SI Prefixes

Multiple	Exponential Form	Prefix	SI Symbol
1 000 000 000	10^9	giga	G
1 000 000	10^6	mega	M
1 000	10^3	kilo	k

Submultiple			
0.001	10^{-3}	milli	m
0.000 001	10^{-6}	micro	μ
0.000 000 001	10^{-9}	nano	n

Conversion Factors (FPS) to (SI)

Quantity	Unit of Measurement (FPS)	Equals	Unit of Measurement (SI)
Force	lb		4.4482 N
Mass	slug		14.5938 kg
Length	ft		0.3048 m

Conversion Factors (FPS)

$$1 \text{ ft} = 12 \text{ in. (inches)}$$
$$1 \text{ mi. (mile)} = 5280 \text{ ft}$$
$$1 \text{ kip (kilopound)} = 1000 \text{ lb}$$
$$1 \text{ ton} = 2000 \text{ lb}$$

ENGINEERING MECHANICS

DYNAMICS

TWELFTH EDITION

R. C. HIBBELER

PRENTICE HALL
Upper Saddle River, NJ 07458

Library of Congress Cataloging-in-Publication Data on File

Vice President and Editorial Director, ECS: *Marcia Horton*
Acquisitions Editor: *Tacy Quinn*
Associate Editor: *Dee Bernhard*
Editorial Assistant: *Bernadette Marciniak*
Managing Editor: *Scott Disanno*
Production Editor: *Rose Kernan*
Art Director, Interior and Cover Designer: *Kenny Beck*
Art Editor: *Gregory Dulles*
Media Editor: *Daniel Sandin*
Operations Specialist: *Lisa McDowell*
Manufacturing Manager: *Alexis Heydt-Long*
Senior Marketing Manager: *Tim Galligan*

About the Cover:
Background image: Helicopter Tail Rotor Blades/Shutterstock/Steve Mann
Inset image: Lightflight Helicopter in Flight/Corbis/George Hall

Pearson Education Ltd., *London*
Pearson Education Australia Pty. Ltd., *Sydney*
Pearson Education Singapore, Pte. Ltd.
Pearson Education North Asia Ltd., *Hong Kong*
Pearson Education Canada, Inc., *Toronto*
Pearson Educación de Mexico, S.A. de C.V.
Pearson Education—Japan, *Tokyo*
Pearson Education Malaysia, Pte. Ltd.
Pearson Education, Inc., *Upper Saddle River, New Jersey*

Prentice Hall
is an imprint of

www.pearsonhighered.com

Printed in the United States of America
10 9 8 7 6 5 4 3 2

ISBN-10: 0-13-607791-9
ISBN-13: 978-0-13-607791-6

To the Student

With the hope that this work will stimulate
an interest in Engineering Mechanics
and provide an acceptable guide to its understanding.

PREFACE

The main purpose of this book is to provide the student with a clear and thorough presentation of the theory and application of engineering mechanics. To achieve this objective, this work has been shaped by the comments and suggestions of hundreds of reviewers in the teaching profession, as well as many of the author's students. The twelfth edition of this book has been significantly enhanced from the previous edition and it is hoped that both the instructor and student will benefit greatly from these improvements.

New Features

Fundamental Problems. These problem sets are located just after the example problems. They offer students simple applications of the concepts and, therefore, provide them with the chance to develop their problem-solving skills before attempting to solve any of the standard problems that follow. You may consider these problems as extended examples since they *all have partial solutions and answers* that are given in the back of the book. Additionally, the fundamental problems offer students an excellent means of studying for exams; and they can be used at a later time as a preparation for the Fundamentals in Engineering Exam.

Rewriting. Each section of the text was carefully reviewed and, in many areas, the material has been redeveloped to better explain the concepts. This has included adding or changing several of the examples in order to provide more emphasis on the applications of the important concepts.

Conceptual Problems. Throughout the text, usually at the end of each chapter, there is a set of problems that involve conceptual situations related to the application of the mechanics principles contained in the chapter. These analysis and design problems are intended to engage the students in thinking through a real-life situation as depicted in a photo. They can be assigned after the students have developed some expertise in the subject matter.

Additional Photos. The relevance of knowing the subject matter is reflected by realistic applications depicted in over 60 new and updated photos placed throughout the book. These photos are generally used to explain how the principles of mechanics apply to real-world situations. In some sections, photographs have been used to show how engineers must first make an idealized model for analysis and then proceed to draw a free-body diagram of this model in order to apply the theory.

New Problems. There are approximately 50%, or about 850, new problems added to this edition including aerospace and petroleum engineering, and biomechanics applications. Also, this new edition now has approximately 17% more problems than in the previous edition.

Hallmark Features

Besides the new features mentioned above, other outstanding features that define the contents of the text include the following.

Organization and Approach. Each chapter is organized into well-defined sections that contain an explanation of specific topics, illustrative example problems, and a set of homework problems. The topics within each section are placed into subgroups defined by boldface titles. The purpose of this is to present a structured method for introducing each new definition or concept and to make the book convenient for later reference and review.

Chapter Contents. Each chapter begins with an illustration demonstrating a broad-range application of the material within the chapter. A bulleted list of the chapter contents is provided to give a general overview of the material that will be covered.

Emphasis on Free-Body Diagrams. Drawing a free-body diagram is particularly important when solving problems, and for this reason this step is strongly emphasized throughout the book. In particular, special sections and examples are devoted to show how to draw free-body diagrams. Specific homework problems have also been added to develop this practice.

Procedures for Analysis. A general procedure for analyzing any mechanical problem is presented at the end of the first chapter. Then this procedure is customized to relate to specific types of problems that are covered throughout the book. This unique feature provides the student with a logical and orderly method to follow when applying the theory. The example problems are solved using this outlined method in order to clarify its numerical application. Realize, however, that once the relevant principles have been mastered and enough confidence and judgment have been obtained, the student can then develop his or her own procedures for solving problems.

Important Points. This feature provides a review or summary of the most important concepts in a section and highlights the most significant points that should be realized when applying the theory to solve problems.

Conceptual Understanding. Through the use of photographs placed throughout the book, theory is applied in a simplified way in order to illustrate some of its more important conceptual features and instill the physical meaning of many of the terms used in the equations. These simplified applications increase interest in the subject matter and better prepare the student to understand the examples and solve problems.

Homework Problems. Apart from the Fundamental and Conceptual type problems mentioned previously, other types of problems contained in the book include the following:

- **Free-Body Diagram Problems.** Some sections of the book contain introductory problems that only require drawing the free-body diagram for the specific problems within a problem set. These assignments will impress upon the student the importance of mastering this skill as a requirement for a complete solution of any equilibrium problem.

• **General Analysis and Design Problems.** The majority of problems in the book depict realistic situations encountered in engineering practice. Some of these problems come from actual products used in industry. It is hoped that this realism will both stimulate the student's interest in engineering mechanics and provide a means for developing the skill to reduce any such problem from its physical description to a model or symbolic representation to which the principles of mechanics may be applied.

Throughout the book, there is an approximate balance of problems using either SI or FPS units. Furthermore, in any set, an attempt has been made to arrange the problems in order of increasing difficulty except for the end of chapter review problems, which are presented in random order.

• **Computer Problems.** An effort has been made to include some problems that may be solved using a numerical procedure executed on either a desktop computer or a programmable pocket calculator. The intent here is to broaden the student's capacity for using other forms of mathematical analysis without sacrificing the time needed to focus on the application of the principles of mechanics. Problems of this type, which either can or must be solved using numerical procedures, are identified by a "square" symbol (■) preceding the problem number.

With so many homework problems in this new edition, they have now been placed in three different categories. Problems that are simply indicated by a problem number have an answer given in the back of the book. If a bullet (•) precedes the problem number, then a suggestion, key equation, or additional numerical result is given along with the answer. Finally, an asterisk (*) before every fourth problem number indicates a problem without an answer.

Accuracy. As with the previous editions, apart from the author, the accuracy of the text and problem solutions has been thoroughly checked by four other parties: Scott Hendricks, Virginia Polytechnic Institute and State University; Karim Nohra, University of South Florida; Kurt Norlin, Laurel Tech Integrated Publishing Services; and finally Kai Beng Yap, a practicing engineer, who in addition to accuracy review provided content development suggestions.

Contents

The book is divided into 11 chapters, in which the principles are applied first to simple, then to more complicated situations.

The kinematics of a particle is discussed in Chapter 12, followed by a discussion of particle kinetics in Chapter 13 (Equation of Motion), Chapter 14 (Work and Energy), and Chapter 15 (Impulse and Momentum). The concepts of particle dynamics contained in these four chapters are then summarized in a "review" section, and the student is given the chance to identify and solve a variety of problems. A similar sequence of presentation is given for the planar motion of a rigid body: Chapter 16 (Planar Kinematics), Chapter 17 (Equations of Motion), Chapter 18 (Work and Energy), and Chapter 19 (Impulse and Momentum), followed by a summary and review set of problems for these chapters.

If time permits, some of the material involving three-dimensional rigid-body motion may be included in the course. The kinematics and kinetics of this motion are discussed in Chapters 20 and 21, respectively. Chapter 22 (Vibrations) may be included if the student has the necessary mathematical background. Sections of the book that are considered to be beyond the scope of the basic dynamics course are indicated by a star (★) and may be omitted. Note that this material also provides a suitable reference for basic principles when it is discussed in more advanced courses. Finally, Appendix A provides a list of mathematical formulas needed to solve the problems in the book, Appendix B provides a brief review of vector analysis, and Appendix C reviews application of the chain rule.

Alternative Coverage. At the discretion of the instructor, it is possible to cover Chapters 12 through 19 in the following order with no loss in continuity: Chapters 12 and 16 (Kinematics), Chapters 13 and 17 (Equations of Motion), Chapter 14 and 18 (Work and Energy), and Chapters 15 and 19 (Impulse and Momentum).

Acknowledgments

The author has endeavored to write this book so that it will appeal to both the student and instructor. Through the years, many people have helped in its development, and I will always be grateful for their valued suggestions and comments. Specifically, I wish to thank the following individuals who have contributed their comments relative to preparing the Twelfth Edition of this work:

Per Reinhall, *University of Washington*
Faissal A. Moslehy, *University of Central Florida*
Richard R. Neptune, *University of Texas at Austin*
Robert Rennaker, *University of Oklahoma*

A particular note of thanks is also given to Professor Will Liddell, Jr., and Henry Kuhlman. In addition, there are a few people that I feel deserve particular recognition. Vince O'Brien, Director of Team-Based Project Management at Pearson Education, and Rose Kernan, my production editor for many years, have both provided me with their encouragement and support. Frankly, without their help, this totally revised and enhanced edition would not be possible. Furthermore a long-time friend and associate, Kai Beng Yap, was of great help to me in checking the entire manuscript and helping to prepare the problem solutions. A special note of thanks also goes to Kurt Norlan of *Laurel Tech Integrated Publishing Services* in this regard. During the production process I am thankful for the assistance of my wife Conny and daughter Mary Ann with the proofreading and typing needed to prepare the manuscript for publication.

Lastly, many thanks are extended to all my students and to members of the teaching profession who have freely taken the time to send me their suggestions and comments. Since this list is too long to mention, it is hoped that those who have given help in this manner will accept this anonymous recognition.

I would greatly appreciate hearing from you if at any time you have any comments, suggestions, or problems related to any matters regarding this edition.

Russell Charles Hibbeler
hibbeler@bellsouth.net

Mastering ENGINEERING™

MasteringEngineering is the most technologically advanced online tutorial and homework system. It tutors students individually while providing instructors with rich teaching diagnostics.

MasteringEngineering is built upon the same platform as MasteringPhysics, the only online physics homework system with research showing that it improves student learning. A wide variety of published papers based on NSF-sponsored research and tests illustrate the benefits of MasteringEngineering. To read these papers, please visit **www.masteringengineering.com**.

MasteringEngineering for Students

MasteringEngineering improves understanding. As an Instructor-assigned homework and tutorial system, MasteringEngineering is designed to provide students with customized coaching and individualized feedback to help improve problem-solving skills. Students complete homework efficiently and effectively with tutorials that provide targeted help.

▼ **Immediate and specific feedback** on wrong answers coach students individually. Specific feedback on common errors helps explain why a particular answer is not correct.

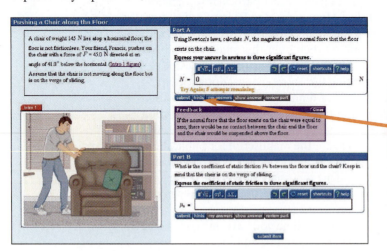

▲ **Hints provide individualized coaching.** Skip the hints you don't need and access only the ones that you need, for the most efficient path to the correct solution.

MasteringEngineering for Instructors

Incorporate dynamic homework into your course with automatic grading and adaptive tutoring. Choose from a wide variety of stimulating problems, including free-body diagram drawing capabilities, algorithmically-generated problem sets, and more.

MasteringEngineering emulates the instructor's office-hour environment, coaching students on problem-solving techniques by asking students simpler sub-questions.

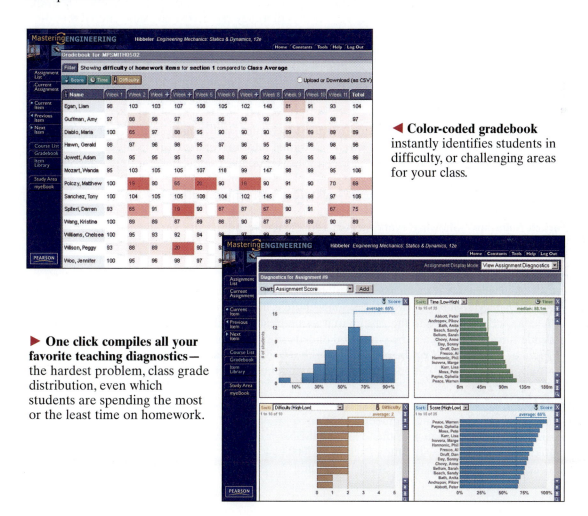

◀ **Color-coded gradebook** instantly identifies students in difficulty, or challenging areas for your class.

▶ **One click compiles all your favorite teaching diagnostics—** the hardest problem, class grade distribution, even which students are spending the most or the least time on homework.

Contact your Pearson Prentice Hall representative for more information.

Resources for Instructors

- **Instructor's Solutions Manual.** This supplement provides complete solutions supported by problem statements and problem figures. The twelfth edition manual was revised to improve readability and was triple accuracy checked.
- **Instructor's Resource CD-ROM.** Visual resources to accompany the text are located on this CD as well as on the Pearson Higher Education website: www.pearsonhighered.com. If you are in need of a login and password for this site, please contact your local Pearson representative. Visual resources include all art from the text, available in PowerPoint slide and JPEG format.
- **Video Solutions.** Developed by Professor Edward Berger, University of Virginia, video solutions are located on the Companion Website for the text and offer step-by-step solution walkthroughs of representative homework problems from each section of the text. Make efficient use of class time and office hours by showing students the complete and concise problem-solving approaches that they can access any time and view at their own pace. The videos are designed to be a flexible resource to be used however each instructor and student prefers. A valuable tutorial resource, the videos are also helpful for student self-evaluation as students can pause the videos to check their understanding and work alongside the video. Access the videos at www.prenhall.com/hibbeler and follow the links for the *Engineering Mechanics: Dynamics*, Twelfth Edition text.

Resources for Students

- **Dynamics Study Pack.** This supplement contains chapter-by-chapter study materials, a Free-Body Diagram Workbook and access to the Companion Website where additional tutorial resources are located.
- **Companion Website.** The Companion Website, located at www.prenhall.com/hibbeler, includes opportunities for practice and review including:
  - **Video Solutions**—Complete, step-by-step solution walkthroughs of representative homework problems from each section. Videos offer:
    - **Fully worked Solutions**—Showing every step of representative homework problems, to help students make vital connections between concepts.
    - **Self-paced Instruction**—Students can navigate each problem and select, play, rewind, fast-forward, stop, and jump-to-sections within each problem's solution.
    - **24/7 Access**—Help whenever students need it with over 20 hours of helpful review. An access code for the *Engineering Mechanics: Dynamics*, Twelfth Edition website is included inside the *Dynamics Study Pack*. To redeem the code and gain access to the site, go to www.prenhall.com/hibbeler and follow the directions on the access code card. Access can also be purchased directly from the site.
- **Dynamics Practice Problems Workbook.** This workbook contains additional worked problems. The problems are partially solved and are designed to help guide students through difficult topics.

Ordering Options

The *Dynamics Study Pack* and MasteringEngineering resources are available as stand-alone items for student purchase and are also available packaged with the texts. The ISBN for each valuepack is as follows:

- *Engineering Mechanics: Dynamics* with Study Pack: 0-13-700239-4
- *Engineering Mechanics: Dynamics* with Study Pack and MasteringEngineering Student Access Card: 0-13-701630-1

Custom Solutions

New options for textbook customization are now available for *Engineering Mechanics*, Twelfth Edition. Please contact your local Pearson/Prentice Hall representative for details.

CONTENTS

Review

16
Planar Kinematics of a Rigid Body   311

17
Planar Kinetics of a Rigid Body: Force and Acceleration   395

18
Planar Kinetics of a Rigid Body: Work and Energy   455

19
Planar Kinetics of a Rigid Body: Impulse and Momentum   495

20
Three-Dimensional Kinematics of a Rigid Body   549

21
Three-Dimensional Kinetics of a Rigid Body   579

22
Vibrations   631

Appendix

Fundamental Problems Partial Solutions and Answers

Answers to Selected Problems

Index

Credits

Chapter 12, The United States Navy Blue Angels perform in an air show as part of San Francisco's Fleet Week celebration. ©Roger Ressmeyer/CORBIS. All Rights Reserved.

Chapter 13, Orange juice factory, elevated view. Getty Images.

Chapter 14, White roller coaster of the Mukogaokayuen ground, Kanagawa. ©Yoshio Kosaka/CORBIS. All Rights Reserved.

Chapter 15, Close-up of golf club hitting a golf ball off the tee. Alamy Images Royalty Free.

Chapter 16, Windmills in Livermore, part of an extensive wind farm, an alternative source of electrical power, California, United States. Brent Winebrenner/Lonely Planet Images/Photo 20-20.

Chapter 17, Burnout drag racing car at Santa Pod Raceway, England. Alamy Images.

Chapter 18, Drilling rig. Getty Images.

Chapter 19, NASA shuttle docking with the International Space Station. Dennis Hallinan/Alamy Images.

Chapter 20, Man watching robotic welding. ©Ted Horowitz/CORBIS. All Rights Reserved.

Chapter 21, A spinning Calypso ride provides a blur of bright colors at Waldameer Park and Water World in Erie, Pennsylvania. Jim Cole/Alamy Images.

Chapter 22, A train track and train wheel give great perspective to the size and power of railway transportation. Joe Belanger/Alamy Images.

Cover 1, Lightflight Helecopter in flight. Lightflight Helecopter is used by Stanford University Hospital Santa Clara Valley Medical Center. ©CORBIS/ All rights reserved.

Cover 2, Helecopter tail rotor blades detail. Steve Mann/Shutterstock.

Other images provided by the author.

DYNAMICS

TWELFTH EDITION

Although each of these planes is rather large, from a distance their motion can be analysed as if each plane were a particle.

Kinematics of a Particle

12

CHAPTER OBJECTIVES

- To introduce the concepts of position, displacement, velocity, and acceleration.
- To study particle motion along a straight line and represent this motion graphically.
- To investigate particle motion along a curved path using different coordinate systems.
- To present an analysis of dependent motion of two particles.
- To examine the principles of relative motion of two particles using translating axes.

12.1 Introduction

Mechanics is a branch of the physical sciences that is concerned with the state of rest or motion of bodies subjected to the action of forces. Engineering mechanics is divided into two areas of study, namely, statics and dynamics. *Statics* is concerned with the equilibrium of a body that is either at rest or moves with constant velocity. Here we will consider *dynamics,* which deals with the accelerated motion of a body. The subject of dynamics will be presented in two parts: *kinematics*, which treats only the geometric aspects of the motion, and *kinetics*, which is the analysis of the forces causing the motion. To develop these principles, the dynamics of a particle will be discussed first, followed by topics in rigid-body dynamics in two and then three dimensions.

Historically, the principles of dynamics developed when it was possible to make an accurate measurement of time. Galileo Galilei (1564–1642) was one of the first major contributors to this field. His work consisted of experiments using pendulums and falling bodies. The most significant contributions in dynamics, however, were made by Isaac Newton (1642–1727), who is noted for his formulation of the three fundamental laws of motion and the law of universal gravitational attraction. Shortly after these laws were postulated, important techniques for their application were developed by Euler, D'Alembert, Lagrange, and others.

There are many problems in engineering whose solutions require application of the principles of dynamics. Typically the structural design of any vehicle, such as an automobile or airplane, requires consideration of the motion to which it is subjected. This is also true for many mechanical devices, such as motors, pumps, movable tools, industrial manipulators, and machinery. Furthermore, predictions of the motions of artificial satellites, projectiles, and spacecraft are based on the theory of dynamics. With further advances in technology, there will be an even greater need for knowing how to apply the principles of this subject.

Problem Solving.

Dynamics is considered to be more involved than statics since both the forces applied to a body and its motion must be taken into account. Also, many applications require using calculus, rather than just algebra and trigonometry. In any case, the most effective way of learning the principles of dynamics is *to solve problems*. To be successful at this, it is necessary to present the work in a logical and orderly manner as suggested by the following sequence of steps:

1. Read the problem carefully and try to correlate the actual physical situation with the theory you have studied.

2. Draw any necessary diagrams and tabulate the problem data.

3. Establish a coordinate system and apply the relevant principles, generally in mathematical form.

4. Solve the necessary equations algebraically as far as practical; then, use a consistent set of units and complete the solution numerically. Report the answer with no more significant figures than the accuracy of the given data.

5. Study the answer using technical judgment and common sense to determine whether or not it seems reasonable.

6. Once the solution has been completed, review the problem. Try to think of other ways of obtaining the same solution.

In applying this general procedure, do the work as neatly as possible. Being neat generally stimulates clear and orderly thinking, and vice versa.

12.2   Rectilinear Kinematics: Continuous Motion

We will begin our study of dynamics by discussing the kinematics of a particle that moves along a rectilinear or straight line path. Recall that a *particle* has a mass but negligible size and shape. Therefore we must limit application to those objects that have dimensions that are of no consequence in the analysis of the motion. In most problems, we will be interested in bodies of finite size, such as rockets, projectiles, or vehicles. Each of these objects can be considered as a particle, as long as the motion is characterized by the motion of its mass center and any rotation of the body is neglected.

Rectilinear Kinematics.   The kinematics of a particle is characterized by specifying, at any given instant, the particle's position, velocity, and acceleration.

Position.   The straight-line path of a particle will be defined using a single coordinate axis s, Fig. 12–1a. The origin O on the path is a fixed point, and from this point the *position coordinate s* is used to specify the location of the particle at any given instant. The magnitude of s is the distance from O to the particle, usually measured in meters (m) or feet (ft), and the sense of direction is defined by the algebraic sign on s. Although the choice is arbitrary, in this case s is positive since the coordinate axis is positive to the right of the origin. Likewise, it is negative if the particle is located to the left of O. Realize that position is a vector quantity since it has both magnitude and direction. Here, however, it is being represented by the algebraic scalar s since the direction always remains along the coordinate axis.

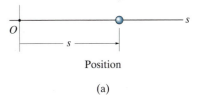

Position

(a)

Displacement.   The *displacement* of the particle is defined as the *change* in its *position*. For example, if the particle moves from one point to another, Fig. 12–1b, the displacement is

$$\Delta s = s' - s$$

In this case Δs is *positive* since the particle's final position is to the *right* of its initial position, i.e., $s' > s$. Likewise, if the final position were to the *left* of its initial position, Δs would be *negative*.

The displacement of a particle is also a *vector quantity*, and it should be distinguished from the distance the particle travels. Specifically, the *distance traveled* is a *positive scalar* that represents the total length of path over which the particle travels.

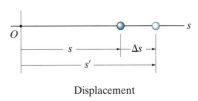

Displacement

(b)

Fig. 12–1

Velocity. If the particle moves through a displacement Δs during the time interval Δt, the *average velocity* of the particle during this time interval is

$$v_{\text{avg}} = \frac{\Delta s}{\Delta t}$$

If we take smaller and smaller values of Δt, the magnitude of Δs becomes smaller and smaller. Consequently, the *instantaneous velocity* is a vector defined as $v = \lim_{\Delta t \to 0}(\Delta s/\Delta t)$, or

$$(\overset{+}{\rightarrow}) \qquad\qquad \boxed{v = \frac{ds}{dt}} \qquad\qquad (12\text{--}1)$$

Since Δt or dt is always positive, the sign used to define the *sense* of the velocity is the same as that of Δs or ds. For example, if the particle is moving to the *right*, Fig. 12–1c, the velocity is *positive;* whereas if it is moving to the *left*, the velocity is *negative*. (This is emphasized here by the arrow written at the left of Eq. 12–1.) The *magnitude* of the velocity is known as the *speed*, and it is generally expressed in units of m/s or ft/s.

Occasionally, the term "average speed" is used. The *average speed* is always a positive scalar and is defined as the total distance traveled by a particle, s_T, divided by the elapsed time Δt; i.e.,

$$(v_{\text{sp}})_{\text{avg}} = \frac{s_T}{\Delta t}$$

For example, the particle in Fig. 12–1d travels along the path of length s_T in time Δt, so its average speed is $(v_{\text{sp}})_{\text{avg}} = s_T/\Delta t$, but its average velocity is $v_{\text{avg}} = -\Delta s/\Delta t$.

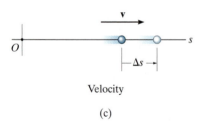

Velocity

(c)

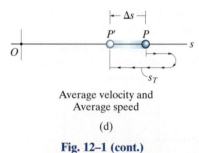

Average velocity and
Average speed

(d)

Fig. 12–1 (cont.)

Acceleration.   Provided the velocity of the particle is known at two points, the *average acceleration* of the particle during the time interval Δt is defined as

$$a_{avg} = \frac{\Delta v}{\Delta t}$$

Here Δv represents the difference in the velocity during the time interval Δt, i.e., $\Delta v = v' - v$, Fig. 12–1e.

The *instantaneous acceleration* at time t is a vector that is found by taking smaller and smaller values of Δt and corresponding smaller and smaller values of Δv, so that $a = \lim_{\Delta t \to 0} (\Delta v / \Delta t)$, or

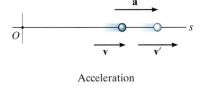

Acceleration

(e)

$$(\overset{+}{\to}) \qquad\qquad a = \frac{dv}{dt} \qquad\qquad (12\text{–}2)$$

Substituting Eq. 12–1 into this result, we can also write

$$(\overset{+}{\to}) \qquad\qquad a = \frac{d^2s}{dt^2}$$

Both the average and instantaneous acceleration can be either positive or negative. In particular, when the particle is *slowing down*, or its speed is decreasing, the particle is said to be *decelerating*. In this case, v' in Fig. 12–1f is *less* than v, and so $\Delta v = v' - v$ will be negative. Consequently, a will also be negative, and therefore it will act to the *left*, in the *opposite sense* to v. Also, note that when the *velocity* is *constant*, the *acceleration is zero* since $\Delta v = v - v = 0$. Units commonly used to express the magnitude of acceleration are m/s^2 or ft/s^2.

Deceleration

(f)

Finally, an important differential relation involving the displacement, velocity, and acceleration along the path may be obtained by eliminating the time differential dt between Eqs. 12–1 and 12–2, which gives

$$(\overset{+}{\to}) \qquad\qquad a \, ds = v \, dv \qquad\qquad (12\text{–}3)$$

Although we have now produced three important kinematic equations, realize that the above equation is not independent of Eqs. 12–1 and 12–2.

Constant Acceleration, $a = a_c$. When the acceleration is constant, each of the three kinematic equations $a_c = dv/dt$, $v = ds/dt$, and $a_c\,ds = v\,dv$ can be integrated to obtain formulas that relate a_c, v, s, and t.

Velocity as a Function of Time.   Integrate $a_c = dv/dt$, assuming that initially $v = v_0$ when $t = 0$.

$$\int_{v_0}^{v} dv = \int_{0}^{t} a_c\,dt$$

$(\overset{+}{\rightarrow})$

$$\boxed{\begin{array}{c} v = v_0 + a_c t \\ \text{Constant Acceleration} \end{array}}$$
(12–4)

Position as a Function of Time.   Integrate $v = ds/dt = v_0 + a_c t$, assuming that initially $s = s_0$ when $t = 0$.

$$\int_{s_0}^{s} ds = \int_{0}^{t} (v_0 + a_c t)\,dt$$

$(\overset{+}{\rightarrow})$

$$\boxed{\begin{array}{c} s = s_0 + v_0 t + \tfrac{1}{2} a_c t^2 \\ \text{Constant Acceleration} \end{array}}$$
(12–5)

Velocity as a Function of Position.   Either solve for t in Eq. 12–4 and substitute into Eq. 12–5, or integrate $v\,dv = a_c\,ds$, assuming that initially $v = v_0$ at $s = s_0$.

$$\int_{v_0}^{v} v\,dv = \int_{s_0}^{s} a_c\,ds$$

$(\overset{+}{\rightarrow})$

$$\boxed{\begin{array}{c} v^2 = v_0^2 + 2a_c(s - s_0) \\ \text{Constant Acceleration} \end{array}}$$
(12–6)

The algebraic signs of s_0, v_0, and a_c, used in the above three equations, are determined from the positive direction of the s axis as indicated by the arrow written at the left of each equation. Remember that these equations are useful *only when the acceleration is constant and when* $t = 0$, $s = s_0$, $v = v_0$. A typical example of constant accelerated motion occurs when a body falls freely toward the earth. If air resistance is neglected and the distance of fall is short, then the *downward* acceleration of the body when it is close to the earth is constant and approximately 9.81 m/s² or 32.2 ft/s². The proof of this is given in Example 13.2.

Important Points

- Dynamics is concerned with bodies that have accelerated motion.
- Kinematics is a study of the geometry of the motion.
- Kinetics is a study of the forces that cause the motion.
- Rectilinear kinematics refers to straight-line motion.
- Speed refers to the magnitude of velocity.
- Average speed is the total distance traveled divided by the total time. This is different from the average velocity, which is the displacement divided by the time.
- A particle that is slowing down is decelerating.
- A particle can have an acceleration and yet have zero velocity.
- The relationship $a\,ds = v\,dv$ is derived from $a = dv/dt$ and $v = ds/dt$, by eliminating dt.

During the time this rocket undergoes rectilinear motion, its altitude as a function of time can be measured and expressed as $s = s(t)$. Its velocity can then be found using $v = ds/dt$, and its acceleration can be determined from $a = dv/dt$.

Procedure for Analysis

Coordinate System.

- Establish a position coordinate s along the path and specify its *fixed origin* and positive direction.
- Since motion is along a straight line, the vector quantities position, velocity, and acceleration can be represented as algebraic scalars. For analytical work the sense of s, v, and a is then defined by their *algebraic signs*.
- The positive sense for each of these scalars can be indicated by an arrow shown alongside each kinematic equation as it is applied.

Kinematic Equations.

- If a relation is known between any *two* of the four variables a, v, s and t, then a third variable can be obtained by using one of the kinematic equations, $a = dv/dt$, $v = ds/dt$ or $a\,ds = v\,dv$, since each equation relates all three variables.*
- Whenever integration is performed, it is important that the position and velocity be known at a given instant in order to evaluate either the constant of integration if an indefinite integral is used, or the limits of integration if a definite integral is used.
- Remember that Eqs. 12–4 through 12–6 have only limited use. These equations apply *only* when the *acceleration is constant* and the initial conditions are $s = s_0$ and $v = v_0$ when $t = 0$.

*Some standard differentiation and integration formulas are given in Appendix A.

12

EXAMPLE | 12.1

The car in Fig. 12–2 moves in a straight line such that for a short time its velocity is defined by $v = (3t^2 + 2t)$ ft/s, where t is in seconds. Determine its position and acceleration when $t = 3$ s. When $t = 0$, $s = 0$.

Fig. 12–2

SOLUTION

Coordinate System. The position coordinate extends from the fixed origin O to the car, positive to the right.

Position. Since $v = f(t)$, the car's position can be determined from $v = ds/dt$, since this equation relates v, s, and t. Noting that $s = 0$ when $t = 0$, we have*

$(\xrightarrow{+})$

$$v = \frac{ds}{dt} = (3t^2 + 2t)$$

$$\int_0^s ds = \int_0^t (3t^2 + 2t)\,dt$$

$$s\Big|_0^s = t^3 + t^2\Big|_0^t$$

$$s = t^3 + t^2$$

When $t = 3$ s,

$$s = (3)^3 + (3)^2 = 36 \text{ ft} \qquad\qquad Ans.$$

Acceleration. Since $v = f(t)$, the acceleration is determined from $a = dv/dt$, since this equation relates a, v, and t.

$(\xrightarrow{+})$

$$a = \frac{dv}{dt} = \frac{d}{dt}(3t^2 + 2t)$$

$$= 6t + 2$$

When $t = 3$ s,

$$a = 6(3) + 2 = 20 \text{ ft/s}^2 \rightarrow \qquad\qquad Ans.$$

NOTE: The formulas for constant acceleration *cannot* be used to solve this problem, because the acceleration is a function of time.

*The *same result* can be obtained by evaluating a constant of integration C rather than using definite limits on the integral. For example, integrating $ds = (3t^2 + 2t)dt$ yields $s = t^3 + t^2 + C$. Using the condition that at $t = 0$, $s = 0$, then $C = 0$.

EXAMPLE | 12.2

A small projectile is fired vertically *downward* into a fluid medium with an initial velocity of 60 m/s. Due to the drag resistance of the fluid the projectile experiences a deceleration of $a = (-0.4v^3) \text{ m/s}^2$, where v is in m/s. Determine the projectile's velocity and position 4 s after it is fired.

SOLUTION

Coordinate System. Since the motion is downward, the position coordinate is positive downward, with origin located at O, Fig. 12–3.

Velocity. Here $a = f(v)$ and so we must determine the velocity as a function of time using $a = dv/dt$, since this equation relates v, a, and t. (Why not use $v = v_0 + a_c t$?) Separating the variables and integrating, with $v_0 = 60$ m/s when $t = 0$, yields

Fig. 12–3

$$(+\downarrow) \qquad a = \frac{dv}{dt} = -0.4v^3$$

$$\int_{60 \text{ m/s}}^{v} \frac{dv}{-0.4v^3} = \int_{0}^{t} dt$$

$$\frac{1}{-0.4}\left(\frac{1}{-2}\right)\frac{1}{v^2}\bigg|_{60}^{v} = t - 0$$

$$\frac{1}{0.8}\left[\frac{1}{v^2} - \frac{1}{(60)^2}\right] = t$$

$$v = \left\{\left[\frac{1}{(60)^2} + 0.8t\right]^{-1/2}\right\} \text{ m/s}$$

Here the positive root is taken, since the projectile will continue to move downward. When $t = 4$ s,

$$v = 0.559 \text{ m/s} \downarrow \qquad\qquad Ans.$$

Position. Knowing $v = f(t)$, we can obtain the projectile's position from $v = ds/dt$, since this equation relates s, v, and t. Using the initial condition $s = 0$, when $t = 0$, we have

$$(+\downarrow) \qquad v = \frac{ds}{dt} = \left[\frac{1}{(60)^2} + 0.8t\right]^{-1/2}$$

$$\int_{0}^{s} ds = \int_{0}^{t}\left[\frac{1}{(60)^2} + 0.8t\right]^{-1/2} dt$$

$$s = \frac{2}{0.8}\left[\frac{1}{(60)^2} + 0.8t\right]^{1/2}\bigg|_{0}^{t}$$

$$s = \frac{1}{0.4}\left\{\left[\frac{1}{(60)^2} + 0.8t\right]^{1/2} - \frac{1}{60}\right\} \text{ m}$$

When $t = 4$ s,

$$s = 4.43 \text{ m} \qquad\qquad Ans.$$

EXAMPLE 12.3

During a test a rocket travels upward at 75 m/s, and when it is 40 m from the ground its engine fails. Determine the maximum height s_B reached by the rocket and its speed just before it hits the ground. While in motion the rocket is subjected to a constant downward acceleration of 9.81 m/s^2 due to gravity. Neglect the effect of air resistance.

SOLUTION

Coordinate System. The origin O for the position coordinate s is taken at ground level with positive upward, Fig. 12–4.

Maximum Height. Since the rocket is traveling *upward*, $v_A = +75$m/s when $t = 0$. At the maximum height $s = s_B$ the velocity $v_B = 0$. For the entire motion, the acceleration is $a_c = -9.81$ m/s^2 (negative since it acts in the *opposite* sense to positive velocity or positive displacement). Since a_c is *constant* the rocket's position may be related to its velocity at the two points A and B on the path by using Eq. 12–6, namely,

$$(+\uparrow) \qquad v_B^2 = v_A^2 + 2a_c(s_B - s_A)$$

$$0 = (75 \text{ m/s})^2 + 2(-9.81 \text{ m/s}^2)(s_B - 40 \text{ m})$$

$$s_B = 327 \text{ m} \qquad \qquad Ans.$$

Velocity. To obtain the velocity of the rocket just before it hits the ground, we can apply Eq. 12–6 between points B and C, Fig. 12–4.

$$(+\uparrow) \qquad \qquad v_C^2 = v_B^2 + 2a_c(s_C - s_B)$$

$$= 0 + 2(-9.81 \text{ m/s}^2)(0 - 327 \text{ m})$$

$$v_C = -80.1 \text{ m/s} = 80.1 \text{ m/s} \downarrow \qquad Ans.$$

The negative root was chosen since the rocket is moving downward. Similarly, Eq. 12–6 may also be applied between points A and C, i.e.,

$$(+\uparrow) \qquad v_C^2 = v_A^2 + 2a_c(s_C - s_A)$$

$$= (75 \text{ m/s})^2 + 2(-9.81 \text{ m/s}^2)(0 - 40 \text{ m})$$

$$v_C = -80.1 \text{ m/s} = 80.1 \text{ m/s} \downarrow \qquad Ans.$$

NOTE: It should be realized that the rocket is subjected to a *deceleration* from A to B of 9.81 m/s^2, and then from B to C it is *accelerated* at this rate. Furthermore, even though the rocket momentarily comes to *rest* at B ($v_B = 0$) the acceleration at B is still 9.81 m/s^2 downward!

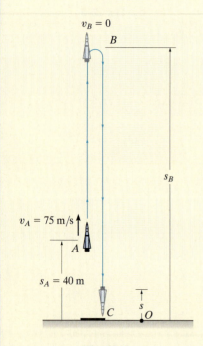

$v_B = 0$

$v_A = 75$ m/s

$s_A = 40$ m

Fig. 12–4

EXAMPLE 12.4

A metallic particle is subjected to the influence of a magnetic field as it travels downward through a fluid that extends from plate A to plate B, Fig. 12–5. If the particle is released from rest at the midpoint C, $s = 100$ mm, and the acceleration is $a = (4s)$ m/s^2, where s is in meters, determine the velocity of the particle when it reaches plate B, $s = 200$ mm, and the time it takes to travel from C to B.

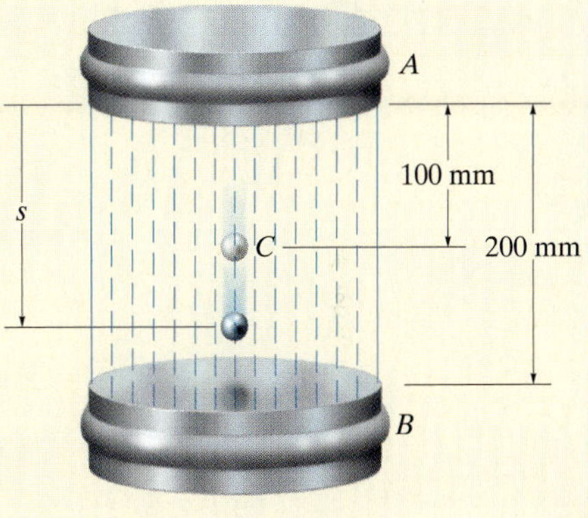

SOLUTION

Coordinate System. As shown in Fig. 12–5, s is positive downward, measured from plate A.

Velocity. Since $a = f(s)$, the velocity as a function of position can be obtained by using $v\, dv = a\, ds$. Realizing that $v = 0$ at $s = 0.1$ m, we have

$(+\downarrow)$

$$v\, dv = a\, ds$$

$$\int_0^v v\, dv = \int_{0.1\text{ m}}^s 4s\, ds$$

$$\frac{1}{2}v^2\Big|_0^v = \frac{4}{2}s^2\Big|_{0.1\text{ m}}^s$$

$$v = 2(s^2 - 0.01)^{1/2} \text{ m/s} \qquad (1)$$

At $s = 200$ mm $= 0.2$ m,

$$v_B = 0.346 \text{ m/s} = 346 \text{ mm/s} \downarrow \qquad \textit{Ans.}$$

The positive root is chosen since the particle is traveling downward, i.e., in the $+s$ direction.

Time. The time for the particle to travel from C to B can be obtained using $v = ds/dt$ and Eq. 1, where $s = 0.1$ m when $t = 0$. From Appendix A,

$(+\downarrow)$

$$ds = v\, dt$$

$$= 2(s^2 - 0.01)^{1/2} dt$$

$$\int_{0.1}^s \frac{ds}{(s^2 - 0.01)^{1/2}} = \int_0^t 2\, dt$$

$$\ln\left(\sqrt{s^2 - 0.01} + s\right)\Big|_{0.1}^s = 2t\Big|_0^t$$

$$\ln\left(\sqrt{s^2 - 0.01} + s\right) + 2.303 = 2t$$

At $s = 0.2$ m,

$$t = \frac{\ln\left(\sqrt{(0.2)^2 - 0.01} + 0.2\right) + 2.303}{2} = 0.658 \text{ s} \qquad \textit{Ans.}$$

Note: The formulas for constant acceleration cannot be used here because the acceleration changes with position, i.e., $a = 4s$.

Fig. 12–5

12

EXAMPLE | 12.5

A particle moves along a horizontal path with a velocity of $v = (3t^2 - 6t)$ m/s, where t is the time in seconds. If it is initially located at the origin O, determine the distance traveled in 3.5 s, and the particle's average velocity and average speed during the time interval.

SOLUTION

Coordinate System. Here positive motion is to the right, measured from the origin O, Fig. 12–6a.

Distance Traveled. Since $v = f(t)$, the position as a function of time may be found by integrating $v = ds/dt$ with $t = 0$, $s = 0$.

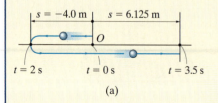

$s = -4.0$ m $s = 6.125$ m

$t = 2$ s $t = 0$ s $t = 3.5$ s

(a)

$$(\xrightarrow{+}) \qquad ds = v\,dt$$

$$= (3t^2 - 6t)dt$$

$$\int_0^s ds = \int_0^t (3t^2 - 6t)\,dt$$

$$s = (t^3 - 3t^2)\text{m} \qquad (1)$$

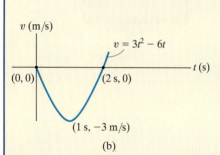

v (m/s)

$v = 3t^2 - 6t$

(0, 0)   (2 s, 0) t (s)

(1 s, −3 m/s)

(b)

Fig. 12–6

In order to determine the distance traveled in 3.5 s, it is necessary to investigate the path of motion. If we consider a graph of the velocity function, Fig. 12–6b, then it reveals that for $0 < t < 2$ s the velocity is *negative*, which means the particle is traveling to the *left*, and for $t > 2$ s the velocity is *positive*, and hence the particle is traveling to the *right*. Also, note that $v = 0$ at $t = 2$ s. The particle's position when $t = 0$, $t = 2$ s, and $t = 3.5$ s can now be determined from Eq. 1. This yields

$$s|_{t=0} = 0 \qquad s|_{t=2\,\text{s}} = -4.0 \text{ m} \qquad s|_{t=3.5\,\text{s}} = 6.125 \text{ m}$$

The path is shown in Fig. 12–6a. Hence, the distance traveled in 3.5 s is

$$s_T = 4.0 + 4.0 + 6.125 = 14.125 \text{ m} = 14.1 \text{ m} \qquad \textit{Ans.}$$

Velocity. The *displacement* from $t = 0$ to $t = 3.5$ s is

$$\Delta s = s|_{t=3.5\,\text{s}} - s|_{t=0} = 6.125 \text{ m} - 0 = 6.125 \text{ m}$$

and so the average velocity is

$$v_{\text{avg}} = \frac{\Delta s}{\Delta t} = \frac{6.125 \text{ m}}{3.5 \text{ s} - 0} = 1.75 \text{ m/s} \rightarrow \qquad \textit{Ans.}$$

The average speed is defined in terms of the *distance traveled* s_T. This positive scalar is

$$(v_{\text{sp}})_{\text{avg}} = \frac{s_T}{\Delta t} = \frac{14.125 \text{ m}}{3.5 \text{ s} - 0} = 4.04 \text{ m/s} \qquad \textit{Ans.}$$

Note: In this problem, the acceleration is $a = dv/dt = (6t - 6)$ m/s^2, which is not constant.

FUNDAMENTAL PROBLEMS

F12–1. Initially, the car travels along a straight road with a speed of 35 m/s. If the brakes are applied and the speed of the car is reduced to 10 m/s in 15 s, determine the constant deceleration of the car.

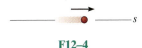

F12–1

F12–2. A ball is thrown vertically upward with a speed of 15 m/s. Determine the time of flight when it returns to its original position.

F12–2

F12–3. A particle travels along a straight line with a velocity of $v = (4t - 3t^2)$ m/s, where t is in seconds. Determine the position of the particle when $t = 4$ s. $s = 0$ when $t = 0$.

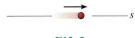

F12–3

F12–4. A particle travels along a straight line with a speed $v = (0.5t^3 - 8t)$ m/s, where t is in seconds. Determine the acceleration of the particle when $t = 2$ s.

F12–4

F12–5. The position of the particle is given by $s = (2t^2 - 8t + 6)$ m, where t is in seconds. Determine the time when the velocity of the particle is zero, and the total distance traveled by the particle when $t = 3$ s.

F12–5

F12–6. A particle travels along a straight line with an acceleration of $a = (10 - 0.2s)$ m/s^2, where s is measured in meters. Determine the velocity of the particle when $s = 10$ m if $v = 5$ m/s at $s = 0$.

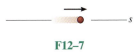

F12–6

F12–7. A particle moves along a straight line such that its acceleration is $a = (4t^2 - 2)$ m/s^2, where t is in seconds. When $t = 0$, the particle is located 2 m to the left of the origin, and when $t = 2$ s, it is 20 m to the left of the origin. Determine the position of the particle when $t = 4$ s.

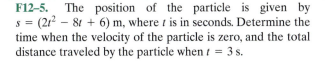

F12–7

F12–8. A particle travels along a straight line with a velocity of $v = (20 - 0.05s^2)$ m/s, where s is in meters. Determine the acceleration of the particle at $s = 15$ m.

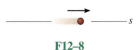

F12–8

PROBLEMS

•12–1. A car starts from rest and with constant acceleration achieves a velocity of 15 m/s when it travels a distance of 200 m. Determine the acceleration of the car and the time required.

12–2. A train starts from rest at a station and travels with a constant acceleration of 1 m/s². Determine the velocity of the train when $t = 30$ s and the distance traveled during this time.

12–3. An elevator descends from rest with an acceleration of 5 ft/s² until it achieves a velocity of 15 ft/s. Determine the time required and the distance traveled.

*12–4. A car is traveling at 15 m/s, when the traffic light 50 m ahead turns yellow. Determine the required constant deceleration of the car and the time needed to stop the car at the light.

•12–5. A particle is moving along a straight line with the acceleration $a = (12t - 3t^{1/2})$ ft/s², where t is in seconds. Determine the velocity and the position of the particle as a function of time. When $t = 0$, $v = 0$ and $s = 15$ ft.

12–6. A ball is released from the bottom of an elevator which is traveling upward with a velocity of 6 ft/s. If the ball strikes the bottom of the elevator shaft in 3 s, determine the height of the elevator from the bottom of the shaft at the instant the ball is released. Also, find the velocity of the ball when it strikes the bottom of the shaft.

12–7. A car has an initial speed of 25 m/s and a constant deceleration of 3 m/s². Determine the velocity of the car when $t = 4$ s. What is the displacement of the car during the 4-s time interval? How much time is needed to stop the car?

*12–8. If a particle has an initial velocity of $v_0 = 12$ ft/s to the right, at $s_0 = 0$, determine its position when $t = 10$ s, if $a = 2$ ft/s² to the left.

•12–9. The acceleration of a particle traveling along a straight line is $a = k/v$, where k is a constant. If $s = 0, v = v_0$ when $t = 0$, determine the velocity of the particle as a function of time t.

12–10. Car A starts from rest at $t = 0$ and travels along a straight road with a constant acceleration of 6 ft/s² until it reaches a speed of 80 ft/s. Afterwards it maintains this speed. Also, when $t = 0$, car B located 6000 ft down the road is traveling towards A at a constant speed of 60 ft/s. Determine the distance traveled by car A when they pass each other.

Prob. 12–10

12–11. A particle travels along a straight line with a velocity $v = (12 - 3t^2)$ m/s, where t is in seconds. When $t = 1$ s, the particle is located 10 m to the left of the origin. Determine the acceleration when $t = 4$ s, the displacement from $t = 0$ to $t = 10$ s, and the distance the particle travels during this time period.

*12–12. A sphere is fired downwards into a medium with an initial speed of 27 m/s. If it experiences a deceleration of $a = (-6t)$ m/s², where t is in seconds, determine the distance traveled before it stops.

•12–13. A particle travels along a straight line such that in 2 s it moves from an initial position $s_A = +0.5$ m to a position $s_B = -1.5$ m. Then in another 4 s it moves from s_B to $s_C = +2.5$ m. Determine the particle's average velocity and average speed during the 6-s time interval.

12–14. A particle travels along a straight-line path such that in 4 s it moves from an initial position $s_A = -8$ m to a position $s_B = +3$ m. Then in another 5 s it moves from s_B to $s_C = -6$ m. Determine the particle's average velocity and average speed during the 9-s time interval.

12–15. Tests reveal that a normal driver takes about 0.75 s before he or she can *react* to a situation to avoid a collision. It takes about 3 s for a driver having 0.1% alcohol in his system to do the same. If such drivers are traveling on a straight road at 30 mph (44 ft/s) and their cars can decelerate at 2 ft/s², determine the shortest stopping distance d for each from the moment they see the pedestrians. *Moral*: If you must drink, please don't drive!

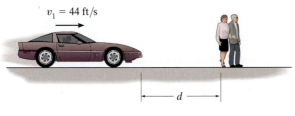

$v_1 = 44$ ft/s

d

Prob. 12–15

*12–16. As a train accelerates uniformly it passes successive kilometer marks while traveling at velocities of 2 m/s and then 10 m/s. Determine the train's velocity when it passes the next kilometer mark and the time it takes to travel the 2-km distance.

•12–17. A ball is thrown with an upward velocity of 5 m/s from the top of a 10-m high building. One second later another ball is thrown vertically from the ground with a velocity of 10 m/s. Determine the height from the ground where the two balls pass each other.

12–18. A car starts from rest and moves with a constant acceleration of 1.5 m/s² until it achieves a velocity of 25 m/s. It then travels with constant velocity for 60 seconds. Determine the average speed and the total distance traveled.

12–19. A car is to be hoisted by elevator to the fourth floor of a parking garage, which is 48 ft above the ground. If the elevator can accelerate at 0.6 ft/s², decelerate at 0.3 ft/s², and reach a maximum speed of 8 ft/s, determine the shortest time to make the lift, starting from rest and ending at rest.

*12–20. A particle is moving along a straight line such that its velocity is defined as $v = (-4s^2)$ m/s, where s is in meters. If $s = 2$ m when $t = 0$, determine the velocity and acceleration as functions of time.

•12–21. Two particles A and B start from rest at the origin $s = 0$ and move along a straight line such that $a_A = (6t - 3)$ ft/s² and $a_B = (12t^2 - 8)$ ft/s², where t is in seconds. Determine the distance between them when $t = 4$ s and the total distance each has traveled in $t = 4$ s.

12–22. A particle moving along a straight line is subjected to a deceleration $a = (-2v^3)$ m/s², where v is in m/s. If it has a velocity $v = 8$ m/s and a position $s = 10$ m when $t = 0$, determine its velocity and position when $t = 4$ s.

12–23. A particle is moving along a straight line such that its acceleration is defined as $a = (-2v)$ m/s², where v is in meters per second. If $v = 20$ m/s when $s = 0$ and $t = 0$, determine the particle's position, velocity, and acceleration as functions of time.

*12–24. A particle starts from rest and travels along a straight line with an acceleration $a = (30 - 0.2v)$ ft/s², where v is in ft/s. Determine the time when the velocity of the particle is $v = 30$ ft/s.

•12–25. When a particle is projected vertically upwards with an initial velocity of v_0, it experiences an acceleration $a = -(g + kv^2)$, where g is the acceleration due to gravity, k is a constant and v is the velocity of the particle. Determine the maximum height reached by the particle.

12–26. The acceleration of a particle traveling along a straight line is $a = (0.02e^t)$ m/s², where t is in seconds. If $v = 0$, $s = 0$ when $t = 0$, determine the velocity and acceleration of the particle at $s = 4$ m.

12–27. A particle moves along a straight line with an acceleration of $a = 5/(3s^{1/3} + s^{5/2})$ m/s², where s is in meters. Determine the particle's velocity when $s = 2$ m, if it starts from rest when $s = 1$ m. Use Simpson's rule to evaluate the integral.

*12–28. If the effects of atmospheric resistance are accounted for, a falling body has an acceleration defined by the equation $a = 9.81[1 - v^2(10^{-4})]$ m/s², where v is in m/s and the positive direction is downward. If the body is released from rest at a very *high altitude*, determine (a) the velocity when $t = 5$ s, and (b) the body's terminal or maximum attainable velocity (as $t \rightarrow \infty$).

•12–29. The position of a particle along a straight line is given by $s = (1.5t^3 - 13.5t^2 + 22.5t)$ ft, where t is in seconds. Determine the position of the particle when $t = 6$ s and the total distance it travels during the 6-s time interval. *Hint:* Plot the path to determine the total distance traveled.

12–30. The velocity of a particle traveling along a straight line is $v = v_0 - ks$, where k is constant. If $s = 0$ when $t = 0$, determine the position and acceleration of the particle as a function of time.

12–31. The acceleration of a particle as it moves along a straight line is given by $a = (2t - 1)$ m/s², where t is in seconds. If $s = 1$ m and $v = 2$ m/s when $t = 0$, determine the particle's velocity and position when $t = 6$ s. Also, determine the total distance the particle travels during this time period.

*12–32. Ball A is thrown vertically upward from the top of a 30-m-high-building with an initial velocity of 5 m/s. At the same instant another ball B is thrown upward from the ground with an initial velocity of 20 m/s. Determine the height from the ground and the time at which they pass.

•12–33. A motorcycle starts from rest at $t = 0$ and travels along a straight road with a constant acceleration of 6 ft/s² until it reaches a speed of 50 ft/s. Afterwards it maintains this speed. Also, when $t = 0$, a car located 6000 ft down the road is traveling toward the motorcycle at a constant speed of 30 ft/s. Determine the time and the distance traveled by the motorcycle when they pass each other.

12–34. A particle moves along a straight line with a velocity $v = (200s)$ mm/s, where s is in millimeters. Determine the acceleration of the particle at $s = 2000$ mm. How long does the particle take to reach this position if $s = 500$ mm when $t = 0$?

■12–35. A particle has an initial speed of 27 m/s. If it experiences a deceleration of $a = (-6t)$ m/s², where t is in seconds, determine its velocity, after it has traveled 10 m. How much time does this take?

*12–36. The acceleration of a particle traveling along a straight line is $a = (8 - 2s)$ m/s², where s is in meters. If $v = 0$ at $s = 0$, determine the velocity of the particle at $s = 2$ m, and the position of the particle when the velocity is maximum.

•12–37. Ball A is thrown vertically upwards with a velocity of v_0. Ball B is thrown upwards from the same point with the same velocity t seconds later. Determine the elapsed time $t < 2v_0/g$ from the instant ball A is thrown to when the balls pass each other, and find the velocity of each ball at this instant.

12–38. As a body is projected to a high altitude above the earth's *surface*, the variation of the acceleration of gravity with respect to altitude y must be taken into account. Neglecting air resistance, this acceleration is determined from the formula $a = -g_0[R^2/(R + y)^2]$, where g_0 is the constant gravitational acceleration at sea level, R is the radius of the earth, and the positive direction is measured upward. If $g_0 = 9.81$ m/s² and $R = 6356$ km, determine the minimum initial velocity (escape velocity) at which a projectile should be shot vertically from the earth's surface so that it does not fall back to the earth. *Hint:* This requires that $v = 0$ as $y \to \infty$.

12–39. Accounting for the variation of gravitational acceleration a with respect to altitude y (see Prob. 12–38), derive an equation that relates the velocity of a freely falling particle to its altitude. Assume that the particle is released from rest at an altitude y_0 from the earth's surface. With what velocity does the particle strike the earth if it is released from rest at an altitude $y_0 = 500$ km? Use the numerical data in Prob. 12–38.

*12–40. When a particle falls through the air, its initial acceleration $a = g$ diminishes until it is zero, and thereafter it falls at a constant or terminal velocity v_f. If this variation of the acceleration can be expressed as $a = (g/v^2_f)(v^2_f - v^2)$, determine the time needed for the velocity to become $v = v_f/2$. Initially the particle falls from rest.

•12–41. A particle is moving along a straight line such that its position from a fixed point is $s = (12 - 15t^2 + 5t^3)$ m, where t is in seconds. Determine the total distance traveled by the particle from $t = 1$ s to $t = 3$ s. Also, find the average speed of the particle during this time interval.

12.3   Rectilinear Kinematics: Erratic Motion

When a particle has erratic or changing motion then its position, velocity, and acceleration *cannot* be described by a single continuous mathematical function along the entire path. Instead, a series of functions will be required to specify the motion at different intervals. For this reason, it is convenient to represent the motion as a graph. If a graph of the motion that relates any two of the variables s, v, a, t can be drawn, then this graph can be used to construct subsequent graphs relating two other variables since the variables are related by the differential relationships $v = ds/dt$, $a = dv/dt$, or $a \, ds = v \, dv$. Several situations occur frequently.

The s–t, v–t, and a–t Graphs.   To construct the v–t graph given the s–t graph, Fig. 12–7a, the equation $v = ds/dt$ should be used, since it relates the variables s and t to v. This equation states that

$$\frac{ds}{dt} = v$$

$$\frac{\text{slope of}}{s\text{–}t \text{ graph}} = \text{velocity}$$

For example, by measuring the slope on the s–t graph when $t = t_1$, the velocity is v_1, which is plotted in Fig. 12–7b. The v–t graph can be constructed by plotting this and other values at each instant.

The a–t graph can be constructed from the v–t graph in a similar manner, Fig. 12–8, since

$$\frac{dv}{dt} = a$$

$$\frac{\text{slope of}}{v\text{–}t \text{ graph}} = \text{acceleration}$$

Examples of various measurements are shown in Fig. 12–8a and plotted in Fig. 12–8b.

If the s–t curve for each interval of motion can be expressed by a mathematical function $s = s(t)$, then the equation of the v–t graph for the same interval can be obtained by differentiating this function with respect to time since $v = ds/dt$. Likewise, the equation of the a–t graph for the same interval can be determined by differentiating $v = v(t)$ since $a = dv/dt$. Since differentiation reduces a polynomial of degree n to that of degree $n - 1$, then if the s–t graph is parabolic (a second-degree curve), the v–t graph will be a sloping line (a first-degree curve), and the a–t graph will be a constant or a horizontal line (a zero-degree curve).

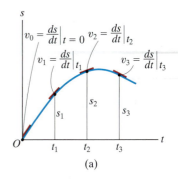

(a)

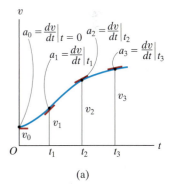

(b)

Fig. 12–7

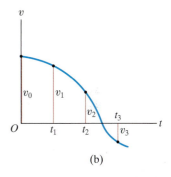

(a)

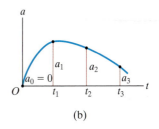

(b)

Fig. 12–8

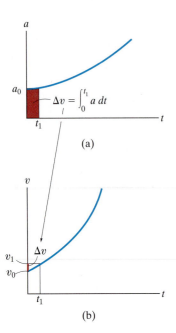

(a)

(b)

Fig. 12–9

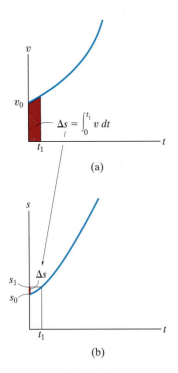

(a)

(b)

Fig. 12–10

If the a–t graph is given, Fig. 12–9a, the v–t graph may be constructed using $a = dv/dt$, written as

$$\Delta v = \int a\, dt$$

$$\begin{array}{c} \text{change in} \\ \text{velocity} \end{array} = \begin{array}{c} \text{area under} \\ a\text{–}t \text{ graph} \end{array}$$

Hence, to construct the v–t graph, we begin with the particle's initial velocity v_0 and then add to this small increments of area (Δv) determined from the a–t graph. In this manner successive points, $v_1 = v_0 + \Delta v$, etc., for the v–t graph are determined, Fig. 12–9b. Notice that an algebraic addition of the area increments of the a–t graph is necessary, since areas lying above the t axis correspond to an increase in v ("positive" area), whereas those lying below the axis indicate a decrease in v ("negative" area).

Similarly, if the v–t graph is given, Fig. 12–10a, it is possible to determine the s–t graph using $v = ds/dt$, written as

$$\Delta s = \int v\, dt$$

$$\text{displacement} = \begin{array}{c} \text{area under} \\ v\text{–}t \text{ graph} \end{array}$$

In the same manner as stated above, we begin with the particle's initial position s_0 and add (algebraically) to this small area increments Δs determined from the v–t graph, Fig. 12–10b.

If segments of the a–t graph can be described by a series of equations, then each of these equations can be *integrated* to yield equations describing the corresponding segments of the v–t graph. In a similar manner, the s–t graph can be obtained by integrating the equations which describe the segments of the v–t graph. As a result, if the a–t graph is linear (a first-degree curve), integration will yield a v–t graph that is parabolic (a second-degree curve) and an s–t graph that is cubic (third-degree curve).

The v–s and a–s Graphs.   If the *a–s* graph can be constructed, then points on the *v–s* graph can be determined by using $v\,dv = a\,ds$. Integrating this equation between the limits $v = v_0$ at $s = s_0$ and $v = v_1$ at $s = s_1$, we have,

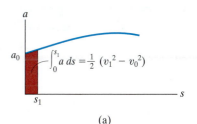

(a)

$$\tfrac{1}{2}(v_1^2 - v_0^2) = \int_{s_0}^{s_1} a\,ds$$

area under
a–s graph

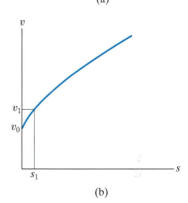

(b)

Fig. 12–11

Therefore, if the red area in Fig. 12–11*a* is determined, and the initial velocity v_0 at $s_0 = 0$ is known, then $v_1 = \left(2\int_{s_0}^{s_1} a\,ds + v_0^2\right)^{1/2}$, Fig. 12–11*b*. Successive points on the *v–s* graph can be constructed in this manner.

If the *v–s* graph is known, the acceleration *a* at any position *s* can be determined using $a\,ds = v\,dv$, written as

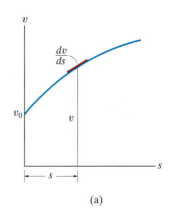

(a)

$$a = v\left(\frac{dv}{ds}\right)$$

velocity times
acceleration = slope of
v–s graph

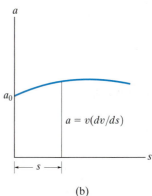

(b)

Fig. 12–12

Thus, at any point (s, v) in Fig. 12–12*a*, the slope dv/ds of the *v–s* graph is measured. Then with *v* and dv/ds known, the value of *a* can be calculated, Fig. 12–12*b*.

The *v–s* graph can also be constructed from the *a–s* graph, or vice versa, by approximating the known graph in various intervals with mathematical functions, $v = f(s)$ or $a = g(s)$, and then using $a\,ds = v\,dv$ to obtain the other graph.

12

EXAMPLE | 12.6

A bicycle moves along a straight road such that its position is described by the graph shown in Fig. 12–13a. Construct the v–t and a–t graphs for $0 \leq t \leq 30$ s.

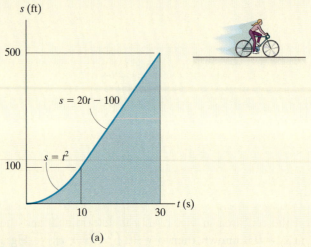

(a)

SOLUTION

v–t Graph. Since $v = ds/dt$, the v–t graph can be determined by differentiating the equations defining the s–t graph, Fig. 12–13a. We have

$$0 \leq t < 10 \text{ s}; \qquad s = (t^2) \text{ ft} \qquad v = \frac{ds}{dt} = (2t) \text{ ft/s}$$

$$10 \text{ s} < t \leq 30 \text{ s}; \qquad s = (20t - 100) \text{ ft} \qquad v = \frac{ds}{dt} = 20 \text{ ft/s}$$

The results are plotted in Fig. 12–13b. We can also obtain specific values of v by measuring the *slope* of the s–t graph at a given instant. For example, at $t = 20$ s, the slope of the s–t graph is determined from the straight line from 10 s to 30 s, i.e.,

$$t = 20 \text{ s}; \qquad v = \frac{\Delta s}{\Delta t} = \frac{500 \text{ ft} - 100 \text{ ft}}{30 \text{ s} - 10 \text{ s}} = 20 \text{ ft/s}$$

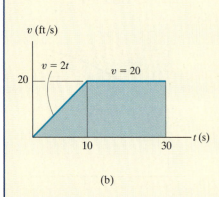

(b)

a–t Graph. Since $a = dv/dt$, the a–t graph can be determined by differentiating the equations defining the lines of the v–t graph. This yields

$$0 \leq t < 10 \text{ s}; \qquad v = (2t) \text{ ft/s} \qquad a = \frac{dv}{dt} = 2 \text{ ft/s}^2$$

$$10 < t \leq 30 \text{ s}; \qquad v = 20 \text{ ft/s} \qquad a = \frac{dv}{dt} = 0$$

The results are plotted in Fig. 12–13c.

NOTE: Show that $a = 2$ ft/s^2 when $t = 5$ s by measuring the slope of the v–t graph.

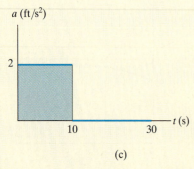

(c)

Fig. 12–13

EXAMPLE 12.7

12

The car in Fig. 12–14a starts from rest and travels along a straight track such that it accelerates at 10 m/s² for 10 s, and then decelerates at 2 m/s². Draw the v–t and s–t graphs and determine the time t' needed to stop the car. How far has the car traveled?

SOLUTION

v–t Graph. Since $dv = a\,dt$, the v–t graph is determined by integrating the straight-line segments of the a–t graph. Using the *initial condition* $v = 0$ when $t = 0$, we have

$$0 \le t < 10 \text{ s}; \quad a = (10) \text{ m/s}^2; \quad \int_0^v dv = \int_0^t 10\,dt, \quad v = 10t$$

When $t = 10$ s, $v = 10(10) = 100$ m/s. Using this as the *initial condition* for the next time period, we have

$$10 \text{ s} < t \le t'; \quad a = (-2) \text{ m/s}^2; \quad \int_{100 \text{ m/s}}^v dv = \int_{10 \text{ s}}^t -2\,dt, \quad v = (-2t + 120) \text{ m/s}$$

When $t = t'$ we require $v = 0$. This yields, Fig. 12–14b,

$$t' = 60 \text{ s} \qquad\qquad Ans.$$

A more direct solution for t' is possible by realizing that the area under the a–t graph is equal to the change in the car's velocity. We require $\Delta v = 0 = A_1 + A_2$, Fig. 12–14a. Thus

$$0 = 10 \text{ m/s}^2(10 \text{ s}) + (-2 \text{ m/s}^2)(t' - 10 \text{ s})$$

$$t' = 60 \text{ s} \qquad\qquad Ans.$$

s–t Graph. Since $ds = v\,dt$, integrating the equations of the v–t graph yields the corresponding equations of the s–t graph. Using the *initial condition* $s = 0$ when $t = 0$, we have

$$0 \le t \le 10 \text{ s}; \quad v = (10t) \text{ m/s}; \quad \int_0^s ds = \int_0^t 10t\,dt, \quad s = (5t^2) \text{ m}$$

When $t = 10$ s, $s = 5(10)^2 = 500$ m. Using this *initial condition*,

$$10 \text{ s} \le t \le 60 \text{ s}; \quad v = (-2t + 120) \text{ m/s}; \quad \int_{500 \text{ m}}^s ds = \int_{10 \text{ s}}^t (-2t + 120)\,dt$$

$$s - 500 = -t^2 + 120t - [-(10)^2 + 120(10)]$$

$$s = (-t^2 + 120t - 600) \text{ m}$$

When $t' = 60$ s, the position is

$$s = -(60)^2 + 120(60) - 600 = 3000 \text{ m} \qquad Ans.$$

The s–t graph is shown in Fig. 12–14c.

NOTE: A direct solution for s is possible when $t' = 60$ s, since the triangular area under the v–t graph would yield the displacement $\Delta s = s - 0$ from $t = 0$ to $t' = 60$ s. Hence,

$$\Delta s = \tfrac{1}{2}(60 \text{ s})(100 \text{ m/s}) = 3000 \text{ m} \qquad Ans.$$

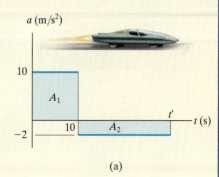

(a)

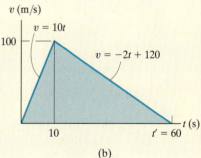

(b)

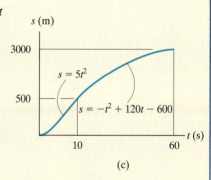

(c)

Fig. 12–14

12

EXAMPLE | 12.8

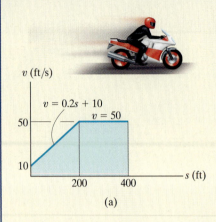

v (ft/s)

$v = 0.2s + 10$

$v = 50$

50

10

200   400 s (ft)

(a)

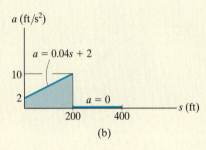

a (ft/s²)

$a = 0.04s + 2$

10

2     $a = 0$

200   400 s (ft)

(b)

Fig. 12–15

The v–s graph describing the motion of a motorcycle is shown in Fig. 12–15a. Construct the a–s graph of the motion and determine the time needed for the motorcycle to reach the position $s = 400$ ft.

SOLUTION

a–s Graph. Since the equations for segments of the v–s graph are given, the a–s graph can be determined using $a\, ds = v\, dv$.

$0 \le s < 200$ ft; $v = (0.2s + 10)$ ft/s

$$a = v\frac{dv}{ds} = (0.2s + 10)\frac{d}{ds}(0.2s + 10) = 0.04s + 2$$

200 ft $< s \le 400$ ft; $v = 50$ ft/s

$$a = v\frac{dv}{ds} = (50)\frac{d}{ds}(50) = 0$$

The results are plotted in Fig. 12–15b.

Time. The time can be obtained using the v–s graph and $v = ds/dt$, because this equation relates v, s, and t. For the first segment of motion, $s = 0$ when $t = 0$, so

$0 \le s < 200$ ft; $v = (0.2s + 10)$ ft/s; $dt = \dfrac{ds}{v} = \dfrac{ds}{0.2s + 10}$

$$\int_0^t dt = \int_0^s \frac{ds}{0.2s + 10}$$

$$t = (5\ln(0.2s + 10) - 5\ln 10)\text{ s}$$

At $s = 200$ ft, $t = 5\ln[0.2(200) + 10] - 5\ln 10 = 8.05$ s. Therefore, using these initial conditions for the second segment of motion,

200 ft $< s \le 400$ ft; $v = 50$ ft/s; $dt = \dfrac{ds}{v} = \dfrac{ds}{50}$

$$\int_{8.05\text{ s}}^t dt = \int_{200\text{ m}}^s \frac{ds}{50};$$

$$t - 8.05 = \frac{s}{50} - 4; \quad t = \left(\frac{s}{50} + 4.05\right)\text{ s}$$

Therefore, at $s = 400$ ft,

$$t = \frac{400}{50} + 4.05 = 12.0\text{ s} \qquad\qquad Ans.$$

NOTE: The graphical results can be checked in part by calculating slopes. For example, at $s = 0$, $a = v(dv/ds) = 10(50 - 10)/200 = 2$ m/s². Also, the results can be checked in part by inspection. The v–s graph indicates the initial increase in velocity (acceleration) followed by constant velocity ($a = 0$).

FUNDAMENTAL PROBLEMS

F12–9. The particle travels along a straight track such that its position is described by the s–t graph. Construct the v–t graph for the same time interval.

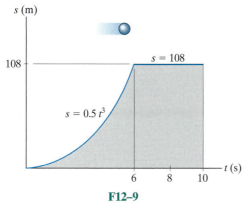

F12–9

F12–10. A van travels along a straight road with a velocity described by the graph. Construct the s–t and a–t graphs during the same period. Take $s = 0$ when $t = 0$.

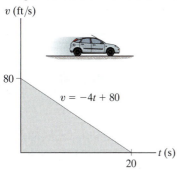

F12–10

F12–11. A bicycle travels along a straight road where its velocity is described by the v–s graph. Construct the a–s graph for the same time interval.

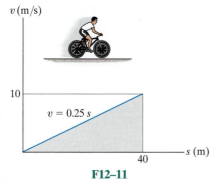

F12–11

F12–12. The sports car travels along a straight road such that its position is described by the graph. Construct the v–t and a–t graphs for the time interval $0 \le t \le 10$ s.

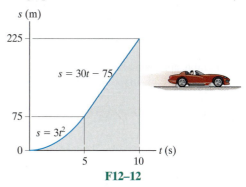

F12–12

F12–13. The dragster starts from rest and has an acceleration described by the graph. Construct the v–t graph for the time interval $0 \le t \le t'$, where t' is the time for the car to come to rest.

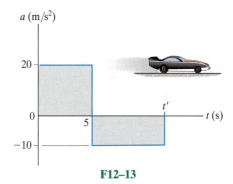

F12–13

F12–14. The dragster starts from rest and has a velocity described by the graph. Construct the s–t graph during the time interval $0 \le t \le 15$ s. Also, determine the total distance traveled during this time interval.

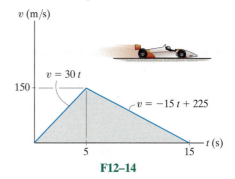

F12–14

PROBLEMS

12–42. The speed of a train during the first minute has been recorded as follows:

t (s)	0	20	40	60
v (m/s)	0	16	21	24

Plot the v–t graph, approximating the curve as straight-line segments between the given points. Determine the total distance traveled.

12–43. A two-stage missile is fired vertically from rest with the acceleration shown. In 15 s the first stage A burns out and the second stage B ignites. Plot the v–t and s–t graphs which describe the two-stage motion of the missile for $0 \le t \le 20$ s.

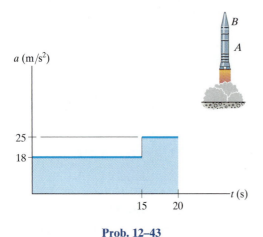

Prob. 12–43

*12–44.** A freight train starts from rest and travels with a constant acceleration of 0.5 ft/s². After a time t' it maintains a constant speed so that when $t = 160$ s it has traveled 2000 ft. Determine the time t' and draw the v–t graph for the motion.

•**12–45.** If the position of a particle is defined by $s = [2 \sin [(\pi/5)t] + 4]$ m, where t is in seconds, construct the s–t, v–t, and a–t graphs for $0 \le t \le 10$ s.

12–46. A train starts from station A and for the first kilometer, it travels with a uniform acceleration. Then, for the next two kilometers, it travels with a uniform speed. Finally, the train decelerates uniformly for another kilometer before coming to rest at station B. If the time for the whole journey is six minutes, draw the v–t graph and determine the maximum speed of the train.

12–47. The particle travels along a straight line with the velocity described by the graph. Construct the a–s graph.

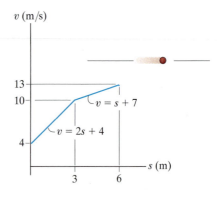

Prob. 12–47

*12–48.** The a–s graph for a jeep traveling along a straight road is given for the first 300 m of its motion. Construct the v–s graph. At $s = 0, v = 0$.

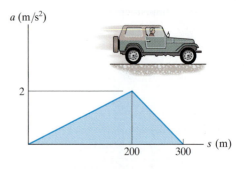

Prob. 12–48

•12–49. A particle travels along a curve defined by the equation $s = (t^3 - 3t^2 + 2t)$ m. where t is in seconds. Draw the $s - t$, $v - t$, and $a - t$ graphs for the particle for $0 \le t \le 3$ s.

12–50. A truck is traveling along the straight line with a velocity described by the graph. Construct the $a-s$ graph for $0 \le s \le 1500$ ft.

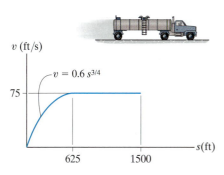

v (ft/s)

$v = 0.6\, s^{3/4}$

75

625   1500 s(ft)

Prob. 12–50

12–51. A car starts from rest and travels along a straight road with a velocity described by the graph. Determine the total distance traveled until the car stops. Construct the $s-t$ and $a-t$ graphs.

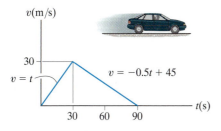

v(m/s)

30

$v = t$ $v = -0.5t + 45$

30   60   90 t(s)

Prob. 12–51

*12–52. A car travels up a hill with the speed shown. Determine the total distance the car travels until it stops ($t = 60$ s). Plot the $a-t$ graph.

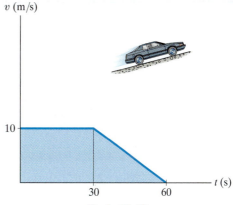

v (m/s)

10

30   60 t (s)

Prob. 12–52

•12–53. The snowmobile moves along a straight course according to the $v-t$ graph. Construct the $s-t$ and $a-t$ graphs for the same 50-s time interval. When $t = 0$, $s = 0$.

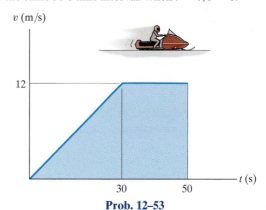

v (m/s)

12

30   50 t (s)

Prob. 12–53

12–54. A motorcyclist at A is traveling at 60 ft/s when he wishes to pass the truck T which is traveling at a constant speed of 60 ft/s. To do so the motorcyclist accelerates at 6 ft/s² until reaching a maximum speed of 85 ft/s. If he then maintains this speed, determine the time needed for him to reach a point located 100 ft in front of the truck. Draw the $v-t$ and $s-t$ graphs for the motorcycle during this time.

$(v_m)_1 = 60$ ft/s $(v_m)_2 = 85$ ft/s

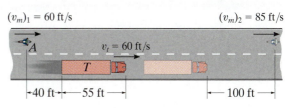

A $v_t = 60$ ft/s

T

40 ft   55 ft   100 ft

Prob. 12–54

12–55. An airplane traveling at 70 m/s lands on a straight runway and has a deceleration described by the graph. Determine the time t' and the distance traveled for it to reach a speed of 5 m/s. Construct the $v-t$ and $s-t$ graphs for this time interval, $0 \leq t \leq t'$.

•12–57. The dragster starts from rest and travels along a straight track with an acceleration-deceleration described by the graph. Construct the $v-s$ graph for $0 \leq s \leq s'$, and determine the distance s' traveled before the dragster again comes to rest.

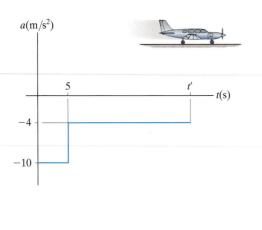

Prob. 12–55

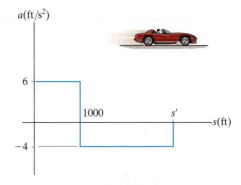

Prob. 12–57

***12–56.** The position of a cyclist traveling along a straight road is described by the graph. Construct the $v-t$ and $a-t$ graphs.

12–58. A sports car travels along a straight road with an acceleration-deceleration described by the graph. If the car starts from rest, determine the distance s' the car travels until it stops. Construct the $v-s$ graph for $0 \leq s \leq s'$.

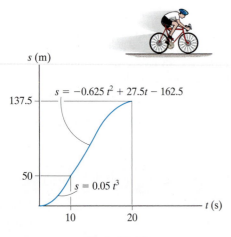

Prob. 12–56

Prob. 12–58

12–59. A missile starting from rest travels along a straight track and for 10 s has an acceleration as shown. Draw the $v-t$ graph that describes the motion and find the distance traveled in 10 s.

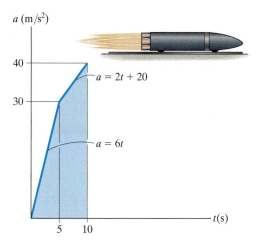

a (m/s²)

40

$a = 2t + 20$

30

$a = 6t$

5   10 t(s)

Prob. 12–59

•12–61. The $v-t$ graph of a car while traveling along a road is shown. Draw the $s-t$ and $a-t$ graphs for the motion.

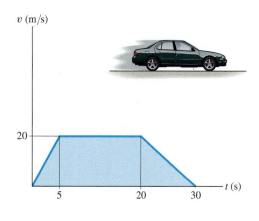

v (m/s)

20

5   20   30 t (s)

Prob. 12–61

***12–60.** A motorcyclist starting from rest travels along a straight road and for 10 s has an acceleration as shown. Draw the $v-t$ graph that describes the motion and find the distance traveled in 10 s.

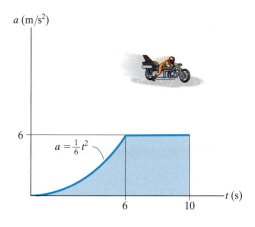

a (m/s²)

6

$a = \frac{1}{6}t^2$

6   10 t (s)

Prob. 12–60

12–62. The boat travels in a straight line with the acceleration described by the $a-s$ graph. If it starts from rest, construct the $v-s$ graph and determine the boat's maximum speed. What distance s' does it travel before it stops?

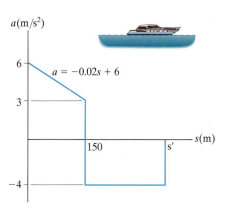

a(m/s²)

6

$a = -0.02s + 6$

3

150   s' s(m)

−4

Prob. 12–62

12–63. The rocket has an acceleration described by the graph. If it starts from rest, construct the v–t and s–t graphs for the motion for the time interval $0 \leq t \leq 14\,\text{s}$.

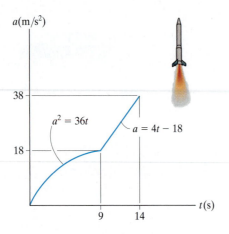

Prob. 12–63

•12–65. The acceleration of the speed boat starting from rest is described by the graph. Construct the v–s graph.

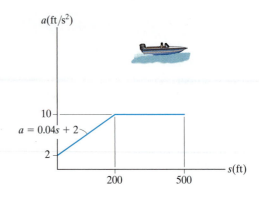

Prob. 12–65

*12–64. The jet bike is moving along a straight road with the speed described by the v–s graph. Construct the a–s graph.

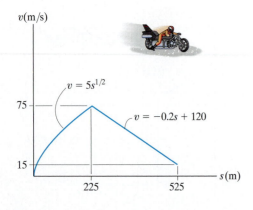

Prob. 12–64

12–66. The boat travels along a straight line with the speed described by the graph. Construct the s–t and a–s graphs. Also, determine the time required for the boat to travel a distance $s = 400\,\text{m}$ if $s = 0$ when $t = 0$.

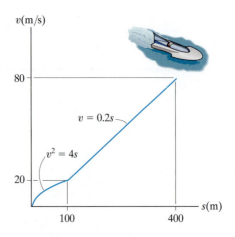

Prob. 12–66

12–67. The *s–t* graph for a train has been determined experimentally. From the data, construct the *v–t* and *a–t* graphs for the motion.

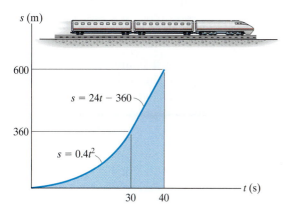

$$s = 24t - 360$$

$$s = 0.4t^2$$

Prob. 12–67

•12–69. The airplane travels along a straight runway with an acceleration described by the graph. If it starts from rest and requires a velocity of 90 m/s to take off, determine the minimum length of runway required and the time *t'* for take off. Construct the *v–t* and *s–t* graphs.

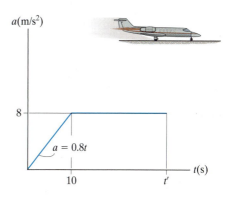

$$a = 0.8t$$

Prob. 12–69

***12–68.** The airplane lands at 250 ft/s on a straight runway and has a deceleration described by the graph. Determine the distance *s'* traveled before its speed is decreased to 25 ft/s. Draw the *s–t* graph.

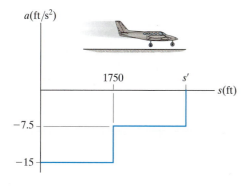

Prob. 12–68

12–70. The *a–t* graph of the bullet train is shown. If the train starts from rest, determine the elapsed time *t'* before it again comes to rest. What is the total distance traveled during this time interval? Construct the *v–t* and *s–t* graphs.

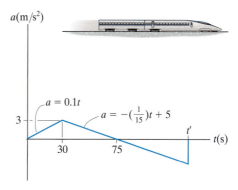

$$a = 0.1t$$

$$a = -\left(\tfrac{1}{15}\right)t + 5$$

Prob. 12–70

12

12.4   General Curvilinear Motion

Curvilinear motion occurs when a particle moves along a curved path. Since this path is often described in three dimensions, vector analysis will be used to formulate the particle's position, velocity, and acceleration.* In this section the general aspects of curvilinear motion are discussed, and in subsequent sections we will consider three types of coordinate systems often used to analyze this motion.

Position.   Consider a particle located at a point on a space curve defined by the path function $s(t)$, Fig. 12–16a. The position of the particle, measured from a fixed point O, will be designated by the *position vector* $\mathbf{r} = \mathbf{r}(t)$. Notice that both the magnitude and direction of this vector will change as the particle moves along the curve.

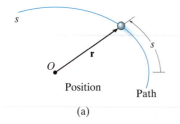

Position
(a)

Displacement.   Suppose that during a small time interval Δt the particle moves a distance Δs along the curve to a new position, defined by $\mathbf{r}' = \mathbf{r} + \Delta\mathbf{r}$, Fig. 12–16b. The *displacement* $\Delta\mathbf{r}$ represents the change in the particle's position and is determined by vector subtraction; i.e., $\Delta\mathbf{r} = \mathbf{r}' - \mathbf{r}$.

Velocity.   During the time Δt, the *average velocity* of the particle is

$$\mathbf{v}_{avg} = \frac{\Delta\mathbf{r}}{\Delta t}$$

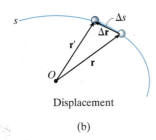

Displacement
(b)

The *instantaneous velocity* is determined from this equation by letting $\Delta t \to 0$, and consequently the direction of $\Delta\mathbf{r}$ *approaches* the *tangent* to the curve. Hence, $\mathbf{v} = \lim_{\Delta t \to 0}(\Delta\mathbf{r}/\Delta t)$ or

$$\mathbf{v} = \frac{d\mathbf{r}}{dt} \qquad (12\text{–}7)$$

Since $d\mathbf{r}$ will be tangent to the curve, the *direction* of $\mathbf{v}$ is also *tangent to the curve*, Fig. 12–16c. The *magnitude* of $\mathbf{v}$, which is called the *speed*, is obtained by realizing that the length of the straight line segment $\Delta\mathbf{r}$ in Fig. 12–16b approaches the arc length Δs as $\Delta t \to 0$, we have $v = \lim_{\Delta t \to 0}(\Delta r/\Delta t) = \lim_{\Delta t \to 0}(\Delta s/\Delta t)$, or

$$v = \frac{ds}{dt} \qquad (12\text{–}8)$$

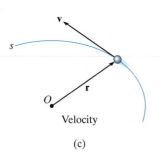

Velocity
(c)

Fig. 12–16

Thus, the *speed* can be obtained by differentiating the path function s with respect to time.

---

*A summary of some of the important concepts of vector analysis is given in Appendix B.

Acceleration.   If the particle has a velocity **v** at time t and a velocity **v′** = **v** + Δ**v** at t + Δt, Fig. 12–16d, then the *average acceleration* of the particle during the time interval Δt is

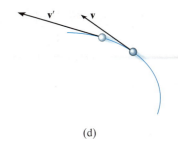

(d)

$$\mathbf{a}_{\text{avg}} = \frac{\Delta \mathbf{v}}{\Delta t}$$

where Δ**v** = **v′** − **v**. To study this time rate of change, the two velocity vectors in Fig. 12–16d are plotted in Fig. 12–16e such that their tails are located at the fixed point O' and their arrowheads touch points on a curve. This curve is called a *hodograph*, and when constructed, it describes the locus of points for the arrowhead of the velocity vector in the same manner as the *path s* describes the locus of points for the arrowhead of the position vector, Fig. 12–16a.

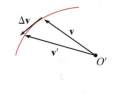

(e)

To obtain the *instantaneous acceleration*, let Δ$t \to 0$ in the above equation. In the limit Δ**v** will approach the *tangent to the hodograph*, and so $\mathbf{a} = \lim_{\Delta t \to 0} (\Delta \mathbf{v}/\Delta t)$, or

$$\mathbf{a} = \frac{d\mathbf{v}}{dt} \qquad (12\text{–}9)$$

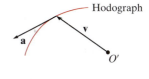

Hodograph

(f)

Substituting Eq. 12–7 into this result, we can also write

$$\mathbf{a} = \frac{d^2\mathbf{r}}{dt^2}$$

By definition of the derivative, **a** acts *tangent to the hodograph*, Fig. 12–16f, and, *in general it is not tangent to the path of motion*, Fig. 12–16g. To clarify this point, realize that Δ**v** and consequently **a** must account for the change made in *both* the magnitude *and* direction of the velocity **v** as the particle moves from one point to the next along the path, Fig. 12–16d. However, in order for the particle to follow any curved path, the directional change always "swings" the velocity vector toward the "inside" or "concave side" of the path, and therefore **a** *cannot* remain tangent to the path. In summary, **v** is always tangent to the *path* and **a** is always tangent to the *hodograph*.

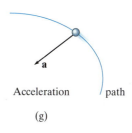

Acceleration   path

(g)

Fig. 12–16 (cont.)

12.5 Curvilinear Motion: Rectangular Components

Occasionally the motion of a particle can best be described along a path that can be expressed in terms of its x, y, z coordinates.

Position. If the particle is at point (x, y, z) on the curved path s shown in Fig. 12–17a, then its location is defined by the *position vector*

$$\mathbf{r} = x\mathbf{i} + y\mathbf{j} + z\mathbf{k} \tag{12–10}$$

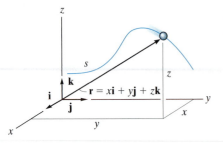

Position

(a)

When the particle moves, the x, y, z components of $\mathbf{r}$ will be functions of time; i.e., $x = x(t)$, $y = y(t)$, $z = z(t)$, so that $\mathbf{r} = \mathbf{r}(t)$.

At any instant the *magnitude* of $\mathbf{r}$ is defined from Eq. C–3 in Appendix C as

$$r = \sqrt{x^2 + y^2 + z^2}$$

And the *direction* of $\mathbf{r}$ is specified by the unit vector $\mathbf{u}_r = \mathbf{r}/r$.

Velocity. The first time derivative of $\mathbf{r}$ yields the velocity of the particle. Hence,

$$\mathbf{v} = \frac{d\mathbf{r}}{dt} = \frac{d}{dt}(x\mathbf{i}) + \frac{d}{dt}(y\mathbf{j}) + \frac{d}{dt}(z\mathbf{k})$$

When taking this derivative, it is necessary to account for changes in *both* the magnitude and direction of each of the vector's components. For example, the derivative of the $\mathbf{i}$ component of $\mathbf{r}$ is

$$\frac{d}{dt}(x\mathbf{i}) = \frac{dx}{dt}\mathbf{i} + x\frac{d\mathbf{i}}{dt}$$

The second term on the right side is zero, provided the x, y, z reference frame is *fixed*, and therefore the *direction* (and the *magnitude*) of $\mathbf{i}$ does not change with time. Differentiation of the $\mathbf{j}$ and $\mathbf{k}$ components may be carried out in a similar manner, which yields the final result,

$$\mathbf{v} = \frac{d\mathbf{r}}{dt} = v_x\mathbf{i} + v_y\mathbf{j} + v_z\mathbf{k} \tag{12–11}$$

where

$$v_x = \dot{x} \quad v_y = \dot{y} \quad v_z = \dot{z} \tag{12–12}$$

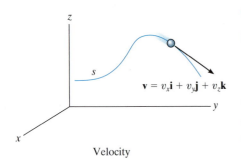

Velocity

(b)

Fig. 12–17

The "dot" notation $\dot{x}$, $\dot{y}$, $\dot{z}$ represents the first time derivatives of $x = x(t)$, $y = y(t)$, $z = z(t)$, respectively.

The velocity has a *magnitude* that is found from

$$v = \sqrt{v_x^2 + v_y^2 + v_z^2}$$

and a *direction* that is specified by the unit vector $\mathbf{u}_v = \mathbf{v}/v$. As discussed in Sec. 12–4, this direction is *always tangent to the path*, as shown in Fig. 12–17b.

Acceleration.   The acceleration of the particle is obtained by taking the first time derivative of Eq. 12–11 (or the second time derivative of Eq. 12–10). We have

$$\mathbf{a} = \frac{d\mathbf{v}}{dt} = a_x\mathbf{i} + a_y\mathbf{j} + a_z\mathbf{k} \qquad (12\text{–}13)$$

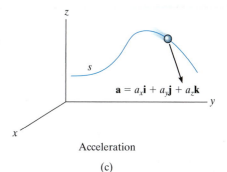

Acceleration

(c)

where

$$\begin{aligned}
a_x &= \dot{v}_x = \ddot{x} \\
a_y &= \dot{v}_y = \ddot{y} \\
a_z &= \dot{v}_z = \ddot{z}
\end{aligned} \qquad (12\text{–}14)$$

Here a_x, a_y, a_z represent, respectively, the first time derivatives of $v_x = v_x(t)$, $v_y = v_y(t)$, $v_z = v_z(t)$, or the second time derivatives of the functions $x = x(t)$, $y = y(t)$, $z = z(t)$.

The acceleration has a *magnitude*

$$a = \sqrt{a_x^2 + a_y^2 + a_z^2}$$

and a *direction* specified by the unit vector $\mathbf{u}_a = \mathbf{a}/a$. Since $\mathbf{a}$ represents the time rate of *change* in both the magnitude and direction of the velocity, in general $\mathbf{a}$ will *not* be tangent to the path, Fig. 12–17c.

12

Important Points

- Curvilinear motion can cause changes in *both* the magnitude and direction of the position, velocity, and acceleration vectors.

- The velocity vector is always directed *tangent* to the path.

- In general, the acceleration vector is *not* tangent to the path, but rather, it is tangent to the hodograph.

- If the motion is described using rectangular coordinates, then the components along each of the axes do not change direction, only their magnitude and sense (algebraic sign) will change.

- By considering the component motions, the change in magnitude and direction of the particle's position and velocity are automatically taken into account.

Procedure for Analysis

Coordinate System.

- A rectangular coordinate system can be used to solve problems for which the motion can conveniently be expressed in terms of its x, y, z components.

Kinematic Quantities.

- Since *rectilinear motion* occurs along *each coordinate axis*, the motion along each axis is found using $v = ds/dt$ and $a = dv/dt$; or in cases where the motion is not expressed as a function of time, the equation $a\, ds = v\, dv$ can be used.

- In two dimensions, the equation of the path $y = f(x)$ can be used to relate the x and y components of velocity and acceleration by applying the chain rule of calculus. A review of this concept is given in Appendix C.

- Once the x, y, z components of **v** and **a** have been determined, the magnitudes of these vectors are found from the Pythagorean theorem, Eq. B–3, and their coordinate direction angles from the components of their unit vectors, Eqs. B–4 and B–5.

EXAMPLE 12.9

At any instant the horizontal position of the weather balloon in Fig. 12–18a is defined by $x = (8t)$ ft, where t is in seconds. If the equation of the path is $y = x^2/10$, determine the magnitude and direction of the velocity and the acceleration when $t = 2$ s.

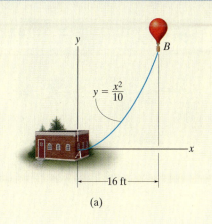

$y = \dfrac{x^2}{10}$

16 ft

(a)

SOLUTION

Velocity. The velocity component in the x direction is

$$v_x = \dot{x} = \frac{d}{dt}(8t) = 8 \text{ ft/s} \rightarrow$$

To find the relationship between the velocity components we will use the chain rule of calculus. (See Appendix A for a full explanation.)

$$v_y = \dot{y} = \frac{d}{dt}(x^2/10) = 2x\dot{x}/10 = 2(16)(8)/10 = 25.6 \text{ ft/s} \uparrow$$

When $t = 2$ s, the magnitude of velocity is therefore

$$v = \sqrt{(8 \text{ ft/s})^2 + (25.6 \text{ ft/s})^2} = 26.8 \text{ ft/s} \qquad \textit{Ans.}$$

The direction is tangent to the path, Fig. 12–18b, where

$$\theta_v = \tan^{-1}\frac{v_y}{v_x} = \tan^{-1}\frac{25.6}{8} = 72.6° \qquad \textit{Ans.}$$

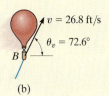

$v = 26.8$ ft/s

$\theta_v = 72.6°$

B

(b)

Acceleration. The relationship between the acceleration components is determined using the chain rule. (See Appendix C.) We have

$$a_x = \dot{v}_x = \frac{d}{dt}(8) = 0$$

$$a_y = \dot{v}_y = \frac{d}{dt}(2x\dot{x}/10) = 2(\dot{x})\dot{x}/10 + 2x(\ddot{x})/10$$

$$= 2(8)^2/10 + 2(16)(0)/10 = 12.8 \text{ ft/s}^2 \uparrow$$

Thus,

$$a = \sqrt{(0)^2 + (12.8)^2} = 12.8 \text{ ft/s}^2 \qquad \textit{Ans.}$$

The direction of **a**, as shown in Fig. 12–18c, is

$$\theta_a = \tan^{-1}\frac{12.8}{0} = 90° \qquad \textit{Ans.}$$

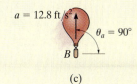

$a = 12.8$ ft/s²

$\theta_a = 90°$

B

(c)

Fig. 12–18

NOTE: It is also possible to obtain v_y and a_y by first expressing $y = f(t) = (8t)^2/10 = 6.4t^2$ and then taking successive time derivatives.

12

EXAMPLE | 12.10

For a short time, the path of the plane in Fig. 12–19a is described by $y = (0.001x^2)$ m. If the plane is rising with a constant velocity of 10 m/s, determine the magnitudes of the velocity and acceleration of the plane when it is at $y = 100$ m.

SOLUTION

When $y = 100$ m, then $100 = 0.001x^2$ or $x = 316.2$ m. Also, since $v_y = 10$ m/s, then

$$y = v_y t; \qquad 100\text{ m} = (10\text{ m/s})\,t \qquad t = 10\text{ s}$$

Velocity. Using the chain rule (see Appendix C) to find the relationship between the velocity components, we have

$$v_y = \dot{y} = \frac{d}{dt}(0.001x^2) = (0.002x)\dot{x} = 0.002xv_x \qquad (1)$$

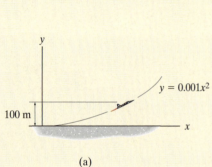

$y = 0.001x^2$

100 m

(a)

Thus

$$10\text{ m/s} = 0.002(316.2\text{ m})(v_x)$$
$$v_x = 15.81\text{ m/s}$$

The magnitude of the velocity is therefore

$$v = \sqrt{v_x^2 + v_y^2} = \sqrt{(15.81\text{ m/s})^2 + (10\text{ m/s})^2} = 18.7\text{ m/s} \qquad \textit{Ans.}$$

Acceleration. Using the chain rule, the time derivative of Eq. (1) gives the relation between the acceleration components.

$$a_y = \dot{v}_y = 0.002\dot{x}v_x + 0.002x\dot{v}_x = 0.002(v_x^2 + xa_x)$$

When $x = 316.2$ m, $v_x = 15.81$ m/s, $\dot{v}_y = a_y = 0$,

$$0 = 0.002((15.81\text{ m/s})^2 + 316.2\text{ m}(a_x))$$
$$a_x = -0.791\text{ m/s}^2$$

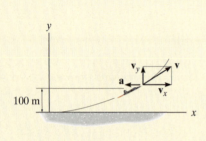

(b)

Fig. 12–19

The magnitude of the plane's acceleration is therefore

$$a = \sqrt{a_x^2 + a_y^2} = \sqrt{(-0.791\text{ m/s}^2)^2 + (0\text{ m/s}^2)^2}$$
$$= 0.791\text{ m/s}^2 \qquad \textit{Ans.}$$

These results are shown in Fig. 12–19b.

12.6   Motion of a Projectile

The free-flight motion of a projectile is often studied in terms of its rectangular components. To illustrate the kinematic analysis, consider a projectile launched at point (x_0, y_0), with an initial velocity of $\mathbf{v}_0$, having components $(\mathbf{v}_0)_x$ and $(\mathbf{v}_0)_y$, Fig. 12–20. When air resistance is neglected, the only force acting on the projectile is its weight, which causes the projectile to have a *constant downward acceleration* of approximately $a_c = g = 9.81 \text{ m/s}^2$ or $g = 32.2 \text{ ft/s}^2$.*

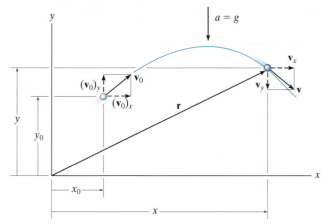

Fig. 12–20

Horizontal Motion.   Since $a_x = 0$, application of the constant acceleration equations, 12–4 to 12–6, yields

$(\overset{+}{\rightarrow})$ $\qquad\qquad v = v_0 + a_c t; \qquad\qquad\qquad v_x = (v_0)_x$

$(\overset{+}{\rightarrow})$ $\qquad\qquad x = x_0 + v_0 t + \frac{1}{2} a_c t^2; \qquad\quad x = x_0 + (v_0)_x t$

$(\overset{+}{\rightarrow})$ $\qquad\qquad v^2 = v_0^2 + 2a_c(x - x_0); \qquad v_x = (v_0)_x$

The first and last equations indicate that *the horizontal component of velocity always remains constant during the motion.*

Vertical Motion.   Since the positive y axis is directed upward, then $a_y = -g$. Applying Eqs. 12–4 to 12–6, we get

$(+\uparrow)$ $\qquad\qquad v = v_0 + a_c t; \qquad\qquad\qquad v_y = (v_0)_y - gt$

$(+\uparrow)$ $\qquad\qquad y = y_0 + v_0 t + \frac{1}{2} a_c t^2; \qquad\quad y = y_0 + (v_0)_y t - \frac{1}{2} g t^2$

$(+\uparrow)$ $\qquad\qquad v^2 = v_0^2 + 2a_c(y - y_0); \qquad v_y^2 = (v_0)_y^2 - 2g(y - y_0)$

Recall that the last equation can be formulated on the basis of eliminating the time t from the first two equations, and therefore *only two of the above three equations are independent of one another.*

* This assumes that the earth's gravitational field does not vary with altitude.

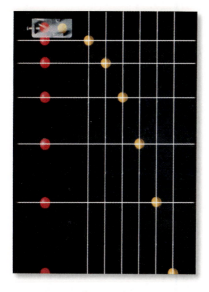

Each picture in this sequence is taken after the same time interval. The red ball falls from rest, whereas the yellow ball is given a horizontal velocity when released. Both balls accelerate downward at the same rate, and so they remain at the same elevation at any instant. This acceleration causes the difference in elevation between the balls to increase between successive photos. Also, note the horizontal distance between successive photos of the yellow ball is constant since the velocity in the horizontal direction remains constant.

EXAMPLE | 12.12

The chipping machine is designed to eject wood chips at $v_O = 25$ ft/s as shown in Fig. 12–22. If the tube is oriented at 30° from the horizontal, determine how high, h, the chips strike the pile if at this instant they land on the pile 20 ft from the tube.

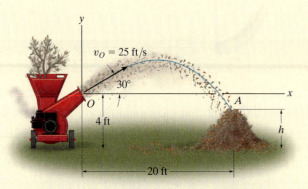

Fig. 12–22

SOLUTION

Coordinate System.   When the motion is analyzed between points O and A, the three unknowns are the height h, time of flight t_{OA}, and vertical component of velocity $(v_A)_y$. [Note that $(v_A)_x = (v_O)_x$.] With the origin of coordinates at O, Fig. 12–22, the initial velocity of a chip has components of

$$(v_O)_x = (25 \cos 30°) \text{ ft/s} = 21.65 \text{ ft/s} \rightarrow$$

$$(v_O)_y = (25 \sin 30°) \text{ ft/s} = 12.5 \text{ ft/s} \uparrow$$

Also, $(v_A)_x = (v_O)_x = 21.65$ ft/s and $a_y = -32.2$ ft/s^2. Since we do not need to determine $(v_A)_y$, we have

Horizontal Motion.

$$(\xrightarrow{+})\qquad x_A = x_O + (v_O)_x t_{OA}$$
$$20 \text{ ft} = 0 + (21.65 \text{ ft/s})t_{OA}$$
$$t_{OA} = 0.9238 \text{ s}$$

Vertical Motion.   Relating t_{OA} to the initial and final elevations of a chip, we have

$$(+\uparrow)\quad y_A = y_O + (v_O)_y t_{OA} + \tfrac{1}{2}a_c t_{OA}^2$$
$$(h - 4 \text{ ft}) = 0 + (12.5 \text{ ft/s})(0.9238 \text{ s}) + \tfrac{1}{2}(-32.2 \text{ ft/s}^2)(0.9238 \text{ s})^2$$
$$h = 1.81 \text{ ft}\qquad\qquad\qquad Ans.$$

NOTE: We can determine $(v_A)_y$ by using $(v_A)_y = (v_O)_y + a_c t_{OA}$.

EXAMPLE 12.13

The track for this racing event was designed so that riders jump off the slope at 30°, from a height of 1 m. During a race it was observed that the rider shown in Fig. 12–23a remained in mid air for 1.5 s. Determine the speed at which he was traveling off the ramp, the horizontal distance he travels before striking the ground, and the maximum height he attains. Neglect the size of the bike and rider.

(a)

SOLUTION

Coordinate System. As shown in Fig. 12–23b, the origin of the coordinates is established at A. Between the end points of the path AB the three unknowns are the initial speed v_A, range R, and the vertical component of velocity $(v_B)_y$.

Vertical Motion. Since the time of flight and the vertical distance between the ends of the path are known, we can determine v_A.

$$(+\uparrow) \qquad y_B = y_A + (v_A)_y t_{AB} + \tfrac{1}{2} a_c t_{AB}^2$$

$$-1 \text{ m} = 0 + v_A \sin 30°(1.5 \text{ s}) + \tfrac{1}{2}(-9.81 \text{ m/s}^2)(1.5 \text{ s})^2$$

$$v_A = 13.38 \text{ m/s} = 13.4 \text{ m/s} \qquad \qquad Ans.$$

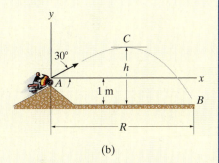

(b)

Fig. 12–23

Horizontal Motion. The range R can now be determined.

$$(\xrightarrow{+}) \qquad x_B = x_A + (v_A)_x t_{AB}$$

$$R = 0 + 13.38 \cos 30° \text{ m/s}(1.5 \text{ s})$$

$$= 17.4 \text{ m} \qquad \qquad Ans.$$

In order to find the maximum height h we will consider the path AC, Fig. 12–23b. Here the three unknowns are the time of flight t_{AC}, the horizontal distance from A to C, and the height h. At the maximum height $(v_C)_y = 0$, and since v_A is known, we can determine h *directly* without considering t_{AC} using the following equation.

$$(v_C)_y^2 = (v_A)_y^2 + 2a_c[y_C - y_A]$$

$$0^2 = (13.38 \sin 30° \text{ m/s})^2 + 2(-9.81 \text{ m/s}^2)[(h - 1 \text{ m}) - 0]$$

$$h = 3.28 \text{ m} \qquad \qquad Ans.$$

NOTE: Show that the bike will strike the ground at B with a velocity having components of

$$(v_B)_x = 11.6 \text{ m/s} \rightarrow, \quad (v_B)_y = 8.02 \text{ m/s} \downarrow$$

12

FUNDAMENTAL PROBLEMS

F12–15. If the x and y components of a particle's velocity are $v_x = (32t)$ m/s and $v_y = 8$ m/s, determine the equation of the path $y = f(x)$. $x = 0$ and $y = 0$ when $t = 0$.

F12–16. A particle is traveling along the straight path. If its position along the x axis is $x = (8t)$ m, where t is in seconds, determine its speed when $t = 2$ s.

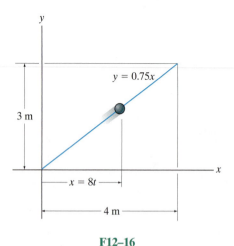

F12–16

F12–17. A particle is constrained to travel along the path. If $x = (4t^4)$ m, where t is in seconds, determine the magnitude of the particle's velocity and acceleration when $t = 0.5$ s.

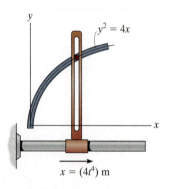

F12–17

F12–18. A particle travels along a straight-line path $y = 0.5x$. If the x component of the particle's velocity is $v_x = (2t^2)$ m/s, where t is in seconds, determine the magnitude of the particle's velocity and acceleration when $t = 4$ s.

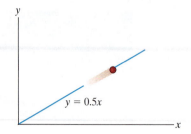

F12–18

F12–19. A particle is traveling along the parabolic path $y = 0.25x^2$. If $x = (2t^2)$ m, where t is in seconds, determine the magnitude of the particle's velocity and acceleration when $t = 2$ s.

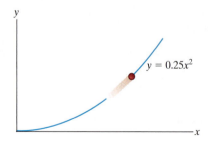

F12–19

F12–20. The position of a box sliding down the ramp can be described by $\mathbf{r} = [2 \sin(2t)\mathbf{i} + 2 \cos t\mathbf{j} - 2t^2\mathbf{k}]$ ft, where t is in seconds and the arguments for the sine and cosine are in radians. Determine the velocity and acceleration of the box when $t = 2$ s.

F12–20

F12–21. The ball is kicked from point A with the initial velocity $v_A = 10$ m/s. Determine the maximum height h it reaches.

F12–22. The ball is kicked from point A with the initial velocity $v_A = 10$ m/s. Determine the range R, and the speed when the ball strikes the ground.

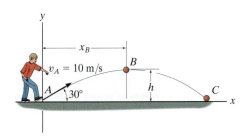

F12–21/22

F12–23. Determine the speed at which the basketball at A must be thrown at the angle of 30° so that it makes it to the basket at B.

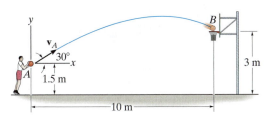

F12–23

F12–24. Water is sprayed at an angle of 90° from the slope at 20 m/s. Determine the range R.

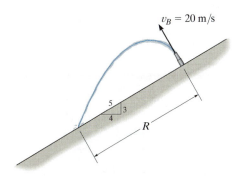

F12–24

F12–25. A ball is thrown from A. If it is required to clear the wall at B, determine the minimum magnitude of its initial velocity $\mathbf{v}_A$.

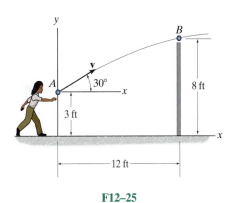

F12–25

F12–26. A projectile is fired with an initial velocity of $v_A = 150$ m/s off the roof of the building. Determine the range R where it strikes the ground at B.

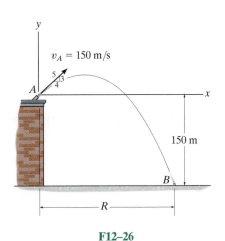

F12–26

PROBLEMS

12–71. The position of a particle is $\mathbf{r} = \{(3t^3 - 2t)\mathbf{i} - (4t^{1/2} + t)\mathbf{j} + (3t^2 - 2)\mathbf{k}\}$ m, where t is in seconds. Determine the magnitude of the particle's velocity and acceleration when $t = 2$ s.

*12–72. The velocity of a particle is $\mathbf{v} = \{3\mathbf{i} + (6 - 2t)\mathbf{j}\}$ m/s, where t is in seconds. If $\mathbf{r} = \mathbf{0}$ when $t = 0$, determine the displacement of the particle during the time interval $t = 1$ s to $t = 3$ s.

•12–73. A particle travels along the parabolic path $y = bx^2$. If its component of velocity along the y axis is $v_y = ct^2$, determine the x and y components of the particle's acceleration. Here b and c are constants.

12–74. The velocity of a particle is given by $\mathbf{v} = \{16t^2\mathbf{i} + 4t^3\mathbf{j} + (5t + 2)\mathbf{k}\}$ m/s, where t is in seconds. If the particle is at the origin when $t = 0$, determine the magnitude of the particle's acceleration when $t = 2$ s. Also, what is the x, y, z coordinate position of the particle at this instant?

12–75. A particle travels along the circular path $x^2 + y^2 = r^2$. If the y component of the particle's velocity is $v_y = 2r \cos 2t$, determine the x and y components of its acceleration at any instant.

*12–76. The box slides down the slope described by the equation $y = (0.05x^2)$ m, where x is in meters. If the box has x components of velocity and acceleration of $v_x = -3$ m/s and $a_x = -1.5$ m/s^2 at $x = 5$ m, determine the y components of the velocity and the acceleration of the box at this instant.

•12–77. The position of a particle is defined by $\mathbf{r} = \{5 \cos 2t\, \mathbf{i} + 4 \sin 2t\, \mathbf{j}\}$ m, where t is in seconds and the arguments for the sine and cosine are given in radians. Determine the magnitudes of the velocity and acceleration of the particle when $t = 1$ s. Also, prove that the path of the particle is elliptical.

12–78. Pegs A and B are restricted to move in the elliptical slots due to the motion of the slotted link. If the link moves with a constant speed of 10 m/s, determine the magnitude of the velocity and acceleration of peg A when $x = 1$ m.

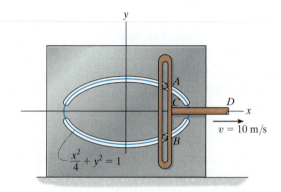

Prob. 12–78

12–79. A particle travels along the path $y^2 = 4x$ with a constant speed of $v = 4$ m/s. Determine the x and y components of the particle's velocity and acceleration when the particle is at $x = 4$ m.

*12–80. The van travels over the hill described by $y = (-1.5(10^{-3})\,x^2 + 15)$ ft. If it has a constant speed of 75 ft/s, determine the x and y components of the van's velocity and acceleration when $x = 50$ ft.

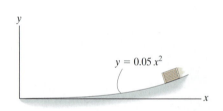

Prob. 12–76

Prob. 12–80

•**12–81.** A particle travels along the circular path from A to B in 1 s. If it takes 3 s for it to go from A to C, determine its *average velocity* when it goes from B to C.

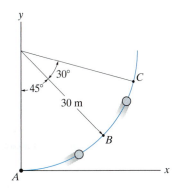

Prob. 12–81

12–82. A car travels east 2 km for 5 minutes, then north 3 km for 8 minutes, and then west 4 km for 10 minutes. Determine the total distance traveled and the magnitude of displacement of the car. Also, what is the magnitude of the average velocity and the average speed?

12–83. The roller coaster car travels down the helical path at constant speed such that the parametric equations that define its position are $x = c \sin kt$, $y = c \cos kt$, $z = h - bt$, where k, c, h, and b are constants. Determine the magnitudes of its velocity and acceleration.

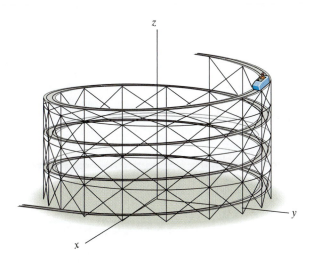

Prob. 12–83

*****12–84.** The path of a particle is defined by $y^2 = 4kx$, and the component of velocity along the y axis is $v_y = ct$, where both k and c are constants. Determine the x and y components of acceleration when $y = y_0$.

•**12–85.** A particle moves along the curve $y = x - (x^2/400)$, where x and y are in ft. If the velocity component in the x direction is $v_x = 2$ ft/s and remains *constant*, determine the magnitudes of the velocity and acceleration when $x = 20$ ft.

12–86. The motorcycle travels with constant speed v_0 along the path that, for a short distance, takes the form of a sine curve. Determine the x and y components of its velocity at any instant on the curve.

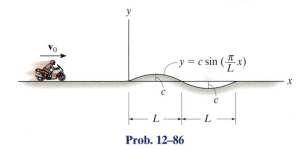

Prob. 12–86

12–87. The skateboard rider leaves the ramp at A with an initial velocity v_A at a 30° angle. If he strikes the ground at B, determine v_A and the time of flight.

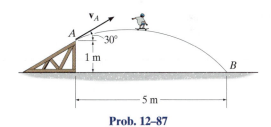

Prob. 12–87

*****12–88.** The pitcher throws the baseball horizontally with a speed of 140 ft/s from a height of 5 ft. If the batter is 60 ft away, determine the time for the ball to arrive at the batter and the height h at which it passes the batter.

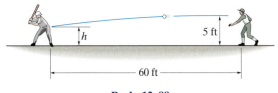

Prob. 12–88

12

•12–89. The ball is thrown off the top of the building. If it strikes the ground at B in 3 s, determine the initial velocity v_A and the inclination angle θ_A at which it was thrown. Also, find the magnitude of the ball's velocity when it strikes the ground.

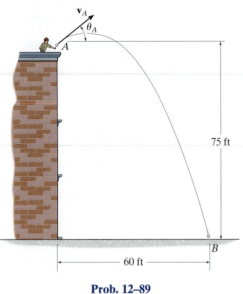

75 ft

60 ft

B

Prob. 12–89

12–90. A projectile is fired with a speed of $v = 60$ m/s at an angle of 60°. A second projectile is then fired with the same speed 0.5 s later. Determine the angle θ of the second projectile so that the two projectiles collide. At what position (x, y) will this happen?

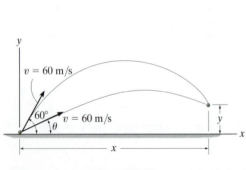

y

$v = 60$ m/s

60°

$v = 60$ m/s

θ

y

x

x

Prob. 12–90

12–91. The fireman holds the hose at an angle $\theta = 30°$ with horizontal, and the water is discharged from the hose at A with a speed of $v_A = 40$ ft/s. If the water stream strikes the building at B, determine his two possible distances s from the building.

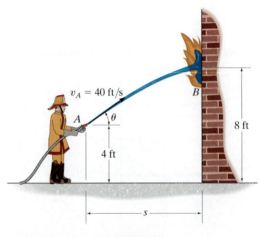

$v_A = 40$ ft/s

A

θ

B

8 ft

4 ft

s

Prob. 12–91

***12–92.** Water is discharged from the hose with a speed of 40 ft/s. Determine the two possible angles θ the fireman can hold the hose so that the water strikes the building at B. Take $s = 20$ ft.

$v_A = 40$ ft/s

A

θ

B

8 ft

4 ft

s

Prob. 12–92

•12–93. The pitching machine is adjusted so that the baseball is launched with a speed of $v_A = 30$ m/s. If the ball strikes the ground at B, determine the two possible angles θ_A at which it was launched.

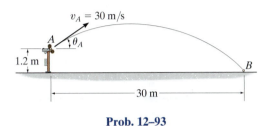

Prob. 12–93

12–94. It is observed that the time for the ball to strike the ground at B is 2.5 s. Determine the speed v_A and angle θ_A at which the ball was thrown.

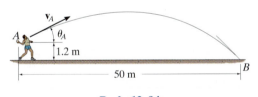

Prob. 12–94

12–95. If the motorcycle leaves the ramp traveling at 110 ft/s, determine the height h ramp B must have so that the motorcycle lands safely.

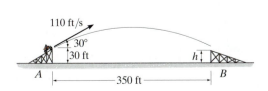

Prob. 12–95

*12–96. The baseball player A hits the baseball with $v_A = 40$ ft/s and $\theta_A = 60°$. When the ball is directly above of player B he begins to run under it. Determine the constant speed v_B and the distance d at which B must run in order to make the catch at the same elevation at which the ball was hit.

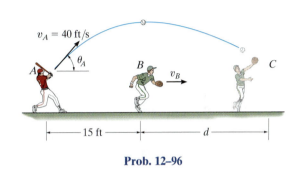

Prob. 12–96

•12–97. A boy throws a ball at O in the air with a speed v_0 at an angle θ_1. If he then throws another ball with the same speed v_0 at an angle $\theta_2 < \theta_1$, determine the time between the throws so that the balls collide in mid air at B.

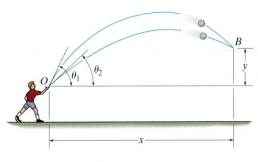

Prob. 12–97

12–98. The golf ball is hit at A with a speed of $v_A = 40\,\text{m/s}$ and directed at an angle of $30°$ with the horizontal as shown. Determine the distance d where the ball strikes the slope at B.

*12–100.** The velocity of the water jet discharging from the orifice can be obtained from $v = \sqrt{2gh}$, where $h = 2\,\text{m}$ is the depth of the orifice from the free water surface. Determine the time for a particle of water leaving the orifice to reach point B and the horizontal distance x where it hits the surface.

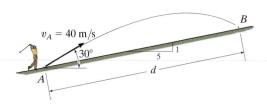

Prob. 12–98

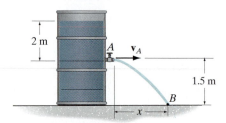

Prob. 12–100

12–99. If the football is kicked at the $45°$ angle, determine its minimum initial speed v_A so that it passes over the goal post at C. At what distance s from the goal post will the football strike the ground at B?

•**12–101.** A projectile is fired from the platform at B. The shooter fires his gun from point A at an angle of $30°$. Determine the muzzle speed of the bullet if it hits the projectile at C.

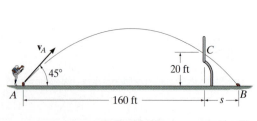

Prob. 12–99

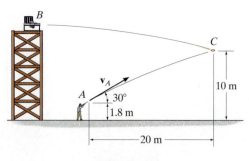

Prob. 12–101

12–102. A golf ball is struck with a velocity of 80 ft/s as shown. Determine the distance d to where it will land.

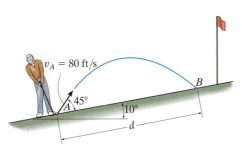

Prob. 12–102

•12–105. The boy at A attempts to throw a ball over the roof of a barn with an initial speed of $v_A = 15$ m/s. Determine the angle θ_A at which the ball must be thrown so that it reaches its maximum height at C. Also, find the distance d where the boy should stand to make the throw.

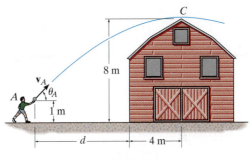

Prob. 12–105

12–103. The football is to be kicked over the goalpost, which is 15 ft high. If its initial speed is $v_A = 80$ ft/s, determine if it makes it over the goalpost, and if so, by how much, h.

*12–104. The football is kicked over the goalpost with an initial velocity of $v_A = 80$ ft/s as shown. Determine the point $B (x, y)$ where it strikes the bleachers.

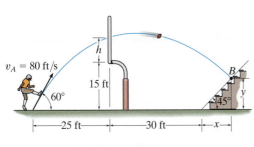

Probs. 12–103/104

12–106. The boy at A attempts to throw a ball over the roof of a barn such that it is launched at an angle $\theta_A = 40°$. Determine the minimum speed v_A at which he must throw the ball so that it reaches its maximum height at C. Also, find the distance d where the boy must stand so that he can make the throw.

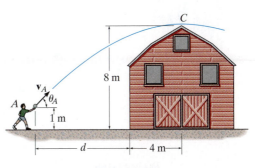

Prob. 12–106

12–107. The fireman wishes to direct the flow of water from his hose to the fire at B. Determine two possible angles θ_1 and θ_2 at which this can be done. Water flows from the hose at $v_A = 80$ ft/s.

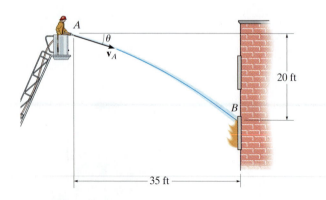

Prob. 12–107

•12–109. Determine the horizontal velocity v_A of a tennis ball at A so that it just clears the net at B. Also, find the distance s where the ball strikes the ground.

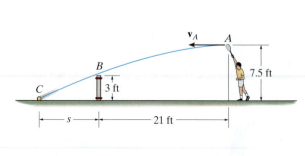

Prob. 12–109

***12–108.** Small packages traveling on the conveyor belt fall off into a l m long loading car. If the conveyor is running at a constant speed of $v_C = 2$ m/s, determine the smallest and largest distance R at which the end A of the car may be placed from the conveyor so that the packages enter the car.

12–110. It is observed that the skier leaves the ramp A at an angle $\theta_A = 25°$ with the horizontal. If he strikes the ground at B, determine his initial speed v_A and the time of flight t_{AB}.

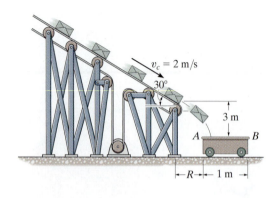

Prob. 12–108

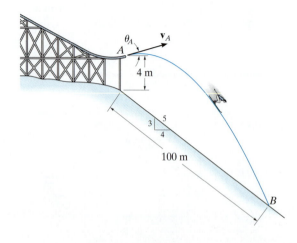

Prob. 12–110

12.7   Curvilinear Motion: Normal and Tangential Components

When the path along which a particle travels is *known*, then it is often convenient to describe the motion using n and t coordinate axes which act normal and tangent to the path, respectively, and at the instant considered have their *origin located at the particle*.

Planar Motion.   Consider the particle shown in Fig. 12–24a, which moves in a plane along a fixed curve, such that at a given instant it is at position s, measured from point O. We will now consider a coordinate system that has its origin at a *fixed point* on the curve, and at the instant considered this origin happens to *coincide* with the location of the particle. The t axis is *tangent* to the curve at the point and is positive in the direction of *increasing s*. We will designate this positive direction with the unit vector $\mathbf{u}_t$. A unique choice for the *normal axis* can be made by noting that geometrically the curve is constructed from a series of differential arc segments ds, Fig. 12–24b. Each segment ds is formed from the arc of an associated circle having a *radius of curvature* ρ (rho) and *center of curvature O′*. The normal axis n is perpendicular to the t axis with its positive sense directed *toward* the center of curvature $O′$, Fig. 12–24a. This positive direction, which is *always* on the concave side of the curve, will be designated by the unit vector $\mathbf{u}_n$. The plane which contains the n and t axes is referred to as the embracing or *osculating plane*, and in this case it is fixed in the plane of motion.*

Velocity.   Since the particle moves, s is a function of time. As indicated in Sec. 12.4, the particle's velocity $\mathbf{v}$ has a *direction* that is *always tangent to the path*, Fig. 12–24c, and a *magnitude* that is determined by taking the time derivative of the path function $s = s(t)$, i.e., $v = ds/dt$ (Eq. 12–8). Hence

$$\mathbf{v} = v\mathbf{u}_t \qquad (12\text{–}15)$$

where

$$v = \dot{s} \qquad (12\text{–}16)$$

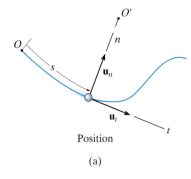

Position

(a)

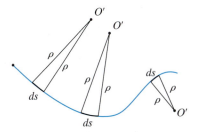

Radius of curvature

(b)

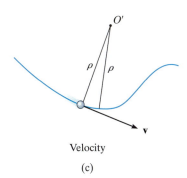

Velocity

(c)

Fig. 12–24

*The osculating plane may also be defined as the plane which has the greatest contact with the curve at a point. It is the limiting position of a plane contacting both the point and the arc segment ds. As noted above, the osculating plane is always coincident with a plane curve; however, each point on a three-dimensional curve has a unique osculating plane.

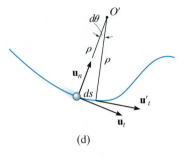

(d)

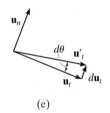

(e)

Acceleration.   The acceleration of the particle is the time rate of change of the velocity. Thus,

$$\mathbf{a} = \dot{\mathbf{v}} = \dot{v}\mathbf{u}_t + v\dot{\mathbf{u}}_t \qquad (12\text{–}17)$$

In order to determine the time derivative $\dot{\mathbf{u}}_t$, note that as the particle moves along the arc ds in time dt, $\mathbf{u}_t$ preserves its magnitude of unity; however, its *direction* changes, and becomes $\mathbf{u}'_t$, Fig. 12–24d. As shown in Fig. 12–24e, we require $\mathbf{u}'_t = \mathbf{u}_t + d\mathbf{u}_t$. Here $d\mathbf{u}_t$ stretches between the arrowheads of $\mathbf{u}_t$ and $\mathbf{u}'_t$, which lie on an infinitesimal arc of radius $u_t = 1$. Hence, $d\mathbf{u}_t$ has a *magnitude* of $du_t = (1)\, d\theta$, and its *direction* is defined by $\mathbf{u}_n$. Consequently, $d\mathbf{u}_t = d\theta\mathbf{u}_n$, and therefore the time derivative becomes $\dot{\mathbf{u}}_t = \dot{\theta}\mathbf{u}_n$. Since $ds = \rho d\theta$, Fig. 12–24d, then $\dot{\theta} = \dot{s}/\rho$, and therefore

$$\dot{\mathbf{u}}_t = \dot{\theta}\mathbf{u}_n = \frac{\dot{s}}{\rho}\mathbf{u}_n = \frac{v}{\rho}\mathbf{u}_n$$

Substituting into Eq. 12–17, $\mathbf{a}$ can be written as the sum of its two components,

$$\boxed{\mathbf{a} = a_t\mathbf{u}_t + a_n\mathbf{u}_n} \qquad (12\text{–}18)$$

where

$$\boxed{a_t = \dot{v}} \qquad \text{or} \qquad \boxed{a_t\, ds = v\, dv} \qquad (12\text{–}19)$$

and

$$\boxed{a_n = \frac{v^2}{\rho}} \qquad (12\text{–}20)$$

These two mutually perpendicular components are shown in Fig. 12–24f. Therefore, the *magnitude* of acceleration is the positive value of

$$a = \sqrt{a_t^2 + a_n^2} \qquad (12\text{–}21)$$

Acceleration

(f)

Fig. 12–24 (cont.)

To better understand these results, consider the following two special cases of motion.

1. If the particle moves along a straight line, then $\rho \to \infty$ and from Eq. 12–20, $a_n = 0$. Thus $a = a_t = \dot{v}$, and we can conclude that the *tangential component of acceleration represents the time rate of change in the magnitude of the velocity.*

2. If the particle moves along a curve with a constant speed, then $a_t = \dot{v} = 0$ and $a = a_n = v^2/\rho$. Therefore, the *normal component of acceleration represents the time rate of change in the direction of the velocity.* Since $\mathbf{a}_n$ *always* acts towards the center of curvature, this component is sometimes referred to as the *centripetal* (or center seeking) *acceleration.*

As a result of these interpretations, a particle moving along the curved path in Fig. 12–25 will have accelerations directed as shown.

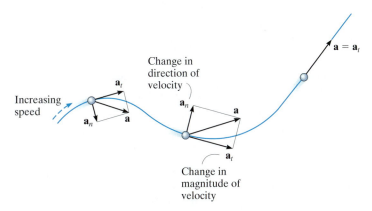

Fig. 12–25

Three-Dimensional Motion.

If the particle moves along a space curve, Fig. 12–26, then at a given instant the t axis is uniquely specified; however, an infinite number of straight lines can be constructed normal to the tangent axis. As in the case of planar motion, we will choose the positive n axis directed toward the path's center of curvature O'. This axis is referred to as the *principal normal* to the curve. With the n and t axes so defined, Eqs. 12–15 through 12–21 can be used to determine $\mathbf{v}$ and $\mathbf{a}$. Since $\mathbf{u}_t$ and $\mathbf{u}_n$ are always perpendicular to one another and lie in the osculating plane, for spatial motion a third unit vector, $\mathbf{u}_b$, defines the *binormal axis b* which is perpendicular to $\mathbf{u}_t$ and $\mathbf{u}_n$, Fig. 12–26.

Since the three unit vectors are related to one another by the vector cross product, e.g., $\mathbf{u}_b = \mathbf{u}_t \times \mathbf{u}_n$, Fig. 12–26, it may be possible to use this relation to establish the direction of one of the axes, if the directions of the other two are known. For example, if no motion occurs in the $\mathbf{u}_b$ direction, and this direction and $\mathbf{u}_t$ are known, then $\mathbf{u}_n$ can be determined, where in this case $\mathbf{u}_n = \mathbf{u}_b \times \mathbf{u}_t$, Fig. 12–26. Remember, though, that $\mathbf{u}_n$ is always on the concave side of the curve.

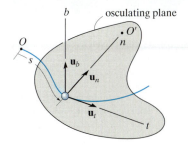

Fig. 12–26

12

Procedure for Analysis

Coordinate System.

- Provided the *path* of the particle is *known*, we can establish a set of n and t coordinates having a *fixed origin,* which is coincident with the particle at the instant considered.
- The positive tangent axis acts in the direction of motion and the positive normal axis is directed toward the path's center of curvature.

Velocity.

- The particle's *velocity* is always tangent to the path.
- The magnitude of velocity is found from the time derivative of the path function.

$$v = \dot{s}$$

Tangential Acceleration.

- The tangential component of acceleration is the result of the time rate of change in the *magnitude* of velocity. This component acts in the positive s direction if the particle's speed is increasing or in the opposite direction if the speed is decreasing.
- The relations between a_t, v, t and s are the same as for rectilinear motion, namely,

$$a_t = \dot{v} \qquad a_t \, ds = v \, dv$$

- If a_t is constant, $a_t = (a_t)_c$, the above equations, when integrated, yield

$$s = s_0 + v_0 t + \tfrac{1}{2}(a_t)_c t^2$$
$$v = v_0 + (a_t)_c t$$
$$v^2 = v_0^2 + 2(a_t)_c(s - s_0)$$

Normal Acceleration.

- The normal component of acceleration is the result of the time rate of change in the *direction* of the velocity. This component is *always* directed toward the center of curvature of the path, i.e., along the positive n axis.
- The magnitude of this component is determined from

$$a_n = \frac{v^2}{\rho}$$

- If the path is expressed as $y = f(x)$, the radius of curvature ρ at any point on the path is determined from the equation

$$\rho = \frac{[1 + (dy/dx)^2]^{3/2}}{|d^2y/dx^2|}$$

The derivation of this result is given in any standard calculus text.

Motorists traveling along this clover-leaf interchange experience a normal acceleration due to the change in direction of their velocity. A tangential component of acceleration occurs when the cars' speed is increased or decreased.

EXAMPLE | **12.14**

When the skier reaches point A along the parabolic path in Fig. 12–27a, he has a speed of 6 m/s which is increasing at 2 m/s². Determine the direction of his velocity and the direction and magnitude of his acceleration at this instant. Neglect the size of the skier in the calculation.

SOLUTION

Coordinate System. Although the path has been expressed in terms of its x and y coordinates, we can still establish the origin of the n, t axes at the fixed point A on the path and determine the components of $\mathbf{v}$ and $\mathbf{a}$ along these axes, Fig. 12–27a.

Velocity. By definition, the velocity is always directed tangent to the path. Since $y = \frac{1}{20}x^2$, $dy/dx = \frac{1}{10}x$, then at $x = 10$ m, $dy/dx = 1$. Hence, at A, $\mathbf{v}$ makes an angle of $\theta = \tan^{-1}1 = 45°$ with the x axis, Fig. 12–27a. Therefore,

$$v_A = 6 \text{ m/s} \qquad 45° \nearrow \qquad\qquad Ans.$$

The acceleration is determined from $\mathbf{a} = \dot{v}\mathbf{u}_t + (v^2/\rho)\mathbf{u}_n$. However, it is first necessary to determine the radius of curvature of the path at A (10 m, 5 m). Since $d^2y/dx^2 = \frac{1}{10}$, then

$$\rho = \frac{[1 + (dy/dx)^2]^{3/2}}{|d^2y/dx^2|} = \left.\frac{\left[1 + \left(\frac{1}{10}x\right)^2\right]^{3/2}}{\left|\frac{1}{10}\right|}\right|_{x=10 \text{ m}} = 28.28 \text{ m}$$

The acceleration becomes

$$\mathbf{a}_A = \dot{v}\mathbf{u}_t + \frac{v^2}{\rho}\mathbf{u}_n$$

$$= 2\mathbf{u}_t + \frac{(6 \text{ m/s})^2}{28.28 \text{ m}}\mathbf{u}_n$$

$$= \{2\mathbf{u}_t + 1.273\mathbf{u}_n\}\text{m/s}^2$$

As shown in Fig. 12–27b,

$$a = \sqrt{(2 \text{ m/s}^2)^2 + (1.273 \text{ m/s}^2)^2} = 2.37 \text{ m/s}^2$$

$$\phi = \tan^{-1}\frac{2}{1.273} = 57.5°$$

Thus, $45° + 90° + 57.5° - 180° = 12.5°$ so that,

$$a = 2.37 \text{ m/s}^2 \qquad 12.5° \nearrow \qquad\qquad Ans.$$

NOTE: By using n, t coordinates, we were able to readily solve this problem through the use of Eq. 12–18, since it accounts for the separate changes in the magnitude and direction of $\mathbf{v}$.

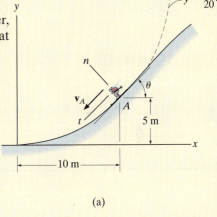

(a)

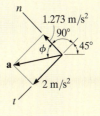

(b)

Fig. 12–27

12

EXAMPLE | 12.15

A race car C travels around the horizontal circular track that has a radius of 300 ft, Fig. 12–28. If the car increases its speed at a constant rate of 7 ft/s², starting from rest, determine the time needed for it to reach an acceleration of 8 ft/s². What is its speed at this instant?

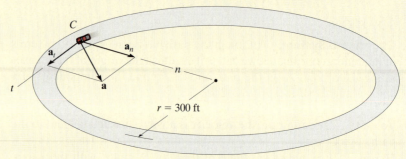

Fig. 12–28

SOLUTION

Coordinate System. The origin of the n and t axes is coincident with the car at the instant considered. The t axis is in the direction of motion, and the positive n axis is directed toward the center of the circle. This coordinate system is selected since the path is known.

Acceleration. The magnitude of acceleration can be related to its components using $a = \sqrt{a_t^2 + a_n^2}$. Here $a_t = 7$ ft/s². Since $a_n = v^2/\rho$, the velocity as a function of time must be determined first.

$$v = v_0 + (a_t)_c t$$
$$v = 0 + 7t$$

Thus

$$a_n = \frac{v^2}{\rho} = \frac{(7t)^2}{300} = 0.163t^2 \text{ ft/s}^2$$

The time needed for the acceleration to reach 8 ft/s² is therefore

$$a = \sqrt{a_t^2 + a_n^2}$$
$$8 \text{ ft/s}^2 = \sqrt{(7 \text{ ft/s}^2)^2 + (0.163t^2)^2}$$

Solving for the positive value of t yields

$$0.163t^2 = \sqrt{(8 \text{ ft/s}^2)^2 - (7 \text{ ft/s}^2)^2}$$
$$t = 4.87 \text{ s} \qquad \qquad Ans.$$

Velocity. The speed at time $t = 4.87$ s is

$$v = 7t = 7(4.87) = 34.1 \text{ ft/s} \qquad Ans.$$

NOTE: Remember the velocity will always be tangent to the path, whereas the acceleration will be directed within the curvature of the path.

•**12–129.** When the
25 m/s, which is inc
magnitude of the ac
instant and the direc

12–130. If the roll
speed increases at
magnitude of its a
$s_B = 40$ m.

12–131. The car is
The driver then appl
the car's speed at th
in m/s. Determine t
reaches point C on
car to travel from A

*12–132. The car
driver applies the
speed at the rate of
Determine the accel
point C on the circ
travel from A to C.

100 m

EXAMPLE 12.16

The boxes in Fig. 12–29a travel along the industrial conveyor. If a box as in Fig. 12–29b starts from rest at A and increases its speed such that $a_t = (0.2t)$ m/s^2, where t is in seconds, determine the magnitude of its acceleration when it arrives at point B.

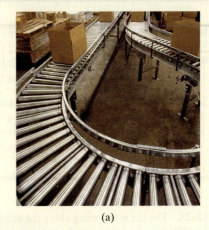

(a)

SOLUTION

Coordinate System. The position of the box at any instant is defined from the fixed point A using the position or path coordinate s, Fig. 12–29b. The acceleration is to be determined at B, so the origin of the n, t axes is at this point.

Acceleration. To determine the acceleration components $a_t = \dot{v}$ and $a_n = v^2/\rho$, it is first necessary to formulate v and $\dot{v}$ so that they may be evaluated at B. Since $v_A = 0$ when $t = 0$, then

$$a_t = \dot{v} = 0.2t \tag{1}$$

$$\int_0^v dv = \int_0^t 0.2t\, dt$$

$$v = 0.1t^2 \tag{2}$$

The time needed for the box to reach point B can be determined by realizing that the position of B is $s_B = 3 + 2\pi(2)/4 = 6.142$ m, Fig. 12–29b, and since $s_A = 0$ when $t = 0$ we have

$$v = \frac{ds}{dt} = 0.1t^2$$

$$\int_0^{6.142\ \text{m}} ds = \int_0^{t_B} 0.1t^2 dt$$

$$6.142\ \text{m} = 0.0333 t_B^3$$

$$t_B = 5.690\text{s}$$

Substituting into Eqs. 1 and 2 yields

$$(a_B)_t = \dot{v}_B = 0.2(5.690) = 1.138\ \text{m/s}^2$$

$$v_B = 0.1(5.69)^2 = 3.238\ \text{m/s}$$

At B, $\rho_B = 2$ m, so that

$$(a_B)_n = \frac{v_B^2}{\rho_B} = \frac{(3.238\ \text{m/s})^2}{2\ \text{m}} = 5.242\ \text{m/s}^2$$

The magnitude of $\mathbf{a}_B$, Fig. 12–29c, is therefore

$$a_B = \sqrt{(1.138\ \text{m/s}^2)^2 + (5.242\ \text{m/s}^2)^2} = 5.36\ \text{m/s}^2 \quad \textit{Ans.}$$

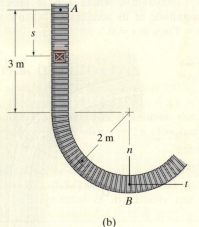

(b)

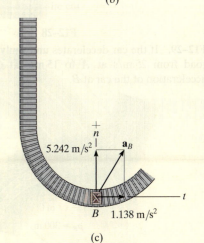

(c)

Fig. 12–29

12

•12–121. The trai
which is decreasi
magnitude of accel

12–122. The train
and begins to de
$a_t = -0.25$ m/s².
acceleration of the
$s_{AB} = 412$ m.

12

•12–148. A spiral transition curve is used on railroads to connect a straight portion of the track with a curved portion. If the spiral is defined by the equation $y = (10^{-6})x^3$, where x and y are in feet, determine the magnitude of the acceleration of a train engine moving with a constant speed of 40 ft/s when it is at point $x = 600$ ft.

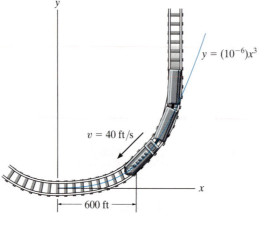

$y = (10^{-6})x^3$

$v = 40$ ft/s

A

600 ft

Prob. 12–148

12–150. Particles A and B are traveling around a circular track at a speed of 8 m/s at the instant shown. If the speed of B is increasing by $(a_t)_B = 4$ m/s², and at the same instant A has an increase in speed of $(a_t)_A = 0.8t$ m/s², determine how long it takes for a collision to occur. What is the magnitude of the acceleration of each particle just before the collision occurs?

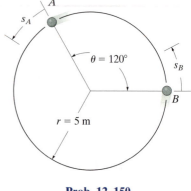

A

s_A

$\theta = 120°$

s_B

B

$r = 5$ m

Prob. 12–150

12–123. The car pa
which its speed
Determine the ma
reaches point B, wh

*12–124. If the ca
and begins to inc
$a_t = 0.5$ m/s², det
acceleration when

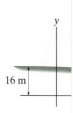

16 m

•12–149. Particles A and B are traveling counter-clockwise around a circular track at a constant speed of 8 m/s. If at the instant shown the speed of A begins to increase by $(a_t)_A = (0.4s_A)$ m/s², where s_A is in meters, determine the distance measured counterclockwise along the track from B to A when $t = 1$ s. What is the magnitude of the acceleration of each particle at this instant?

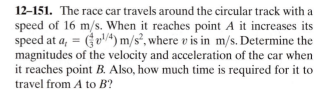

A

s_A

$\theta = 120°$

s_B

B

$r = 5$ m

Prob. 12–149

12–151. The race car travels around the circular track with a speed of 16 m/s. When it reaches point A it increases its speed at $a_t = (\frac{4}{3}v^{1/4})$ m/s², where v is in m/s. Determine the magnitudes of the velocity and acceleration of the car when it reaches point B. Also, how much time is required for it to travel from A to B?

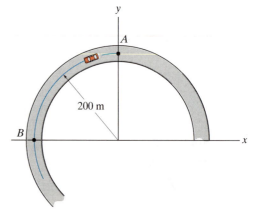

y

A

200 m

B

x

Prob. 12–151

*12–152. A particle travels along the path $y = a + bx + cx^2$, where a, b, c are constants. If the speed of the particle is constant, $v = v_0$, determine the x and y components of velocity and the normal component of acceleration when $x = 0$.

•12–153. The ball is kicked with an initial speed $v_A = 8$ m/s at an angle $\theta_A = 40°$ with the horizontal. Find the equation of the path, $y = f(x)$, and then determine the normal and tangential components of its acceleration when $t = 0.25$ s.

12–154. The motion of a particle is defined by the equations $x = (2t + t^2)$ m and $y = (t^2)$ m, where t is in seconds. Determine the normal and tangential components of the particle's velocity and acceleration when $t = 2$ s.

12–155. The motorcycle travels along the elliptical track at a constant speed v. Determine the greatest magnitude of the acceleration if $a > b$.

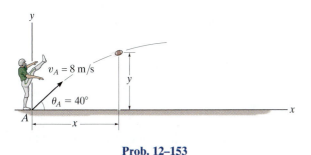

Prob. 12–153

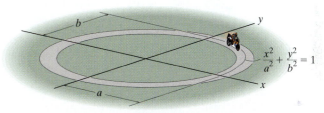

Prob. 12–155

12.8 Curvilinear Motion: Cylindrical Components

Sometimes the motion of the particle is constrained on a path that is best described using cylindrical coordinates. If motion is restricted to the plane, then polar coordinates are used.

Polar Coordinates. We can specify the location of the particle shown in Fig. 12–30a using a *radial coordinate r*, which extends outward from the fixed origin O to the particle, and a *transverse coordinate* θ, which is the counterclockwise angle between a fixed reference line and the r axis. The angle is generally measured in degrees or radians, where 1 rad = $180°/\pi$. The positive directions of the r and θ coordinates are defined by the unit vectors $\mathbf{u}_r$ and $\mathbf{u}_\theta$, respectively. Here $\mathbf{u}_r$ is in the direction of increasing r when θ is held fixed, and $\mathbf{u}_\theta$ is in a direction of increasing θ when r is held fixed. Note that these directions are perpendicular to one another.

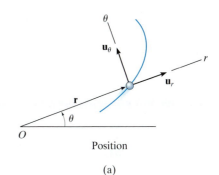

Position

(a)

Fig. 12–30

12

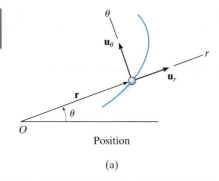

Position

(a)

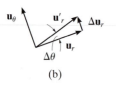

(b)

Position.   At any instant the position of the particle, Fig. 12–30a, is defined by the position vector

$$\mathbf{r} = r\mathbf{u}_r \qquad (12\text{–}22)$$

Velocity.   The instantaneous velocity $\mathbf{v}$ is obtained by taking the time derivative of $\mathbf{r}$. Using a dot to represent the time derivative, we have

$$\mathbf{v} = \dot{\mathbf{r}} = \dot{r}\mathbf{u}_r + r\dot{\mathbf{u}}_r$$

To evaluate $\dot{\mathbf{u}}_r$, notice that $\mathbf{u}_r$ only changes its direction with respect to time, since by definition the magnitude of this vector is always one unit. Hence, during the time Δt, a change Δr will not cause a change in the direction of $\mathbf{u}_r$; however, a change $\Delta \theta$ will cause $\mathbf{u}_r$ to become $\mathbf{u}_r'$, where $\mathbf{u}_r' = \mathbf{u}_r + \Delta\mathbf{u}_r$, Fig. 12–30b. The time change in $\mathbf{u}_r$ is then $\Delta\mathbf{u}_r$. For small angles $\Delta\theta$ this vector has a magnitude $\Delta u_r \approx 1(\Delta\theta)$ and acts in the $\mathbf{u}_\theta$ direction. Therefore, $\Delta\mathbf{u}_r = \Delta\theta\mathbf{u}_\theta$, and so

$$\dot{\mathbf{u}}_r = \lim_{\Delta t \to 0} \frac{\Delta\mathbf{u}_r}{\Delta t} = \left(\lim_{\Delta t \to 0} \frac{\Delta\theta}{\Delta t} \right)\mathbf{u}_\theta$$

$$\dot{\mathbf{u}}_r = \dot{\theta}\mathbf{u}_\theta \qquad (12\text{–}23)$$

Substituting into the above equation, the velocity can be written in component form as

$$\mathbf{v} = v_r\mathbf{u}_r + v_\theta\mathbf{u}_\theta \qquad (12\text{–}24)$$

where

$$\begin{aligned} v_r &= \dot{r} \\ v_\theta &= r\dot{\theta} \end{aligned} \qquad (12\text{–}25)$$

These components are shown graphically in Fig. 12–30c. The *radial component* v_r is a measure of the rate of increase or decrease in the length of the radial coordinate, i.e., $\dot{r}$; whereas the *transverse component* v_θ can be interpreted as the rate of motion along the circumference of a circle having a radius r. In particular, the term $\dot{\theta} = d\theta/dt$ is called the *angular velocity*, since it indicates the time rate of change of the angle θ. Common units used for this measurement are rad/s.

Since v_r and v_θ are mutually perpendicular, the *magnitude* of velocity or speed is simply the positive value of

$$v = \sqrt{(\dot{r})^2 + (r\dot{\theta})^2} \qquad (12\text{–}26)$$

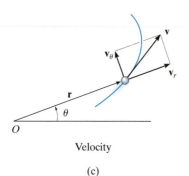

Velocity

(c)

Fig. 12–30 (cont.)

and the *direction* of $\mathbf{v}$ is, of course, tangent to the path, Fig. 12–30c.

Acceleration. Taking the time derivatives of Eq. 12–24, using Eqs. 12–25, we obtain the particle's instantaneous acceleration,

$$\mathbf{a} = \dot{\mathbf{v}} = \ddot{r}\mathbf{u}_r + \dot{r}\dot{\mathbf{u}}_r + \dot{r}\dot{\theta}\mathbf{u}_\theta + r\ddot{\theta}\mathbf{u}_\theta + r\dot{\theta}\dot{\mathbf{u}}_\theta$$

To evaluate $\dot{\mathbf{u}}_\theta$, it is necessary only to find the change in the direction of $\mathbf{u}_\theta$ since its magnitude is always unity. During the time Δt, a change Δr will not change the direction of $\mathbf{u}_\theta$, however, a change $\Delta\theta$ will cause $\mathbf{u}_\theta$ to become $\mathbf{u}'_\theta$, where $\mathbf{u}'_\theta = \mathbf{u}_\theta + \Delta\mathbf{u}_\theta$, Fig. 12–30d. The time change in $\mathbf{u}_\theta$ is thus $\Delta\mathbf{u}_\theta$. For small angles this vector has a magnitude $\Delta u_\theta \approx 1(\Delta\theta)$ and acts in the $-\mathbf{u}_r$, direction; i.e., $\Delta\mathbf{u}_\theta = -\Delta\theta\mathbf{u}_r$. Thus,

$$\dot{\mathbf{u}}_\theta = \lim_{\Delta t \to 0} \frac{\Delta\mathbf{u}_\theta}{\Delta t} = -\left(\lim_{\Delta t \to 0} \frac{\Delta\theta}{\Delta t}\right)\mathbf{u}_r$$

$$\dot{\mathbf{u}}_\theta = -\dot{\theta}\mathbf{u}_r \qquad (12\text{–}27)$$

Substituting this result and Eq. 12–23 into the above equation for $\mathbf{a}$, we can write the acceleration in component form as

$$\boxed{\mathbf{a} = a_r\mathbf{u}_r + a_\theta\mathbf{u}_\theta} \qquad (12\text{–}28)$$

where

$$\boxed{\begin{aligned} a_r &= \ddot{r} - r\dot{\theta}^2 \\ a_\theta &= r\ddot{\theta} + 2\dot{r}\dot{\theta} \end{aligned}} \qquad (12\text{–}29)$$

The term $\ddot{\theta} = d^2\theta/dt^2 = d/dt(d\theta/dt)$ is called the *angular acceleration* since it measures the change made in the angular velocity during an instant of time. Units for this measurement are rad/s^2.

Since $\mathbf{a}_r$, and $\mathbf{a}_\theta$ are always perpendicular, the *magnitude* of acceleration is simply the positive value of

$$a = \sqrt{(\ddot{r} - r\dot{\theta}^2)^2 + (r\ddot{\theta} + 2\dot{r}\dot{\theta})^2} \qquad (12\text{–}30)$$

The *direction* is determined from the vector addition of its two components. In general, $\mathbf{a}$ will *not* be tangent to the path, Fig. 12–30e.

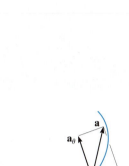

Acceleration

(e)

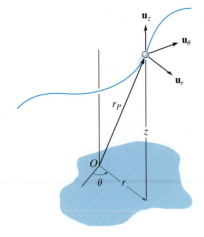

Fig. 12–31

Cylindrical Coordinates.

Cylindrical Coordinates. If the particle moves along a space curve as shown in Fig. 12–31, then its location may be specified by the three *cylindrical coordinates, r, θ, z.* The z coordinate is identical to that used for rectangular coordinates. Since the unit vector defining its direction, $\mathbf{u}_z$, is constant, the time derivatives of this vector are zero, and therefore the position, velocity, and acceleration of the particle can be written in terms of its cylindrical coordinates as follows:

$$\mathbf{r}_P = r\mathbf{u}_r + z\mathbf{u}_z$$

$$\mathbf{v} = \dot{r}\mathbf{u}_r + r\dot{\theta}\mathbf{u}_\theta + \dot{z}\mathbf{u}_z \tag{12–31}$$

$$\mathbf{a} = (\ddot{r} - r\dot{\theta}^2)\mathbf{u}_r + (r\ddot{\theta} + 2\dot{r}\dot{\theta})\mathbf{u}_\theta + \ddot{z}\mathbf{u}_z \tag{12–32}$$

Time Derivatives.

Time Derivatives. The above equations require that we obtain the time derivatives $\dot{r}, \ddot{r}, \dot{\theta}$, and $\ddot{\theta}$ in order to evaluate the r and θ components of $\mathbf{v}$ and $\mathbf{a}$. Two types of problems generally occur:

1. If the polar coordinates are specified as time parametric equations, $r = r(t)$ and $\theta = \theta(t)$, then the time derivatives can be found directly.

2. If the time-parametric equations are not given, then the path $r = f(\theta)$ must be known. Using the chain rule of calculus we can then find the relation between $\dot{r}$ and $\dot{\theta}$, and between $\ddot{r}$ and $\ddot{\theta}$. Application of the chain rule, along with some examples, is explained in Appendix C.

The spiral motion of this boy can be followed by using cylindrical components. Here the radial coordinate r is constant, the transverse coordinate θ will increase with time as the boy rotates about the vertical, and his altitude z will decrease with time.

Procedure for Analysis

Coordinate System.

- Polar coordinates are a suitable choice for solving problems when data regarding the angular motion of the radial coordinate r is given to describe the particle's motion. Also, some paths of motion can conveniently be described in terms of these coordinates.

- To use polar coordinates, the origin is established at a fixed point, and the radial line r is directed to the particle.

- The transverse coordinate θ is measured from a fixed reference line to the radial line.

Velocity and Acceleration.

- Once r and the four time derivatives $\dot{r}, \ddot{r}, \dot{\theta}$, and $\ddot{\theta}$ have been evaluated at the instant considered, their values can be substituted into Eqs. 12–25 and 12–29 to obtain the radial and transverse components of $\mathbf{v}$ and $\mathbf{a}$.

- If it is necessary to take the time derivatives of $r = f(\theta)$, then the chain rule of calculus must be used. See Appendix C.

- Motion in three dimensions requires a simple extension of the above procedure to include $\dot{z}$ and $\ddot{z}$.

EXAMPLE 12.17

The amusement park ride shown in Fig. 12–32a consists of a chair that is rotating in a horizontal circular path of radius r such that the arm OB has an angular velocity $\dot{\theta}$ and angular acceleration $\ddot{\theta}$. Determine the radial and transverse components of velocity and acceleration of the passenger. Neglect his size in the calculation.

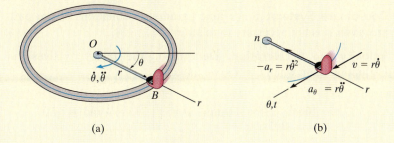

(a)   (b)

Fig. 12–32

SOLUTION

Coordinate System. Since the angular motion of the arm is reported, polar coordinates are chosen for the solution, Fig. 12–32a. Here θ is not related to r, since the radius is constant for all θ.

Velocity and Acceleration. It is first necessary to specify the first and second time derivatives of r and θ. Since r is *constant*, we have

$$r = r \qquad \dot{r} = 0 \qquad \ddot{r} = 0$$

Thus,

$$v_r = \dot{r} = 0 \qquad\qquad\qquad\qquad Ans.$$
$$v_\theta = r\dot{\theta} \qquad\qquad\qquad\qquad Ans.$$
$$a_r = \ddot{r} - r\dot{\theta}^2 = -r\dot{\theta}^2 \qquad\qquad Ans.$$
$$a_\theta = r\ddot{\theta} + 2\dot{r}\dot{\theta} = r\ddot{\theta} \qquad\qquad Ans.$$

These results are shown in Fig. 12–32b.

NOTE: The n, t axes are also shown in Fig. 12–32b, which in this special case of circular motion happen to be *collinear* with the r and θ axes, respectively. Since $v = v_\theta = v_t = r\dot{\theta}$, then by comparison,

$$-a_r = a_n = \frac{v^2}{\rho} = \frac{(r\dot{\theta})^2}{r} = r\dot{\theta}^2$$

$$a_\theta = a_t = \frac{dv}{dt} = \frac{d}{dt}(r\dot{\theta}) = \frac{dr}{dt}\dot{\theta} + r\frac{d\dot{\theta}}{dt} = 0 + r\ddot{\theta}$$

EXAMPLE | 12.18

The rod OA in Fig. 12–33a rotates in the horizontal plane such that $\theta = (t^3)$ rad. At the same time, the collar B is sliding outward along OA so that $r = (100t^2)$ mm. If in both cases t is in seconds, determine the velocity and acceleration of the collar when $t = 1$ s.

SOLUTION

Coordinate System. Since time-parametric equations of the path are given, it is not necessary to relate r to θ.

Velocity and Acceleration. Determining the time derivatives and evaluating them when $t = 1$ s, we have

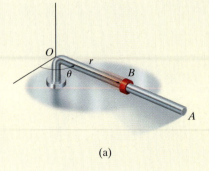

(a)

$$r = 100t^2 \Big|_{t=1\,\text{s}} = 100\,\text{mm} \qquad \theta = t^3 \Big|_{t=1\,\text{s}} = 1\,\text{rad} = 57.3°$$

$$\dot{r} = 200t \Big|_{t=1\,\text{s}} = 200\,\text{mm/s} \qquad \dot{\theta} = 3t^2 \Big|_{t=1\,\text{s}} = 3\,\text{rad/s}$$

$$\ddot{r} = 200 \Big|_{t=1\,\text{s}} = 200\,\text{mm/s}^2 \qquad \ddot{\theta} = 6t \Big|_{t=1\,\text{s}} = 6\,\text{rad/s}^2.$$

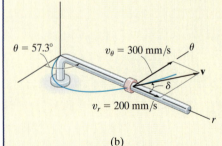

(b)

As shown in Fig. 12–33b,

$$\mathbf{v} = \dot{r}\mathbf{u}_r + r\dot{\theta}\mathbf{u}_\theta$$
$$= 200\mathbf{u}_r + 100(3)\mathbf{u}_\theta = \{200\mathbf{u}_r + 300\mathbf{u}_\theta\}\,\text{mm/s}$$

The magnitude of $\mathbf{v}$ is

$$v = \sqrt{(200)^2 + (300)^2} = 361\,\text{mm/s} \qquad\qquad Ans.$$

$$\delta = \tan^{-1}\!\left(\frac{300}{200}\right) = 56.3° \qquad \delta + 57.3° = 114° \qquad Ans.$$

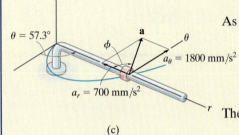

(c)

Fig. 12–33

As shown in Fig. 12–33c,

$$\mathbf{a} = (\ddot{r} - r\dot{\theta}^2)\mathbf{u}_r + (r\ddot{\theta} + 2\dot{r}\dot{\theta})\mathbf{u}_\theta$$
$$= [200 - 100(3)^2]\mathbf{u}_r + [100(6) + 2(200)3]\mathbf{u}_\theta$$
$$= \{-700\mathbf{u}_r + 1800\mathbf{u}_\theta\}\,\text{mm/s}^2$$

The magnitude of $\mathbf{a}$ is

$$a = \sqrt{(700)^2 + (1800)^2} = 1930\,\text{mm/s}^2 \qquad\qquad Ans.$$

$$\phi = \tan^{-1}\!\left(\frac{1800}{700}\right) = 68.7° \qquad (180° - \phi) + 57.3° = 169° \qquad Ans.$$

NOTE: The velocity is tangent to the path; however, the acceleration is directed within the curvature of the path, as expected.

EXAMPLE 12.19

12

The searchlight in Fig. 12–34a casts a spot of light along the face of a wall that is located 100 m from the searchlight. Determine the magnitudes of the velocity and acceleration at which the spot appears to travel across the wall at the instant $\theta = 45°$. The searchlight rotates at a constant rate of $\dot\theta = 4$ rad/s.

SOLUTION

Coordinate System. Polar coordinates will be used to solve this problem since the angular rate of the searchlight is given. To find the necessary time derivatives it is first necessary to relate r to θ. From Fig. 12–34a,

$$r = 100/\cos\theta = 100\sec\theta$$

Velocity and Acceleration. Using the chain rule of calculus, noting that $d(\sec\theta) = \sec\theta\tan\theta\,d\theta$, and $d(\tan\theta) = \sec^2\theta\,d\theta$, we have

$$\dot r = 100(\sec\theta\tan\theta)\dot\theta$$
$$\ddot r = 100(\sec\theta\tan\theta)\dot\theta(\tan\theta)\dot\theta + 100\sec\theta(\sec^2\theta)\dot\theta(\dot\theta)$$
$$\quad + 100\sec\theta\tan\theta(\ddot\theta)$$
$$= 100\sec\theta\tan^2\theta\,(\dot\theta)^2 + 100\sec^3\theta\,(\dot\theta)^2 + 100(\sec\theta\tan\theta)\ddot\theta$$

Since $\dot\theta = 4$ rad/s = constant, then $\ddot\theta = 0$, and the above equations, when $\theta = 45°$, become

$$r = 100\sec 45° = 141.4$$
$$\dot r = 400\sec 45°\tan 45° = 565.7$$
$$\ddot r = 1600\,(\sec 45°\tan^2 45° + \sec^3 45°) = 6788.2$$

As shown in Fig. 12–34b,

$$\mathbf{v} = \dot r\mathbf{u}_r + r\dot\theta\mathbf{u}_\theta$$
$$= 565.7\mathbf{u}_r + 141.4(4)\mathbf{u}_\theta$$
$$= \{565.7\mathbf{u}_r + 565.7\mathbf{u}_\theta\}\ \text{m/s}$$
$$v = \sqrt{v_r^2 + v_\theta^2} = \sqrt{(565.7)^2 + (565.7)^2}$$
$$= 800\ \text{m/s} \qquad\qquad\qquad Ans.$$

As shown in Fig. 12–34c,

$$\mathbf{a} = (\ddot r - r\dot\theta^2)\mathbf{u}_r + (r\ddot\theta + 2\dot r\dot\theta)\mathbf{u}_\theta$$
$$= [6788.2 - 141.4(4)^2]\mathbf{u}_r + [141.4(0) + 2(565.7)4]\mathbf{u}_\theta$$
$$= \{4525.5\mathbf{u}_r + 4525.5\mathbf{u}_\theta\}\ \text{m/s}^2$$
$$a = \sqrt{a_r^2 + a_\theta^2} = \sqrt{(4525.5)^2 + (4525.5)^2}$$
$$= 6400\ \text{m/s}^2 \qquad\qquad\qquad Ans.$$

NOTE: It is also possible to find a without having to calculate $\ddot r$ (or a_r). As shown in Fig. 12–34d, since $a_\theta = 4525.5$ m/s^2, then by vector resolution, $a = 4525.5/\cos 45° = 6400$ m/s^2.

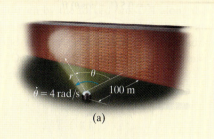

(a)

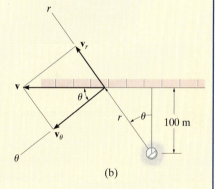

(b)

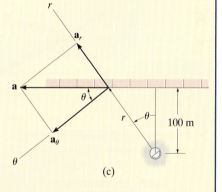

(c)

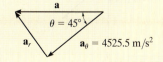

(d)

Fig. 12–34

12–174. The airplane on the amusement park ride moves along a path defined by the equations $r = 4$ m, $\theta = (0.2t)$ rad, and $z = (0.5 \cos \theta)$ m, where t is in seconds. Determine the cylindrical components of the velocity and acceleration of the airplane when $t = 6$ s.

•**12–177.** The driver of the car maintains a constant speed of 40 m/s. Determine the angular velocity of the camera tracking the car when $\theta = 15°$.

12–178. When $\theta = 15°$, the car has a speed of 50 m/s which is increasing at 6 m/s². Determine the angular velocity of the camera tracking the car at this instant.

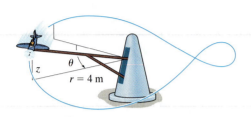

Prob. 12–174

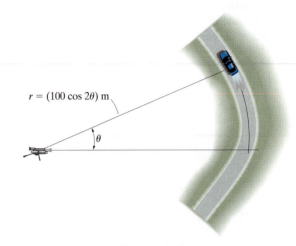

$r = (100 \cos 2\theta)$ m

Probs. 12–177/178

12–175. The motion of peg P is constrained by the lemniscate curved slot in OB and by the slotted arm OA. If OA rotates counterclockwise with a constant angular velocity of $\dot\theta = 3$ rad/s, determine the magnitudes of the velocity and acceleration of peg P at $\theta = 30°$.

*****12–176.** The motion of peg P is constrained by the lemniscate curved slot in OB and by the slotted arm OA. If OA rotates counterclockwise with an angular velocity of $\dot\theta = (3t^{3/2})$ rad/s, where t is in seconds, determine the magnitudes of the velocity and acceleration of peg P at $\theta = 30°$. When $t = 0, \theta = 0°$.

12–179. If the cam rotates clockwise with a constant angular velocity of $\dot\theta = 5$ rad/s, determine the magnitudes of the velocity and acceleration of the follower rod AB at the instant $\theta = 30°$. The surface of the cam has a shape of limaçon defined by $r = (200 + 100 \cos \theta)$ mm.

*****12–180.** At the instant $\theta = 30°$, the cam rotates with a clockwise angular velocity of $\dot\theta = 5$ rad/s and and angular acceleration of $\ddot\theta = 6$ rad/s². Determine the magnitudes of the velocity and acceleration of the follower rod AB at this instant. The surface of the cam has a shape of a limaçon defined by $r = (200 + 100 \cos \theta)$ mm.

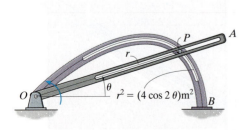

Probs. 12–175/176

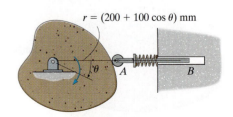

$r = (200 + 100 \cos \theta)$ mm

Probs. 12–179/180

•12–181. The automobile travels from a parking deck down along a cylindrical spiral ramp at a constant speed of $v = 1.5$ m/s. If the ramp descends a distance of 12 m for every full revolution, $\theta = 2\pi$ rad, determine the magnitude of the car's acceleration as it moves along the ramp, $r = 10$ m. *Hint:* For part of the solution, note that the tangent to the ramp at any point is at an angle of $\phi = \tan^{-1}(12/[2\pi(10)]) = 10.81°$ from the horizontal. Use this to determine the velocity components v_θ and v_z, which in turn are used to determine $\dot{\theta}$ and $\dot{z}$.

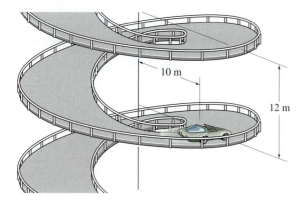

Prob. 12–181

12–182. The box slides down the helical ramp with a constant speed of $v = 2$ m/s. Determine the magnitude of its acceleration. The ramp descends a vertical distance of 1 m for every full revolution. The mean radius of the ramp is $r = 0.5$ m.

12–183. The box slides down the helical ramp such that by $r = 0.5$ m, $\theta = (0.5t^3)$ rad, and $z = (2 - 0.2t^2)$ m, where t is in seconds. Determine the magnitudes of the velocity and acceleration of the box at the instant $\theta = 2\pi$ rad.

Probs. 12–182/183

*12–184. Rod *OA* rotates counterclockwise with a constant angular velocity of $\dot{\theta} = 6$ rad/s. Through mechanical means collar *B* moves along the rod with a speed of $\dot{r} = (4t^2)$ m/s, where t is in seconds. If $r = 0$ when $t = 0$, determine the magnitudes of velocity and acceleration of the collar when $t = 0.75$ s.

•12–185. Rod *OA* is rotating counterclockwise with an angular velocity of $\dot{\theta} = (2t^2)$ rad/s. Through mechanical means collar *B* moves along the rod with a speed of $\dot{r} = (4t^2)$ m/s. If $\theta = 0$ and $r = 0$ when $t = 0$, determine the magnitudes of velocity and acceleration of the collar at $\theta = 60°$.

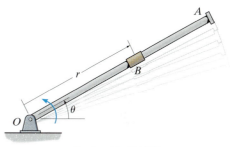

Probs. 12–184/185

12–186. The slotted arm *AB* drives pin *C* through the spiral groove described by the equation $r = a\theta$. If the angular velocity is constant at $\dot{\theta}$, determine the radial and transverse components of velocity and acceleration of the pin.

12–187. The slotted arm *AB* drives pin *C* through the spiral groove described by the equation $r = (1.5\,\theta)$ ft, where θ is in radians. If the arm starts from rest when $\theta = 60°$ and is driven at an angular velocity of $\dot{\theta} = (4t)$ rad/s, where t is in seconds, determine the radial and transverse components of velocity and acceleration of the pin *C* when $t = 1$ s.

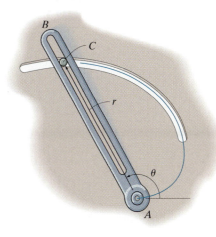

Probs. 12–186/187

*12–188. The partial surface of the cam is that of a logarithmic spiral $r = (40e^{0.05\theta})$ mm, where θ is in radians. If the cam rotates at a constant angular velocity of $\dot{\theta} = 4$ rad/s, determine the magnitudes of the velocity and acceleration of the point on the cam that contacts the follower rod at the instant $\theta = 30°$.

•12–189. Solve Prob. 12–188, if the cam has an angular acceleration of $\ddot{\theta} = 2$ rad/s² when its angular velocity is $\dot{\theta} = 4$ rad/s at $\theta = 30°$.

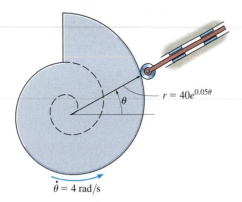

$r = 40e^{0.05\theta}$

θ

$\dot{\theta} = 4$ rad/s

Probs. 12–188/189

12–190. A particle moves along an Archimedean spiral $r = (8\theta)$ ft, where θ is given in radians. If $\dot{\theta} = 4$ rad/s (constant), determine the radial and transverse components of the particle's velocity and acceleration at the instant $\theta = \pi/2$ rad. Sketch the curve and show the components on the curve.

12–191. Solve Prob. 12–190 if the particle has an angular acceleration $\ddot{\theta} = 5$ rad/s² when $\dot{\theta} = 4$ rad/s at $\theta = \pi/2$ rad.

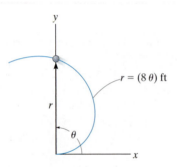

y

$r = (8\,\theta)$ ft

r

θ

x

Probs. 12–190/191

*12–192. The boat moves along a path defined by $r^2 = [10(10^3) \cos 2\theta]$ ft², where θ is in radians. If $\theta = (0.4t^2)$ rad, where t is in seconds, determine the radial and transverse components of the boat's velocity and acceleration at the instant $t = 1$ s.

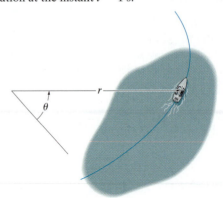

r

θ

Prob. 12–192

•12–193. A car travels along a road, which for a short distance is defined by $r = (200/\theta)$ ft, where θ is in radians. If it maintains a constant speed of $v = 35$ ft/s, determine the radial and transverse components of its velocity when $\theta = \pi/3$ rad.

12–194. For a short time the jet plane moves along a path in the shape of a lemniscate, $r^2 = (2500 \cos 2\theta)$ km². At the instant $\theta = 30°$, the radar tracking device is rotating at $\dot{\theta} = 5(10^{-3})$ rad/s with $\ddot{\theta} = 2(10^{-3})$ rad/s². Determine the radial and transverse components of velocity and acceleration of the plane at this instant.

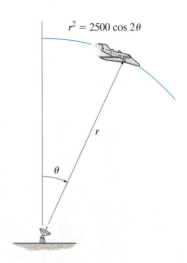

$r^2 = 2500 \cos 2\theta$

r

θ

Prob. 12–194

12.9 Absolute Dependent Motion Analysis of Two Particles

In some types of problems the motion of one particle will *depend* on the corresponding motion of another particle. This dependency commonly occurs if the particles, here represented by blocks, are interconnected by inextensible cords which are wrapped around pulleys. For example, the movement of block A downward along the inclined plane in Fig. 12–36 will cause a corresponding movement of block B up the other incline. We can show this mathematically by first specifying the location of the blocks using *position coordinates* s_A and s_B. Note that each of the coordinate axes is (1) measured from a *fixed* point (O) or *fixed* datum line, (2) measured along each inclined plane *in the direction of motion* of each block, and (3) has a positive sense from C to A and D to B. If the total cord length is l_T, the two position coordinates are related by the equation

$$s_A + l_{CD} + s_B = l_T$$

Here l_{CD} is the length of the cord passing over arc CD. Taking the time derivative of this expression, realizing that l_{CD} and l_T *remain constant*, while s_A and s_B measure the segments of the cord that change in length. We have

$$\frac{ds_A}{dt} + \frac{ds_B}{dt} = 0 \quad \text{or} \quad v_B = -v_A$$

The negative sign indicates that when block A has a velocity downward, i.e., in the direction of positive s_A, it causes a corresponding upward velocity of block B; i.e., B moves in the negative s_B direction.

In a similar manner, time differentiation of the velocities yields the relation between the accelerations, i.e.,

$$a_B = -a_A$$

A more complicated example is shown in Fig. 12–37a. In this case, the position of block A is specified by s_A, and the position of the *end* of the cord from which block B is suspended is defined by s_B. As above, we have chosen position coordinates which (1) have their origin at fixed points or datums, (2) are measured in the direction of motion of each block, and (3) are positive to the right for s_A and positive downward for s_B. During the motion, the length of the red colored segments of the cord in Fig. 12–37a *remains constant*. If l represents the total length of cord minus these segments, then the position coordinates can be related by the equation

$$2s_B + h + s_A = l$$

Since l and h are constant during the motion, the two time derivatives yield

$$2v_B = -v_A \quad 2a_B = -a_A$$

Hence, when B moves downward ($+s_B$), A moves to the left ($-s_A$) with twice the motion.

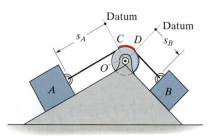

Fig. 12–36

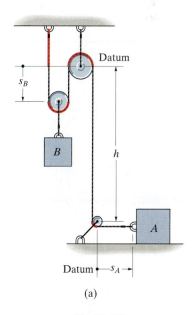

(a)

Fig. 12–37

12

12

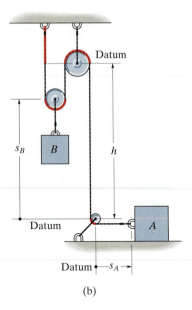

Datum

s_B B   h

Datum

Datum s_A

(b)

Fig. 12–37 (cont.)

The motion of the traveling block on this oil rig depends upon the motion of the cable connected to the winch which operates it. It is important to be able to relate these motions in order to determine the power requirements of the winch and the force in the cable caused by any accelerated motion.

This example can also be worked by defining the position of block B from the center of the bottom pulley (a fixed point), Fig. 12–37b. In this case

$$2(h - s_B) + h + s_A = l$$

Time differentiation yields

$$2v_B = v_A \qquad 2a_B = a_A$$

Here the signs are the same. Why?

Procedure for Analysis

The above method of relating the dependent motion of one particle to that of another can be performed using algebraic scalars or position coordinates provided each particle moves along a rectilinear path. When this is the case, only the magnitudes of the velocity and acceleration of the particles will change, not their line of direction.

Position-Coordinate Equation.

- Establish each position coordinate with an origin located at a *fixed* point or datum.

- It is *not necessary* that the *origin* be the *same* for each of the coordinates; however, it is *important* that each coordinate axis selected be directed along the *path of motion* of the particle.

- Using geometry or trigonometry, relate the position coordinates to the total length of the cord, l_T, or to that portion of cord, l, which *excludes* the segments that do not change length as the particles move—such as arc segments wrapped over pulleys.

- If a problem involves a *system* of two or more cords wrapped around pulleys, then the position of a point on one cord must be related to the position of a point on another cord using the above procedure. Separate equations are written for a fixed length of each cord of the system and the positions of the two particles are then related by these equations (see Examples 12.22 and 12.23).

Time Derivatives.

- Two successive time derivatives of the position-coordinate equations yield the required velocity and acceleration equations which relate the motions of the particles.

- The signs of the terms in these equations will be consistent with those that specify the positive and negative sense of the position coordinates.

EXAMPLE 12.21

Determine the speed of block A in Fig. 12–38 if block B has an upward speed of 6 ft/s.

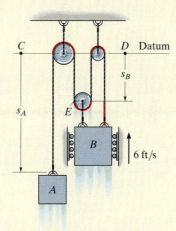

Fig. 12–38

SOLUTION

Position-Coordinate Equation.   There is *one cord* in this system having segments which change length. Position coordinates s_A and s_B will be used since each is measured from a fixed point (C or D) and extends along each block's *path of motion*. In particular, s_B is directed to point E since motion of B and E is the *same*.

The red colored segments of the cord in Fig. 12–38 remain at a constant length and do not have to be considered as the blocks move. The remaining length of cord, l, is also constant and is related to the changing position coordinates s_A and s_B by the equation

$$s_A + 3s_B = l$$

Time Derivative.   Taking the time derivative yields

$$v_A + 3v_B = 0$$

so that when $v_B = -6$ ft/s (upward),

$$v_A = 18 \text{ ft/s} \downarrow \qquad\qquad Ans.$$

12

EXAMPLE | 12.22

Determine the speed of A in Fig. 12–39 if B has an upward speed of 6 ft/s.

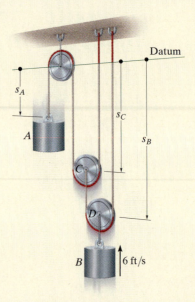

Fig. 12–39

SOLUTION

Position-Coordinate Equation. As shown, the positions of blocks A and B are defined using coordinates s_A and s_B. Since the system has *two cords* with segments that change length, it will be necessary to use a third coordinate, s_C, in order to relate s_A to s_B. In other words, the length of one of the cords can be expressed in terms of s_A and s_C, and the length of the other cord can be expressed in terms of s_B and s_C.

The red colored segments of the cords in Fig. 12–39 do not have to be considered in the analysis. Why? For the remaining cord lengths, say l_1 and l_2, we have

$$s_A + 2s_C = l_1 \qquad s_B + (s_B - s_C) = l_2$$

Time Derivative. Taking the time derivative of these equations yields

$$v_A + 2v_C = 0 \qquad 2v_B - v_C = 0$$

Eliminating v_C produces the relationship between the motions of each cylinder.

$$v_A + 4v_B = 0$$

so that when $v_B = -6$ ft/s (upward),

$$v_A = +24 \text{ ft/s} = 24 \text{ ft/s} \downarrow \qquad\qquad \textit{Ans.}$$

EXAMPLE | 12.23

Determine the speed of block B in Fig. 12–40 if the end of the cord at A is pulled down with a speed of 2 m/s.

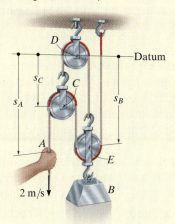

Fig. 12–40

SOLUTION

Position-Coordinate Equation.   The position of point A is defined by s_A, and the position of block B is specified by s_B since point E on the pulley will have the *same motion* as the block. Both coordinates are measured from a horizontal datum passing through the *fixed* pin at pulley D. Since the system consists of *two* cords, the coordinates s_A and s_B cannot be related directly. Instead, by establishing a third position coordinate, s_C, we can now express the length of one of the cords in terms of s_B and s_C, and the length of the other cord in terms of s_A, s_B, and s_C.

Excluding the red colored segments of the cords in Fig. 12–40, the remaining constant cord lengths l_1 and l_2 (along with the hook and link dimensions) can be expressed as

$$s_C + s_B = l_1$$
$$(s_A - s_C) + (s_B - s_C) + s_B = l_2$$

Time Derivative.   The time derivative of each equation gives

$$v_C + v_B = 0$$
$$v_A - 2v_C + 2v_B = 0$$

Eliminating v_C, we obtain

$$v_A + 4v_B = 0$$

so that when $v_A = 2$ m/s (downward),

$$v_B = -0.5 \text{ m/s} = 0.5 \text{ m/s} \uparrow \qquad\qquad Ans.$$

EXAMPLE | 12.24

A man at A is hoisting a safe S as shown in Fig. 12–41 by walking to the right with a constant velocity $v_A = 0.5$ m/s. Determine the velocity and acceleration of the safe when it reaches the elevation of 10 m. The rope is 30 m long and passes over a small pulley at D.

SOLUTION

Position-Coordinate Equation. This problem is unlike the previous examples since rope segment DA changes *both direction and magnitude*. However, the ends of the rope, which define the positions of S and A, are specified by means of the x and y coordinates since they must be measured from a fixed point and *directed along the paths of motion* of the ends of the rope.

The x and y coordinates may be related since the rope has a fixed length $l = 30$ m, which at all times is equal to the length of segment DA plus CD. Using the Pythagorean theorem to determine l_{DA}, we have $l_{DA} = \sqrt{(15)^2 + x^2}$; also, $l_{CD} = 15 - y$. Hence,

$$l = l_{DA} + l_{CD}$$

$$30 = \sqrt{(15)^2 + x^2} + (15 - y)$$

$$y = \sqrt{225 + x^2} - 15 \tag{1}$$

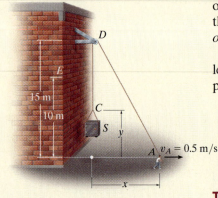

Fig. 12–41

Time Derivatives. Taking the time derivative, using the chain rule (see Appendix C), where $v_S = dy/dt$ and $v_A = dx/dt$, yields

$$v_S = \frac{dy}{dt} = \left[\frac{1}{2} \frac{2x}{\sqrt{225 + x^2}} \right] \frac{dx}{dt}$$

$$= \frac{x}{\sqrt{225 + x^2}} v_A \tag{2}$$

At $y = 10$ m, x is determined from Eq. 1, i.e., $x = 20$ m. Hence, from Eq. 2 with $v_A = 0.5$ m/s,

$$v_S = \frac{20}{\sqrt{225 + (20)^2}}(0.5) = 0.4\,\text{m/s} = 400\,\text{mm/s} \uparrow \quad Ans.$$

The acceleration is determined by taking the time derivative of Eq. 2. Since v_A is constant, then $a_A = dv_A/dt = 0$, and we have

$$a_S = \frac{d^2y}{dt^2} = \left[\frac{-x(dx/dt)}{(225 + x^2)^{3/2}} \right] x v_A + \left[\frac{1}{\sqrt{225 + x^2}} \right]\left(\frac{dx}{dt}\right) v_A + \left[\frac{1}{\sqrt{225 + x^2}} \right] x \frac{dv_A}{dt} = \frac{225 v_A^2}{(225 + x^2)^{3/2}}$$

At $x = 20$ m, with $v_A = 0.5$ m/s, the acceleration becomes

$$a_S = \frac{225(0.5 \text{ m/s})^2}{[225 + (20 \text{ m})^2]^{3/2}} = 0.00360 \text{ m/s}^2 = 3.60 \text{ mm/s}^2 \uparrow Ans.$$

NOTE: The constant velocity at A causes the other end C of the rope to have an acceleration since $\mathbf{v}_A$ causes segment DA to change its direction as well as its length.

12.10   Relative-Motion of Two Particles Using Translating Axes

Throughout this chapter the absolute motion of a particle has been determined using a single fixed reference frame. There are many cases, however, where the path of motion for a particle is complicated, so that it may be easier to analyze the motion in parts by using two or more frames of reference. For example, the motion of a particle located at the tip of an airplane propeller, while the plane is in flight, is more easily described if one observes first the motion of the airplane from a fixed reference and then superimposes (vectorially) the circular motion of the particle measured from a reference attached to the airplane.

In this section *translating frames of reference* will be considered for the analysis. Relative-motion analysis of particles using rotating frames of reference will be treated in Secs. 16.8 and 20.4, since such an analysis depends on prior knowledge of the kinematics of line segments.

Position.   Consider particles A and B, which move along the arbitrary paths shown in Fig. 12–42. The *absolute position* of each particle, $\mathbf{r}_A$ and $\mathbf{r}_B$, is measured from the common origin O of the *fixed x, y, z* reference frame. The origin of a second frame of reference x', y', z' is attached to and moves with particle A. The axes of this frame are *only permitted to translate* relative to the fixed frame. The position of B measured relative to A is denoted by the *relative-position vector* $\mathbf{r}_{B/A}$. Using vector addition, the three vectors shown in Fig. 12–42 can be related by the equation

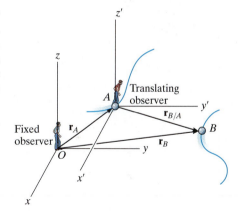

$$\mathbf{r}_B = \mathbf{r}_A + \mathbf{r}_{B/A} \qquad (12\text{–}33)$$

Velocity.   An equation that relates the velocities of the particles is determined by taking the time derivative of the above equation; i.e.,

$$\mathbf{v}_B = \mathbf{v}_A + \mathbf{v}_{B/A} \qquad (12\text{–}34)$$

Fig. 12–42

Here $\mathbf{v}_B = d\mathbf{r}_B/dt$ and $\mathbf{v}_A = d\mathbf{r}_A/dt$ refer to *absolute velocities*, since they are observed from the fixed frame; whereas the *relative velocity* $\mathbf{v}_{B/A} = d\mathbf{r}_{B/A}/dt$ is observed from the translating frame. It is important to note that since the x', y', z' axes translate, the *components* of $\mathbf{r}_{B/A}$ will *not* change direction and therefore the time derivative of these components will only have to account for the change in their magnitudes. Equation 12–34 therefore states that the velocity of B is equal to the velocity of A plus (vectorially) the velocity of "B with respect to A," as measured by the *translating observer* fixed in the x', y', z' reference frame.

12

Acceleration. The time derivative of Eq. 12–34 yields a similar vector relation between the *absolute* and *relative accelerations* of particles A and B.

$$\mathbf{a}_B = \mathbf{a}_A + \mathbf{a}_{B/A} \qquad (12\text{–}35)$$

Here $\mathbf{a}_{B/A}$ is the acceleration of B as seen by the observer located at A and translating with the x', y', z' reference frame.*

Procedure For Analysis

- When applying the relative velocity and acceleration equations, it is first necessary to specify the particle A that is the origin for the translating x', y', z' axes. Usually this point has a *known* velocity or acceleration.

- Since vector addition forms a triangle, there can be at most *two unknowns*, represented by the magnitudes and/or directions of the vector quantities.

- These unknowns can be solved for either graphically, using trigonometry (law of sines, law of cosines), or by resolving each of the three vectors into rectangular or Cartesian components, thereby generating a set of scalar equations.

The pilots of these jet planes flying close to one another must be aware of their relative positions and velocities at all times in order to avoid a collision.

* An easy way to remember the setup of these equations, is to note the "cancellation" of the subscript A between the two terms, e.g., $\mathbf{a}_B = \mathbf{a}_{\cancel{A}} + \mathbf{a}_{B/\cancel{A}}$.

EXAMPLE 12.25

A train travels at a constant speed of 60 mi/h, crosses over a road as shown in Fig. 12–43a. If the automobile A is traveling at 45 mi/h along the road, determine the magnitude and direction of the velocity of the train relative to the automobile.

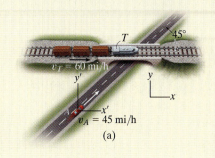
(a)

SOLUTION I

Vector Analysis. The relative velocity $\mathbf{v}_{T/A}$ is measured from the translating x', y' axes attached to the automobile, Fig. 12–43a. It is determined from $\mathbf{v}_T = \mathbf{v}_A + \mathbf{v}_{T/A}$. Since $\mathbf{v}_T$ and $\mathbf{v}_A$ are known in *both* magnitude and direction, the unknowns become the x and y components of $\mathbf{v}_{T/A}$. Using the x, y axes in Fig. 12–43a, we have

$$\mathbf{v}_T = \mathbf{v}_A + \mathbf{v}_{T/A}$$

$$60\mathbf{i} = (45 \cos 45°\mathbf{i} + 45 \sin 45°\mathbf{j}) + \mathbf{v}_{T/A}$$

$$\mathbf{v}_{T/A} = \{28.2\mathbf{i} - 31.8\mathbf{j}\} \text{ mi/h} \qquad \textit{Ans.}$$

The magnitude of $\mathbf{v}_{T/A}$ is thus

$$v_{T/A} = \sqrt{(28.2)^2 + (-31.8)^2} = 42.5 \text{ mi/h} \qquad \textit{Ans.}$$

From the direction of each component, Fig. 12–43b, the direction of $\mathbf{v}_{T/A}$ is

$$\tan \theta = \frac{(v_{T/A})_y}{(v_{T/A})_x} = \frac{31.8}{28.2}$$

$$\theta = 48.5° \quad \nwarrow \qquad \textit{Ans.}$$

Note that the vector addition shown in Fig. 12–43b indicates the correct sense for $\mathbf{v}_{T/A}$. This figure anticipates the answer and can be used to check it.

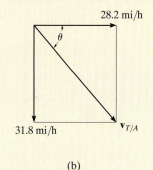
(b)

SOLUTION II

Scalar Analysis. The unknown components of $\mathbf{v}_{T/A}$ can also be determined by applying a scalar analysis. We will assume these components act in the *positive* x and y directions. Thus,

$$\mathbf{v}_T = \mathbf{v}_A + \mathbf{v}_{T/A}$$

$$\begin{bmatrix} 60 \text{ mi/h} \\ \rightarrow \end{bmatrix} = \begin{bmatrix} 45 \text{ mi/h} \\ \nearrow^{45°} \end{bmatrix} + \begin{bmatrix} (v_{T/A})_x \\ \rightarrow \end{bmatrix} + \begin{bmatrix} (v_{T/A})_y \\ \uparrow \end{bmatrix}$$

Resolving each vector into its x and y components yields

$(\xrightarrow{+})$ $60 = 45 \cos 45° + (v_{T/A})_x + 0$

$(+\uparrow)$ $0 = 45 \sin 45° + 0 + (v_{T/A})_y$

Solving, we obtain the previous results,

$$(v_{T/A})_x = 28.2 \text{ mi/h} = 28.2 \text{ mi/h} \rightarrow$$

$$(v_{T/A})_y = -31.8 \text{ mi/h} = 31.8 \text{ mi/h} \downarrow$$

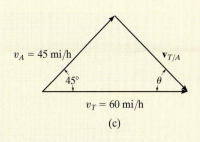
(c)

Fig. 12–43

12

EXAMPLE | 12.26

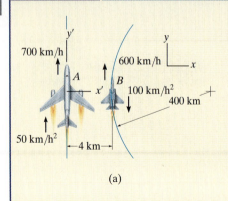

(a)

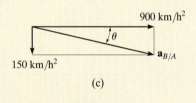

(b)

Plane A in Fig. 12–44a is flying along a straight-line path, whereas plane B is flying along a circular path having a radius of curvature of $\rho_B = 400$ km. Determine the velocity and acceleration of B as measured by the pilot of A.

SOLUTION

Velocity. The origin of the x and y axes are located at an arbitrary fixed point. Since the motion relative to plane A is to be determined, the *translating frame of reference* x', y' is attached to it, Fig. 12–44a. Applying the relative-velocity equation in scalar form since the velocity vectors of both planes are parallel at the instant shown, we have

$$(+\uparrow) \qquad\qquad v_B = v_A + v_{B/A}$$

$$600 \text{ km/h} = 700 \text{ km/h} + v_{B/A}$$

$$v_{B/A} = -100 \text{ km/h} = 100 \text{ km/h} \downarrow \qquad Ans.$$

The vector addition is shown in Fig. 12–44b.

Acceleration. Plane B has both tangential and normal components of acceleration since it is flying along a *curved path*. From Eq. 12–20, the magnitude of the normal component is

$$(a_B)_n = \frac{v_B^2}{\rho} = \frac{(600 \text{ km/h})^2}{400 \text{ km}} = 900 \text{ km/h}^2$$

Applying the relative-acceleration equation gives

$$\mathbf{a}_B = \mathbf{a}_A + \mathbf{a}_{B/A}$$

$$900\mathbf{i} - 100\mathbf{j} = 50\mathbf{j} + \mathbf{a}_{B/A}$$

Thus,

$$\mathbf{a}_{B/A} = \{900\mathbf{i} - 150\mathbf{j}\} \text{ km/h}^2$$

From Fig. 12–44c, the magnitude and direction of $\mathbf{a}_{B/A}$ are therefore

$$a_{B/A} = 912 \text{ km/h}^2 \quad \theta = \tan^{-1}\frac{150}{900} = 9.46° \searrow \qquad Ans.$$

NOTE: The solution to this problem was possible using a translating frame of reference, since the pilot in plane A is "translating." Observation of the motion of plane A with respect to the pilot of plane B, however, must be obtained using a *rotating* set of axes attached to plane B. (This assumes, of course, that the pilot of B is fixed in the rotating frame, so he does not turn his eyes to follow the motion of A.) The analysis for this case is given in Example 16.21.

(c)

Fig. 12–44

EXAMPLE 12.27

At the instant shown in Fig. 12–45a, cars A and B are traveling with speeds of 18 m/s and 12 m/s, respectively. Also at this instant, A has a decrease in speed of 2 m/s², and B has an increase in speed of 3 m/s². Determine the velocity and acceleration of B with respect to A.

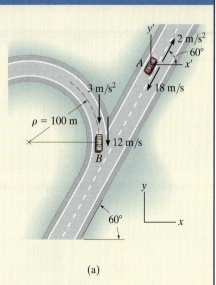

SOLUTION

Velocity. The fixed x, y axes are established at an arbitrary point on the ground and the translating x', y' axes are attached to car A, Fig. 12–45a. Why? The relative velocity is determined from $\mathbf{v}_B = \mathbf{v}_A + \mathbf{v}_{B/A}$. What are the two unknowns? Using a Cartesian vector analysis, we have

$$\mathbf{v}_B = \mathbf{v}_A + \mathbf{v}_{B/A}$$

$$-12\mathbf{j} = (-18\cos 60°\mathbf{i} - 18\sin 60°\mathbf{j}) + \mathbf{v}_{B/A}$$

$$\mathbf{v}_{B/A} = \{9\mathbf{i} + 3.588\mathbf{j}\} \text{ m/s}$$

Thus,

$$v_{B/A} = \sqrt{(9)^2 + (3.588)^2} = 9.69 \text{ m/s} \qquad \textit{Ans.}$$

Noting that $\mathbf{v}_{B/A}$ has $+\mathbf{i}$ and $+\mathbf{j}$ components, Fig. 12–45b, its direction is

$$\tan\theta = \frac{(v_{B/A})_y}{(v_{B/A})_x} = \frac{3.588}{9}$$

$$\theta = 21.7° \qquad \textit{Ans.}$$

(a)

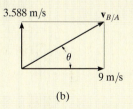

(b)

Acceleration. Car B has both tangential and normal components of acceleration. Why? The magnitude of the normal component is

$$(a_B)_n = \frac{v_B^2}{\rho} = \frac{(12 \text{ m/s})^2}{100 \text{ m}} = 1.440 \text{ m/s}^2$$

Applying the equation for relative acceleration yields

$$\mathbf{a}_B = \mathbf{a}_A + \mathbf{a}_{B/A}$$

$$(-1.440\mathbf{i} - 3\mathbf{j}) = (2\cos 60°\mathbf{i} + 2\sin 60°\mathbf{j}) + \mathbf{a}_{B/A}$$

$$\mathbf{a}_{B/A} = \{-2.440\mathbf{i} - 4.732\mathbf{j}\} \text{ m/s}^2$$

Here $\mathbf{a}_{B/A}$ has $-\mathbf{i}$ and $-\mathbf{j}$ components. Thus, from Fig. 12–45c,

$$a_{B/A} = \sqrt{(2.440)^2 + (4.732)^2} = 5.32 \text{ m/s}^2 \qquad \textit{Ans.}$$

$$\tan\phi = \frac{(a_{B/A})_y}{(a_{B/A})_x} = \frac{4.732}{2.440}$$

$$\phi = 62.7° \qquad \textit{Ans.}$$

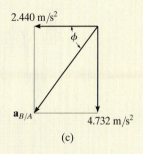

(c)

Fig. 12–45

NOTE: Is it possible to obtain the relative acceleration of $\mathbf{a}_{A/B}$ using this method? Refer to the comment made at the end of Example 12.26.

FUNDAMENTAL PROBLEMS

F12–39. Determine the speed of block D if end A of the rope is pulled down with a speed of $v_A = 3$ m/s.

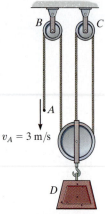

$v_A = 3$ m/s

F12–39

F12–40. Determine the speed of block A if end B of the rope is pulled down with a speed of 6 m/s.

6 m/s

F12–40

F12–41. Determine the speed of block A if end B of the rope is pulled down with a speed of 1.5 m/s.

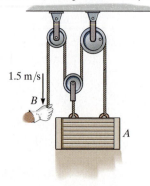

1.5 m/s

F12–41

F12–42. Determine the speed of block A if end F of the rope is pulled down with a speed of $v_F = 3$ m/s.

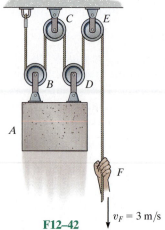

$v_F = 3$ m/s

F12–42

F12–43. Determine the speed of car A if point P on the cable has a speed of 4 m/s when the motor M winds the cable in.

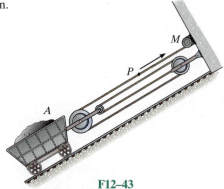

F12–43

F12–44. Determine the speed of cylinder B if cylinder A moves downward with a speed of $v_A = 4$ ft/s.

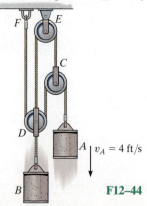

$v_A = 4$ ft/s

F12–44

F12–45.   Car *A* is traveling with a constant speed of 80 km/h due north, while car *B* is traveling with a constant speed of 100 km/h due east. Determine the velocity of car *B* relative to car *A*.

F12–47.   The boats *A* and *B* travel with constant speeds of $v_A = 15$ m/s and $v_B = 10$ m/s when they leave the pier at *O* at the same time. Determine the distance between them when $t = 4$ s.

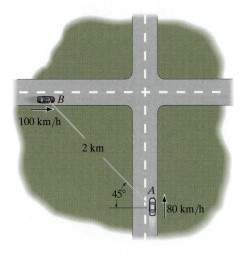

F12–45

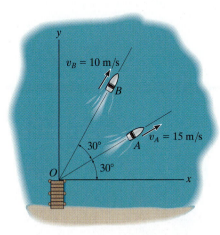

F12–47

F12–46.   Two planes *A* and *B* are traveling with the constant velocities shown. Determine the magnitude and direction of the velocity of plane *B* relative to plane *A*.

F12–48.   At the instant shown, cars *A* and *B* are traveling at the speeds shown. If *B* is accelerating at 1200 km/h² while *A* maintains a constant speed, determine the velocity and acceleration of *A* with respect to *B*.

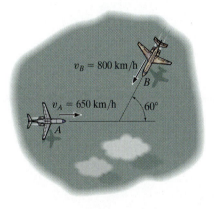

F12–46

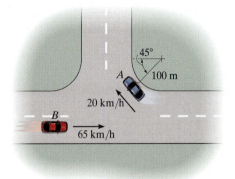

F12–48

12

PROBLEMS

12–195. The mine car C is being pulled up the incline using the motor M and the rope-and-pulley arrangement shown. Determine the speed v_P at which a point P on the cable must be traveling toward the motor to move the car up the plane with a constant speed of $v = 2$ m/s.

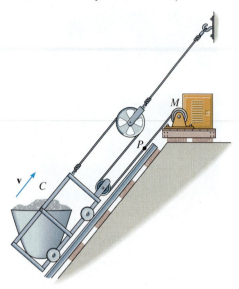

Prob. 12–195

*12–196. Determine the displacement of the log if the truck at C pulls the cable 4 ft to the right.

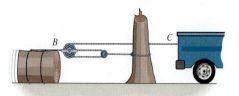

Prob. 12–196

•12–197. If the hydraulic cylinder H draws in rod BC at 2 ft/s, determine the speed of slider A.

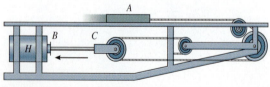

Prob. 12–197

12–198. If end A of the rope moves downward with a speed of 5 m/s, determine the speed of cylinder B.

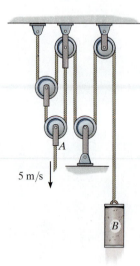

5 m/s

Prob. 12–198

12–199. Determine the speed of the elevator if each motor draws in the cable with a constant speed of 5 m/s.

Prob. 12–199

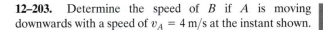

*12–200. Determine the speed of cylinder A, if the rope is drawn towards the motor M at a constant rate of 10 m/s.

•12–201. If the rope is drawn towards the motor M at a speed of $v_M = (5t^{3/2})$ m/s, where t is in seconds, determine the speed of cylinder A when $t = 1$ s.

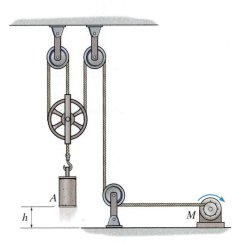

Probs. 12–200/201

12–203. Determine the speed of B if A is moving downwards with a speed of $v_A = 4$ m/s at the instant shown.

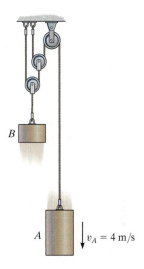

Prob. 12–203

12–202. If the end of the cable at A is pulled down with a speed of 2 m/s, determine the speed at which block B rises.

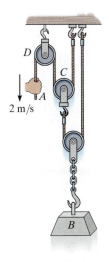

Prob. 12–202

*12–204. The crane is used to hoist the load. If the motors at A and B are drawing in the cable at a speed of 2 ft/s and 4 ft/s, respectively, determine the speed of the load.

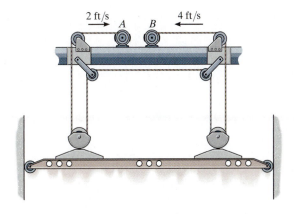

Prob. 12–204

•12–205. The cable at B is pulled downwards at 4 ft/s, and the speed is decreasing at 2 ft/s². Determine the velocity and acceleration of block A at this instant.

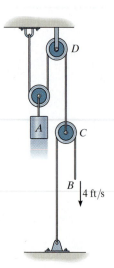

Prob. 12–205

12–206. If block A is moving downward with a speed of 4 ft/s while C is moving up at 2 ft/s, determine the speed of block B.

12–207. If block A is moving downward at 6 ft/s while block C is moving down at 18 ft/s, determine the speed of block B.

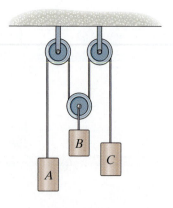

Probs. 12–206/207

*12–208. If the end of the cable at A is pulled down with a speed of 2 m/s, determine the speed at which block E rises.

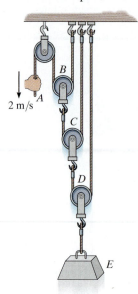

Prob. 12–208

•12–209. If motors at A and B draw in their attached cables with an acceleration of $a = (0.2t)$ m/s², where t is in seconds, determine the speed of the block when it reaches a height of $h = 4$ m, starting from rest at $h = 0$. Also, how much time does it take to reach this height?

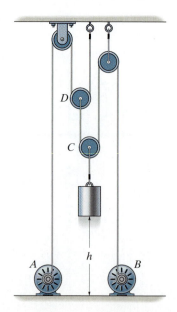

Prob. 12–209

12–210. The motor at C pulls in the cable with an acceleration $a_C = (3t^2)$ m/s^2, where t is in seconds. The motor at D draws in its cable at $a_D = 5$ m/s^2. If both motors start at the same instant from rest when $d = 3$ m, determine (a) the time needed for $d = 0$, and (b) the velocities of blocks A and B when this occurs.

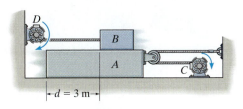

Prob. 12–210

12–211. The motion of the collar at A is controlled by a motor at B such that when the collar is at $s_A = 3$ ft it is moving upwards at 2 ft/s and decreasing at 1 ft/s^2. Determine the velocity and acceleration of a point on the cable as it is drawn into the motor B at this instant.

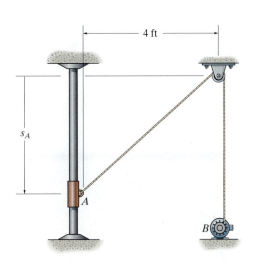

Prob. 12–211

*12–212. The man pulls the boy up to the tree limb C by walking backward at a constant speed of 1.5 m/s. Determine the speed at which the boy is being lifted at the instant $x_A = 4$ m. Neglect the size of the limb. When $x_A = 0$, $y_B = 8$ m, so that A and B are coincident, i.e., the rope is 16 m long.

•12–213. The man pulls the boy up to the tree limb C by walking backward. If he starts from rest when $x_A = 0$ and moves backward with a constant acceleration $a_A = 0.2$ m/s^2, determine the speed of the boy at the instant $y_B = 4$ m. Neglect the size of the limb. When $x_A = 0$, $y_B = 8$ m, so that A and B are coincident, i.e., the rope is 16 m long.

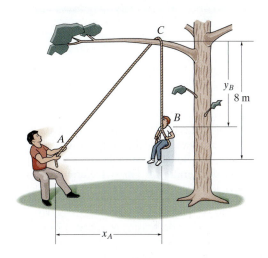

Probs. 12–212/213

12–214. If the truck travels at a constant speed of $v_T = 6$ ft/s, determine the speed of the crate for any angle θ of the rope. The rope has a length of 100 ft and passes over a pulley of negligible size at A. *Hint:* Relate the coordinates x_T and x_C to the length of the rope and take the time derivative. Then substitute the trigonometric relation between x_C and θ.

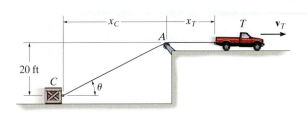

Prob. 12–214

12–215. At the instant shown, car A travels along the straight portion of the road with a speed of 25 m/s. At this same instant car B travels along the circular portion of the road with a speed of 15 m/s. Determine the velocity of car B relative to car A.

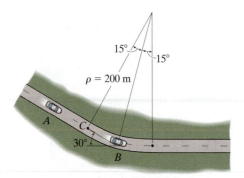

Prob. 12–215

***12–216.** Car A travels along a straight road at a speed of 25 m/s while accelerating at 1.5 m/s². At this same instant car C is traveling along the straight road with a speed of 30 m/s while decelerating at 3 m/s². Determine the velocity and acceleration of car A relative to car C.

•**12–217.** Car B is traveling along the curved road with a speed of 15 m/s while decreasing its speed at 2 m/s². At this same instant car C is traveling along the straight road with a speed of 30 m/s while decelerating at 3 m/s². Determine the velocity and acceleration of car B relative to car C.

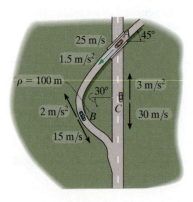

Probs. 12–216/217

12–218. The ship travels at a constant speed of $v_s = 20$ m/s and the wind is blowing at a speed of $v_w = 10$ m/s, as shown. Determine the magnitude and direction of the horizontal component of velocity of the smoke coming from the smoke stack as it appears to a passenger on the ship.

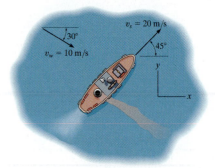

Prob. 12–218

12–219. The car is traveling at a constant speed of 100 km/h. If the rain is falling at 6 m/s in the direction shown, determine the velocity of the rain as seen by the driver.

Prob. 12–219

***12–220.** The man can row the boat in still water with a speed of 5 m/s. If the river is flowing at 2 m/s, determine the speed of the boat and the angle θ he must direct the boat so that it travels from A to B.

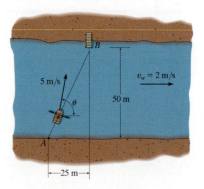

Prob. 12–220

•**12–221.** At the instant shown, cars A and B travel at speeds of 30 mi/h and 20 mi/h, respectively. If B is increasing its speed by 1200 mi/h^2, while A maintains a constant speed, determine the velocity and acceleration of B with respect to A.

12–222. At the instant shown, cars A and B travel at speeds of 30 m/h and 20 mi/h, respectively. If A is increasing its speed at 400 mi/h^2 whereas the speed of B is decreasing at 800 mi/h^2, determine the velocity and acceleration of B with respect to A.

***12–224.** At the instant shown, cars A and B travel at speeds of 70 mi/h and 50 mi/h, respectively. If B is increasing its speed by 1100 mi/h^2, while A maintains a constant speed, determine the velocity and acceleration of B with respect to A. Car B moves along a curve having a radius of curvature of 0.7 mi.

•**12–225.** At the instant shown, cars A and B travel at speeds of 70 mi/h and 50 mi/h, respectively. If B is decreasing its speed at 1400 mi/h^2 while A is increasing its speed at 800 mi/h^2, determine the acceleration of B with respect to A. Car B moves along a curve having a radius of curvature of 0.7 mi.

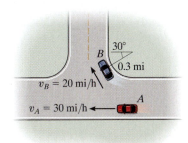

Probs. 12–221/222

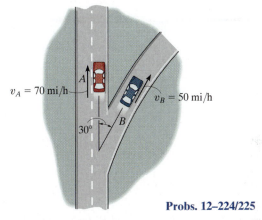

Probs. 12–224/225

12–223. Two boats leave the shore at the same time and travel in the directions shown. If $v_A = 20$ ft/s and $v_B = 15$ ft/s, determine the velocity of boat A with respect to boat B. How long after leaving the shore will the boats be 800 ft apart?

12–226. An aircraft carrier is traveling forward with a velocity of 50 km/h. At the instant shown, the plane at A has just taken off and has attained a forward horizontal air speed of 200 km/h, measured from still water. If the plane at B is traveling along the runway of the carrier at 175 km/h in the direction shown, determine the velocity of A with respect to B.

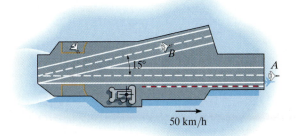

Prob. 12–226

12–227. A car is traveling north along a straight road at 50 km/h. An instrument in the car indicates that the wind is directed towards the east. If the car's speed is 80 km/h, the instrument indicates that the wind is directed towards the north-east. Determine the speed and direction of the wind.

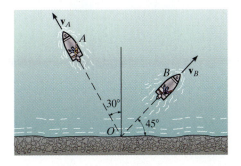

Prob. 12–223

12

*12–228. At the instant shown car A is traveling with a velocity of 30 m/s and has an acceleration of 2 m/s^2 along the highway. At the same instant B is traveling on the trumpet interchange curve with a speed of 15 m/s, which is decreasing at 0.8 m/s^2. Determine the relative velocity and relative acceleration of B with respect to A at this instant.

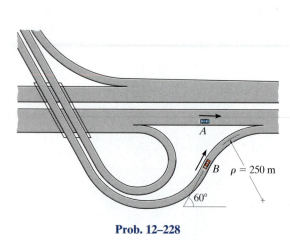

Prob. 12–228

•12–229. Two cyclists A and B travel at the same constant speed v. Determine the velocity of A with respect to B if A travels along the circular track, while B travels along the diameter of the circle.

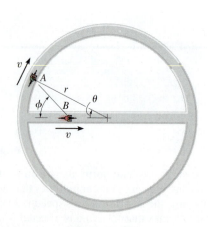

Prob. 12–229

12–230. A man walks at 5 km/h in the direction of a 20-km/h wind. If raindrops fall vertically at 7 km/h in *still air*, determine the direction in which the drops appear to fall with respect to the man. Assume the horizontal speed of the raindrops is equal to that of the wind.

Prob. 12–230

12–231. A man can row a boat at 5 m/s in still water. He wishes to cross a 50-m-wide river to point B, 50 m downstream. If the river flows with a velocity of 2 m/s, determine the speed of the boat and the time needed to make the crossing.

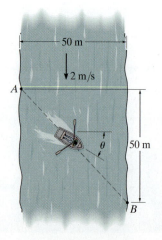

Prob. 12–231

CONCEPTUAL PROBLEMS

P12–1. If you measured the time it takes for the construction elevator to go from A to B, then B to C, and then C to D, and you also know the distance between each of the points, how could you determine the average velocity and average acceleration of the elevator as it ascends from A to D? Use numerical values to explain how this can be done.

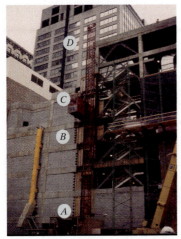

P12–1

P12–2. If the sprinkler at A is 1 m from the ground, then scale the necessary measurements from the photo to determine the approximate velocity of the water jet as it flows from the nozzle of the sprinkler.

P12–2

P12–3. The basketball was thrown at an angle measured from the horizontal to the man's outstretched arms. If the basket is 10 ft from the ground, make appropriate measurements in the photo and determine if the ball located as shown will pass through the basket.

P12–3

P12–4. The pilot tells you the wingspan of her plane and her constant airspeed. How would you determine the acceleration of the plane at the moment shown? Use numerical values and take any necessary measurements from the photo.

P12–4

12

CHAPTER REVIEW

Rectilinear Kinematics

Rectilinear kinematics refers to motion along a straight line. A position coordinate s specifies the location of the particle on the line, and the displacement Δs is the change in this position.

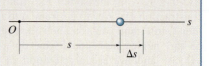

The average velocity is a vector quantity, defined as the displacement divided by the time interval.

$$v_{avg} = \frac{-\Delta s}{\Delta t}$$

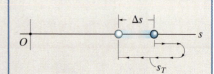

The average speed is a scalar, and is the total distance traveled divided by the time of travel.

$$(v_{sp})_{avg} = \frac{s_T}{\Delta t}$$

The time, position, velocity, and acceleration are related by three differential equations.

$$a = \frac{dv}{dt}, \qquad v = \frac{ds}{dt}, \qquad a\, ds = v\, dv$$

If the acceleration is known to be constant, then the differential equations relating time, position, velocity, and acceleration can be integrated.

$$v = v_0 + a_c t$$
$$s = s_0 + v_0 t + \tfrac{1}{2} a_c t^2$$
$$v^2 = v_0^2 + 2a_c(s - s_0)$$

Graphical Solutions

If the motion is erratic, then it can be described by a graph. If one of these graphs is given, then the others can be established using the differential relations between a, v, s, and t.

$$a = \frac{dv}{dt},$$
$$v = \frac{ds}{dt},$$
$$a\, ds = v\, dv$$

Curvilinear Motion, x, y, z

Curvilinear motion along the path can be resolved into rectilinear motion along the x, y, z axes. The equation of the path is used to relate the motion along each axis.

$$v_x = \dot{x} \qquad a_x = \dot{v}_x$$
$$v_y = \dot{y} \qquad a_y = \dot{v}_y$$
$$v_z = \dot{z} \qquad a_z = \dot{v}_z$$

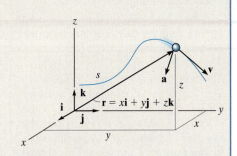

Projectile Motion

Free-flight motion of a projectile follows a parabolic path. It has a constant velocity in the horizontal direction, and a constant downward acceleration of $g = 9.81 \text{ m/s}^2$ or 32.2 ft/s^2 in the vertical direction. Any two of the three equations for constant acceleration apply in the vertical direction, and in the horizontal direction only one equation applies.

$$(+\uparrow) \quad v_y = (v_0)_y + a_c t$$
$$(+\uparrow) \quad y = y_0 + (v_0)_y t + \tfrac{1}{2} a_c t^2$$
$$(+\uparrow) \quad v_y^2 = (v_0)_y^2 + 2a_c(y - y_0)$$
$$(\xrightarrow{+}) \quad x = x_0 + (v_0)_x t$$

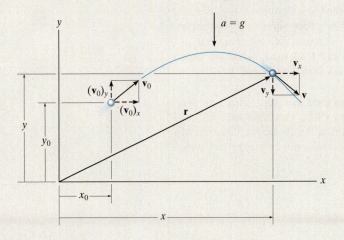

Curvilinear Motion *n, t*

If normal and tangential axes are used for the analysis, then **v** is always in the positive *t* direction.

The acceleration has two components. The tangential component, $\mathbf{a}_t$, accounts for the change in the magnitude of the velocity; a slowing down is in the negative *t* direction, and a speeding up is in the positive *t* direction. The normal component $\mathbf{a}_n$ accounts for the change in the direction of the velocity. This component is always in the positive *n* direction.

$$a_t = \dot{v} \qquad \text{or} \qquad a_t\,ds = v\,dv$$

$$a_n = \frac{v^2}{\rho}$$

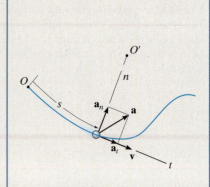

Curvilinear Motion *r, θ*

If the path of motion is expressed in polar coordinates, then the velocity and acceleration components can be related to the time derivatives of *r* and *θ*.

To apply the time-derivative equations, it is necessary to determine $r, \dot{r}, \ddot{r}, \theta, \dot{\theta}, \ddot{\theta}$ at the instant considered. If the path $r = f(\theta)$ is given, then the chain rule of calculus must be used to obtain time derivatives. (See Appendix C.)

Once the data are substituted into the equations, then the algebraic sign of the results will indicate the direction of the components of **v** or **a** along each axis.

$$v_r = \dot{r}$$
$$v_\theta = r\dot{\theta}$$

$$a_r = \ddot{r} - r\dot{\theta}^2$$
$$a_\theta = r\ddot{\theta} + 2\dot{r}\dot{\theta}$$

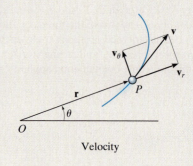

Velocity

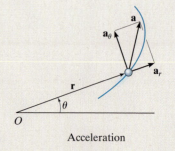

Acceleration

Absolute Dependent Motion of Two Particles

The dependent motion of blocks that are suspended from pulleys and cables can be related by the geometry of the system. This is done by first establishing position coordinates, measured from a fixed origin to each block. Each coordinate must be directed along the line of motion of a block.

Using geometry and/or trigonometry, the coordinates are then related to the cable length in order to formulate a position coordinate equation.

$$2s_B + h + s_A = l$$

The first time derivative of this equation gives a relationship between the velocities of the blocks, and a second time derivative gives the relation between their accelerations.

$$2v_B = -v_A$$
$$2a_B = -a_A$$

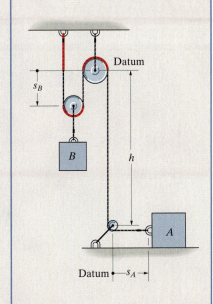

Relative-Motion Analysis Using Translating Axes

If two particles A and B undergo independent motions, then these motions can be related to their relative motion using a *translating set of axes* attached to one of the particles (A).

$$\mathbf{r}_B = \mathbf{r}_A + \mathbf{r}_{B/A}$$

For planar motion, each vector equation produces two scalar equations, one in the x, and the other in the y direction. For solution, the vectors can be expressed in Cartesian form, or the x and y scalar components can be written directly.

$$\mathbf{v}_B = \mathbf{v}_A + \mathbf{v}_{B/A}$$
$$\mathbf{a}_B = \mathbf{a}_A + \mathbf{a}_{B/A}$$

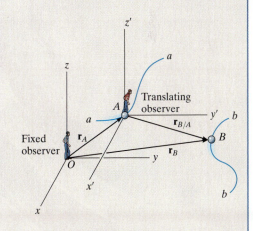

The design of conveyors for a bottling plant requires knowledge of the forces that act on them and the ability to predict the motion of the bottles they transport.

Kinetics of a Particle: Force and Acceleration

13

CHAPTER OBJECTIVES

- To state Newton's Second Law of Motion and to define mass and weight.

- To analyze the accelerated motion of a particle using the equation of motion with different coordinate systems.

- To investigate central-force motion and apply it to problems in space mechanics.

13.1 Newton's Second Law of Motion

Kinetics is a branch of dynamics that deals with the relationship between the change in motion of a body and the forces that cause this change. The basis for kinetics is Newton's second law, which states that when an *unbalanced force* acts on a particle, the particle will *accelerate* in the direction of the force with a magnitude that is proportional to the force.

This law can be verified experimentally by applying a known unbalanced force **F** to a particle, and then measuring the acceleration **a**. Since the force and acceleration are directly proportional, the constant of proportionality, m, may be determined from the ratio $m = F/a$. This positive scalar m is called the *mass* of the particle. Being constant during any acceleration, m provides a quantitative measure of the resistance of the particle to a change in its velocity, that is its inertia.

13

13.4   Equations of Motion: Rectangular Coordinates

When a particle moves relative to an inertial x, y, z frame of reference, the forces acting on the particle, as well as its acceleration, can be expressed in terms of their $\mathbf{i}, \mathbf{j}, \mathbf{k}$ components, Fig. 13–5. Applying the equation of motion, we have

$$\Sigma\mathbf{F} = m\mathbf{a}; \quad \Sigma F_x\mathbf{i} + \Sigma F_y\mathbf{j} + \Sigma F_z\mathbf{k} = m(a_x\mathbf{i} + a_y\mathbf{j} + a_z\mathbf{k})$$

For this equation to be satisfied, the respective $\mathbf{i}, \mathbf{j}, \mathbf{k}$ components on the left side must equal the corresponding components on the right side. Consequently, we may write the following three scalar equations:

$$\boxed{\begin{aligned} \Sigma F_x &= ma_x \\ \Sigma F_y &= ma_y \\ \Sigma F_z &= ma_z \end{aligned}} \qquad (13\text{–}7)$$

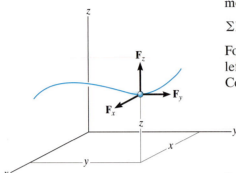

Fig. 13–5

In particular, if the particle is constrained to move only in the x–y plane, then the first two of these equations are used to specify the motion.

Procedure for Analysis

The equations of motion are used to solve problems which require a relationship between the forces acting on a particle and the accelerated motion they cause.

Free-Body Diagram.

- Select the inertial coordinate system. Most often, rectangular or x, y, z coordinates are chosen to analyze problems for which the particle has *rectilinear motion*.

- Once the coordinates are established, draw the particle's free-body diagram. Drawing this diagram is *very important* since it provides a graphical representation that accounts for *all the forces* ($\Sigma\mathbf{F}$) which act on the particle, and thereby makes it possible to resolve these forces into their x, y, z components.

- The direction and sense of the particle's acceleration $\mathbf{a}$ should also be established. If the sense is unknown, for mathematical convenience assume that the sense of each acceleration component acts in the *same direction* as its *positive* inertial coordinate axis.

- The acceleration may be represented as the $m\mathbf{a}$ vector on the kinetic diagram.*

- Identify the unknowns in the problem.

---

*It is a convention in this text always to use the kinetic diagram as a graphical aid when developing the proofs and theory. The particle's acceleration or its components will be shown as blue colored vectors near the free-body diagram in the examples.

Equations of Motion.

- If the forces can be resolved directly from the free-body diagram, apply the equations of motion in their scalar component form.

- If the geometry of the problem appears complicated, which often occurs in three dimensions, Cartesian vector analysis can be used for the solution.

- *Friction.* If a moving particle contacts a rough surface, it may be necessary to use the *frictional equation*, which relates the frictional and normal forces $\mathbf{F}_f$ and $\mathbf{N}$ acting at the surface of contact by using the coefficient of kinetic friction, i.e., $F_f = \mu_k N$. Remember that $\mathbf{F}_f$ always acts on the free-body diagram such that it opposes the motion of the particle relative to the surface it contacts. If the particle is *on the verge* of relative motion, then the coefficient of static friction should be used.

- *Spring.* If the particle is connected to an *elastic spring* having negligible mass, the spring force F_s can be related to the deformation of the spring by the equation $F_s = ks$. Here k is the spring's stiffness measured as a force per unit length, and s is the stretch or compression defined as the difference between the deformed length l and the undeformed length l_0, i.e., $s = l - l_0$.

Kinematics.

- If the velocity or position of the particle is to be found, it will be necessary to apply the necessary kinematic equations once the particle's acceleration is determined from $\Sigma \mathbf{F} = m\mathbf{a}$.

- If *acceleration* is a function of time, use $a = dv/dt$ and $v = ds/dt$ which, when integrated, yield the particle's velocity and position, respectively.

- If *acceleration* is a function of displacement, integrate $a\,ds = v\,dv$ to obtain the velocity as a function of position.

- If *acceleration is constant*, use $v = v_0 + a_c t$, $s = s_0 + v_0 t + \frac{1}{2}a_c t^2$, $v^2 = v_0^2 + 2a_c(s - s_0)$ to determine the velocity or position of the particle.

- If the problem involves the dependent motion of several particles, use the method outlined in Sec. 12.9 to relate their accelerations. In all cases, make sure the positive inertial coordinate directions used for writing the kinematic equations are the same as those used for writing the equations of motion; otherwise, simultaneous solution of the equations will result in errors.

- If the solution for an unknown vector component yields a negative scalar, it indicates that the component acts in the direction opposite to that which was assumed.

13

EXAMPLE 13.1

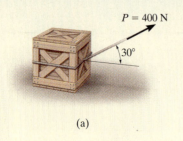

$P = 400$ N

30°

(a)

The 50-kg crate shown in Fig. 13–6a rests on a horizontal surface for which the coefficient of kinetic friction is $\mu_k = 0.3$. If the crate is subjected to a 400-N towing force as shown, determine the velocity of the crate in 3 s starting from rest.

SOLUTION

Using the equations of motion, we can relate the crate's acceleration to the force causing the motion. The crate's velocity can then be determined using kinematics.

Free-Body Diagram. The weight of the crate is $W = mg = 50$ kg $(9.81$ m/s$^2) = 490.5$ N. As shown in Fig. 13–6b, the frictional force has a magnitude $F = \mu_k N_C$ and acts to the left, since it opposes the motion of the crate. The acceleration **a** is assumed to act horizontally, in the positive x direction. There are two unknowns, namely N_C and a.

Equations of Motion. Using the data shown on the free-body diagram, we have

$$\xrightarrow{+} \Sigma F_x = ma_x; \qquad 400 \cos 30° - 0.3 N_C = 50a \qquad (1)$$
$$+\uparrow \Sigma F_y = ma_y; \qquad N_C - 490.5 + 400 \sin 30° = 0 \qquad (2)$$

Solving Eq. 2 for N_C, substituting the result into Eq. 1, and solving for a yields

$$N_C = 290.5 \text{ N}$$
$$a = 5.185 \text{ m/s}^2$$

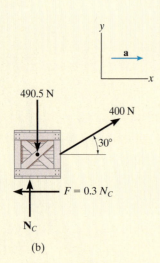

490.5 N

400 N

30°

$F = 0.3\, N_C$

$\mathbf{N}_C$

(b)

Fig. 13–6

Kinematics. Notice that the acceleration is *constant*, since the applied force **P** is constant. Since the initial velocity is zero, the velocity of the crate in 3 s is

$$(\xrightarrow{+}) \qquad v = v_0 + a_c t = 0 + 5.185(3)$$
$$= 15.6 \text{ m/s} \rightarrow \qquad \qquad \textit{Ans.}$$

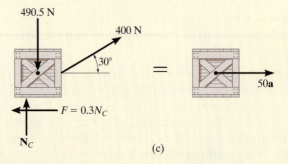

(c)

NOTE: We can also use the alternative procedure of drawing the crate's free-body *and* kinetic diagrams, Fig. 13–6c, prior to applying the equations of motion.

EXAMPLE | 13.2

A 10-kg projectile is fired vertically upward from the ground, with an initial velocity of 50 m/s, Fig. 13–7a. Determine the maximum height to which it will travel if (a) atmospheric resistance is neglected; and (b) atmospheric resistance is measured as $F_D = (0.01v^2)$ N, where v is the speed of the projectile at any instant, measured in m/s.

SOLUTION

In both cases the known force on the projectile can be related to its acceleration using the equation of motion. Kinematics can then be used to relate the projectile's acceleration to its position.

Part (a) Free-Body Diagram. As shown in Fig. 13–7b, the projectile's weight is $W = mg = 10(9.81) = 98.1$ N. We will assume the unknown acceleration **a** acts upward in the *positive z* direction.

Equation of Motion.

$$+\uparrow \Sigma F_z = ma_z; \qquad -98.1 = 10a, \qquad a = -9.81 \text{ m/s}^2$$

The result indicates that the projectile, like every object having free-flight motion near the earth's surface, is subjected to a *constant* downward acceleration of 9.81 m/s².

Kinematics. Initially, $z_0 = 0$ and $v_0 = 50$ m/s, and at the maximum height $z = h$, $v = 0$. Since the acceleration is *constant*, then

$$(+\uparrow) \qquad v^2 = v_0^2 + 2a_c(z - z_0)$$
$$0 = (50)^2 + 2(-9.81)(h - 0)$$
$$h = 127 \text{ m} \qquad\qquad Ans.$$

Part (b) Free-Body Diagram. Since the force $F_D = (0.01v^2)$ N tends to retard the upward motion of the projectile, it acts downward as shown on the free-body diagram, Fig. 13–7c.

Equation of Motion.

$$+\uparrow \Sigma F_z = ma_z; \qquad -0.01v^2 - 98.1 = 10a, \qquad a = -(0.001v^2 + 9.81)$$

Kinematics. Here the acceleration is *not constant* since F_D depends on the velocity. Since $a = f(v)$, we can relate a to position using

$$(+\uparrow) \, a\,dz = v\,dv; \qquad -(0.001v^2 + 9.81)\,dz = v\,dv$$

Separating the variables and integrating, realizing that initially $z_0 = 0$, $v_0 = 50$ m/s (positive upward), and at $z = h$, $v = 0$, we have

$$\int_0^h dz = -\int_{50}^0 \frac{v\,dv}{0.001v^2 + 9.81} = -500 \ln(v^2 + 9810) \Big|_{50 \text{ m/s}}^0$$

$$h = 114 \text{ m} \qquad\qquad Ans.$$

NOTE: The answer indicates a lower elevation than that obtained in part (a) due to atmospheric resistance or drag.

(a)

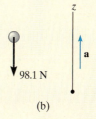

(b)

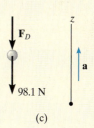

(c)

Fig. 13–7

13

EXAMPLE | 13.3

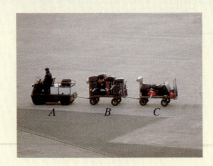

The baggage truck A shown in the photo has a weight of 900 lb and tows a 550-lb cart B and a 325-lb cart C. For a short time the driving frictional force developed at the wheels of the truck is $F_A = (40t)$ lb, where t is in seconds. If the truck starts from rest, determine its speed in 2 seconds. Also, what is the horizontal force acting on the coupling between the truck and cart B at this instant? Neglect the size of the truck and carts.

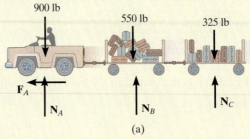

(a)

SOLUTION

Free-Body Diagram.   As shown in Fig. 13–8a, it is the frictional driving force that gives both the truck and carts an acceleration. Here we have considered all three vehicles as a single system.

Equation of Motion.   Only motion in the horizontal direction has to be considered.

$$\xleftarrow{+} \Sigma F_x = ma_x; \qquad 40t = \left(\frac{900 + 550 + 325}{32.2} \right) a$$

$$a = 0.7256t$$

Kinematics.   Since the acceleration is a function of time, the velocity of the truck is obtained using $a = dv/dt$ with the initial condition that $v_0 = 0$ at $t = 0$. We have

$$\int_0^v dv = \int_0^{2\,s} 0.7256t \, dt; \quad v = 0.3628t^2 \Big|_0^{2\,s} = 1.45 \text{ ft/s} \qquad \textit{Ans.}$$

Free-Body Diagram.   In order to determine the force between the truck and cart B, we will consider a free-body diagram of the truck so that we can "expose" the coupling force **T** as external to the free-body diagram, Fig. 13–8b.

Equation of Motion.   When $t = 2$ s, then

$$\xleftarrow{+} \Sigma F_x = ma_x; \qquad 40(2) - T = \left(\frac{900}{32.2} \right)[0.7256(2)]$$

$$T = 39.4 \text{ lb} \qquad \textit{Ans.}$$

NOTE: Try and obtain this same result by considering a free-body diagram of carts B and C as a single system.

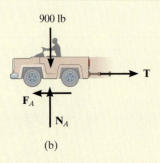

(b)

Fig. 13–8

EXAMPLE 13.4

A smooth 2-kg collar C, shown in Fig. 13–9a, is attached to a spring having a stiffness $k = 3$ N/m and an unstretched length of 0.75 m. If the collar is released from rest at A, determine its acceleration and the normal force of the rod on the collar at the instant $y = 1$ m.

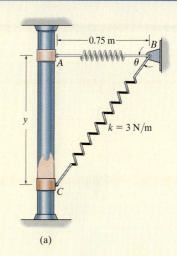

(a)

SOLUTION

Free-Body Diagram.   The free-body diagram of the collar when it is located at the arbitrary position y is shown in Fig. 13–9b. Furthermore, the collar is *assumed* to be accelerating so that "**a**" acts downward in the *positive* y direction. There are four unknowns, namely, N_C, F_s, a, and θ.

Equations of Motion.

$$\xrightarrow{+} \Sigma F_x = ma_x; \qquad -N_C + F_s \cos \theta = 0 \qquad (1)$$

$$+\downarrow \Sigma F_y = ma_y; \qquad 19.62 - F_s \sin \theta = 2a \qquad (2)$$

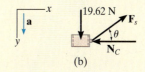

(b)

Fig. 13–9

From Eq. 2 it is seen that the acceleration depends on the magnitude and direction of the spring force. Solution for N_C and a is possible once F_s and θ are known.

The magnitude of the spring force is a function of the stretch s of the spring; i.e., $F_s = ks$. Here the unstretched length is $AB = 0.75$ m, Fig. 13–9a; therefore, $s = CB - AB = \sqrt{y^2 + (0.75)^2} - 0.75$. Since $k = 3$ N/m, then

$$F_s = ks = 3\left(\sqrt{y^2 + (0.75)^2} - 0.75\right) \qquad (3)$$

From Fig. 13–9a, the angle θ is related to y by trigonometry.

$$\tan \theta = \frac{y}{0.75} \qquad (4)$$

Substituting $y = 1$ m into Eqs. 3 and 4 yields $F_s = 1.50$ N and $\theta = 53.1°$. Substituting these results into Eqs. 1 and 2, we obtain

$$N_C = 0.900 \text{ N} \qquad \qquad Ans.$$

$$a = 9.21 \text{ m/s}^2 \downarrow \qquad \qquad Ans.$$

NOTE: This is not a case of constant acceleration, since the spring force changes both its magnitude and direction as the collar moves downward.

EXAMPLE 13.5

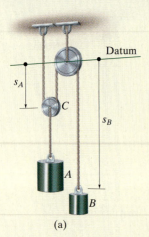

s_A

s_B

Datum

C

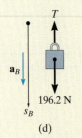

A

B

(a)

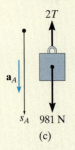

T T

2T

(b)

2T

a_A

s_A 981 N

(c)

T

a_B

196.2 N

s_B

(d)

Fig. 13–10

The 100-kg block A shown in Fig. 13–10a is released from rest. If the masses of the pulleys and the cord are neglected, determine the speed of the 20-kg block B in 2 s.

SOLUTION

Free-Body Diagrams.   Since the mass of the pulleys is *neglected*, then for pulley C, $ma = 0$ and we can apply $\Sigma F_y = 0$ as shown in Fig. 13–10b. The free-body diagrams for blocks A and B are shown in Fig. 13–10c and d, respectively. Notice that for A to remain stationary $T = 490.5$ N, whereas for B to remain static $T = 196.2$ N. Hence A will move down while B moves up. Although this is the case, we will assume both blocks accelerate downward, in the direction of $+s_A$ and $+s_B$. The three unknowns are T, a_A, and a_B.

Equations of Motion.   Block A,

$$+\downarrow \Sigma F_y = ma_y; \qquad 981 - 2T = 100a_A \qquad (1)$$

Block B,

$$+\downarrow \Sigma F_y = ma_y; \qquad 196.2 - T = 20a_B \qquad (2)$$

Kinematics.   The necessary third equation is obtained by relating a_A to a_B using a dependent motion analysis, discussed in Sect. 12.9. The coordinates s_A and s_B in Fig. 13–10a measure the positions of A and B from the fixed datum. It is seen that

$$2s_A + s_B = l$$

where l is constant and represents the total vertical length of cord. Differentiating this expression twice with respect to time yields

$$2a_A = -a_B \qquad (3)$$

Notice that when writing Eqs. 1 to 3, the *positive direction was always assumed downward*. It is very important to be *consistent* in this assumption since we are seeking a simultaneous solution of equations. The results are

$$T = 327.0 \text{ N}$$
$$a_A = 3.27 \text{ m/s}^2$$
$$a_B = -6.54 \text{ m/s}^2$$

Hence when block A accelerates *downward*, block B accelerates *upward* as expected. Since a_B is constant, the velocity of block B in 2 s is thus

$$(+\downarrow) \qquad\qquad v = v_0 + a_B t$$
$$= 0 + (-6.54)(2)$$
$$= -13.1 \text{ m/s} \qquad\qquad Ans.$$

The negative sign indicates that block B is moving upward.

FUNDAMENTAL PROBLEMS

F13–1. The motor winds in the cable with a constant acceleration, such that the 20-kg crate moves a distance $s = 6$ m in 3 s, starting from rest. Determine the tension developed in the cable. The coefficient of kinetic friction between the crate and the plane is $\mu_k = 0.3$.

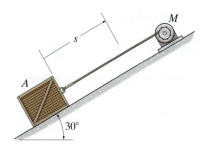

F13–1

F13–2. If motor M exerts a force of $F = (10t^2 + 100)$ N on the cable, where t is in seconds, determine the velocity of the 25-kg crate when $t = 4$ s. The coefficients of static and kinetic friction between the crate and the plane are $\mu_s = 0.3$ and $\mu_k = 0.25$, respectively. The crate is initially at rest.

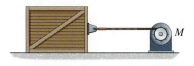

F13–2

F13–3. A spring of stiffness $k = 500$ N/m is mounted against the 10-kg block. If the block is subjected to the force of $F = 500$ N, determine its velocity at $s = 0.5$ m. When $s = 0$, the block is at rest and the spring is uncompressed. The contact surface is smooth.

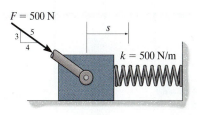

F13–3

F13–4. The 2-Mg car is being towed by a winch. If the winch exerts a force of $T = (100s)$ N on the cable, where s is the displacement of the car in meters, determine the speed of the car when $s = 10$ m, starting from rest. Neglect rolling resistance of the car.

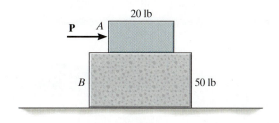

F13–4

F13–5. The spring has a stiffness $k = 200$ N/m and is unstretched when the 25-kg block is at A. Determine the acceleration of the block when $s = 0.4$ m. The contact surface between the block and the plane is smooth.

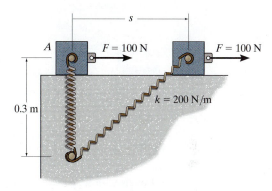

F13–5

F13–6. Block B rests upon a smooth surface. If the coefficients of static and kinetic friction between A and B are $\mu_s = 0.4$ and $\mu_k = 0.3$, respectively, determine the acceleration of each block if $P = 6$ lb.

F13–6

13

PROBLEMS

•13–1. The casting has a mass of 3 Mg. Suspended in a vertical position and initially at rest, it is given an upward speed of 200 mm/s in 0.3 s using a crane hook H. Determine the tension in cables AC and AB during this time interval if the acceleration is constant.

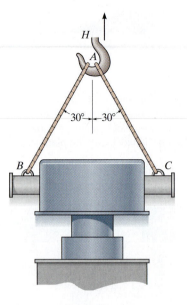

Prob. 13–1

13–2. The 160-Mg train travels with a speed of 80 km/h when it starts to climb the slope. If the engine exerts a traction force $\mathbf{F}$ of $1/20$ of the weight of the train and the rolling resistance $\mathbf{F}_D$ is equal to $1/500$ of the weight of the train, determine the deceleration of the train.

13–3. The 160-Mg train starts from rest and begins to climb the slope as shown. If the engine exerts a traction force $\mathbf{F}$ of $1/8$ of the weight of the train, determine the speed of the train when it has traveled up the slope a distance of 1 km. Neglect rolling resistance.

Probs. 13–2/3

*13–4. The 2-Mg truck is traveling at 15 m/s when the brakes on all its wheels are applied, causing it to skid for a distance of 10 m before coming to rest. Determine the constant horizontal force developed in the coupling C, and the frictional force developed between the tires of the truck and the road during this time. The total mass of the boat and trailer is 1 Mg.

Prob. 13–4

•13–5. If blocks A and B of mass 10 kg and 6 kg, respectively, are placed on the inclined plane and released, determine the force developed in the link. The coefficients of kinetic friction between the blocks and the inclined plane are $\mu_A = 0.1$ and $\mu_B = 0.3$. Neglect the mass of the link.

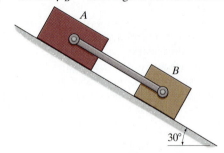

Prob. 13–5

13–6. Motors A and B draw in the cable with the accelerations shown. Determine the acceleration of the 300-lb crate C and the tension developed in the cable. Neglect the mass of all the pulleys.

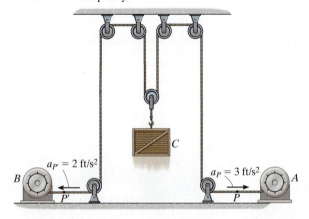

Prob. 13–6

13–7.   The van is traveling at 20 km/h when the coupling of the trailer at *A* fails. If the trailer has a mass of 250 kg and coasts 45 m before coming to rest, determine the constant horizontal force *F* created by rolling friction which causes the trailer to stop.

Prob. 13–7

*13–8.**   If the 10-lb block *A* slides down the plane with a constant velocity when $\theta = 30°$, determine the acceleration of the block when $\theta = 45°$.

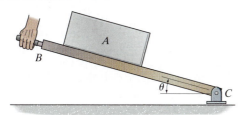

Prob. 13–8

•**13–9.**   Each of the three barges has a mass of 30 Mg, whereas the tugboat has a mass of 12 Mg. As the barges are being pulled forward with a constant velocity of 4 m/s, the tugboat must overcome the frictional resistance of the water, which is 2 kN for each barge and 1.5 kN for the tugboat. If the cable between *A* and *B* breaks, determine the acceleration of the tugboat.

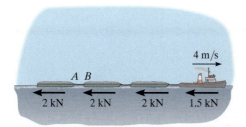

Prob. 13–9

13–10.   The crate has a mass of 80 kg and is being towed by a chain which is always directed at 20° from the horizontal as shown. If the magnitude of **P** is increased until the crate begins to slide, determine the crate's initial acceleration if the coefficient of static friction is $\mu_s = 0.5$ and the coefficient of kinetic friction is $\mu_k = 0.3$.

13–11.   The crate has a mass of 80 kg and is being towed by a chain which is always directed at 20° from the horizontal as shown. Determine the crate's acceleration in $t = 2$ s if the coefficient of static friction is $\mu_s = 0.4$, the coefficient of kinetic friction is $\mu_k = 0.3$, and the towing force is $P = (90t^2)$ N, where t is in seconds.

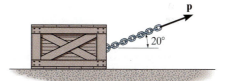

Probs. 13–10/11

*13–12.**   Determine the acceleration of the system and the tension in each cable. The inclined plane is smooth, and the coefficient of kinetic friction between the horizontal surface and block *C* is $(\mu_k)_C = 0.2$.

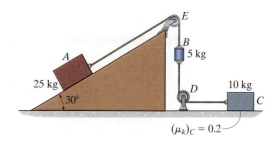

Prob. 13–12

13

•**13–13.** The two boxcars A and B have a weight of 20 000 lb and 30 000 lb, respectively. If they coast freely down the incline when the brakes are applied to all the wheels of car A causing it to skid, determine the force in the coupling C between the two cars. The coefficient of kinetic friction between the wheels of A and the tracks is $\mu_k = 0.5$. The wheels of car B are free to roll. Neglect their mass in the calculation. *Suggestion:* Solve the problem by representing single resultant normal forces acting on A and B, respectively.

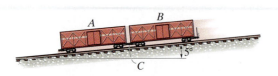

Prob. 13–13

13–14. The 3.5-Mg engine is suspended from a spreader beam AB having a negligible mass and is hoisted by a crane which gives it an acceleration of 4 m/s^2 when it has a velocity of 2 m/s. Determine the force in chains CA and CB during the lift.

13–15. The 3.5-Mg engine is suspended from a spreader beam having a negligible mass and is hoisted by a crane which exerts a force of 40 kN on the hoisting cable. Determine the distance the engine is hoisted in 4 s, starting from rest.

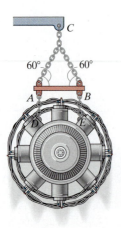

Probs. 13–14/15

*****13–16.** The man pushes on the 60-lb crate with a force **F**. The force is always directed down at 30° from the horizontal as shown, and its magnitude is increased until the crate begins to slide. Determine the crate's initial acceleration if the coefficient of static friction is $\mu_s = 0.6$ and the coefficient of kinetic friction is $\mu_k = 0.3$.

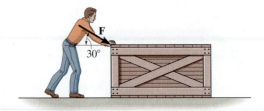

Prob. 13–16

•**13–17.** A force of $F = 15$ lb is applied to the cord. Determine how high the 30-lb block A rises in 2 s starting from rest. Neglect the weight of the pulleys and cord.

13–18. Determine the constant force **F** which must be applied to the cord in order to cause the 30-lb block A to have a speed of 12 ft/s when it has been displaced 3 ft upward starting from rest. Neglect the weight of the pulleys and cord.

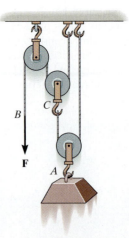

Probs. 13–17/18

13–19. The 800-kg car at B is connected to the 350-kg car at A by a spring coupling. Determine the stretch in the spring if (a) the wheels of both cars are free to roll and (b) the brakes are applied to all four wheels of car B, causing the wheels to skid. Take $(\mu_k)_B = 0.4$. Neglect the mass of the wheels.

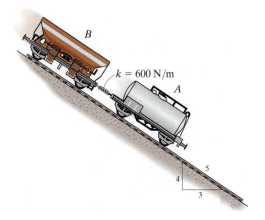

Prob. 13–19

•13–21. Block B has a mass m and is released from rest when it is on top of cart A, which has a mass of $3m$. Determine the tension in cord CD needed to hold the cart from moving while B slides down A. Neglect friction.

13–22. Block B has a mass m and is released from rest when it is on top of cart A, which has a mass of $3m$. Determine the tension in cord CD needed to hold the cart from moving while B slides down A. The coefficient of kinetic friction between A and B is μ_k.

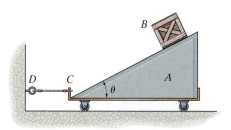

Probs. 13–21/22

*13–20. The 10-lb block A travels to the right at $v_A = 2$ ft/s at the instant shown. If the coefficient of kinetic friction is $\mu_k = 0.2$ between the surface and A, determine the velocity of A when it has moved 4 ft. Block B has a weight of 20 lb.

Prob. 13–20

13–23. The 2-kg shaft CA passes through a smooth journal bearing at B. Initially, the springs, which are coiled loosely around the shaft, are unstretched when no force is applied to the shaft. In this position $s = s' = 250$ mm and the shaft is at rest. If a horizontal force of $F = 5$ kN is applied, determine the speed of the shaft at the instant $s = 50$ mm, $s' = 450$ mm. The ends of the springs are attached to the bearing at B and the caps at C and A.

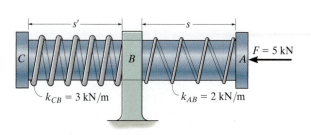

Prob. 13–23

*13–24. If the force of the motor M on the cable is shown in the graph, determine the velocity of the cart when $t = 3$ s. The load and cart have a mass of 200 kg and the car starts from rest.

13–26. A freight elevator, including its load, has a mass of 500 kg. It is prevented from rotating by the track and wheels mounted along its sides. When $t = 2$ s, the motor M draws in the cable with a speed of 6 m/s, *measured relative to the elevator.* If it starts from rest, determine the constant acceleration of the elevator and the tension in the cable. Neglect the mass of the pulleys, motor, and cables.

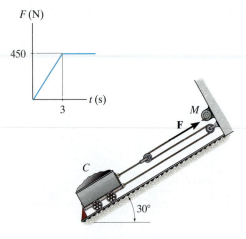

Prob. 13–24

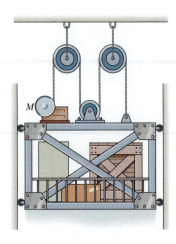

Prob. 13–26

•13–25. If the motor draws in the cable with an acceleration of 3 m/s^2, determine the reactions at the supports A and B. The beam has a uniform mass of 30 kg/m, and the crate has a mass of 200 kg. Neglect the mass of the motor and pulleys.

13–27. Determine the required mass of block A so that when it is released from rest it moves the 5-kg block B a distance of 0.75 m up along the smooth inclined plane in $t = 2$ s. Neglect the mass of the pulleys and cords.

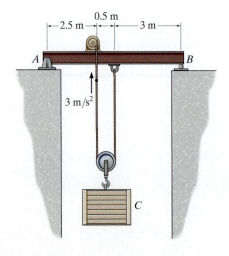

Prob. 13–25

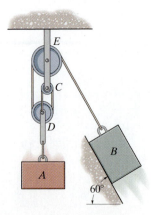

Prob. 13–27

*13–28.   Blocks A and B have a mass of m_A and m_B, where $m_A > m_B$. If pulley C is given an acceleration of $\mathbf{a}_0$, determine the acceleration of the blocks. Neglect the mass of the pulley.

Prob. 13–28

•13–29.   The tractor is used to lift the 150-kg load B with the 24-m-long rope, boom, and pulley system. If the tractor travels to the right at a constant speed of 4 m/s, determine the tension in the rope when $s_A = 5$ m. When $s_A = 0$, $s_B = 0$.

13–30.   The tractor is used to lift the 150-kg load B with the 24-m-long rope, boom, and pulley system. If the tractor travels to the right with an acceleration of 3 m/s² and has a velocity of 4 m/s at the instant $s_A = 5$ m, determine the tension in the rope at this instant. When $s_A = 0$, $s_B = 0$.

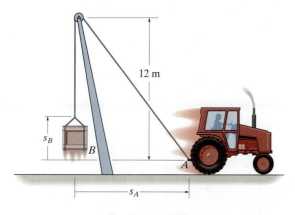

Probs. 13–29/30

13–31.   The 75-kg man climbs up the rope with an acceleration of 0.25 m/s², measured relative to the rope. Determine the tension in the rope and the acceleration of the 80-kg block.

Prob. 13–31

*13–32.   Motor M draws in the cable with an acceleration of 4 ft/s², measured relative to the 200-lb mine car. Determine the acceleration of the car and the tension in the cable. Neglect the mass of the pulleys.

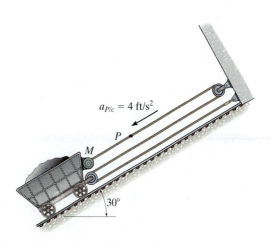

Prob. 13–32

13

•13–33. The 2-lb collar C fits loosely on the smooth shaft. If the spring is unstretched when $s = 0$ and the collar is given a velocity of 15 ft/s, determine the velocity of the collar when $s = 1$ ft.

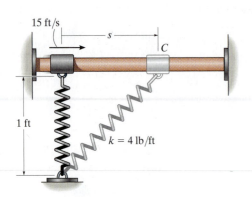

Prob. 13–33

13–34. In the cathode-ray tube, electrons having a mass m are emitted from a source point S and begin to travel horizontally with an initial velocity $\mathbf{v}_0$. While passing between the grid plates a distance l, they are subjected to a vertical force having a magnitude eV/w, where e is the charge of an electron, V the applied voltage acting across the plates, and w the distance between the plates. After passing clear of the plates, the electrons then travel in straight lines and strike the screen at A. Determine the deflection d of the electrons in terms of the dimensions of the voltage plate and tube. Neglect gravity which causes a slight vertical deflection when the electron travels from S to the screen, and the slight deflection between the plates.

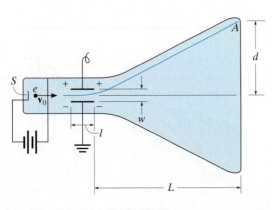

Prob. 13–34

13–35. The 2-kg collar C is free to slide along the smooth shaft AB. Determine the acceleration of collar C if (a) the shaft is fixed from moving, (b) collar A, which is fixed to shaft AB, moves to the left at constant velocity along the horizontal guide, and (c) collar A is subjected to an acceleration of 2 m/s² to the left. In all cases, the motion occurs in the vertical plane.

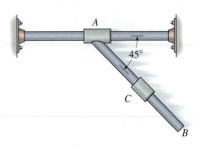

Prob. 13–35

*13–36. Blocks A and B each have a mass m. Determine the largest horizontal force $\mathbf{P}$ which can be applied to B so that A will not move relative to B. All surfaces are smooth.

•13–37. Blocks A and B each have a mass m. Determine the largest horizontal force $\mathbf{P}$ which can be applied to B so that A will not slip on B. The coefficient of static friction between A and B is μ_s. Neglect any friction between B and C.

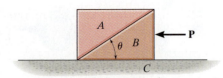

Probs. 13–36/37

13–38. If a force $F = 200$ N is applied to the 30-kg cart, show that the 20-kg block A will slide on the cart. Also determine the time for block A to move on the cart 1.5 m. The coefficients of static and kinetic friction between the block and the cart are $\mu_s = 0.3$ and $\mu_k = 0.25$. Both the cart and the block start from rest.

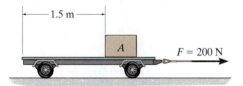

Prob. 13–38

13–39. Suppose it is possible to dig a smooth tunnel through the earth from a city at A to a city at B as shown. By the theory of gravitation, any vehicle C of mass m placed within the tunnel would be subjected to a gravitational force which is always directed toward the center of the earth D. This force $\mathbf{F}$ has a magnitude that is directly proportional to its distance r from the earth's center. Hence, if the vehicle has a weight of $W = mg$ when it is located on the earth's surface, then at an arbitrary location r the magnitude of force $\mathbf{F}$ is $F = (mg/R)r$, where $R = 6328$ km, the radius of the earth. If the vehicle is released from rest when it is at B, $x = s = 2$ Mm, determine the time needed for it to reach A, and the maximum velocity it attains. Neglect the effect of the earth's rotation in the calculation and assume the earth has a constant density. *Hint:* Write the equation of motion in the x direction, noting that $r \cos \theta = x$. Integrate, using the kinematic relation $v \, dv = a \, dx$, then integrate the result using $v = dx/dt$.

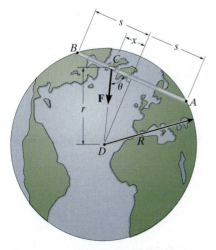

Prob. 13–39

13–40. The 30-lb crate is being hoisted upward with a constant acceleration of 6 ft/s². If the uniform beam AB has a weight of 200 lb, determine the components of reaction at the fixed support A. Neglect the size and mass of the pulley at B. *Hint:* First find the tension in the cable, then analyze the forces in the beam using statics.

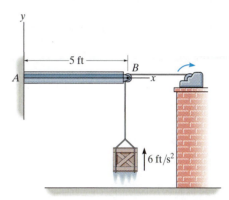

Prob. 13–40

•13–41. If a horizontal force of $P = 10$ lb is applied to block A, determine the acceleration of block B. Neglect friction. *Hint:* Show that $a_B = a_A \tan 15°$.

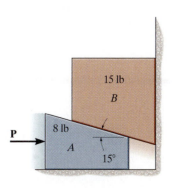

Prob. 13–41

13–42. Block A has a mass m_A and is attached to a spring having a stiffness k and unstretched length l_0. If another block B, having a mass m_B, is pressed against A so that the spring deforms a distance d, determine the distance both blocks slide on the smooth surface before they begin to separate. What is their velocity at this instant?

13–43. Block A has a mass m_A and is attached to a spring having a stiffness k and unstretched length l_0. If another block B, having a mass m_B, is pressed against A so that the spring deforms a distance d, show that for separation to occur it is necessary that $d > 2\mu_k g(m_A + m_B)/k$, where μ_k is the coefficient of kinetic friction between the blocks and the ground. Also, what is the distance the blocks slide on the surface before they separate?

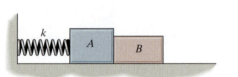

Probs. 13–42/43

*13–44. The 600-kg dragster is traveling with a velocity of 125 m/s when the engine is shut off and the braking parachute is deployed. If air resistance imposed on the dragster due to the parachute is $F_D = (6000 + 0.9v^2)\,\text{N}$, where v is in m/s, determine the time required for the dragster to come to rest.

Prob. 13–44

•13–45. The buoyancy force on the 500-kg balloon is $F = 6\,\text{kN}$, and the air resistance is $F_D = (100v)\,\text{N}$, where v is in m/s. Determine the terminal or maximum velocity of the balloon if it starts from rest.

Prob. 13–45

13–46. The parachutist of mass m is falling with a velocity of v_0 at the instant he opens the parachute. If air resistance is $F_D = Cv^2$, determine her maximum velocity (terminal velocity) during the descent.

Prob. 13–46

13–47. The weight of a particle varies with altitude such that $W = m(gr_0^2)/r^2$, where r_0 is the radius of the earth and r is the distance from the particle to the earth's center. If the particle is fired vertically with a velocity v_0 from the earth's surface, determine its velocity as a function of position r. What is the smallest velocity v_0 required to escape the earth's gravitational field, what is r_{max}, and what is the time required to reach this altitude?

13.5   Equations of Motion: Normal and Tangential Coordinates

When a particle moves along a curved path which is known, the equation of motion for the particle may be written in the tangential, normal, and binormal directions, Fig. 13–11. Note that there is no motion of the particle in the binormal direction, since the particle is constrained to move along the path. We have

$$\Sigma \mathbf{F} = m\mathbf{a}$$

$$\Sigma F_t \mathbf{u}_t + \Sigma F_n \mathbf{u}_n + \Sigma F_b \mathbf{u}_b = m\mathbf{a}_t + m\mathbf{a}_n$$

This equation is satisfied provided

$$
\begin{aligned}
\Sigma F_t &= ma_t \\
\Sigma F_n &= ma_n \\
\Sigma F_b &= 0
\end{aligned}
\tag{13–8}
$$

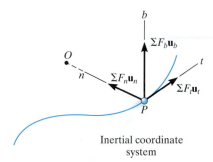

Inertial coordinate system

Fig. 13–11

Recall that $a_t\ (= dv/dt)$ represents the time rate of change in the magnitude of velocity. So if $\Sigma \mathbf{F}_t$ acts in the direction of motion, the particle's speed will increase, whereas if it acts in the opposite direction, the particle will slow down. Likewise, $a_n\ (= v^2/\rho)$ represents the time rate of change in the velocity's direction. It is caused by $\Sigma \mathbf{F}_n$, which *always* acts in the positive n direction, i.e., toward the path's center of curvature. From this reason it is often referred to as the *centripetal force*.

The centrifuge is used to subject a passenger to a very large normal acceleration caused by rapid rotation. Realize that this acceleration is *caused by* the unbalanced normal force exerted on the passenger by the seat of the centrifuge.

Procedure for Analysis

When a problem involves the motion of a particle along a *known curved path*, normal and tangential coordinates should be considered for the analysis since the acceleration components can be readily formulated. The method for applying the equations of motion, which relate the forces to the acceleration, has been outlined in the procedure given in Sec. 13.4. Specifically, for *t, n, b* coordinates it may be stated as follows:

Free-Body Diagram.

- Establish the inertial *t, n, b* coordinate system at the particle and draw the particle's free-body diagram.

- The particle's normal acceleration $\mathbf{a}_n$ *always* acts in the positive *n* direction.

- If the tangential acceleration $\mathbf{a}_t$ is unknown, assume it acts in the positive *t* direction.

- There is no acceleration in the *b* direction.

- Identify the unknowns in the problem.

Equations of Motion.

- Apply the equations of motion, Eqs. 13–8.

Kinematics.

- Formulate the tangential and normal components of acceleration; i.e., $a_t = dv/dt$ or $a_t = v\,dv/ds$ and $a_n = v^2/\rho$.

- If the path is defined as $y = f(x)$, the radius of curvature at the point where the particle is located can be obtained from $\rho = [1 + (dy/dx)^2]^{3/2}/|d^2y/dx^2|$.

13

EXAMPLE 13.6

Determine the banking angle θ for the race track so that the wheels of the racing cars shown in Fig. 13–12a will not have to depend upon friction to prevent any car from sliding up or down the track. Assume the cars have negligible size, a mass m, and travel around the curve of radius ρ with a constant speed v.

(a)

SOLUTION

Before looking at the following solution, give some thought as to why it should be solved using t, n, b coordinates.

Free-Body Diagram.   As shown in Fig. 13–12b, and as stated in the problem, no frictional force acts on the car. Here N_C represents the *resultant* of the ground on all four wheels. Since a_n can be calculated, the unknowns are N_C and θ.

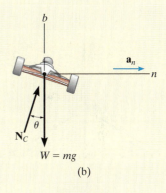

(b)

Fig. 13–12

Equations of Motion.   Using the n, b axes shown,

$$\xrightarrow{+} \Sigma F_n = ma_n; \qquad N_C \sin \theta = m\frac{v^2}{\rho} \qquad (1)$$

$$+\uparrow \Sigma F_b = 0; \qquad N_C \cos \theta - mg = 0 \qquad (2)$$

Eliminating N_C and m from these equations by dividing Eq. 1 by Eq. 2, we obtain

$$\tan \theta = \frac{v^2}{g\rho}$$

$$\theta = \tan^{-1}\left(\frac{v^2}{g\rho}\right) \qquad \textit{Ans.}$$

NOTE: The result is independent of the mass of the car. Also, a force summation in the tangential direction is of no consequence to the solution. If it were considered, then $a_t = dv/dt = 0$, since the car moves with *constant speed*. A further analysis of this problem is discussed in Prob. 21–47.

EXAMPLE 13.7

The 3-kg disk D is attached to the end of a cord as shown in Fig. 13–13a. The other end of the cord is attached to a ball-and-socket joint located at the center of a platform. If the platform rotates rapidly, and the disk is placed on it and released from rest as shown, determine the time it takes for the disk to reach a speed great enough to break the cord. The maximum tension the cord can sustain is 100 N, and the coefficient of kinetic friction between the disk and the platform is $\mu_k = 0.1$.

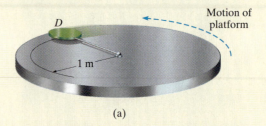

(a)

SOLUTION

Free-Body Diagram. The frictional force has a magnitude $F = \mu_k N_D = 0.1 N_D$ and a sense of direction that opposes the *relative motion* of the disk with respect to the platform. It is this force that gives the disk a tangential component of acceleration causing v to increase, thereby causing T to increase until it reaches 100 N. The weight of the disk is $W = 3(9.81) = 29.43$ N. Since a_n can be related to v, the unknowns are N_D, a_t, and v.

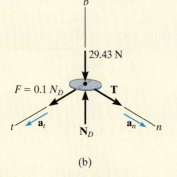

(b)

Fig. 13–13

Equations of Motion.

$$\Sigma F_n = ma_n; \qquad\qquad\qquad T = 3\left(\frac{v^2}{1}\right) \qquad (1)$$

$$\Sigma F_t = ma_t; \qquad\qquad\qquad 0.1 N_D = 3a_t \qquad (2)$$

$$\Sigma F_b = 0; \qquad\qquad\qquad N_D - 29.43 = 0 \qquad (3)$$

Setting $T = 100$ N, Eq. 1 can be solved for the critical speed v_{cr} of the disk needed to break the cord. Solving all the equations, we obtain

$$N_D = 29.43 \text{ N}$$
$$a_t = 0.981 \text{ m/s}^2$$
$$v_{cr} = 5.77 \text{ m/s}$$

Kinematics. Since a_t is *constant*, the time needed to break the cord is

$$v_{cr} = v_0 + a_t t$$
$$5.77 = 0 + (0.981)t$$
$$t = 5.89 \text{ s} \qquad\qquad\qquad Ans.$$

EXAMPLE 13.8

Design of the ski jump shown in the photo requires knowing the type of forces that will be exerted on the skier and her approximate trajectory. If in this case the jump can be approximated by the parabola shown in Fig. 13–14a, determine the normal force on the 150-lb skier the instant she arrives at the end of the jump, point A, where her velocity is 65 ft/s. Also, what is her acceleration at this point?

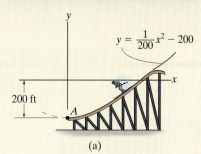

SOLUTION
Why consider using n, t coordinates to solve this problem?

Free-Body Diagram. Since $dy/dx = x/100 \,|_{x=0} = 0$, the slope at A is horizontal. The free-body diagram of the skier when she is at A is shown in Fig. 13–14b. Since the path is *curved*, there are two components of acceleration, $\mathbf{a}_n$ and $\mathbf{a}_t$. Since a_n can be calculated, the unknowns are a_t and N_A.

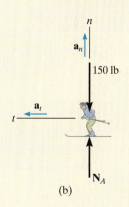

Equations of Motion.

$$+\uparrow \Sigma F_n = ma_n; \qquad N_A - 150 = \frac{150}{32.2}\left(\frac{(65)^2}{\rho}\right) \qquad (1)$$

$$\xleftarrow{+} \Sigma F_t = ma_t; \qquad 0 = \frac{150}{32.2}a_t \qquad (2)$$

The radius of curvature ρ for the path must be determined at point $A(0, -200$ ft). Here $y = \frac{1}{200}x^2 - 200$, $dy/dx = \frac{1}{100}x$, $d^2y/dx^2 = \frac{1}{100}$, so that at $x = 0$,

$$\rho = \frac{[1 + (dy/dx)^2]^{3/2}}{|d^2y/dx^2|}\bigg|_{x=0} = \frac{[1 + (0)^2]^{3/2}}{\left|\frac{1}{100}\right|} = 100 \text{ ft}$$

Substituting this into Eq. 1 and solving for N_A, we obtain

$$N_A = 347 \text{ lb} \qquad\qquad Ans.$$

Kinematics. From Eq. 2,

$$a_t = 0$$

Thus,

$$a_n = \frac{v^2}{\rho} = \frac{(65)^2}{100} = 42.2 \text{ ft/s}^2$$

$$a_A = a_n = 42.2 \text{ ft/s}^2 \uparrow \qquad\qquad Ans.$$

NOTE: Apply the equation of motion in the y direction and show that when the skier is in midair her acceleration is 32.2 ft/s².

Fig. 13–14

13

EXAMPLE | 13.9

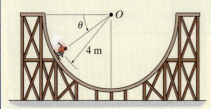

(a)

The 60-kg skateboarder in Fig.13–15a coasts down the circular track. If he starts from rest when $\theta = 0°$, determine the magnitude of the normal reaction the track exerts on him when $\theta = 60°$. Neglect his size for the calculation.

SOLUTION

Free-Body Diagram. The free-body diagram of the skateboarder when he is at an *arbitrary position* θ is shown in Fig. 13–15b. At $\theta = 60°$ there are three unknowns, N_s, a_t, and a_n (or v).

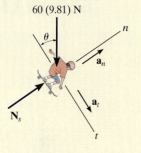

(b)

Equations of Motion.

$$\downarrow \Sigma F_n = ma_n; \quad N_S - [60(9.81)\text{N}] \sin \theta = (60 \text{ kg})\left(\frac{v^2}{4\text{m}}\right) \qquad (1)$$

$$\downarrow \Sigma F_t = ma_t; \quad [60(9.81)\text{N}] \cos \theta = (60 \text{ kg}) a_t$$

$$a_t = 9.81 \cos \theta$$

Kinematics. Since a_t is expressed in terms of θ, the equation $v \, dv = a_t \, ds$ must be used to determine the speed of the skateboarder when $\theta = 60°$. Using the geometric relation $s = \theta r$, where $ds = r \, d\theta = (4 \text{ m}) \, d\theta$, Fig. 13–15c, and the initial condition $v = 0$ at $\theta = 0°$, we have,

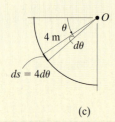

(c)

Fig. 13–15

$$v \, dv = a_t \, ds$$

$$\int_0^v v \, dv = \int_0^{60°} 9.81 \cos \theta (4 \, d\theta)$$

$$\left.\frac{v^2}{2}\right|_0^v = 39.24 \sin \theta \Big|_0^{60°}$$

$$\frac{v^2}{2} - 0 = 39.24(\sin 60° - 0)$$

$$v^2 = 67.97 \text{ m}^2/\text{s}^2$$

Substituting this result and $\theta = 60°$ into Eq. (1), yields

$$N_s = 1529.23 \text{ N} = 1.53 \text{ kN} \qquad \textit{Ans.}$$

FUNDAMENTAL PROBLEMS

F13–7. The block rests at a distance of 2 m from the center of the platform. If the coefficient of static friction between the block and the platform is $\mu_s = 0.3$, determine the maximum speed which the block can attain before it begins to slip. Assume the angular motion of the disk is slowly increasing.

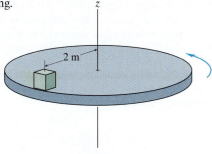

F13–7

F13–8. Determine the maximum speed that the jeep can travel over the crest of the hill and not lose contact with the road.

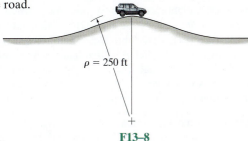

$\rho = 250$ ft

F13–8

F13–9. A pilot weighs 150 lb and is traveling at a constant speed of 120 ft/s. Determine the normal force he exerts on the seat of the plane when he is upside down at A. The loop has a radius of curvature of 400 ft.

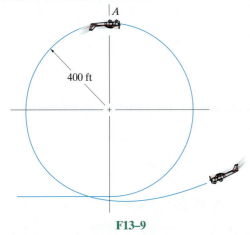

F13–9

F13–10. The sports car is traveling along a 30° banked road having a radius of curvature of $\rho = 500$ ft. If the coefficient of static friction between the tires and the road is $\mu_s = 0.2$, determine the maximum safe speed so no slipping occurs. Neglect the size of the car.

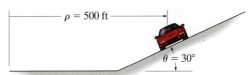

$\rho = 500$ ft

$\theta = 30°$

F13–10

F13–11. If the 10-kg ball has a velocity of 3 m/s when it is at the position A, along the vertical path, determine the tension in the cord and the increase in the speed of the ball at this position.

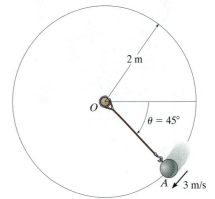

2 m

O

$\theta = 45°$

A 3 m/s

F13–11

F13–12. The motorcycle has a mass of 0.5 Mg and a negligible size. It passes point A traveling with a speed of 15 m/s, which is increasing at a constant rate of 1.5 m/s². Determine the resultant frictional force exerted by the road on the tires at this instant.

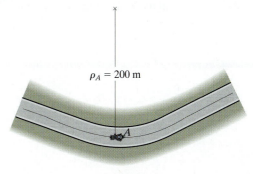

$\rho_A = 200$ m

A

F13–12

PROBLEMS

13

*13–48. The 2-kg block B and 15-kg cylinder A are connected to a light cord that passes through a hole in the center of the smooth table. If the block is given a speed of $v = 10 \, \text{m/s}$, determine the radius r of the circular path along which it travels.

•13–49. The 2-kg block B and 15-kg cylinder A are connected to a light cord that passes through a hole in the center of the smooth table. If the block travels along a circular path of radius $r = 1.5 \, \text{m}$, determine the speed of the block.

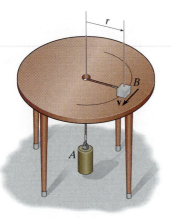

Probs. 13–48/49

13–50. At the instant shown, the 50-kg projectile travels in the vertical plane with a speed of $v = 40 \, \text{m/s}$. Determine the tangential component of its acceleration and the radius of curvature ρ of its trajectory at this instant.

13–51. At the instant shown, the radius of curvature of the vertical trajectory of the 50-kg projectile is $\rho = 200 \, \text{m}$. Determine the speed of the projectile at this instant.

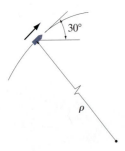

Probs. 13–50/51

*13–52. Determine the mass of the sun, knowing that the distance from the earth to the sun is $149.6(10^6)$ km. *Hint:* Use Eq. 13–1 to represent the force of gravity acting on the earth.

•13–53. The sports car, having a mass of 1700 kg, travels horizontally along a 20° banked track which is circular and has a radius of curvature of $\rho = 100 \, \text{m}$. If the coefficient of static friction between the tires and the road is $\mu_s = 0.2$, determine the *maximum constant speed* at which the car can travel without sliding up the slope. Neglect the size of the car.

13–54. Using the data in Prob. 13–53, determine the *minimum speed* at which the car can travel around the track without sliding down the slope.

Probs. 13–53/54

13–55. The device shown is used to produce the experience of weightlessness in a passenger when he reaches point A, $\theta = 90°$, along the path. If the passenger has a mass of 75 kg, determine the minimum speed he should have when he reaches A so that he does not exert a normal reaction on the seat. The chair is pin-connected to the frame BC so that he is always seated in an upright position. During the motion his speed remains constant.

*13–56. A man having the mass of 75 kg sits in the chair which is pin-connected to the frame BC. If the man is always seated in an upright position, determine the horizontal and vertical reactions of the chair on the man at the instant $\theta = 45°$. At this instant he has a speed of 6 m/s, which is increasing at $0.5 \, \text{m/s}^2$.

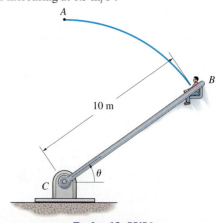

Probs. 13–55/56

•13–57. Determine the tension in wire CD just after wire AB is cut. The small bob has a mass m.

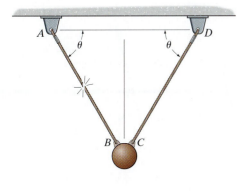

Prob. 13–57

13–59. An acrobat has a weight of 150 lb and is sitting on a chair which is perched on top of a pole as shown. If by a mechanical drive the pole rotates downward at a constant rate from $\theta = 0°$, such that the acrobat's center of mass G maintains a *constant speed* of $v_a = 10$ ft/s, determine the angle θ at which he begins to "fly" out of the chair. Neglect friction and assume that the distance from the pivot O to G is $\rho = 15$ ft.

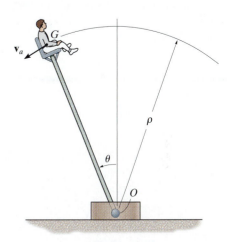

Prob. 13–59

13–58. Determine the time for the satellite to complete its orbit around the earth. The orbit has a radius r measured from the center of the earth. The masses of the satellite and the earth are m_s and M_e, respectively.

Prob. 13–58

*13–60. A spring, having an unstretched length of 2 ft, has one end attached to the 10-lb ball. Determine the angle θ of the spring if the ball has a speed of 6 ft/s tangent to the horizontal circular path.

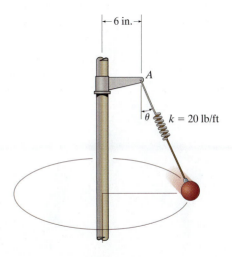

Prob. 13–60

•13–61. If the ball has a mass of 30 kg and a speed $v = 4$ m/s at the instant it is at its lowest point, $\theta = 0°$, determine the tension in the cord at this instant. Also, determine the angle θ to which the ball swings and momentarily stops. Neglect the size of the ball.

13–62. The ball has a mass of 30 kg and a speed $v = 4$ m/s at the instant it is at its lowest point, $\theta = 0°$. Determine the tension in the cord and the rate at which the ball's speed is decreasing at the instant $\theta = 20°$. Neglect the size of the ball.

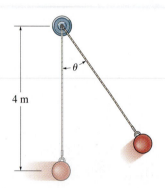

Probs. 13–61/62

13–63. The vehicle is designed to combine the feel of a motorcycle with the comfort and safety of an automobile. If the vehicle is traveling at a constant speed of 80 km/h along a circular curved road of radius 100 m, determine the tilt angle θ of the vehicle so that only a normal force from the seat acts on the driver. Neglect the size of the driver.

Prob. 13–63

*13–64. The ball has a mass m and is attached to the cord of length l. The cord is tied at the top to a swivel and the ball is given a velocity $\mathbf{v}_0$. Show that the angle θ which the cord makes with the vertical as the ball travels around the circular path must satisfy the equation $\tan \theta \sin \theta = v_0^2/gl$. Neglect air resistance and the size of the ball.

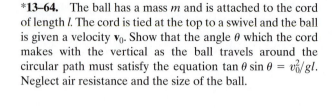

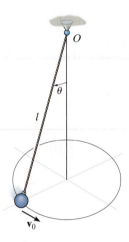

Prob. 13–64

•13–65. The smooth block B, having a mass of 0.2 kg, is attached to the vertex A of the right circular cone using a light cord. If the block has a speed of 0.5 m/s around the cone, determine the tension in the cord and the reaction which the cone exerts on the block. Neglect the size of the block.

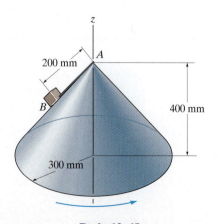

Prob. 13–65

13–66. Determine the minimum coefficient of static friction between the tires and the road surface so that the 1.5-Mg car does not slide as it travels at 80 km/h on the curved road. Neglect the size of the car.

13–67. If the coefficient of static friction between the tires and the road surface is $\mu_s = 0.25$, determine the maximum speed of the 1.5-Mg car without causing it to slide when it travels on the curve. Neglect the size of the car.

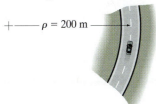

$\rho = 200$ m

Probs. 13–66/67

*13–68. At the instant shown, the 3000-lb car is traveling with a speed of 75 ft/s, which is increasing at a rate of 6 ft/s². Determine the magnitude of the resultant frictional force the road exerts on the tires of the car. Neglect the size of the car.

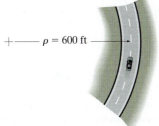

$\rho = 600$ ft

Prob. 13–68

•13–69. Determine the maximum speed at which the car with mass m can pass over the top point A of the vertical curved road and still maintain contact with the road. If the car maintains this speed, what is the normal reaction the road exerts on the car when it passes the lowest point B on the road?

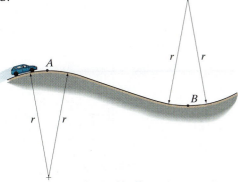

Prob. 13–69

13–70. A 5-Mg airplane is flying at a constant speed of 350 km/h along a horizontal circular path of radius $r = 3000$ m. Determine the uplift force **L** acting on the airplane and the banking angle θ. Neglect the size of the airplane.

13–71. A 5-Mg airplane is flying at a constant speed of 350 km/h along a horizontal circular path. If the banking angle $\theta = 15°$, determine the uplift force **L** acting on the airplane and the radius r of the circular path. Neglect the size of the airplane.

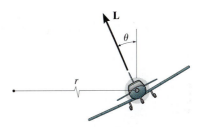

Probs. 13–70/71

*13–72. The 0.8-Mg car travels over the hill having the shape of a parabola. If the driver maintains a constant speed of 9 m/s, determine both the resultant normal force and the resultant frictional force that all the wheels of the car exert on the road at the instant it reaches point A. Neglect the size of the car.

•13–73. The 0.8-Mg car travels over the hill having the shape of a parabola. When the car is at point A, it is traveling at 9 m/s and increasing its speed at 3 m/s². Determine both the resultant normal force and the resultant frictional force that all the wheels of the car exert on the road at this instant. Neglect the size of the car.

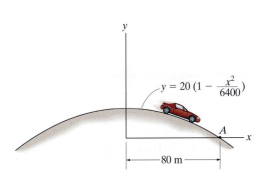

$$y = 20\left(1 - \frac{x^2}{6400}\right)$$

80 m

Probs. 13–72/73

13–74. The 6-kg block is confined to move along the smooth parabolic path. The attached spring restricts the motion and, due to the roller guide, always remains horizontal as the block descends. If the spring has a stiffness of $k = 10$ N/m, and unstretched length of 0.5 m, determine the normal force of the path on the block at the instant $x = 1$ m when the block has a speed of 4 m/s. Also, what is the rate of increase in speed of the block at this point? Neglect the mass of the roller and the spring.

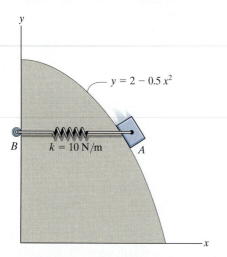

Prob. 13–74

13–75. Prove that if the block is released from rest at point B of a smooth path of *arbitrary shape*, the speed it attains when it reaches point A is equal to the speed it attains when it falls freely through a distance h; i.e., $v = \sqrt{2gh}$.

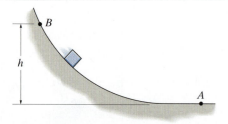

Prob. 13–75

*13–76.** A toboggan and rider of total mass 90 kg travel down along the (smooth) slope defined by the equation $y = 0.08x^2$. At the instant $x = 10$ m, the toboggan's speed is 5 m/s. At this point, determine the rate of increase in speed and the normal force which the slope exerts on the toboggan. Neglect the size of the toboggan and rider for the calculation.

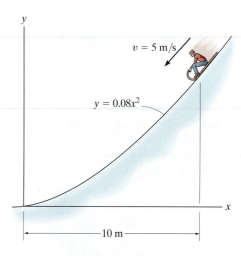

Prob. 13–76

•**13–77.** The skier starts from rest at $A(10 \text{ m}, 0)$ and descends the smooth slope, which may be approximated by a parabola. If she has a mass of 52 kg, determine the normal force the ground exerts on the skier at the instant she arrives at point B. Neglect the size of the skier. *Hint:* Use the result of Prob. 13–75.

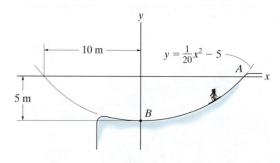

Prob. 13–77

13–78.   The 5-lb box is projected with a speed of 20 ft/s at *A* up the vertical circular smooth track. Determine the angle θ when the box leaves the track.

13–79.   Determine the minimum speed that must be given to the 5-lb box at *A* in order for it to remain in contact with the circular path. Also, determine the speed of the box when it reaches point *B*.

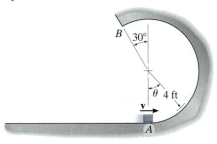

Probs. 13–78/79

*****13–80.**   The 800-kg motorbike travels with a constant speed of 80 km/h up the hill. Determine the normal force the surface exerts on its wheels when it reaches point *A*. Neglect its size.

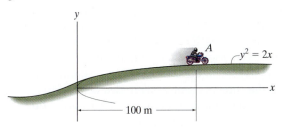

Prob. 13–80

•13–81.   The 1.8-Mg car travels up the incline at a constant speed of 80 km/h. Determine the normal reaction of the road on the car when it reaches point *A*. Neglect its size.

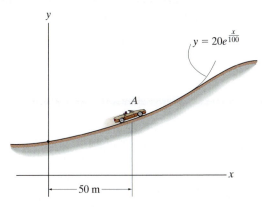

Prob. 13–81

13–82.   Determine the maximum speed the 1.5-Mg car can have and still remain in contact with the road when it passes point *A*. If the car maintains this speed, what is the normal reaction of the road on it when it passes point *B*? Neglect the size of the car.

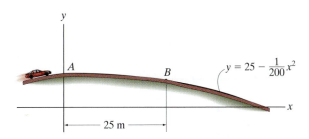

Prob. 13–82

13–83.   The 5-lb collar slides on the smooth rod, so that when it is at *A* it has a speed of 10 ft/s. If the spring to which it is attached has an unstretched length of 3 ft and a stiffness of $k = 10$ lb/ft, determine the normal force on the collar and the acceleration of the collar at this instant.

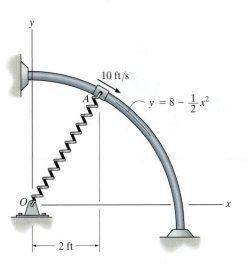

Prob. 13–83

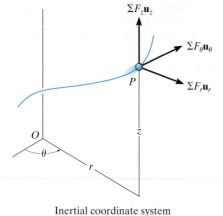

Inertial coordinate system

Fig. 13–16

13.6   Equations of Motion: Cylindrical Coordinates

When all the forces acting on a particle are resolved into cylindrical components, i.e., along the unit-vector directions $\mathbf{u}_r$, $\mathbf{u}_\theta$, $\mathbf{u}_z$, Fig. 13–16, the equation of motion can be expressed as

$$\Sigma \mathbf{F} = m\mathbf{a}$$
$$\Sigma F_r \mathbf{u}_r + \Sigma F_\theta \mathbf{u}_\theta + \Sigma F_z \mathbf{u}_z = ma_r \mathbf{u}_r + ma_\theta \mathbf{u}_\theta + ma_z \mathbf{u}_z$$

To satisfy this equation, we require

$$\Sigma F_r = ma_r$$
$$\Sigma F_\theta = ma_\theta \qquad (13\text{–}9)$$
$$\Sigma F_z = ma_z$$

If the particle is constrained to move only in the r–θ plane, then only the first two of Eqs. 13–9 are used to specify the motion.

Tangential and Normal Forces.
The most straightforward type of problem involving cylindrical coordinates requires the determination of the resultant force components ΣF_r, ΣF_θ, ΣF_z which cause a particle to move with a *known* acceleration. If, however, the particle's accelerated motion is not completely specified at the given instant, then some information regarding the directions or magnitudes of the forces acting on the particle must be known or computed in order to solve Eqs. 13–9. For example, the force $\mathbf{P}$ causes the particle in Fig. 13–17a to move along a path $r = f(\theta)$. The *normal force* $\mathbf{N}$ which the path exerts on the particle is always *perpendicular to the tangent of the path*, whereas the frictional force $\mathbf{F}$ always acts along the tangent in the opposite direction of motion. The *directions* of $\mathbf{N}$ and $\mathbf{F}$ can be specified relative to the radial coordinate by using the angle ψ (psi), Fig. 13–17b, which is defined between the *extended* radial line and the tangent to the curve.

As the car of weight W descends the spiral track, the resultant normal force which the track exerts on the car can be represented by its three cylindrical components. $-\mathbf{N}_r$ creates a radial acceleration $-\mathbf{a}_r$, $\mathbf{N}_\theta$ creates a transverse acceleration $\mathbf{a}_\theta$, and the difference $\mathbf{W} - \mathbf{N}_z$ creates an azimuthal acceleration $-\mathbf{a}_z$.

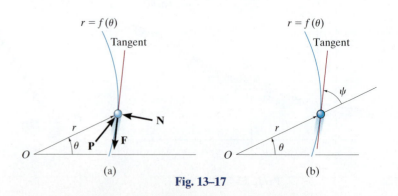

Fig. 13–17

This angle can be obtained by noting that when the particle is displaced a distance ds along the path, Fig. 13–17c, the component of displacement in the radial direction is dr and the component of displacement in the transverse direction is $r\,d\theta$. Since these two components are mutually perpendicular, the angle ψ can be determined from $\tan\psi = r\,d\theta/dr$, or

$$\tan\psi = \frac{r}{dr/d\theta} \qquad (13\text{–}10)$$

If ψ is calculated as a positive quantity, it is measured from the *extended radial line* to the tangent in a counterclockwise sense or in the positive direction of θ. If it is negative, it is measured in the opposite direction to positive θ. For example, consider the cardioid $r = a(1 + \cos\theta)$, shown in Fig. 13–18. Because $dr/d\theta = -a\sin\theta$, then when $\theta = 30°$, $\tan\psi = a(1 + \cos 30°)/(-a\sin 30°) = -3.732$, or $\psi = -75°$, measured clockwise, opposite to $+\theta$ as shown in the figure.

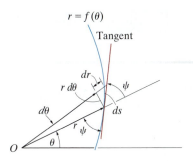

(c)

Fig. 13–17

Procedure for Analysis

Cylindrical or polar coordinates are a suitable choice for the analysis of a problem for which data regarding the angular motion of the radial line r are given, or in cases where the path can be conveniently expressed in terms of these coordinates. Once these coordinates have been established, the equations of motion can then be applied in order to relate the forces acting on the particle to its acceleration components. The method for doing this has been outlined in the procedure for analysis given in Sec. 13.4. The following is a summary of this procedure.

Free-Body Diagram.

- Establish the r, θ, z inertial coordinate system and draw the particle's free-body diagram.
- Assume that $\mathbf{a}_r, \mathbf{a}_\theta, \mathbf{a}_z$ act in the *positive directions* of r, θ, z if they are unknown.
- Identify all the unknowns in the problem.

Equations of Motion.

- Apply the equations of motion, Eqs. 13–9.

Kinematics.

- Use the methods of Sec. 12.8 to determine r and the time derivatives $\dot{r}$, $\ddot{r}$, $\dot{\theta}$, $\ddot{\theta}$, $\ddot{z}$, and then evaluate the acceleration components $a_r = \ddot{r} - r\dot{\theta}^2$, $a_\theta = r\ddot{\theta} + 2\dot{r}\dot{\theta}$, $a_z = \ddot{z}$.
- If any of the acceleration components is computed as a negative quantity, it indicates that it acts in its negative coordinate direction.
- When taking the time derivatives of $r = f(\theta)$, it is very important to use the chain rule of calculus, which is discussed at the end of Appendix C.

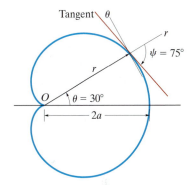

Fig. 13–18

EXAMPLE 13.10

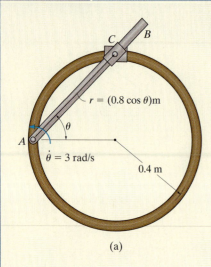

(a)

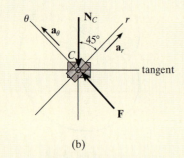

(b)

Fig. 13–19

The smooth 0.5-kg double-collar in Fig. 13–19a can freely slide on arm AB and the circular guide rod. If the arm rotates with a constant angular velocity of $\dot{\theta} = 3$ rad/s, determine the force the arm exerts on the collar at the instant $\theta = 45°$. Motion is in the horizontal plane.

SOLUTION

Free-Body Diagram.   The normal reaction $\mathbf{N}_C$ of the circular guide rod and the force $\mathbf{F}$ of arm AB act on the collar in the plane of motion, Fig. 13–19b. Note that $\mathbf{F}$ acts perpendicular to the axis of arm AB, that is, in the direction of the θ axis, while $\mathbf{N}_C$ acts perpendicular to the tangent of the circular path at $\theta = 45°$. The four unknowns are N_C, F, a_r, a_θ.

Equations of Motion.

$$+\nearrow \Sigma F_r = ma_r; \qquad -N_C \cos 45° = (0.5 \text{ kg}) a_r \qquad (1)$$

$$+\nwarrow \Sigma F_\theta = ma_\theta; \qquad F - N_C \sin 45° = (0.5 \text{ kg}) a_\theta \qquad (2)$$

Kinematics.   Using the chain rule (see Appendix C), the first and second time derivatives of r when $\theta = 45°$, $\dot{\theta} = 3$ rad/s, $\ddot{\theta} = 0$, are

$$r = 0.8 \cos \theta = 0.8 \cos 45° = 0.5657 \text{ m}$$

$$\dot{r} = -0.8 \sin \theta \, \dot{\theta} = -0.8 \sin 45°(3) = -1.6971 \text{ m/s}$$

$$\ddot{r} = -0.8 \left[\sin \theta \, \ddot{\theta} + \cos \theta \, \dot{\theta}^2 \right]$$

$$= -0.8[\sin 45°(0) + \cos 45°(3^2)] = -5.091 \text{ m/s}^2$$

We have

$$a_r = \ddot{r} - r\dot{\theta}^2 = -5.091 \text{ m/s}^2 - (0.5657 \text{ m})(3 \text{ rad/s})^2 = -10.18 \text{ m/s}^2$$

$$a_\theta = r\ddot{\theta} + 2\dot{r}\dot{\theta} = (0.5657 \text{ m})(0) + 2(-1.6971 \text{ m/s})(3 \text{ rad/s})$$

$$= -10.18 \text{ m/s}^2$$

Substituting these results into Eqs. (1) and (2) and solving, we get

$$N_C = 7.20 \text{ N}$$

$$F = 0 \qquad\qquad Ans.$$

EXAMPLE 13.11

The smooth 2-kg cylinder C in Fig. 13–20a has a pin P through its center which passes through the slot in arm OA. If the arm is forced to rotate in the *vertical plane* at a constant rate $\dot\theta = 0.5$ rad/s, determine the force that the arm exerts on the peg at the instant $\theta = 60°$.

SOLUTION
Why is it a good idea to use polar coordinates to solve this problem?

Free-Body Diagram.   The free-body diagram for the cylinder is shown in Fig. 13–20b. The force on the peg, $\mathbf{F}_P$, acts perpendicular to the slot in the arm. As usual, $\mathbf{a}_r$ and $\mathbf{a}_\theta$ are assumed to act in the directions of *positive r* and θ, respectively. Identify the four unknowns.

Equations of Motion.   Using the data in Fig. 13–20b, we have

$$+\nearrow\Sigma F_r = ma_r; \qquad 19.62\sin\theta - N_C\sin\theta = 2a_r \qquad (1)$$

$$+\searrow\Sigma F_\theta = ma_\theta; \quad 19.62\cos\theta + F_P - N_C\cos\theta = 2a_\theta \qquad (2)$$

Kinematics.   From Fig. 13–20a, r can be related to θ by the equation

$$r = \frac{0.4}{\sin\theta} = 0.4\csc\theta$$

Since $d(\csc\theta) = -(\csc\theta\cot\theta)\,d\theta$ and $d(\cot\theta) = -(\csc^2\theta)\,d\theta$, then r and the necessary time derivatives become

$$\dot\theta = 0.5 \qquad r = 0.4\csc\theta$$

$$\ddot\theta = 0 \qquad \dot r = -0.4(\csc\theta\cot\theta)\dot\theta$$

$$= -0.2\csc\theta\cot\theta$$

$$\ddot r = -0.2(-\csc\theta\cot\theta)(\dot\theta)\cot\theta - 0.2\csc\theta(-\csc^2\theta)\dot\theta$$

$$= 0.1\csc\theta(\cot^2\theta + \csc^2\theta)$$

Evaluating these formulas at $\theta = 60°$, we get

$$\dot\theta = 0.5 \qquad r = 0.462$$

$$\ddot\theta = 0 \qquad \dot r = -0.133$$

$$\ddot r = 0.192$$

$$a_r = \ddot r - r\dot\theta^2 = 0.192 - 0.462(0.5)^2 = 0.0770$$

$$a_\theta = r\ddot\theta + 2\dot r\dot\theta = 0 + 2(-0.133)(0.5) = -0.133$$

Substituting these results into Eqs. 1 and 2 with $\theta = 60°$ and solving yields

$$N_C = 19.5\text{ N} \qquad\qquad F_P = -0.356\text{ N} \qquad\qquad Ans.$$

The negative sign indicates that $\mathbf{F}_P$ acts opposite to the direction shown in Fig. 13–20b.

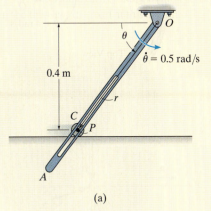

(a)

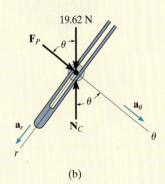

(b)

Fig. 13–20

PROBLEMS

13

*13–84. The path of motion of a 5-lb particle in the horizontal plane is described in terms of polar coordinates as $r = (2t + 1)$ ft and $\theta = (0.5t^2 - t)$ rad, where t is in seconds. Determine the magnitude of the resultant force acting on the particle when $t = 2$ s.

•13–85. Determine the magnitude of the resultant force acting on a 5-kg particle at the instant $t = 2$ s, if the particle is moving along a horizontal path defined by the equations $r = (2t + 10)$ m and $\theta = (1.5t^2 - 6t)$ rad, where t is in seconds.

13–86. A 2-kg particle travels along a horizontal smooth path defined by

$$r = \left(\frac{1}{4}t^3 + 2\right)\text{m}, \theta = \left(\frac{t^2}{4}\right)\text{rad},$$

where t is in seconds. Determine the radial and transverse components of force exerted on the particle when $t = 2$ s.

13–87. A 2-kg particle travels along a path defined by

$$r = (3 + 2t^2)\,\text{m}, \theta = \left(\frac{1}{3}t^3 + 2\right)\text{rad}$$

and $z = (5 - 2t^2)$ m, where t is in seconds. Determine the r, θ, z components of force that the path exerts on the particle at the instant $t = 1$ s.

*13–88. If the coefficient of static friction between the block of mass m and the turntable is μ_s, determine the maximum constant angular velocity of the platform without causing the block to slip.

•13–89. The 0.5-kg collar C can slide freely along the smooth rod AB. At a given instant, rod AB is rotating with an angular velocity of $\dot\theta = 2$ rad/s and has an angular acceleration of $\ddot\theta = 2$ rad/s^2. Determine the normal force of rod AB and the radial reaction of the end plate B on the collar at this instant. Neglect the mass of the rod and the size of the collar.

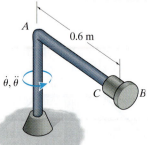

Prob. 13–89

13–90. The 2-kg rod AB moves up and down as its end slides on the smooth contoured surface of the cam, where $r = 0.1$ m and $z = (0.02 \sin \theta)$ m. If the cam is rotating with a constant angular velocity of 5 rad/s, determine the force on the roller A when $\theta = 90°$. Neglect friction at the bearing C and the mass of the roller.

13–91. The 2-kg rod AB moves up and down as its end slides on the smooth contoured surface of the cam, where $r = 0.1$ m and $z = (0.02 \sin \theta)$ m. If the cam is rotating at a constant angular velocity of 5 rad/s, determine the maximum and minimum force the cam exerts on the roller at A. Neglect friction at the bearing C and the mass of the roller.

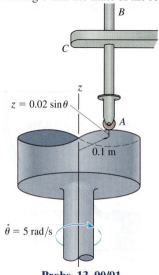

Probs. 13–90/91

Prob. 13–88

*13–92. If the coefficient of static friction between the conical surface and the block of mass m is $\mu_s = 0.2$, determine the minimum constant angular velocity $\dot{\theta}$ so that the block does not slide downwards.

•13–93. If the coefficient of static friction between the conical surface and the block is $\mu_s = 0.2$, determine the maximum constant angular velocity $\dot{\theta}$ without causing the block to slide upwards.

13–95. The mechanism is rotating about the vertical axis with a constant angular velocity of $\dot{\theta} = 6\,\text{rad/s}$. If rod AB is smooth, determine the constant position r of the 3-kg collar C. The spring has an unstretched length of 400 mm. Neglect the mass of the rod and the size of the collar.

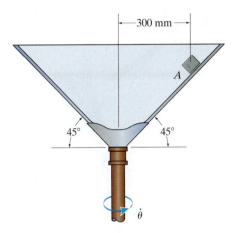

Probs. 13–92/93

Prob. 13–95

13–94. If the position of the 3-kg collar C on the smooth rod AB is held at $r = 720\,\text{mm}$, determine the constant angular velocity $\dot{\theta}$ at which the mechanism is rotating about the vertical axis. The spring has an unstretched length of 400 mm. Neglect the mass of the rod and the size of the collar.

*13–96. Due to the constraint, the 0.5-kg cylinder C travels along the path described by $r = (0.6 \cos \theta)\,\text{m}$. If arm OA rotates counterclockwise with an angular velocity of $\dot{\theta} = 2\,\text{rad/s}$ and an angular acceleration of $\ddot{\theta} = 0.8\,\text{rad/s}^2$ at the instant $\theta = 30°$, determine the force exerted by the arm on the cylinder at this instant. The cylinder is in contact with only one edge of the smooth slot, and the motion occurs in the horizontal plane.

Prob. 13–94

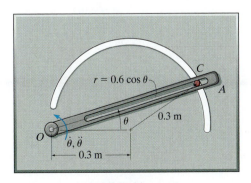

Prob. 13–96

•13–97. The 0.75-lb smooth can is guided along the circular path using the arm guide. If the arm has an angular velocity $\dot{\theta} = 2$ rad/s and an angular acceleration $\ddot{\theta} = 0.4$ rad/s^2 at the instant $\theta = 30°$, determine the force of the guide on the can. Motion occurs in the *horizontal plane*.

13–98. Solve Prob. 13–97 if motion occurs in the *vertical plane*.

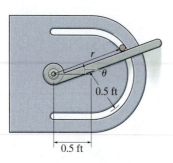

Probs. 13–97/98

13–99. The forked rod is used to move the smooth 2-lb particle around the horizontal path in the shape of a limaçon, $r = (2 + \cos \theta)$ ft. If at all times $\dot{\theta} = 0.5$ rad/s, determine the force which the rod exerts on the particle at the instant $\theta = 90°$. The fork and path contact the particle on only one side.

*13–100. Solve Prob. 13–99 at the instant $\theta = 60°$.

•13–101. The forked rod is used to move the smooth 2-lb particle around the horizontal path in the shape of a limaçon, $r = (2 + \cos \theta)$ ft. If $\theta = (0.5t^2)$ rad, where t is in seconds, determine the force which the rod exerts on the particle at the instant $t = 1$ s. The fork and path contact the particle on only one side.

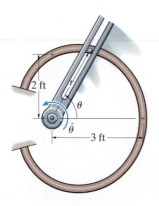

Probs. 13–99/100/101

13–102. The amusement park ride rotates with a constant angular velocity of $\dot{\theta} = 0.8$ rad/s. If the path of the ride is defined by $r = (3 \sin \theta + 5)$ m and $z = (3 \cos \theta)$ m, determine the r, θ, and z components of force exerted by the seat on the 20-kg boy when $\theta = 120°$.

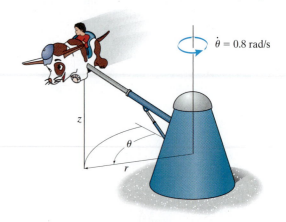

Prob. 13–102

13–103. The airplane executes the vertical loop defined by $r^2 = [810(10^3)\cos 2\theta]$ m^2. If the pilot maintains a constant speed $v = 120$ m/s along the path, determine the normal force the seat exerts on him at the instant $\theta = 0°$. The pilot has a mass of 75 kg.

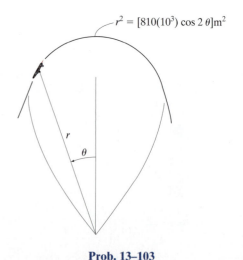

Prob. 13–103

*13–104. A boy standing firmly spins the girl sitting on a circular "dish" or sled in a circular path of radius $r_0 = 3$ m such that her angular velocity is $\dot\theta_0 = 0.1$ rad/s. If the attached cable OC is drawn inward such that the radial coordinate r changes with a constant speed of $\dot r = -0.5$ m/s, determine the tension it exerts on the sled at the instant $r = 2$ m. The sled and girl have a total mass of 50 kg. Neglect the size of the girl and sled and the effects of friction between the sled and ice. *Hint:* First show that the equation of motion in the θ direction yields $a_\theta = r\ddot\theta + 2\dot r\dot\theta = (1/r)\,d/dt(r^2\dot\theta) = 0$. When integrated, $r^2\dot\theta = C$, where the constant C is determined from the problem data.

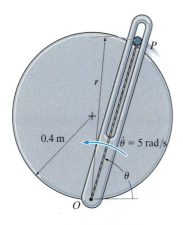

Prob. 13–104

13–105. The smooth particle has a mass of 80 g. It is attached to an elastic cord extending from O to P and due to the slotted arm guide moves along the *horizontal* circular path $r = (0.8 \sin \theta)$ m. If the cord has a stiffness $k = 30$ N/m and an unstretched length of 0.25 m, determine the force of the guide on the particle when $\theta = 60°$. The guide has a constant angular velocity $\dot\theta = 5$ rad/s.

13–106. Solve Prob. 13–105 if $\ddot\theta = 2$ rad/s² when $\dot\theta = 5$ rad/s and $\theta = 60°$.

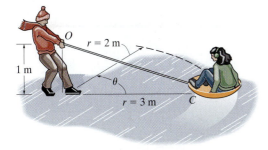

Probs. 13–105/106

13–107. The 1.5-kg cylinder C travels along the path described by $r = (0.6 \sin \theta)$ m. If arm OA rotates counterclockwise with a constant angular velocity of $\dot\theta = 3$ rad/s, determine the force exerted by the smooth slot in arm OA on the cylinder at the instant $\theta = 60°$. The spring has a stiffness of 100 N/m and is unstretched when $\theta = 30°$. The cylinder is in contact with only one edge of the slotted arm. Neglect the size of the cylinder. Motion occurs in the horizontal plane.

*13–108. The 1.5-kg cylinder C travels along the path described by $r = (0.6 \sin \theta)$ m. If arm OA is rotating counterclockwise with an angular velocity of $\dot\theta = 3$ rad/s, determine the force exerted by the smooth slot in arm OA on the cylinder at the instant $\theta = 60°$. The spring has a stiffness of 100 N/m and is unstretched when $\theta = 30°$. The cylinder is in contact with only one edge of the slotted arm. Neglect the size of the cylinder. Motion occurs in the vertical plane.

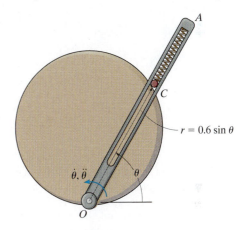

Probs. 13–107/108

•13–109. Using air pressure, the 0.5-kg ball is forced to move through the tube lying in the horizontal plane and having the shape of a logarithmic spiral. If the tangential force exerted on the ball due to air pressure is 6 N, determine the rate of increase in the ball's speed at the instant $\theta = \pi/2$. Also, what is the angle ψ from the extended radial coordinate r to the line of action of the 6-N force?

Prob. 13–109

13–110. The tube rotates in the horizontal plane at a constant rate of $\dot{\theta} = 4$ rad/s. If a 0.2-kg ball B starts at the origin O with an initial radial velocity of $\dot{r} = 1.5$ m/s and moves outward through the tube, determine the radial and transverse components of the ball's velocity at the instant it leaves the outer end at C, $r = 0.5$ m. *Hint:* Show that the equation of motion in the r direction is $\ddot{r} - 16r = 0$. The solution is of the form $r = Ae^{-4t} + Be^{4t}$. Evaluate the integration constants A and B, and determine the time t when $r = 0.5$ m. Proceed to obtain v_r and v_θ.

*13–112. The 0.5-lb ball is guided along the vertical circular path $r = 2r_c \cos\theta$ using the arm OA. If the arm has an angular velocity $\dot{\theta} = 0.4$ rad/s and an angular acceleration $\ddot{\theta} = 0.8$ rad/s^2 at the instant $\theta = 30°$, determine the force of the arm on the ball. Neglect friction and the size of the ball. Set $r_c = 0.4$ ft.

•13–113. The ball of mass m is guided along the vertical circular path $r = 2r_c \cos\theta$ using the arm OA. If the arm has a constant angular velocity $\dot{\theta}_0$, determine the angle $\theta \le 45°$ at which the ball starts to leave the surface of the semicylinder. Neglect friction and the size of the ball.

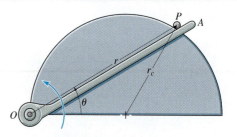

Probs. 13–112/113

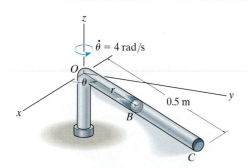

Prob. 13–110

13–111. The pilot of an airplane executes a vertical loop which in part follows the path of a cardioid, $r = 600(1 + \cos\theta)$ ft. If his speed at A $(\theta = 0°)$ is a constant $v_P = 80$ ft/s, determine the vertical force the seat belt must exert on him to hold him to his seat when the plane is upside down at A. He weighs 150 lb.

13–114. The ball has a mass of 1 kg and is confined to move along the smooth vertical slot due to the rotation of the smooth arm OA. Determine the force of the rod on the ball and the normal force of the slot on the ball when $\theta = 30°$. The rod is rotating with a constant angular velocity $\dot{\theta} = 3$ rad/s. Assume the ball contacts only one side of the slot at any instant.

13–115. Solve Prob. 13–114 if the arm has an angular acceleration of $\ddot{\theta} = 2$ rad/s^2 when $\dot{\theta} = 3$ rad/s at $\theta = 30°$.

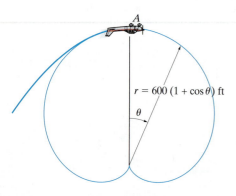

Prob. 13–111

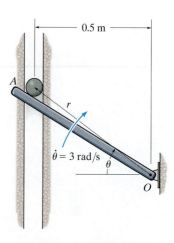

Probs. 13–114/115

*13.7   Central-Force Motion and Space Mechanics

If a particle is moving only under the influence of a force having a line of action which is always directed toward a fixed point, the motion is called *central-force motion*. This type of motion is commonly caused by electrostatic and gravitational forces.

In order to analyze the motion, we will consider the particle P shown in Fig. 13–22a, which has a mass m and is acted upon only by the central force **F**. The free-body diagram for the particle is shown in Fig. 13–22b. Using polar coordinates (r, θ), the equations of motion, Eqs. 13–9, become

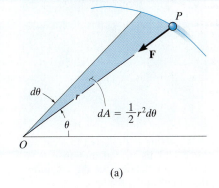

(a)

$$\Sigma F_r = ma_r; \qquad -F = m\left[\frac{d^2r}{dt^2} - r\left(\frac{d\theta}{dt}\right)^2\right]$$

$$\text{(13–11)}$$

$$\Sigma F_\theta = ma_\theta; \qquad 0 = m\left(r\frac{d^2\theta}{dt^2} + 2\frac{dr}{dt}\frac{d\theta}{dt}\right)$$

The second of these equations may be written in the form

$$\frac{1}{r}\left[\frac{d}{dt}\left(r^2\frac{d\theta}{dt}\right)\right] = 0$$

so that integrating yields

$$r^2\frac{d\theta}{dt} = h \qquad \text{(13–12)}$$

Here h is the constant of integration.

From Fig. 13–22a notice that the shaded area described by the radius r, as r moves through an angle $d\theta$, is $dA = \frac{1}{2}r^2\,d\theta$. If the *areal velocity* is defined as

$$\frac{dA}{dt} = \frac{1}{2}r^2\frac{d\theta}{dt} = \frac{h}{2} \qquad \text{(13–13)}$$

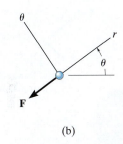

(b)

Fig. 13–22

then it is seen that the areal velocity for a particle subjected to central-force motion is *constant*. In other words, the particle will sweep out equal segments of area per unit of time as it travels along the path. To obtain the *path of motion*, $r = f(\theta)$, the independent variable t must be eliminated from Eqs. 13–11. Using the chain rule of calculus and Eq. 13–12, the time derivatives of Eqs. 13–11 may be replaced by

$$\frac{dr}{dt} = \frac{dr}{d\theta}\frac{d\theta}{dt} = \frac{h}{r^2}\frac{dr}{d\theta}$$

$$\frac{d^2r}{dt^2} = \frac{d}{dt}\left(\frac{h}{r^2}\frac{dr}{d\theta}\right) = \frac{d}{d\theta}\left(\frac{h}{r^2}\frac{dr}{d\theta}\right)\frac{d\theta}{dt} = \left[\frac{d}{d\theta}\left(\frac{h}{r^2}\frac{dr}{d\theta}\right)\right]\frac{h}{r^2}$$

13

This satellite is subjected to a central force and its orbital motion can be closely predicted using the equations developed in this section.

Substituting a new dependent variable (xi) $\xi = 1/r$ into the second equation, we have

$$\frac{d^2r}{dt^2} = -h^2\xi^2\frac{d^2\xi}{d\theta^2}$$

Also, the square of Eq. 13–12 becomes

$$\left(\frac{d\theta}{dt}\right)^2 = h^2\xi^4$$

Substituting these two equations into the first of Eqs. 13–11 yields

$$-h^2\xi^2\frac{d^2\xi}{d\theta^2} - h^2\xi^3 = -\frac{F}{m}$$

or

$$\frac{d^2\xi}{d\theta^2} + \xi = \frac{F}{mh^2\xi^2} \qquad (13\text{–}14)$$

This differential equation defines the path over which the particle travels when it is subjected to the central force **F**.*

For application, the force of gravitational attraction will be considered. Some common examples of central-force systems which depend on gravitation include the motion of the moon and artificial satellites about the earth, and the motion of the planets about the sun. As a typical problem in space mechanics, consider the trajectory of a space satellite or space vehicle launched into free-flight orbit with an initial velocity $\mathbf{v}_0$, Fig. 13–23. It will be assumed that this velocity is initially *parallel* to the tangent at the surface of the earth, as shown in the figure.† Just after the satellite is released into free flight, the only force acting on it is the gravitational force of the earth. (Gravitational attractions involving other bodies such as the moon or sun will be neglected, since for orbits close to the earth their effect is small in comparison with the earth's gravitation.) According to Newton's law of gravitation, force **F** will always act between the mass centers of the earth and the satellite, Fig. 13–23. From Eq. 13–1, this force of attraction has a magnitude of

$$F = G\frac{M_e m}{r^2}$$

where M_e and m represent the mass of the earth and the satellite, respectively, G is the gravitational constant, and r is the distance between

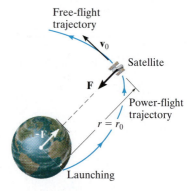

Free-flight
trajectory

$\mathbf{v}_0$

 Satellite

F

Power-flight
trajectory

$r = r_0$

Launching

Fig. 13–23

---

*In the derivation, **F** is considered positive when it is directed toward point O. If **F** is oppositely directed, the right side of Eq. 13–14 should be negative.

†The case where $\mathbf{v}_0$ acts at some initial angle θ to the tangent is best described using the conservation of angular momentum (see Prob. 15–100).

the mass centers. To obtain the orbital path, we set $\xi = 1/r$ in the foregoing equation and substitute the result into Eq. 13–14. We obtain

$$\frac{d^2\xi}{d\theta^2} + \xi = \frac{GM_e}{h^2} \qquad (13\text{–}15)$$

This second-order differential equation has constant coefficients and is nonhomogeneous. The solution is the sum of the complementary and particular solutions given by

$$\xi = \frac{1}{r} = C \cos(\theta - \phi) + \frac{GM_e}{h^2} \qquad (13\text{–}16)$$

This equation represents the *free-flight trajectory* of the satellite. It is the equation of a conic section expressed in terms of polar coordinates.

A geometric interpretation of Eq. 13–16 requires knowledge of the equation for a conic section. As shown in Fig. 13–24, a conic section is defined as the locus of a point P that moves in such a way that the ratio of its distance to a *focus*, or fixed point F, to its perpendicular distance to a fixed line DD called the *directrix*, is constant. This constant ratio will be denoted as e and is called the *eccentricity*. By definition

$$e = \frac{FP}{PA}$$

From Fig. 13–24,

$$FP = r = e(PA) = e[p - r\cos(\theta - \phi)]$$

or

$$\frac{1}{r} = \frac{1}{p}\cos(\theta - \phi) + \frac{1}{ep}$$

Comparing this equation with Eq. 13–16, it is seen that the fixed distance from the focus to the directrix is

$$p = \frac{1}{C} \qquad (13\text{–}17)$$

And the eccentricity of the conic section for the trajectory is

$$e = \frac{Ch^2}{GM_e} \qquad (13\text{–}18)$$

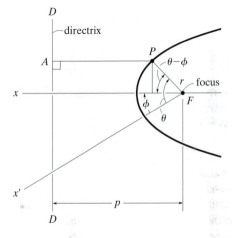

Fig. 13–24

13

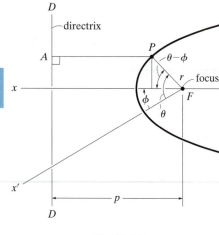

x

Fig. 13–24

Provided the polar angle θ is measured from the x axis (an axis of symmetry since it is perpendicular to the directrix), the angle ϕ is zero, Fig. 13–24, and therefore Eq. 13–16 reduces to

$$\frac{1}{r} = C \cos \theta + \frac{GM_e}{h^2} \qquad (13\text{–}19)$$

The constants h and C are determined from the data obtained for the position and velocity of the satellite at the end of the *power-flight trajectory*. For example, if the initial height or distance to the space vehicle is r_0, measured from the center of the earth, and its initial speed is v_0 at the beginning of its free flight, Fig. 13–25, then the constant h may be obtained from Eq. 13–12. When $\theta = \phi = 0°$, the velocity $\mathbf{v}_0$ has no radial component; therefore, from Eq. 12–25, $v_0 = r_0(d\theta/dt)$, so that

$$h = r_0^2 \frac{d\theta}{dt}$$

or

$$h = r_0 v_0 \qquad (13\text{–}20)$$

To determine C, use Eq. 13–19 with $\theta = 0°, r = r_0$, and substitute Eq. 13–20 for h:

$$C = \frac{1}{r_0}\left(1 - \frac{GM_e}{r_0 v_0^2}\right) \qquad (13\text{–}21)$$

The equation for the free-flight trajectory therefore becomes

$$\frac{1}{r} = \frac{1}{r_0}\left(1 - \frac{GM_e}{r_0 v_0^2}\right)\cos \theta + \frac{GM_e}{r_0^2 v_0^2} \qquad (13\text{–}22)$$

The type of path traveled by the satellite is determined from the value of the eccentricity of the conic section as given by Eq. 13–18. If

$$
\begin{array}{ll}
e = 0 & \text{free-flight trajectory is a circle} \\
e = 1 & \text{free-flight trajectory is a parabola} \\
e < 1 & \text{free-flight trajectory is an ellipse} \\
e > 1 & \text{free-flight trajectory is a hyperbola}
\end{array}
\qquad (13\text{–}23)
$$

Parabolic Path.   Each of these trajectories is shown in Fig. 13–25. From the curves it is seen that when the satellite follows a parabolic path, it is "on the border" of never returning to its initial starting point. The initial launch velocity, $\mathbf{v}_0$, required for the satellite to follow a parabolic path is called the *escape velocity*. The speed, v_e, can be determined by using the second of Eqs. 13–23, $e = 1$, with Eqs. 13–18, 13–20, and 13–21. It is left as an exercise to show that

$$v_e = \sqrt{\frac{2GM_e}{r_0}} \qquad (13\text{–}24)$$

Circular Orbit.   The speed v_c required to launch a satellite into a *circular orbit* can be found using the first of Eqs. 13–23, $e = 0$. Since e is related to h and C, Eq. 13–18, C must be zero to satisfy this equation (from Eq. 13–20, h cannot be zero); and therefore, using Eq. 13–21, we have

$$v_c = \sqrt{\frac{GM_e}{r_0}} \qquad (13\text{–}25)$$

Provided r_0 represents a minimum height for launching, in which frictional resistance from the atmosphere is neglected, speeds at launch which are less than v_c will cause the satellite to reenter the earth's atmosphere and either burn up or crash, Fig. 13–25.

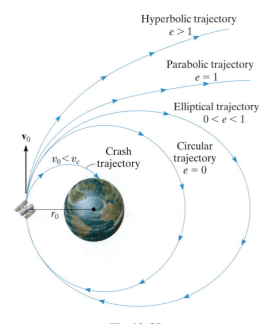

Fig. 13–25

13

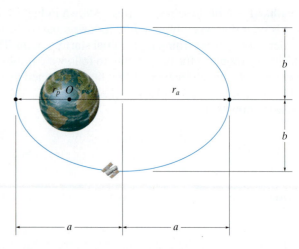

Fig. 13–26

Elliptical Orbit. All the trajectories attained by planets and most satellites are elliptical, Fig. 13–26. For a satellite's orbit about the earth, the *minimum distance* from the orbit to the center of the earth O (which is located at one of the foci of the ellipse) is r_p and can be found using Eq. 13–22 with $\theta = 0°$. Therefore;

$$r_p = r_0 \tag{13–26}$$

This minimum distance is called the *perigee* of the orbit. The *apogee* or maximum distance r_a can be found using Eq. 13–22 with $\theta = 180°$.* Thus,

$$r_a = \frac{r_0}{(2GM_e/r_0v_0^2) - 1} \tag{13–27}$$

With reference to Fig. 13–26, the half length of the major axis of the ellipse is

$$a = \frac{r_p + r_a}{2} \tag{13–28}$$

Using analytical geometry, it can be shown that the half length of the minor axis is determined from the equation

$$b = \sqrt{r_p r_a} \tag{13–29}$$

*Actually, the terminology perigee and apogee pertains only to orbits about the *earth*. If any other heavenly body is located at the focus of an elliptical orbit, the minimum and maximum distances are referred to respectively as the *periapsis* and *apoapsis* of the orbit.

Furthermore, by direct integration, the area of an ellipse is

$$A = \pi ab = \frac{\pi}{2}(r_p + r_a)\sqrt{r_p r_a} \qquad (13\text{--}30)$$

The areal velocity has been defined by Eq. 13–13, $dA/dt = h/2$. Integrating yields $A = hT/2$, where T is the *period* of time required to make one orbital revolution. From Eq. 13–30, the period is

$$T = \frac{\pi}{h}(r_p + r_a)\sqrt{r_p r_a} \qquad (13\text{--}31)$$

In addition to predicting the orbital trajectory of earth satellites, the theory developed in this section is valid, to a surprisingly close approximation, at predicting the actual motion of the planets traveling around the sun. In this case the mass of the sun, M_s, should be substituted for M_e when the appropriate formulas are used.

The fact that the planets do indeed follow elliptic orbits about the sun was discovered by the German astronomer Johannes Kepler in the early seventeenth century. His discovery was made *before* Newton had developed the laws of motion and the law of gravitation, and so at the time it provided important proof as to the validity of these laws. Kepler's laws, developed after 20 years of planetary observation, are summarized as follows:

1. Every planet travels in its orbit such that the line joining it to the center of the sun sweeps over equal areas in equal intervals of time, whatever the line's length.

2. The orbit of every planet is an ellipse with the sun placed at one of its foci.

3. The square of the period of any planet is directly proportional to the cube of the major axis of its orbit.

A mathematical statement of the first and second laws is given by Eqs. 13–13 and 13–22, respectively. The third law can be shown from Eq. 13–31 using Eqs. 13–19, 13–28, and 13–29. (See Prob. 13–116.)

CONCEPTUAL PROBLEMS

P13–1. If the box is released from rest at A, use numerical values to show how you would estimate the time for it to arrive at B. Also, list the assumptions for your analysis.

P13–1

P13–2. The tugboat has a known mass and its propeller provides a known maximum thrust. When the tug is fully powered you observe the time it takes for the tug to reach a speed of known value starting from rest. Show how you could determine the mass of the barge. Neglect the drag force of the water on the tug. Use numerical values to explain you answer.

P13–2

P13–3. Determine the smallest speed of each car A and B so that the passengers do not lose contact with the seat while the arms turn at a constant rate. What is the largest normal force of the seat on each passenger? Use numerical values to explain your answer.

P13–3

P13–4. Each car is pin connected at its ends to the rim of the wheel which turns at a constant speed. Using numerical values, show how to determine the resultant force the seat exerts on the passenger located in the top car A. The passengers are seated towards the center of the wheel. Also, list the assumptions for your analysis.

P13–4

CHAPTER REVIEW

Kinetics

Kinetics is the study of the relation between forces and the acceleration they cause. This relation is based on Newton's second law of motion, expressed mathematically as $\Sigma \mathbf{F} = m\mathbf{a}$.

Before applying the equation of motion, it is important to first draw the particle's *free-body diagram* in order to account for all of the forces that act on the particle. Graphically, this diagram is equal to the *kinetic diagram*, which shows the result of the forces, that is, the $m\mathbf{a}$ vector.

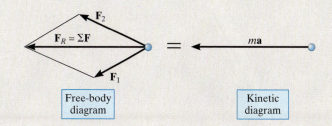

Free-body diagram

Kinetic diagram

Inertial Coordinate Systems

When applying the equation of motion, it is important to measure the acceleration from an inertial coordinate system. This system has axes that do not rotate but are either fixed or translate with a constant velocity. Various types of inertial coordinate systems can be used to apply $\Sigma \mathbf{F} = m\mathbf{a}$ in component form.

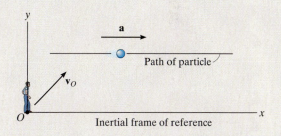

Inertial frame of reference

Rectangular x, y, z axes are used to describe rectilinear motion along each of the axes.

$$\Sigma F_x = ma_x, \quad \Sigma F_y = ma_y, \quad \Sigma F_z = ma_z$$

Normal and tangential n, t axes are often used when the path is known. Recall that $\mathbf{a}_n$ is always directed in the $+n$ direction. It indicates the change in the velocity direction. Also recall that $\mathbf{a}_t$ is tangent to the path. It indicates the change in the velocity magnitude.

$$\Sigma F_t = ma_t, \quad \Sigma F_n = ma_n, \quad \Sigma F_b = 0$$
$$a_t = dv/dt \quad \text{or} \quad a_t = v\, dv/ds$$
$$a_n = v^2/\rho \quad \text{where} \quad \rho = \frac{[1 + (dy/dx)^2]^{3/2}|}{|d^2y/dx^2|}$$

Cylindrical coordinates are useful when angular motion of the radial line r is specified or when the path can conveniently be described with these coordinates.

$$\Sigma F_r = m(\ddot{r} - r\dot{\theta}^2)$$
$$\Sigma F_\theta = m(r\ddot{\theta} + 2\dot{r}\dot{\theta})$$
$$\Sigma F_z = m\ddot{z}$$

Central-Force Motion

When a single force acts upon a particle, such as during the free-flight trajectory of a satellite in a gravitational field, then the motion is referred to as central-force motion. The orbit depends upon the eccentricity e; and as a result, the trajectory can either be circular, parabolic, elliptical, or hyperbolic.

In order to properly design the loop of this roller coaster it is necessary to ensure that the cars have enough energy to be able to make the loop without leaving the tracks.

Kinetics of a Particle: Work and Energy

14

CHAPTER OBJECTIVES

- To develop the principle of work and energy and apply it to solve problems that involve force, velocity, and displacement.

- To study problems that involve power and efficiency.

- To introduce the concept of a conservative force and apply the theorem of conservation of energy to solve kinetic problems.

14.1   The Work of a Force

In this chapter, we will analyze motion of a particle using the concepts of work and energy. The resulting equation will be useful for solving problems that involve force, velocity, and displacement. Before we do this, however, we must first define the work of a force. Specifically, a force **F** will do *work* on a particle only when the particle undergoes a *displacement in the direction of the force*. For example, if the force **F** in Fig. 14–1 causes the particle to move along the path *s* from position **r** to a new position **r**′, the displacement is then $d\mathbf{r} = \mathbf{r}' - \mathbf{r}$. The magnitude of $d\mathbf{r}$ is ds, the length of the differential segment along the path. If the angle between the tails of $d\mathbf{r}$ and **F** is θ, Fig. 14–1, then the work done by **F** is a *scalar quantity*, defined by

$$dU = F\, ds \cos \theta$$

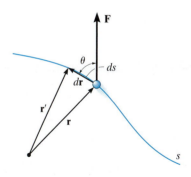

Fig. 14–1

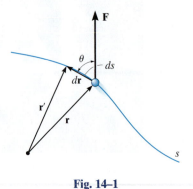

Fig. 14–1

By definition of the dot product (see Eq. B–14) this equation can also be written as

$$dU = \mathbf{F} \cdot d\mathbf{r}$$

This result may be interpreted in one of two ways: either as the product of F and the component of displacement $ds \cos \theta$ in the direction of the force, or as the product of ds and the component of force, $F \cos \theta$, in the direction of displacement. Note that if $0° \le \theta < 90°$, then the force component and the displacement have the *same sense* so that the work is *positive;* whereas if $90° < \theta \le 180°$, these vectors will have *opposite sense*, and therefore the work is *negative*. Also, $dU = 0$ if the force is *perpendicular* to displacement, since $\cos 90° = 0$, or if the force is applied at a *fixed point*, in which case the displacement is zero.

The unit of work in SI units is the joule (J), which is the amount of work done by a one-newton force when it moves through a distance of one meter in the direction of the force $(1\,J = 1\,N \cdot m)$. In the FPS system, work is measured in units of foot-pounds $(ft \cdot lb)$, which is the work done by a one-pound force acting through a distance of one foot in the direction of the force.*

Work of a Variable Force.   If the particle acted upon by the force $\mathbf{F}$ undergoes a finite displacement along its path from $\mathbf{r}_1$ to $\mathbf{r}_2$ or s_1 to s_2, Fig. 14–2a, the work of force $\mathbf{F}$ is determined by integration. Provided $\mathbf{F}$ and θ can be expressed as a function of position, then

$$U_{1-2} = \int_{\mathbf{r}_1}^{\mathbf{r}_2} \mathbf{F} \cdot d\mathbf{r} = \int_{s_1}^{s_2} F \cos \theta \, ds \qquad (14\text{–}1)$$

Sometimes, this relation may be obtained by using experimental data to plot a graph of $F \cos \theta$ vs. s. Then the *area* under this graph bounded by s_1 and s_2 represents the total work, Fig. 14–2b.

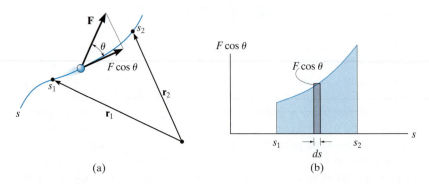

(a)   (b)

Fig. 14–2

*By convention, the units for the moment of a force or torque are written as $lb \cdot ft$, to distinguish them from those used to signify work, $ft \cdot lb$.

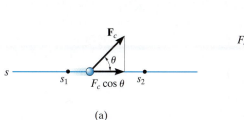

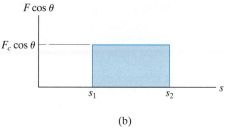

(a)   (b)

Fig. 14–3

Work of a Constant Force Moving Along a Straight Line.

If the force $\mathbf{F}_c$ has a constant magnitude and acts at a constant angle θ from its straight-line path, Fig. 14–3a, then the component of $\mathbf{F}_c$ in the direction of displacement is always $F_c \cos \theta$. The work done by $\mathbf{F}_c$ when the particle is displaced from s_1 to s_2 is determined from Eq. 14–1, in which case

$$U_{1-2} = F_c \cos \theta \int_{s_1}^{s_2} ds$$

or

$$\boxed{U_{1-2} = F_c \cos \theta (s_2 - s_1)} \qquad (14\text{–}2)$$

Here the work of $\mathbf{F}_c$ represents the *area of the rectangle* in Fig. 14–3b.

Work of a Weight.

Consider a particle of weight $\mathbf{W}$, which moves up along the path s shown in Fig. 14–4 from position s_1 to position s_2. At an intermediate point, the displacement $d\mathbf{r} = dx\mathbf{i} + dy\mathbf{j} + dz\mathbf{k}$. Since $\mathbf{W} = -W\mathbf{j}$, applying Eq. 14–1 we have

$$U_{1-2} = \int \mathbf{F} \cdot d\mathbf{r} = \int_{\mathbf{r}_1}^{\mathbf{r}_2} (-W\mathbf{j}) \cdot (dx\mathbf{i} + dy\mathbf{j} + dz\mathbf{k})$$

$$= \int_{y_1}^{y_2} -W \, dy = -W(y_2 - y_1)$$

or

$$\boxed{U_{1-2} = -W \, \Delta y} \qquad (14\text{–}3)$$

Thus, the work is independent of the path and is equal to the magnitude of the particle's weight times its vertical displacement. In the case shown in Fig. 14–4 the work is *negative*, since W is downward and Δy is upward. Note, however, that if the particle is displaced *downward* $(-\Delta y)$, the work of the weight is *positive*. Why?

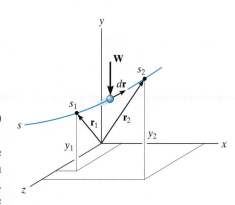

Fig. 14–4

Work of a Spring Force. If an elastic spring is elongated a distance ds, Fig. 14–5a, then the work done by the force that acts on the attached particle is $dU = -F_s ds = -ks\, ds$. The work is *negative* since $\mathbf{F}_s$ acts in the opposite sense to ds. If the particle displaces from s_1 to s_2, the work of $\mathbf{F}_s$ is then

$$U_{1-2} = \int_{s_1}^{s_2} F_s\, ds = \int_{s_1}^{s_2} -ks\, ds$$

$$U_{1-2} = -\left(\tfrac{1}{2}ks_2^2 - \tfrac{1}{2}ks_1^2\right) \qquad (14\text{–}4)$$

This work represents the trapezoidal area under the line $F_s = ks$, Fig. 14–5b.

A mistake in sign can be avoided when applying this equation if one simply notes the direction of the spring force acting on the particle and compares it with the sense of direction of displacement of the particle— if both are in the *same sense, positive work* results; if they are *opposite* to one another, the *work is negative.*

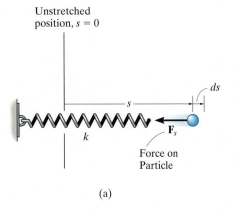

(a)

(b)

Fig. 14–5

The forces acting on the cart as it is pulled a distance s up the incline, are shown on its free-body diagram. The constant towing force $\mathbf{T}$ does positive work of $U_T = (T \cos \phi)s$, the weight does negative work of $U_W = -(W \sin \theta)s$, and the normal force $\mathbf{N}$ does no work since there is no displacement of this force along its line of action.

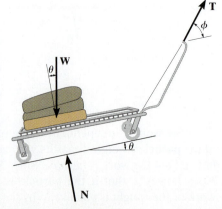

EXAMPLE 14.1

The 10-kg block shown in Fig. 14–6a rests on the smooth incline. If the spring is originally stretched 0.5 m, determine the total work done by all the forces acting on the block when a horizontal force $P = 400$ N pushes the block up the plane $s = 2$ m.

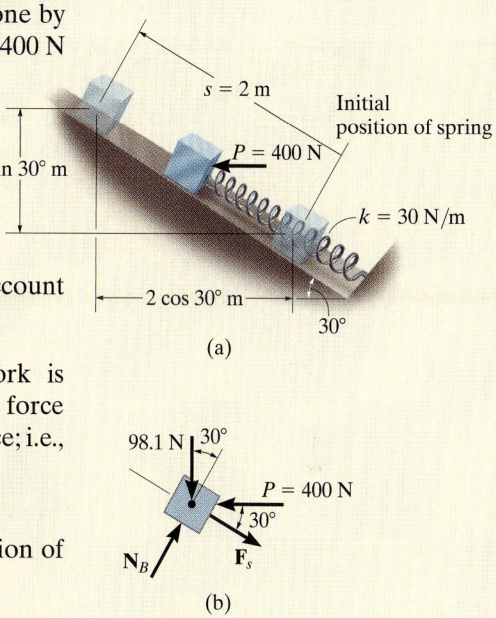

(a)

SOLUTION
First the free-body diagram of the block is drawn in order to account for all the forces that act on the block, Fig. 14–6b.

Horizontal Force P. Since this force is *constant*, the work is determined using Eq. 14–2. The result can be calculated as the force times the component of displacement in the direction of the force; i.e.,

$$U_P = 400 \text{ N } (2 \text{ m cos } 30°) = 692.8 \text{ J}$$

or the displacement times the component of force in the direction of displacement, i.e.,

$$U_P = 400 \text{ N cos } 30°(2 \text{ m}) = 692.8 \text{ J}$$

(b)

Fig. 14–6

Spring Force F$_s$. In the initial position the spring is stretched $s_1 = 0.5$ m and in the final position it is stretched $s_2 = 0.5$ m + 2 m = 2.5 m. We require the work to be negative since the force and displacement are opposite to each other. The work of $\mathbf{F}_s$ is thus

$$U_s = -\left[\tfrac{1}{2}(30 \text{ N/m})(2.5 \text{ m})^2 - \tfrac{1}{2}(30 \text{ N/m})(0.5 \text{ m})^2\right] = -90 \text{ J}$$

Weight W. Since the weight acts in the opposite sense to its vertical displacement, the work is negative; i.e.,

$$U_W = -(98.1 \text{ N}) (2 \text{ m sin } 30°) = -98.1 \text{ J}$$

Note that it is also possible to consider the component of weight in the direction of displacement; i.e.,

$$U_W = -(98.1 \sin 30° \text{ N}) (2 \text{ m}) = -98.1 \text{ J}$$

Normal Force N$_B$. This force does *no work* since it is *always* perpendicular to the displacement.

Total Work. The work of all the forces when the block is displaced 2 m is therefore

$$U_T = 692.8 \text{ J} - 90 \text{ J} - 98.1 \text{ J} = 505 \text{ J} \qquad Ans.$$

14

14.2   Principle of Work and Energy

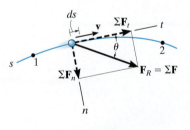

Fig. 14–7

Consider the particle in Fig. 14–7, which is located on the path defined relative to an inertial coordinate system. If the particle has a mass m and is subjected to a system of external forces represented by the resultant $\mathbf{F}_R = \Sigma\mathbf{F}$, then the equation of motion for the particle in the tangential direction is $\Sigma F_t = ma_t$. Applying the kinematic equation $a_t = v \, dv/ds$ and integrating both sides, assuming initially that the particle has a position $s = s_1$ and a speed $v = v_1$, and later at $s = s_2$, $v = v_2$, we have

$$\Sigma \int_{s_1}^{s_2} F_t \, ds = \int_{v_1}^{v_2} mv \, dv$$

$$\Sigma \int_{s_1}^{s_2} F_t \, ds = \tfrac{1}{2}mv_2^2 - \tfrac{1}{2}mv_1^2 \qquad (14\text{–}5)$$

From Fig. 14–7, note that $\Sigma F_t = \Sigma F \cos \theta$, and since work is defined from Eq. 14–1, the final result can be written as

$$\Sigma U_{1-2} = \tfrac{1}{2}mv_2^2 - \tfrac{1}{2}mv_1^2 \qquad (14\text{–}6)$$

This equation represents the *principle of work and energy* for the particle. The term on the left is the sum of the work done by *all* the forces acting on the particle as the particle moves from point 1 to point 2. The two terms on the right side, which are of the form $T = \tfrac{1}{2}mv^2$, define the particle's final and initial *kinetic energy*, respectively. Like work, kinetic energy is a *scalar* and has units of joules (J) and ft · lb. However, unlike work, which can be either positive or negative, the kinetic energy is *always positive*, regardless of the direction of motion of the particle.

When Eq. 14–6 is applied, it is often expressed in the form

$$\boxed{T_1 + \Sigma U_{1-2} = T_2} \qquad (14\text{–}7)$$

which states that the particle's initial kinetic energy plus the work done by all the forces acting on the particle as it moves from its initial to its final position is equal to the particle's final kinetic energy.

As noted from the derivation, the principle of work and energy represents an integrated form of $\Sigma F_t = ma_t$, obtained by using the kinematic equation $a_t = v \, dv/ds$. As a result, this principle will provide a convenient *substitution* for $\Sigma F_t = ma_t$ when solving those types of kinetic problems which involve *force*, *velocity*, and *displacement* since these quantities are involved in Eq. 14–7. For application, it is suggested that the following procedure be used.

Procedure for Analysis

Work (Free-Body Diagram).

- Establish the inertial coordinate system and draw a free-body diagram of the particle in order to account for all the forces that do work on the particle as it moves along its path.

Principle of Work and Energy.

- Apply the principle of work and energy, $T_1 + \Sigma U_{1-2} = T_2$.

- The kinetic energy at the initial and final points is *always positive*, since it involves the speed squared $\left(T = \frac{1}{2}mv^2\right)$.

- A force does work when it moves through a displacement in the direction of the force.

- Work is *positive* when the force component is in the *same sense of direction* as its displacement, otherwise it is negative.

- Forces that are functions of displacement must be integrated to obtain the work. Graphically, the work is equal to the area under the force-displacement curve.

- The work of a weight is the product of the weight magnitude and the vertical displacement, $U_W = \pm Wy$. It is positive when the weight moves downwards.

- The work of a spring is of the form $U_s = \frac{1}{2}ks^2$, where k is the spring stiffness and s is the stretch or compression of the spring.

Numerical application of this procedure is illustrated in the examples following Sec. 14.3.

If an oncoming car strikes these crash barrels, the car's kinetic energy will be transformed into work, which causes the barrels, and to some extent the car, to be deformed. By knowing the amount of energy that can be absorbed by each barrel it is possible to design a crash cushion such as this.

14.3   Principle of Work and Energy for a System of Particles

The principle of work and energy can be extended to include a system of particles isolated within an enclosed region of space as shown in Fig. 14–8. Here the arbitrary *i*th particle, having a mass m_i, is subjected to a resultant external force $\mathbf{F}_i$ and a resultant internal force $\mathbf{f}_i$ which all the other particles exert on the *i*th particle. If we apply the principle of work and energy to this and each of the other particles in the system, then since work and energy are scalar quantities, the equations can be summed algebraically, which gives

$$\Sigma T_1 + \Sigma U_{1-2} = \Sigma T_2 \tag{14–8}$$

In this case, the initial kinetic energy of the system plus the work done by all the external and internal forces acting on the system is equal to the final kinetic energy of the system.

If the system represents a *translating rigid body*, or a series of connected translating bodies, then all the particles in each body will undergo the *same displacement*. Therefore, the work of all the internal forces will occur in equal but opposite collinear pairs and so it will cancel out. On the other hand, if the body is assumed to be *nonrigid*, the particles of the body may be displaced along *different paths*, and some of the energy due to force interactions would be given off and lost as heat or stored in the body if permanent deformations occur. We will discuss these effects briefly at the end of this section and in Sec. 15.4. Throughout this text, however, the principle of work and energy will be applied to problems where direct accountability of such energy losses does not have to be considered.

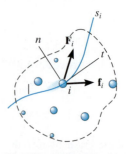

Inertial coordinate system

Fig. 14–8

Work of Friction Caused by Sliding.

A special class of problems will now be investigated which requires a careful application of Eq. 14–8. These problems involve cases where a body slides over the surface of another body in the presence of friction. Consider, for example, a block which is translating a distance s over a rough surface as shown in Fig. 14–9a. If the applied force **P** just balances the *resultant* frictional force $\mu_k N$, Fig. 14–9b, then due to equilibrium a constant velocity **v** is maintained, and one would expect Eq. 14–8 to be applied as follows:

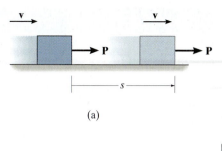

(a)

$$\tfrac{1}{2}mv^2 + Ps - \mu_k Ns = \tfrac{1}{2}mv^2$$

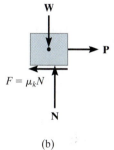

(b)

Indeed this equation is satisfied if $P = \mu_k N$; however, as one realizes from experience, the sliding motion will *generate heat*, a form of energy which seems not to be accounted for in the work-energy equation. In order to explain this paradox and thereby more closely represent the nature of friction, we should actually model the block so that the surfaces of contact are *deformable* (nonrigid).* Recall that the rough portions at the bottom of the block act as "teeth," and when the block slides these teeth *deform slightly* and either break off or vibrate as they pull away from "teeth" at the contacting surface, Fig. 14–9c. As a result, frictional forces that act on the block at these points are displaced slightly, due to the localized deformations, and later they are replaced by other frictional forces as other points of contact are made. At any instant, the *resultant* **F** of all these frictional forces remains essentially constant, i.e., $\mu_k N$; however, due to the many *localized deformations*, the actual displacement s' of $\mu_k N$ is *not* the same as the displacement s of the applied force **P**. Instead, s' will be *less* than s $(s' < s)$, and therefore the *external work* done by the resultant frictional force will be $\mu_k N s'$ and not $\mu_k N s$. The remaining amount of work, $\mu_k N(s - s')$, manifests itself as an increase in *internal energy*, which in fact causes the block's temperature to rise.

(c)

Fig. 14–9

In summary then, Eq. 14–8 can be applied to problems involving sliding friction; however, it should be fully realized that the work of the resultant frictional force is not represented by $\mu_k N s$; instead, this term represents *both* the external work of friction ($\mu_k N s'$) *and* internal work $[\mu_k N(s - s')]$ which is converted into various forms of internal energy, such as heat.†

---

*See Chapter 8 of *Engineering Mechanics: Statics.*
†See B. A. Sherwood and W. H. Bernard, "Work and Heat Transfer in the Presence of Sliding Friction," *Am. J. Phys.* 52, 1001 (1984).

EXAMPLE 14.2

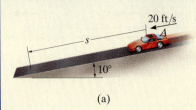

20 ft/s

A

s

$10°$

(a)

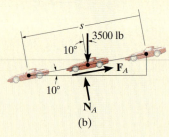

s

3500 lb

$10°$

F_A

$10°$

N_A

(b)

Fig. 14–10

The 3500-lb automobile shown in Fig. 14–10a travels down the $10°$ inclined road at a speed of 20 ft/s. If the driver jams on the brakes, causing his wheels to lock, determine how far s the tires skid on the road. The coefficient of kinetic friction between the wheels and the road is $\mu_k = 0.5$.

SOLUTION

This problem can be solved using the principle of work and energy, since it involves force, velocity, and displacement.

Work (Free-Body Diagram). As shown in Fig. 14–10b, the normal force $\mathbf{N}_A$ does no work since it never undergoes displacement along its line of action. The weight, 3500 lb, is displaced $s \sin 10°$ and does positive work. Why? The frictional force $\mathbf{F}_A$ does both external and internal work when it undergoes a displacement s. This work is negative since it is in the opposite sense of direction to the displacement. Applying the equation of equilibrium normal to the road, we have

$$+\nwarrow\Sigma F_n = 0; \quad N_A - 3500 \cos 10° \text{ lb} = 0 \quad N_A = 3446.8 \text{ lb}$$

Thus,

$$F_A = \mu_k N_A = 0.5 \, (3446.8 \text{ lb}) = 1723.4 \text{ lb}$$

Principle of Work and Energy.

$$T_1 + \Sigma U_{1\text{-}2} = T_2$$

$$\frac{1}{2}\left(\frac{3500 \text{ lb}}{32.2 \text{ ft/s}^2}\right)(20 \text{ ft/s})^2 + 3500 \text{ lb}(s \sin 10°) - (1723.4 \text{ lb})s = 0$$

Solving for s yields

$$s = 19.5 \text{ ft} \qquad\qquad Ans.$$

NOTE: If this problem is solved by using the equation of motion, *two steps* are involved. First, from the free-body diagram, Fig. 14–10b, the equation of motion is applied along the incline. This yields

$$+\swarrow\Sigma F_s = ma_s; \quad 3500 \sin 10° \text{ lb} - 1723.4 \text{ lb} = \frac{3500 \text{ lb}}{32.2 \text{ ft/s}^2}a$$

$$a = -10.3 \text{ ft/s}^2$$

Then, since a is constant, we have

$$(+\swarrow) \quad v^2 = v_0^2 + 2a_c(s - s_0);$$

$$(0)^2 = (20 \text{ ft/s})^2 + 2(-10.3 \text{ ft/s}^2)(s - 0)$$

$$s = 19.5 \text{ ft} \qquad\qquad Ans.$$

EXAMPLE 14.3

For a short time the crane in Fig. 14–11a lifts the 2.50-Mg beam with a force of $F = (28 + 3s^2)$ kN. Determine the speed of the beam when it has risen $s = 3$ m. Also, how much time does it take to attain this height starting from rest?

(a)

SOLUTION

We can solve part of this problem using the principle of work and energy since it involves force, velocity, and displacement. Kinematics must be used to determine the time. Note that at $s = 0$, $F = 28(10^3)$N $> W = 2.50(10^3)(9.81)$N, so motion will occur.

Work (Free-Body Diagram). As shown on the free-body diagram, Fig. 14–11b, the lifting force **F** does positive work, which must be determined by integration since this force is a variable. Also, the weight is constant and will do negative work since the displacement is upwards.

Principles of Work and Energy.

$$T_1 + \Sigma U_{1-2} = T_2$$

$$0 + \int_0^s (28 + 3s^2)(10^3)\,ds - (2.50)(10^3)(9.81)s = \tfrac{1}{2}(2.50)(10^3)v^2$$

$$28(10^3)s + (10^3)s^3 - 24.525(10^3)s = 1.25(10^3)v^2$$

$$v = (2.78s + 0.8s^3)^{\frac{1}{2}} \qquad (1)$$

When $s = 3$ m,

$$v = 5.47 \text{ m/s} \qquad \textit{Ans.}$$

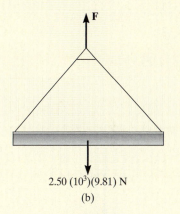

2.50 $(10^3)(9.81)$ N

(b)

Fig. 14–11

Kinematics. Since we were able to express the velocity as a function of displacement, the time can be determined using $v = ds/dt$. In this case,

$$(2.78s + 0.8s^3)^{\frac{1}{2}} = \frac{ds}{dt}$$

$$t = \int_0^3 \frac{ds}{(2.78s + 0.8s^3)^{\frac{1}{2}}}$$

The integration can be performed numerically using a pocket calculator. The result is

$$t = 1.79 \text{ s} \qquad \textit{Ans.}$$

NOTE: The acceleration of the beam can be determined by integrating Eq. (1) using $v\,dv = a\,ds$, or more directly, by applying the equation of motion, $\Sigma F = ma$.

EXAMPLE 14.4

The platform P, shown in Fig. 14–12a, has negligible mass and is tied down so that the 0.4-m-long cords keep a 1-m-long spring compressed 0.6 m when *nothing* is on the platform. If a 2-kg block is placed on the platform and released from rest after the platform is pushed down 0.1 m, Fig. 14–12b, determine the maximum height h the block rises in the air, measured from the ground.

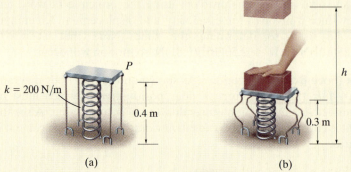

(a)   (b)

Fig. 14–12

SOLUTION

Work (Free-Body Diagram). Since the block is released from rest and later reaches its maximum height, the initial and final velocities are zero. The free-body diagram of the block when it is still in contact with the platform is shown in Fig. 14–12c. Note that the weight does negative work and the spring force does positive work. Why? In particular, the *initial compression* in the spring is $s_1 = 0.6$ m + 0.1 m = 0.7 m. Due to the cords, the spring's *final compression* is $s_2 = 0.6$ m (after the block leaves the platform). The bottom of the block rises from a height of $(0.4 \text{ m} - 0.1 \text{ m}) = 0.3$ m to a final height h.

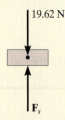

19.62 N

$\mathbf{F}_s$

(c)

Principle of Work and Energy.

$$T_1 + \Sigma U_{1\text{-}2} = T_2$$

$$\tfrac{1}{2}mv_1^2 + \left\{-\left(\tfrac{1}{2}ks_2^2 - \tfrac{1}{2}ks_1^2\right) - W\,\Delta y\right\} = \tfrac{1}{2}mv_2^2$$

Note that here $s_1 = 0.7$ m $> s_2 = 0.6$ m and so the work of the spring as determined from Eq. 14–4 will indeed be positive once the calculation is made. Thus,

$$0 + \left\{-\left[\tfrac{1}{2}(200 \text{ N/m})(0.6 \text{ m})^2 - \tfrac{1}{2}(200 \text{ N/m})(0.7 \text{ m})^2\right]\right.$$

$$\left. - (19.62 \text{ N})[h - (0.3 \text{ m})]\right\} = 0$$

Solving yields

$$h = 0.963 \text{ m} \qquad\qquad Ans.$$

EXAMPLE 14.5

The 40-kg boy in Fig. 14–13a slides down the smooth water slide. If he starts from rest at A, determine his speed when he reaches B and the normal reaction the slide exerts on the boy at this position.

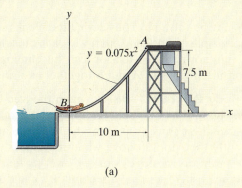

(a)

SOLUTION

Work (Free-Body Diagram).   As shown on the free-body diagram, Fig. 14–13b, there are two forces acting on the boy as he goes down the slide. Note that the normal force does no work.

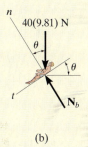

(b)

Principle of Work and Energy.

$$T_A + \Sigma U_{A-B} = T_B$$

$$0 + (40(9.81)\text{N})(7.5 \text{ m}) = \tfrac{1}{2}(40 \text{ kg})v_B^2$$

$$v_B = 12.13 \text{ m/s} = 12.1 \text{ m/s} \qquad Ans.$$

Equation of Motion.   Referring to the free-body diagram of the boy when he is at B, Fig. 14–13c, the normal reaction N_B can now be obtained by applying the equation of motion along the n axis. Here the radius of curvature of the path is

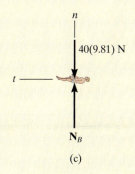

(c)

$$\rho_B = \frac{\left[1 + \left(\dfrac{dy}{dx}\right)^2\right]^{3/2}}{|d^2y/dx^2|} = \frac{[1 + (0.15x)^2]^{3/2}}{|0.15|}\bigg|_{x=0} = 6.667 \text{ m}$$

Thus,

$$+\uparrow \Sigma F_n = ma_n; \quad N_B - 40(9.81) \text{ N} = 40 \text{ kg}\left(\frac{(12.13 \text{ m/s})^2}{6.667 \text{ m}}\right)$$

$$N_B = 1275.3 \text{ N} = 1.28 \text{ kN} \qquad Ans.$$

Fig. 14–13

14

EXAMPLE 14.6

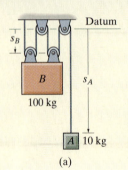

(a)

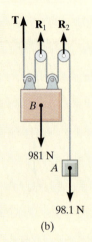

(b)

Fig. 14–14

Blocks A and B shown in Fig. 14–14a have a mass of 10 kg and 100 kg, respectively. Determine the distance B travels when it is released from rest to the point where its speed becomes 2 m/s.

SOLUTION

This problem may be solved by considering the blocks separately and applying the principle of work and energy to each block. However, the work of the (unknown) cable tension can be eliminated from the analysis by considering blocks A and B together as a *single system*.

Work (Free-Body Diagram). As shown on the free-body diagram of the system, Fig. 14–14b, the cable force $\mathbf{T}$ and reactions $\mathbf{R}_1$ and $\mathbf{R}_2$ do *no work*, since these forces represent the reactions at the supports and consequently they do not move while the blocks are displaced. The weights both do positive work if we *assume* both move downward, in the positive sense of direction of s_A and s_B.

Principle of Work and Energy. Realizing the blocks are released from rest, we have

$$\Sigma T_1 + \Sigma U_{1\text{-}2} = \Sigma T_2$$

$$\{\tfrac{1}{2}m_A(v_A)_1^2 + \tfrac{1}{2}m_B(v_B)_1^2\} + \{W_A \, \Delta s_A + W_B \, \Delta s_B\} =$$
$$\{\tfrac{1}{2}m_A(v_A)_2^2 + \tfrac{1}{2}m_B(v_B)_2^2\}$$

$$\{0 + 0\} + \{98.1 \text{ N } (\Delta s_A) + 981 \text{ N } (\Delta s_B)\} =$$
$$\{\tfrac{1}{2}(10 \text{ kg})(v_A)_2^2 + \tfrac{1}{2}(100 \text{ kg})(2 \text{ m/s})^2\} \qquad (1)$$

Kinematics. Using the methods of kinematics discussed in Sec. 12.9, it may be seen from Fig. 14–14a that the total length l of all the vertical segments of cable may be expressed in terms of the position coordinates s_A and s_B as

$$s_A + 4s_B = l$$

Hence, a change in position yields the displacement equation

$$\Delta s_A + 4 \, \Delta s_B = 0$$

$$\Delta s_A = -4 \, \Delta s_B$$

Here we see that a downward displacement of one block produces an upward displacement of the other block. Note that Δs_A and Δs_B must have the *same* sign convention in both Eqs. 1 and 2. Taking the time derivative yields

$$v_A = -4v_B = -4(2 \text{ m/s}) = -8 \text{ m/s} \qquad (2)$$

Retaining the negative sign in Eq. 2 and substituting into Eq. 1 yields

$$\Delta s_B = 0.883 \text{ m} \downarrow \qquad \qquad \textit{Ans.}$$

FUNDAMENTAL PROBLEMS

F14–1. The spring is placed between the wall and the 10-kg block. If the block is subjected to a force of $F = 500$ N, determine its velocity when $s = 0.5$ m. When $s = 0$, the block is at rest and the spring is uncompressed. The contact surface is smooth.

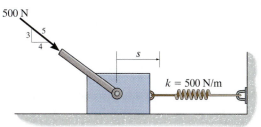

F14–1

F14–2. If the motor exerts a constant force of 300 N on the cable, determine the speed of the 20-kg crate when it travels $s = 10$ m up the plane, starting from rest. The coefficient of kinetic friction between the crate and the plane is $\mu_k = 0.3$.

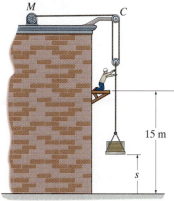

F14–2

F14–3. If the motor exerts a force of $F = (600 + 2s^2)$ N on the cable, determine the speed of the 100-kg crate when it rises to $s = 15$ m. The crate is initially at rest on the ground.

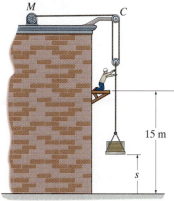

F14–3

F14–4. The 1.8-Mg dragster is traveling at 125 m/s when the engine is shut off and the parachute is released. If the drag force of the parachute can be approximated by the graph, determine the speed of the dragster when it has traveled 400 m.

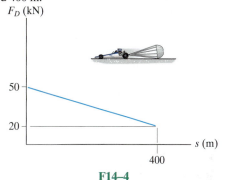

F14–4

F14–5. When $s = 0.6$ m, the spring is unstretched and the 10-kg block has a speed of 5 m/s down the smooth plane. Determine the distance s when the block stops.

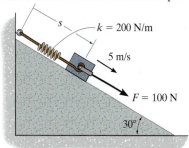

F14–5

F14–6. The 5-lb collar is pulled by a cord that passes around a small peg at C. If the cord is subjected to a constant force of $F = 10$ lb, and the collar is at rest when it is at A, determine its speed when it reaches B. Neglect friction.

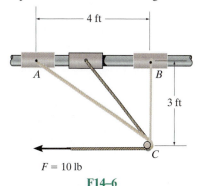

F14–6

PROBLEMS

•**14–1.** A 1500-lb crate is pulled along the ground with a constant speed for a distance of 25 ft, using a cable that makes an angle of 15° with the horizontal. Determine the tension in the cable and the work done by this force. The coefficient of kinetic friction between the ground and the crate is $\mu_k = 0.55$.

14–2. The motion of a 6500-lb boat is arrested using a bumper which provides a resistance as shown in the graph. Determine the maximum distance the boat dents the bumper if its approaching speed is 3 ft/s.

***14–4.** When a 7-kg projectile is fired from a cannon barrel that has a length of 2 m, the explosive force exerted on the projectile, while it is in the barrel, varies in the manner shown. Determine the approximate muzzle velocity of the projectile at the instant it leaves the barrel. Neglect the effects of friction inside the barrel and assume the barrel is horizontal.

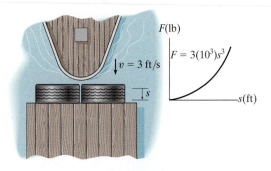

Prob. 14–2

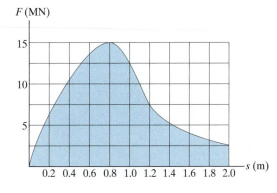

Prob. 14–4

14–3. The smooth plug has a weight of 20 lb and is pushed against a series of Belleville spring washers so that the compression in the spring is $s = 0.05$ ft. If the force of the spring on the plug is $F = (3s^{1/3})$ lb, where s is given in feet, determine the speed of the plug after it moves away from the spring. Neglect friction.

•**14–5.** The 1.5-kg block slides along a smooth plane and strikes a *nonlinear spring* with a speed of $v = 4$ m/s. The spring is termed "nonlinear" because it has a resistance of $F_s = ks^2$, where $k = 900$ N/m². Determine the speed of the block after it has compressed the spring $s = 0.2$ m.

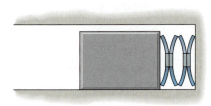

Prob. 14–3

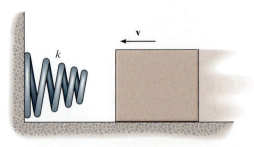

Prob. 14–5

14–6. When the driver applies the brakes of a light truck traveling 10 km/h, it skids 3 m before stopping. How far will the truck skid if it is traveling 80 km/h when the brakes are applied?

Prob. 14–6

14–7. The 6-lb block is released from rest at A and slides down the smooth parabolic surface. Determine the maximum compression of the spring.

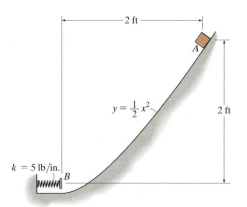

Prob. 14–7

****14–8.** The spring in the toy gun has an unstretched length of 100 mm. It is compressed and locked in the position shown. When the trigger is pulled, the spring unstretches 12.5 mm, and the 20-g ball moves along the barrel. Determine the speed of the ball when it leaves the gun. Neglect friction.

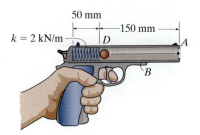

Prob. 14–8

•**14–9.** Springs AB and CD have a stiffness of $k = 300$ N/m and $k' = 200$ N/m, respectively, and both springs have an unstretched length of 600 mm. If the 2-kg smooth collar starts from rest when the springs are unstretched, determine the speed of the collar when it has moved 200 mm.

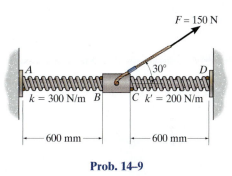

Prob. 14–9

14–10. The 2-Mg car has a velocity of $v_1 = 100$ km/h when the driver sees an obstacle in front of the car. If it takes 0.75 s for him to react and lock the brakes, causing the car to skid, determine the distance the car travels before it stops. The coefficient of kinetic friction between the tires and the road is $\mu_k = 0.25$.

14–11. The 2-Mg car has a velocity of $v_1 = 100$ km/h when the driver sees an obstacle in front of the car. It takes 0.75 s for him to react and lock the brakes, causing the car to skid. If the car stops when it has traveled a distance of 175 m, determine the coefficient of kinetic friction between the tires and the road.

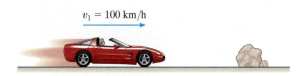

Probs. 14–10/11

*14–12. The 10-lb block is released from rest at A. Determine the compression of each of the springs after the block strikes the platform and is brought momentarily to rest. Initially both springs are unstretched. Assume the platform has a negligible mass.

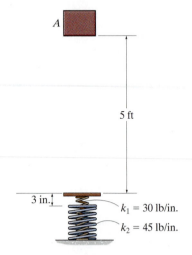

5 ft

3 in.

$k_1 = 30$ lb/in.

$k_2 = 45$ lb/in.

Prob. 14–12

14–13. Determine the velocity of the 60-lb block A if the two blocks are released from rest and the 40-lb block B moves 2 ft up the incline. The coefficient of kinetic friction between both blocks and the inclined planes is $\mu_k = 0.10$.

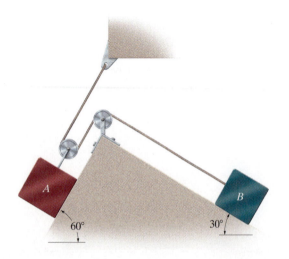

60°   30°

Prob. 14–13

14–14. The force $\mathbf{F}$, acting in a constant direction on the 20-kg block, has a magnitude which varies with the position s of the block. Determine how far the block slides before its velocity becomes 5 m/s. When $s = 0$ the block is moving to the right at 2 m/s. The coefficient of kinetic friction between the block and surface is $\mu_k = 0.3$.

14–15. The force $\mathbf{F}$, acting in a constant direction on the 20-kg block, has a magnitude which varies with position s of the block. Determine the speed of the block after it slides 3 m. When $s = 0$ the block is moving to the right at 2 m/s. The coefficient of kinetic friction between the block and surface is $\mu_k = 0.3$.

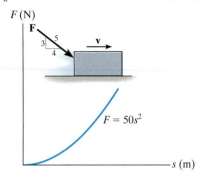

F (N)

$F = 50s^2$

s (m)

Prob. 14–14/15

14–16. A rocket of mass m is fired vertically from the surface of the earth, i.e., at $r = r_1$. Assuming no mass is lost as it travels upward, determine the work it must do against gravity to reach a distance r_2. The force of gravity is $F = GM_em/r^2$ (Eq. 13–1), where M_e is the mass of the earth and r the distance between the rocket and the center of the earth.

r_2

r

r_1

Prob. 14–16

•14–17. The cylinder has a weight of 20 lb and is pushed against a series of Belleville spring washers so that the compression in the spring is $s = 0.05$ ft. If the force of the spring on the cylinder is $F = (100s^{1/3})$ lb, where s is given in feet, determine the speed of the cylinder just after it moves away from the spring, i.e., at $s = 0$.

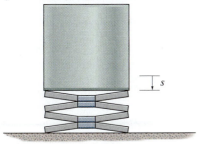

Prob. 14–17

14–18. The collar has a mass of 20 kg and rests on the smooth rod. Two springs are attached to it and the ends of the rod as shown. Each spring has an uncompressed length of 1 m. If the collar is displaced $s = 0.5$ m and released from rest, determine its velocity at the instant it returns to the point $s = 0$.

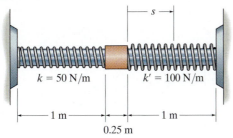

Prob. 14–18

14–19. Determine the height h of the incline D to which the 200-kg roller coaster car will reach, if it is launched at B with a speed just sufficient for it to round the top of the loop at C without leaving the track. The radius of curvature at C is $\rho_c = 25$ m.

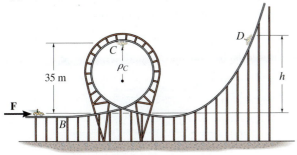

Prob. 14–19

*14–20. Packages having a weight of 15 lb are transferred horizontally from one conveyor to the next using a ramp for which $\mu_k = 0.15$. The top conveyor is moving at 6 ft/s and the packages are spaced 3 ft apart. Determine the required speed of the bottom conveyor so no sliding occurs when the packages come horizontally in contact with it. What is the spacing s between the packages on the bottom conveyor?

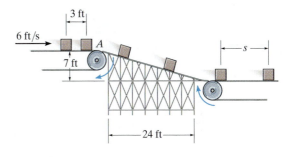

Prob. 14–20

•14–21. The 0.5-kg ball of negligible size is fired up the smooth vertical circular track using the spring plunger. The plunger keeps the spring compressed 0.08 m when $s = 0$. Determine how far s it must be pulled back and released so that the ball will begin to leave the track when $\theta = 135°$.

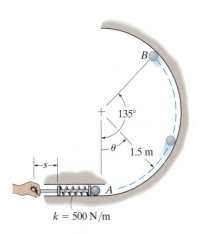

Prob. 14–21

14–22. The 2-lb box slides on the smooth circular ramp. If the box has a velocity of 30 ft/s at A, determine the velocity of the box and normal force acting on the ramp when the box is located at B and C. Assume the radius of curvature of the path at C is still 5 ft.

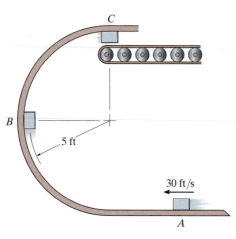

Prob. 14–22

14–23. Packages having a weight of 50 lb are delivered to the chute at $v_A = 3$ ft/s using a conveyor belt. Determine their speeds when they reach points B, C, and D. Also calculate the normal force of the chute on the packages at B and C. Neglect friction and the size of the packages.

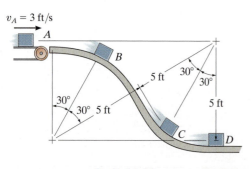

Prob. 14–23

*14–24. The 2-lb block slides down the smooth parabolic surface, such that when it is at A it has a speed of 10 ft/s. Determine the magnitude of the block's velocity and acceleration when it reaches point B, and the maximum height y_{max} reached by the block.

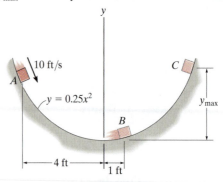

Prob. 14–24

•14–25. The skier starts from rest at A and travels down the ramp. If friction and air resistance can be neglected, determine his speed v_B when he reaches B. Also, find the distance s to where he strikes the ground at C, if he makes the jump traveling horizontally at B. Neglect the skier's size. He has a mass of 70 kg.

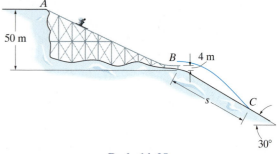

Prob. 14–25

14–26. The crate, which has a mass of 100 kg, is subjected to the action of the two forces. If it is originally at rest, determine the distance it slides in order to attain a speed of 6 m/s. The coefficient of kinetic friction between the crate and the surface is $\mu_k = 0.2$.

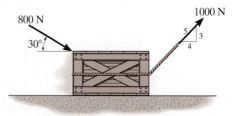

Prob. 14–26

14–27. The 2-lb brick slides down a smooth roof, such that when it is at A it has a velocity of 5 ft/s. Determine the speed of the brick just before it leaves the surface at B, the distance d from the wall to where it strikes the ground, and the speed at which it hits the ground.

•14–29. The 120-lb man acts as a human cannonball by being "fired" from the spring-loaded cannon shown. If the greatest acceleration he can experience is $a = 10g = 322 \text{ ft/s}^2$, determine the required stiffness of the spring which is compressed 2 ft at the moment of firing. With what velocity will he exit the cannon barrel, $d = 8$ ft, when the cannon is fired? When the spring is compressed $s = 2$ ft then $d = 8$ ft. Neglect friction and assume the man holds himself in a rigid position throughout the motion.

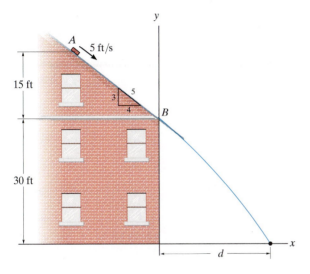

Prob. 14–27

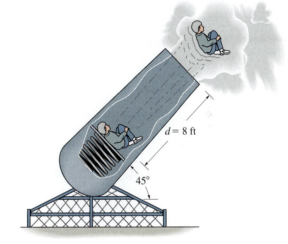

Prob. 14–29

***14–28.** Roller coasters are designed so that riders will not experience a normal force that is more than 3.5 times their weight against the seat of the car. Determine the smallest radius of curvature ρ of the track at its lowest point if the car has a speed of 5 ft/s at the crest of the drop. Neglect friction.

14–30. If the track is to be designed so that the passengers of the roller coaster do not experience a normal force equal to zero or more than 4 times their weight, determine the limiting heights h_A and h_C so that this does not occur. The roller coaster starts from rest at position A. Neglect friction.

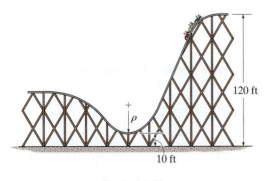

Prob. 14–28

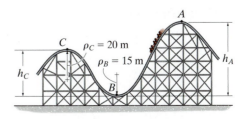

Prob. 14–30

14–31. Marbles having a mass of 5 g fall from rest at A through the glass tube and accumulate in the can at C. Determine the placement R of the can from the end of the tube and the speed at which the marbles fall into the can. Neglect the size of the can.

•14–33. If the coefficient of kinetic friction between the 100-kg crate and the plane is $\mu_k = 0.25$, determine the compression x of the spring required to bring the crate momentarily to rest. Initially the spring is unstretched and the crate is at rest.

14–34. If the coefficient of kinetic friction between the 100-kg crate and the plane is $\mu_k = 0.25$, determine the speed of the crate at the instant the compression of the spring is $x = 1.5$ m. Initially the spring is unstretched and the crate is at rest.

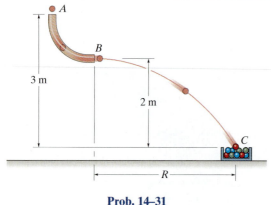

Prob. 14–31

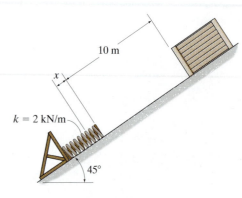

Probs. 14–33/34

*14–32. The ball has a mass of 0.5 kg and is suspended from a rubber band having an unstretched length of 1 m and a stiffness $k = 50$ N/m. If the support at A to which the rubber band is attached is 2 m from the floor, determine the greatest speed the ball can have at A so that it does not touch the floor when it reaches its lowest point B. Neglect the size of the ball and the mass of the rubber band.

14–35. A 2-lb block rests on the smooth semicylindrical surface. An elastic cord having a stiffness $k = 2$ lb/ft is attached to the block at B and to the base of the semicylinder at point C. If the block is released from rest at $A(\theta = 0°)$, determine the unstretched length of the cord so that the block begins to leave the semicylinder at the instant $\theta = 45°$. Neglect the size of the block.

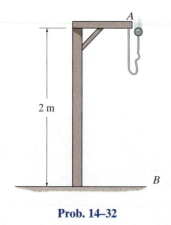

Prob. 14–32

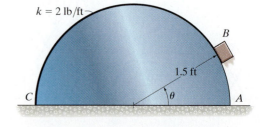

Prob. 14–35

*14–36.   The 50-kg stone has a speed of $v_A = 8$ m/s when it reaches point A. Determine the normal force it exerts on the incline when it reaches point B. Neglect friction and the stone's size.

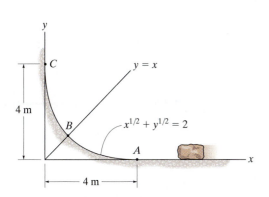

Prob. 14–36

•14–37.   If the 75-kg crate starts from rest at A, determine its speed when it reaches point B. The cable is subjected to a constant force of $F = 300$ N. Neglect friction and the size of the pulley.

14–38.   If the 75-kg crate starts from rest at A, and its speed is 6 m/s when it passes point B, determine the constant force $\mathbf{F}$ exerted on the cable. Neglect friction and the size of the pulley.

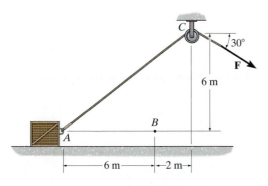

Probs. 14–37/38

14–39.   If the 60-kg skier passes point A with a speed of 5 m/s, determine his speed when he reaches point B. Also find the normal force exerted on him by the slope at this point. Neglect friction.

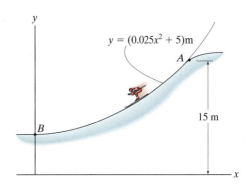

Prob. 14–39

*14–40.   The 150-lb skater passes point A with a speed of 6 ft/s. Determine his speed when he reaches point B and the normal force exerted on him by the track at this point. Neglect friction.

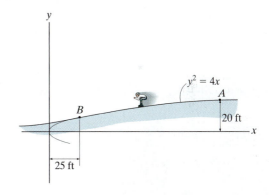

Prob. 14–40

•14–41.   A small box of mass m is given a speed of $v = \sqrt{\frac{1}{48}gr}$ at the top of the smooth half cylinder. Determine the angle θ at which the box leaves the cylinder.

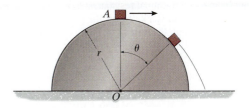

Prob. 14–41

14

14.4   Power and Efficiency

Power.   The term "power" provides a useful basis for chosing the type of motor or machine which is required to do a certain amount of work in a given time. For example, two pumps may each be able to empty a reservoir if given enough time; however, the pump having the larger power will complete the job sooner.

The *power* generated by a machine or engine that performs an amount of work dU within the time interval dt is therefore

$$P = \frac{dU}{dt} \tag{14–9}$$

If the work dU is expressed as $dU = \mathbf{F} \cdot d\mathbf{r}$, then

$$P = \frac{dU}{dt} = \frac{\mathbf{F} \cdot d\mathbf{r}}{dt} = \mathbf{F} \cdot \frac{d\mathbf{r}}{dt}$$

or

$$P = \mathbf{F} \cdot \mathbf{v} \tag{14–10}$$

Hence, power is a *scalar*, where in this formulation $\mathbf{v}$ represents the velocity of the paticle which is acted upon by the force $\mathbf{F}$.

The basic units of power used in the SI and FPS systems are the watt (W) and horsepower (hp), respectively. These units are defined as

$$1 \text{ W} = 1 \text{ J/s} = 1 \text{ N} \cdot \text{m/s}$$

$$1 \text{ hp} = 550 \text{ ft} \cdot \text{lb/s}$$

For conversion between the two systems of units, 1 hp = 746 W.

The power output of this locomotive comes from the driving frictional force **F** developed at its wheels. It is this force that overcomes the frictional resistance of the cars in tow and is able to lift the weight of the train up the grade.

Efficiency.   The *mechanical efficiency* of a machine is defined as the ratio of the output of useful power produced by the machine to the input of power supplied to the machine. Hence,

$$\epsilon = \frac{\text{power output}}{\text{power input}} \tag{14–11}$$

If energy supplied to the machine occurs during the *same time interval* at which it is drawn, then the efficiency may also be expressed in terms of the ratio

$$\epsilon = \frac{\text{energy output}}{\text{energy input}} \qquad (14\text{--}12)$$

Since machines consist of a series of moving parts, frictional forces will always be developed within the machine, and as a result, extra energy or power is needed to overcome these forces. Consequently, power output will be less than power input and so *the efficiency of a machine is always less than 1*.

The power supplied to a body can be determined using the following procedure.

Procedure for Analysis

- First determine the external force **F** acting on the body which causes the motion. This force is usually developed by a machine or engine placed either within or external to the body.

- If the body is accelerating, it may be necessary to draw its free-body diagram and apply the equation of motion ($\Sigma \mathbf{F} = m\mathbf{a}$) to determine **F**.

- Once **F** and the velocity **v** of the particle where **F** is applied have been found, the power is determined by multiplying the force magnitude with the component of velocity acting in the direction of **F**, (i.e., $P = \mathbf{F} \cdot \mathbf{v} = Fv \cos \theta$).

- In some problems the power may be found by calculating the work done by **F** per unit of time ($P_{\text{avg}} = \Delta U / \Delta t$,).

The power requirements of this elevator depend upon the vertical force **F** that acts on the elevator and causes it to move upwards. If the velocity of the elevator is **v**, then the power output is $P = \mathbf{F} \cdot \mathbf{v}$.

EXAMPLE 14.7

The man in Fig. 14–15a pushes on the 50-kg crate with a force of $F = 150$ N. Determine the power supplied by the man when $t = 4$ s. The coefficient of kinetic friction between the floor and the crate is $\mu_k = 0.2$. Initially the create is at rest.

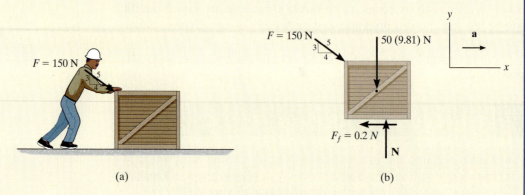

Fig. 14–15

SOLUTION

To determine the power developed by the man, the velocity of the 150-N force must be obtained first. The free-body diagram of the crate is shown in Fig. 14–15b. Applying the equation of motion,

$$+\uparrow \Sigma F_y = ma_y; \quad N - \left(\tfrac{3}{5}\right)150 \text{ N} - 50(9.81) \text{ N} = 0$$
$$N = 580.5 \text{ N}$$
$$\overset{+}{\rightarrow} \Sigma F_x = ma_x; \quad \left(\tfrac{4}{5}\right)150 \text{ N} - 0.2(580.5 \text{ N}) = (50 \text{ kg})a$$
$$a = 0.078 \text{ m/s}^2$$

The velocity of the crate when $t = 4$ s is therefore

$$(\overset{+}{\rightarrow}) \qquad\qquad\qquad v = v_0 + a_c t$$
$$v = 0 + (0.078 \text{ m/s}^2)(4 \text{ s}) = 0.312 \text{ m/s}$$

The power supplied to the crate by the man when $t = 4$ s is therefore

$$P = \mathbf{F} \cdot \mathbf{v} = F_x v = \left(\tfrac{4}{5}\right)(150 \text{ N})(0.312 \text{ m/s})$$

$$= 37.4 \text{ W} \qquad\qquad\qquad Ans.$$

EXAMPLE 14.8

The motor M of the hoist shown in Fig. 14–16a lifts the 75-lb crate C so that the acceleration of point P is 4 ft/s². Determine the power that must be supplied to the motor at the instant P has a velocity of 2 ft/s. Neglect the mass of the pulley and cable and take $\epsilon = 0.85$.

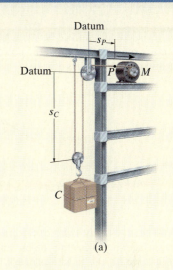

(a)

SOLUTION

In order to find the power output of the motor, it is first necessary to determine the tension in the cable since this force is developed by the motor.

From the free-body diagram, Fig. 14–16b, we have

$$+\downarrow\ \Sigma F_y = ma_y; \qquad -2T + 75\ \text{lb} = \frac{75\ \text{lb}}{32.2\ \text{ft/s}^2}a_C \qquad (1)$$

The acceleration of the crate can be obtained by using kinematics to relate it to the known acceleration of point P, Fig. 14–16a. Using the methods of Sec. 12.9, the coordinates s_C and s_P can be related to a constant portion of cable length l which is changing in the vertical and horizontal directions. We have $2s_C + s_P = l$. Taking the second time derivative of this equation yields

$$2a_C = -a_P \qquad (2)$$

Since $a_P = +4$ ft/s², then $a_C = -(4\ \text{ft/s}^2)/2 = -2$ ft/s². What does the negative sign indicate? Substituting this result into Eq. 1 and *retaining* the negative sign since the acceleration in *both* Eq. 1 and Eq. 2 was considered positive downward, we have

$$-2T + 75\ \text{lb} = \left(\frac{75\ \text{lb}}{32.2\ \text{ft/s}^2}\right)(-2\ \text{ft/s}^2)$$

$$T = 39.83\ \text{lb}$$

(b)

Fig. 14–16

The power output, measured in units of horsepower, required to draw the cable in at a rate of 2 ft/s is therefore

$$P = \mathbf{T} \cdot \mathbf{v} = (39.83\ \text{lb})(2\ \text{ft/s})[1\ \text{hp}/(550\ \text{ft} \cdot \text{lb/s})]$$

$$= 0.1448\ \text{hp}$$

This *power output* requires that the motor provide a *power input* of

$$\text{power input} = \frac{1}{\epsilon}(\text{power output})$$

$$= \frac{1}{0.85}(0.1448\ \text{hp}) = 0.170\ \text{hp} \qquad Ans.$$

NOTE: Since the velocity of the crate is constantly changing, the power requirement is *instantaneous*.

FUNDAMENTAL PROBLEMS

F14–7. If the contact surface between the 20-kg block and the ground is smooth, determine the power of force **F** when $t = 4$ s. Initially, the block is at rest.

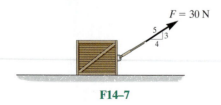

F14–7

F14–8. If $F = (10s)$ N, where s is in meters, and the contact surface between the block and the ground is smooth, determine the power of force **F** when $s = 5$ m. Initially, the 20-kg block is at rest.

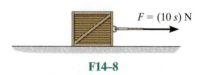

F14–8

F14–9. If the motor winds in the cable with a constant speed of $v = 3$ ft/s, determine the power supplied to the motor. The load weighs 100 lb and the efficiency of the motor is $\epsilon = 0.8$. Neglect the mass of the pulleys.

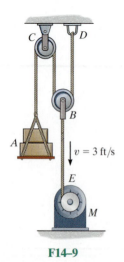

F14–9

F14–10. The coefficient of kinetic friction between the 20-kg block and the inclined plane is $\mu_k = 0.2$. If the block is traveling up the inclined plane with a constant velocity $v = 5$ m/s, determine the power of force **F**.

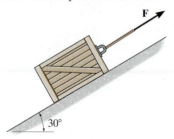

F14–10

F14–11. If the 50-kg load A is hoisted by motor M so that the load has a constant velocity of 1.5 m/s, determine the power input of the motor, which operates at an efficiency $\epsilon = 0.8$.

F14–11

F14–12. At the instant shown, point P on the cable has a velocity $v_P = 12$ m/s, which is increasing at a rate of $a_P = 6$ m/s². Determine the power input of motor M at this instant if it operates with an efficiency $\epsilon = 0.8$. The mass of block A is 50 kg.

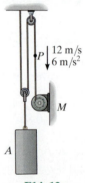

F14–12

PROBLEMS

14–42. The diesel engine of a 400-Mg train increases the train's speed uniformly from rest to 10 m/s in 100 s along a horizontal track. Determine the average power developed.

14–43. Determine the power input for a motor necessary to lift 300 lb at a constant rate of 5 ft/s. The efficiency of the motor is $\epsilon = 0.65$.

*14–44. An electric streetcar has a weight of 15 000 lb and accelerates along a horizontal straight road from rest so that the power is always 100 hp. Determine how far it must travel to reach a speed of 40 ft/s.

•14–45. The Milkin Aircraft Co. manufactures a turbojet engine that is placed in a plane having a weight of 13000 lb. If the engine develops a constant thrust of 5200 lb, determine the power output of the plane when it is just ready to take off with a speed of 600 mi/h.

14–46. The engine of the 3500-lb car is generating a constant power of 50 hp while the car is traveling up the slope with a constant speed. If the engine is operating with an efficiency of $\epsilon = 0.8$, determine the speed of the car. Neglect drag and rolling resistance.

Prob. 14–46

14–47. A loaded truck weighs $16(10^3)$ lb and accelerates uniformly on a level road from 15 ft/s to 30 ft/s during 4 s. If the frictional resistance to motion is 325 lb, determine the maximum power that must be delivered to the wheels.

*14–48. An automobile having a weight of 3500 lb travels up a 7° slope at a constant speed of $v = 40$ ft/s. If friction and wind resistance are neglected, determine the power developed by the engine if the automobile has a mechanical efficiency of $\epsilon = 0.65$.

•14–49. An escalator step moves with a constant speed of 0.6 m/s. If the steps are 125 mm high and 250 mm in length, determine the power of a motor needed to lift an average mass of 150 kg per step. There are 32 steps.

14–50. The man having the weight of 150 lb is able to run up a 15-ft-high flight of stairs in 4 s. Determine the power generated. How long would a 100-W light bulb have to burn to expend the same amount of energy? *Conclusion:* Please turn off the lights when they are not in use!

15 ft

Prob. 14–50

14–51. The material hoist and the load have a total mass of 800 kg and the counterweight C has a mass of 150 kg. At a given instant, the hoist has an upward velocity of 2 m/s and an acceleration of 1.5 m/s^2. Determine the power generated by the motor M at this instant if it operates with an efficiency of $\epsilon = 0.8$.

*14–52. The material hoist and the load have a total mass of 800 kg and the counterweight C has a mass of 150 kg. If the upward speed of the hoist increases uniformly from 0.5 m/s to 1.5 m/s in 1.5 s, determine the average power generated by the motor M during this time. The motor operates with an efficiency of $\epsilon = 0.8$.

M

C

Probs. 14–51/52

14

•**14–53.** The 2-Mg car increases its speed uniformly from rest to 25 m/s in 30 s up the inclined road. Determine the maximum power that must be supplied by the engine, which operates with an efficiency of $\epsilon = 0.8$. Also, find the average power supplied by the engine.

Prob. 14–53

14–54. Determine the velocity of the 200-lb crate if the motor operates with an efficiency of $\epsilon = 0.8$. The power input to the motor is 2.5 hp and the crate is moving at 10 ft/s.

Prob. 14–54

14–55. A constant power of 1.5 hp is supplied to the motor while it operates with an efficiency of $\epsilon = 0.8$. Determine the velocity of the 200-lb crate in 15 seconds, starting from rest. Neglect friction.

Prob. 14–55

14–56. The fluid transmission of a 30 000-lb truck allows the engine to deliver constant power to the rear wheels. Determine the distance required for the truck traveling on a level road to increase its speed from 35 ft/s to 60 ft/s if 90 hp is delivered to the rear wheels. Neglect drag and rolling resistance.

•**14–57.** If the engine of a 1.5-Mg car generates a constant power of 15 kW, determine the speed of the car after it has traveled a distance of 200 m on a level road starting from rest. Neglect friction.

14–58. The 1.2-Mg mine car is being pulled by the winch M mounted on the car. If the winch exerts a force of $F = (150t^{3/2})\,\text{N}$ on the cable, where t is in seconds, determine the power output of the winch when $t = 5$ s, starting from rest.

14–59. The 1.2-Mg mine car is being pulled by the winch M mounted on the car. If the winch generates a constant power output of 30 kW, determine the speed of the car at the instant it has traveled a distance of 30 m, starting from rest.

Probs. 14–58/59

14–60. The 1.2-Mg mine car is being pulled by winch M mounted on the car. If the winch generates a constant power output of 30 kW, and the car starts from rest, determine the speed of the car when $t = 5$ s.

Prob. 14–60

•14–61. The 50-lb crate is hoisted by the motor M. If the crate starts from rest and by constant acceleration attains a speed of 12 ft/s after rising $s = 10$ ft, determine the power that must be supplied to the motor at the instant $s = 10$ ft. The motor has an efficiency $\epsilon = 0.65$. Neglect the mass of the pulley and cable.

14–63. If the jet on the dragster supplies a constant thrust of $T = 20$ kN, determine the power generated by the jet as a function of time. Neglect drag and rolling resistance, and the loss of fuel. The dragster has a mass of 1 Mg and starts from rest.

Prob. 14–63

*14–64. Sand is being discharged from the silo at A to the conveyor and transported to the storage deck at the rate of 360 000 lb/h. An electric motor is attached to the conveyor to maintain the speed of the belt at 3 ft/s. Determine the average power generated by the motor.

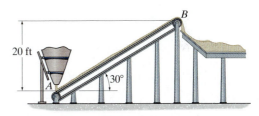

Prob. 14–64

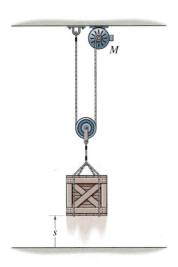

Prob. 14–61

14–62. A motor hoists a 60-kg crate at a constant velocity to a height of $h = 5$ m in 2 s. If the indicated power of the motor is 3.2 kW, determine the motor's efficiency.

14–65. The 500-kg elevator starts from rest and travels upward with a constant acceleration $a_c = 2$ m/s^2. Determine the power output of the motor M when $t = 3$ s. Neglect the mass of the pulleys and cable.

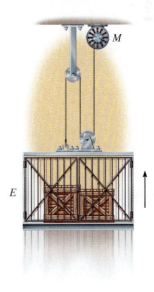

Prob. 14–65

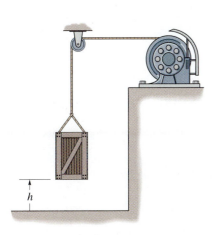

Prob. 14–62

14–66. A rocket having a total mass of 8 Mg is fired vertically from rest. If the engines provide a constant thrust of $T = 300$ kN, determine the power output of the engines as a function of time. Neglect the effect of drag resistance and the loss of fuel mass and weight.

$T = 300$ kN

Prob. 14–66

14–67. The crate has a mass of 150 kg and rests on a surface for which the coefficients of static and kinetic friction are $\mu_s = 0.3$ and $\mu_k = 0.2$, respectively. If the motor M supplies a cable force of $F = (8t^2 + 20)$ N, where t is in seconds, determine the power output developed by the motor when $t = 5$ s.

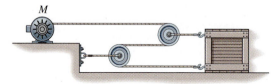

Prob. 14–67

*14–68. The 50-lb block rests on the rough surface for which the coefficient of kinetic friction is $\mu_k = 0.2$. A force $F = (40 + s^2)$ lb, where s is in ft, acts on the block in the direction shown. If the spring is originally unstretched $(s = 0)$ and the block is at rest, determine the power developed by the force the instant the block has moved $s = 1.5$ ft.

Prob. 14–68

•14–69. Using the biomechanical power curve shown, determine the maximum speed attained by the rider and his bicycle, which have a total mass of 92 kg, as the rider ascends the 20° slope starting from rest.

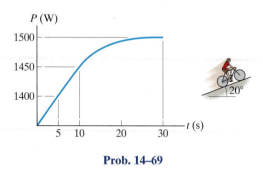

Prob. 14–69

14–70. The 50-kg crate is hoisted up the 30° incline by the pulley system and motor M. If the crate starts from rest and, by constant acceleration, attains a speed of 4 m/s after traveling 8 m along the plane, determine the power that must be supplied to the motor at the instant the crate has moved 8 m. Neglect friction along the plane. The motor has an efficiency of $\epsilon = 0.74$.

14–71. Solve Prob. 14–70 if the coefficient of kinetic friction between the plane and the crate is $\mu_k = 0.3$.

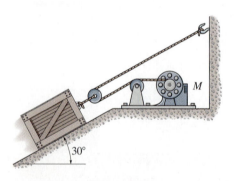

Probs. 14–70/71

14.5   Conservative Forces and Potential Energy

Conservative Force.   If the work of a force is *independent of the path* and depends only on the force's initial and final positions on the path, then we can classify this force as a *conservative force*. Examples of conservative forces are the weight of a particle and the force developed by a spring. The work done by the weight depends *only* on the *vertical displacement* of the weight, and the work done by a spring force depends *only* on the spring's *elongation* or *compression*.

In contrast to a conservative force, consider the force of friction exerted *on a sliding object* by a fixed surface. The work done by the frictional force *depends on the path*—the longer the path, the greater the work. Consequently, *frictional forces are nonconservative*. The work is dissipated from the body in the form of heat.

Energy.   Energy is defined as the capacity for doing work. For example, if a particle is originally at rest, then the principle of work and energy states that $\Sigma U_{1\rightarrow2} = T_2$. In other words, the kinetic energy is equal to the work that must be done on the particle to bring it from a state of rest to a speed v. Thus, the *kinetic energy* is a measure of the particle's *capacity to do work*, which is associated with the *motion* of the particle. When energy comes from the *position* of the particle, measured from a fixed datum or reference plane, it is called potential energy. Thus, *potential energy* is a measure of the amount of work a conservative force will do when it moves from a given position to the datum. In mechanics, the potential energy created by gravity (weight) or an elastic spring is important.

Gravitational Potential Energy.   If a particle is located a distance y *above* an arbitrarily selected datum, as shown in Fig. 14–17, the particle's weight **W** has positive *gravitational potential energy*, V_g, since **W** has the capacity of doing positive work when the particle is moved back down to the datum. Likewise, if the particle is located a distance y *below* the datum, V_g is negative since the weight does negative work when the particle is moved back up to the datum. At the datum $V_g = 0$.

In general, if y is *positive upward*, the gravitational potential energy of the particle of weight W is*

$$V_g = Wy \qquad (14\text{–}13)$$

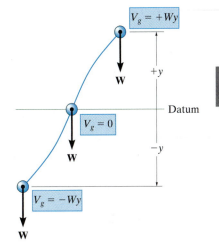

Gravitational potential energy

Fig. 14–17

---

*Here the weight is assumed to be *constant*. This assumption is suitable for small differences in elevation Δy. If the elevation change is significant, however, a variation of weight with elevation must be taken into account (see Prob. 14–16).

Elastic Potential Energy.   When an elastic spring is elongated or compressed a distance s from its unstretched position, elastic potential energy V_e can be stored in the spring. This energy is

$$V_e = +\tfrac{1}{2}ks^2 \qquad (14\text{–}14)$$

Here V_e is *always positive* since, in the deformed position, the force of the spring has the *capacity* or "potential" for always doing positive work on the particle when the spring is returned to its unstretched position, Fig. 14–18.

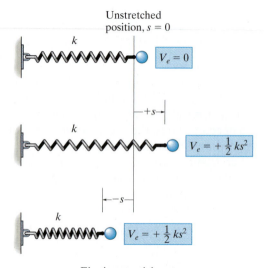

Elastic potential energy

Fig. 14–18

The weight of the sacks resting on this platform causes potential energy to be stored in the supporting springs. As each sack is removed, the platform will *rise* slightly since some of the potential energy within the springs will be transformed into an increase in gravitational potential energy of the remaining sacks. Such a device is useful for removing the sacks without having to bend over to pick them up as they are unloaded.

Potential Function.   In the general case, if a particle is subjected to both gravitational and elastic forces, the particle's potential energy can be expressed as a *potential function*, which is the algebraic sum

$$V = V_g + V_e \qquad\qquad (14\text{–}15)$$

Measurement of V depends on the location of the particle with respect to a selected datum in accordance with Eqs. 14–13 and 14–14.

The work done by a conservative force in moving the particle from one point to another point is measured by the *difference* of this function, i.e.,

$$U_{1-2} = V_1 - V_2 \qquad\qquad (14\text{–}16)$$

For example, the potential function for a particle of weight W suspended from a spring can be expressed in terms of its position, s, measured from a datum located at the unstretched length of the spring, Fig. 14–19. We have

$$V = V_g + V_e$$

$$= -Ws + \tfrac{1}{2}ks^2$$

If the particle moves from s_1 to a lower position s_2, then applying Eq. 14–16 it can be seen that the work of $\mathbf{W}$ and $\mathbf{F}_s$ is

$$U_{1-2} = V_1 - V_2 = \left(-Ws_1 + \tfrac{1}{2}ks_1^2\right) - \left(-Ws_2 + \tfrac{1}{2}ks_2^2\right)$$

$$= W(s_2 - s_1) - \left(\tfrac{1}{2}ks_2^2 - \tfrac{1}{2}ks_1^2\right)$$

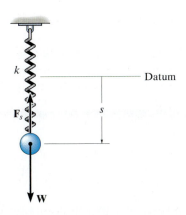

Fig. 14–19

When the displacement along the path is infinitesimal, i.e., from point (x, y, z) to $(x + dx, y + dy, z + dz)$, Eq. 14–16 becomes

$$dU = V(x, y, z) - V(x + dx, y + dy, z + dz)$$
$$= -dV(x, y, z) \tag{14–17}$$

If we represent both the force and its displacement as Cartesian vectors, then the work can also be expressed as

$$dU = \mathbf{F} \cdot d\mathbf{r} = (F_x\mathbf{i} + F_y\mathbf{j} + F_z\mathbf{k}) \cdot (dx\mathbf{i} + dy\mathbf{j} + dz\mathbf{k})$$
$$= F_x \, dx + F_y \, dy + F_z \, dz$$

Substituting this result into Eq. 14–17 and expressing the differential $dV(x, y, z)$ in terms of its partial derivatives yields

$$F_x \, dx + F_y \, dy + F_z \, dz = -\left(\frac{\partial V}{\partial x}dx + \frac{\partial V}{\partial y}dy + \frac{\partial V}{\partial z}dz\right)$$

Since changes in x, y, and z are all independent of one another, this equation is satisfied provided

$$F_x = -\frac{\partial V}{\partial x}, \qquad F_y = -\frac{\partial V}{\partial y}, \qquad F_z = -\frac{\partial V}{\partial z} \tag{14–18}$$

Thus,

$$\mathbf{F} = -\frac{\partial V}{\partial x}\mathbf{i} - \frac{\partial V}{\partial y}\mathbf{j} - \frac{\partial V}{\partial z}\mathbf{k}$$

$$= -\left(\frac{\partial}{\partial x}\mathbf{i} + \frac{\partial}{\partial y}\mathbf{j} + \frac{\partial}{\partial z}\mathbf{k}\right)V$$

or

$$\mathbf{F} = -\nabla V \tag{14–19}$$

where ∇ (del) represents the vector operator $\nabla = (\partial/\partial x)\mathbf{i} + (\partial/\partial y)\mathbf{j} + (\partial/\partial z)\mathbf{k}$.

Equation 14–19 relates a force $\mathbf{F}$ to its potential function V and thereby provides a mathematical criterion for proving that $\mathbf{F}$ is conservative. For example, the gravitational potential function for a weight located a distance y above a datum is $V_g = Wy$. To prove that $\mathbf{W}$ is conservative, it is necessary to show that it satisfies Eq. 14–18 (or Eq. 14–19), in which case

$$F_y = -\frac{\partial V}{\partial y}; \qquad\qquad F_y = -\frac{\partial}{\partial y}(Wy) = -W$$

The negative sign indicates that $\mathbf{W}$ acts downward, opposite to positive y, which is upward.

14.6   Conservation of Energy

When a particle is acted upon by a system of *both* conservative and nonconservative forces, the portion of the work done by the *conservative forces* can be written in terms of the difference in their potential energies using Eq. 14–16, i.e., $(\Sigma U_{1-2})_{cons.} = V_1 - V_2$. As a result, the principle of work and energy can be written as

$$T_1 + V_1 + (\Sigma U_{1-2})_{noncons.} = T_2 + V_2 \qquad (14\text{–}20)$$

Here $(\Sigma U_{1-2})_{noncons.}$ represents the work of the nonconservative forces acting on the particle. If *only conservative forces* do work then we have

$$\boxed{T_1 + V_1 = T_2 + V_2} \qquad (14\text{–}21)$$

This equation is referred to as the *conservation of mechanical energy* or simply the *conservation of energy*. It states that during the motion the sum of the particle's kinetic and potential energies remains *constant*. For this to occur, kinetic energy must be transformed into potential energy, and vice versa. For example, if a ball of weight **W** is dropped from a height h above the ground (datum), Fig. 14–20, the potential energy of the ball is maximum before it is dropped, at which time its kinetic energy is zero. The total mechanical energy of the ball in its initial position is thus

$$E = T_1 + V_1 = 0 + Wh = Wh$$

When the ball has fallen a distance $h/2$, its speed can be determined by using $v^2 = v_0^2 + 2a_c(y - y_0)$, which yields $v = \sqrt{2g(h/2)} = \sqrt{gh}$. The energy of the ball at the mid-height position is therefore

$$E = T_2 + V_2 = \frac{1}{2}\frac{W}{g}\left(\sqrt{gh}\right)^2 + W\left(\frac{h}{2}\right) = Wh$$

Just before the ball strikes the ground, its potential energy is zero and its speed is $v = \sqrt{2gh}$. Here, again, the total energy of the ball is

$$E = T_3 + V_3 = \frac{1}{2}\frac{W}{g}\left(\sqrt{2gh}\right)^2 + 0 = Wh$$

Note that when the ball comes in contact with the ground, it deforms somewhat, and provided the ground is hard enough, the ball will rebound off the surface, reaching a new height h', which will be *less* than the height h from which it was first released. Neglecting air friction, the difference in height accounts for an energy loss, $E_l = W(h - h')$, which occurs during the collision. Portions of this loss produce noise, localized deformation of the ball and ground, and heat.

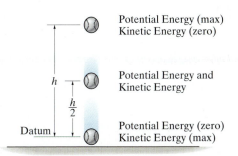

Potential Energy (max)
Kinetic Energy (zero)

h

$\dfrac{h}{2}$

Potential Energy and
Kinetic Energy

Datum

Potential Energy (zero)
Kinetic Energy (max)

Fig. 14–20

14

System of Particles. If a system of particles is *subjected only to conservative forces*, then an equation similar to Eq. 14–21 can be written for the particles. Applying the ideas of the preceding discussion, Eq. 14–8 $(\Sigma T_1 + \Sigma U_{1-2} = \Sigma T_2)$ becomes

$$\Sigma T_1 + \Sigma V_1 = \Sigma T_2 + \Sigma V_2 \tag{14–22}$$

Here, the sum of the system's initial kinetic and potential energies is equal to the sum of the system's final kinetic and potential energies. In other words, $\Sigma T + \Sigma V = $ const.

Procedure for Analysis

The conservation of energy equation can be used to solve problems involving *velocity, displacement*, and *conservative force systems*. It is generally *easier to apply* than the principle of work and energy because this equation requires specifying the particle's kinetic and potential energies at only *two points* along the path, rather than determining the work when the particle moves through a *displacement*. For application it is suggested that the following procedure be used.

Potential Energy.

- Draw two diagrams showing the particle located at its initial and final points along the path.

- If the particle is subjected to a vertical displacement, establish the fixed horizontal datum from which to measure the particle's gravitational potential energy V_g.

- Data pertaining to the elevation y of the particle from the datum and the stretch or compression s of any connecting springs can be determined from the geometry associated with the two diagrams.

- Recall $V_g = Wy$, where y is positive upward from the datum and negative downward from the datum; also for a spring, $V_e = \frac{1}{2}ks^2$, which is *always positive*.

Conservation of Energy.

- Apply the equation $T_1 + V_1 = T_2 + V_2$.

- When determining the kinetic energy, $T = \frac{1}{2}mv^2$, remember that the particle's speed v must be measured from an inertial reference frame.

EXAMPLE 14.9

The gantry structure in the photo is used to test the response of an airplane during a crash. As shown in Fig. 14–21a, the plane, having a mass of 8 Mg, is hoisted back until $\theta = 60°$, and then the pull-back cable AC is released when the plane is at rest. Determine the speed of the plane just before it crashes into the ground, $\theta = 15°$. Also, what is the maximum tension developed in the supporting cable during the motion? Neglect the size of the airplane and the effect of lift caused by the wings during the motion.

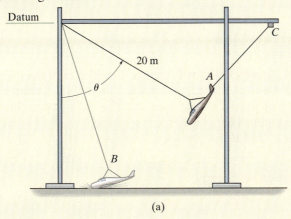

(a)

SOLUTION

Since the force of the cable does *no work* on the plane, it must be obtained using the equation of motion. First, however, we must determine the plane's speed at B.

Potential Energy. For convenience, the datum has been established at the top of the gantry, Fig. 14–21a.

(b)

Fig. 14–21

Conservation of Energy.

$$T_A + V_A = T_B + V_B$$

$$0 - 8000 \text{ kg } (9.81 \text{ m/s}^2)(20 \cos 60° \text{ m}) =$$

$$\tfrac{1}{2}(8000 \text{ kg})v_B^2 - 8000 \text{ kg } (9.81 \text{ m/s}^2)(20 \cos 15° \text{ m})$$

$$v_B = 13.52 \text{ m/s} = 13.5 \text{ m/s} \qquad Ans.$$

Equation of Motion. From the free-body diagram when the plane is at B, Fig. 14–21b, we have

$$+\nwarrow \ \Sigma F_n = ma_n;$$

$$T - (8000(9.81) \text{ N}) \cos 15° = (8000 \text{ kg})\frac{(13.52 \text{ m/s})^2}{20 \text{ m}}$$

$$T = 149 \text{ kN} \qquad Ans.$$

EXAMPLE 14.10

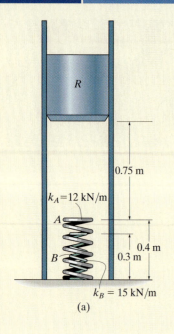

(a)

The ram R shown in Fig. 14–22a has a mass of 100 kg and is released from rest 0.75 m from the top of a spring, A, that has a stiffness $k_A = 12$ kN/m. If a second spring B, having a stiffness $k_B = 15$ kN/m, is "nested" in A, determine the maximum displacement of A needed to stop the downward motion of the ram. The unstretched length of each spring is indicated in the figure. Neglect the mass of the springs.

SOLUTION

Potential Energy. We will *assume* that the ram compresses *both* springs at the instant it comes to rest. The datum is located through the center of gravity of the ram at its initial position, Fig. 14–22b. When the kinetic energy is reduced to zero $(v_2 = 0)$, A is compressed a distance s_A and B compresses $s_B = s_A - 0.1$ m.

Conservation of Energy.

$$T_1 + V_1 = T_2 + V_2$$

$$0 + 0 = 0 + \left\{\tfrac{1}{2}k_A s_A^2 + \tfrac{1}{2}k_B(s_A - 0.1)^2 - Wh\right\}$$

$$0 + 0 = 0 + \left\{\tfrac{1}{2}(12\,000\text{ N/m})s_A^2 + \tfrac{1}{2}(15\,000\text{ N/m})(s_A - 0.1\text{ m})^2\right.$$
$$\left. - 981\text{ N }(0.75\text{ m} + s_A)\right\}$$

Rearranging the terms,

$$13\,500 s_A^2 - 2481 s_A - 660.75 = 0$$

Using the quadratic formula and solving for the positive root, we have

$$s_A = 0.331\text{ m} \qquad\qquad\qquad Ans.$$

Since $s_B = 0.331$ m $- 0.1$ m $= 0.231$ m, which is positive, the assumption that *both* springs are compressed by the ram is correct.

NOTE: The second root, $s_A = -0.148$ m, does not represent the physical situation. Since positive s is measured downward, the negative sign indicates that spring A would have to be "extended" by an amount of 0.148 m to stop the ram.

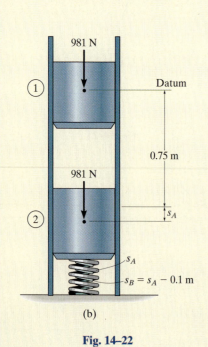

(b)

Fig. 14–22

EXAMPLE 14.11

A smooth 2-kg collar, shown in Fig. 14–23a, fits loosely on the vertical shaft. If the spring is unstretched when the collar is in the position A, determine the speed at which the collar is moving when $y = 1$ m, if (a) it is released from rest at A, and (b) it is released at A with an *upward* velocity $v_A = 2$ m/s.

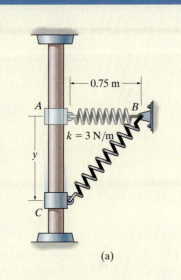

(a)

SOLUTION

Part (a) Potential Energy.   For convenience, the datum is established through AB, Fig. 14–23b. When the collar is at C, the gravitational potential energy is $-(mg)y$, since the collar is *below* the datum, and the elastic potential energy is $\frac{1}{2}ks_{CB}^2$. Here $s_{CB} = 0.5$ m, which represents the *stretch* in the spring as shown in the figure.

Conservation of Energy.

$$T_A + V_A = T_C + V_C$$
$$0 + 0 = \tfrac{1}{2}mv_C^2 + \left\{\tfrac{1}{2}ks_{CB}^2 - mgy\right\}$$
$$0 + 0 = \left\{\tfrac{1}{2}(2 \text{ kg})v_C^2\right\} + \left\{\tfrac{1}{2}(3 \text{ N/m})(0.5 \text{ m})^2 - 2(9.81) \text{ N } (1 \text{ m})\right\}$$
$$v_C = 4.39 \text{ m/s} \downarrow \qquad\qquad Ans.$$

This problem can also be solved by using the equation of motion or the principle of work and energy. Note that for *both* of these methods the variation of the magnitude and direction of the spring force must be taken into account (see Example 13.4). Here, however, the above solution is clearly advantageous since the calculations depend *only* on data calculated at the initial and final points of the path.

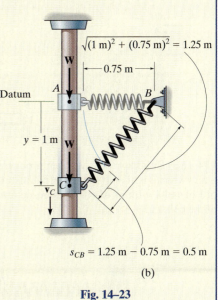

$s_{CB} = 1.25 \text{ m} - 0.75 \text{ m} = 0.5 \text{ m}$

(b)

Fig. 14–23

Part (b) Conservation of Energy.   If $v_A = 2$ m/s, using the data in Fig. 14–23b, we have

$$T_A + V_A = T_C + V_C$$
$$\tfrac{1}{2}mv_A^2 + 0 = \tfrac{1}{2}mv_C^2 + \left\{\tfrac{1}{2}ks_{CB}^2 - mgy\right\}$$
$$\tfrac{1}{2}(2 \text{ kg})(2 \text{ m/s})^2 + 0 = \tfrac{1}{2}(2 \text{ kg})v_C^2 + \left\{\tfrac{1}{2}(3 \text{ N/m})(0.5 \text{ m})^2\right.$$
$$\left. - 2(9.81) \text{ N } (1 \text{ m})\right\}$$
$$v_C = 4.82 \text{ m/s} \downarrow \qquad\qquad Ans.$$

NOTE: The kinetic energy of the collar depends only on the *magnitude* of velocity, and therefore it is immaterial if the collar is moving up or down at 2 m/s when released at A.

FUNDAMENTAL PROBLEMS

F14–13. The 2-kg pendulum bob is released from rest when it is at A. Determine the speed of the bob and the tension in the cord when the bob passes through its lowest position, B.

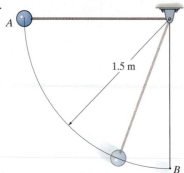

F14–13

F14–14. The 2-kg package leaves the conveyor belt at A with a speed of $v_A = 1$ m/s and slides down the smooth ramp. Determine the required speed of the conveyor belt at B so that the package can be delivered without slipping on the belt. Also, find the normal reaction the curved portion of the ramp exerts on the package at B if $\rho_B = 2$ m.

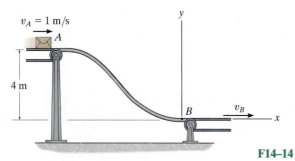

F14–14

F14–15. The 2-kg collar is given a downward velocity of 4 m/s when it is at A. If the spring has an unstretched length of 1 m and a stiffness of $k = 30$ N/m, determine the velocity of the collar at $s = 1$ m.

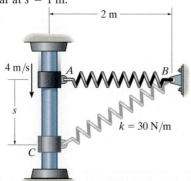

F14–15

F14–16. The 5-lb collar is released from rest at A and travels along the frictionless guide. Determine the speed of the collar when it strikes the stop B. The spring has an unstretched length of 0.5 ft.

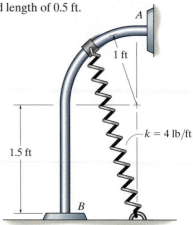

F14–16

F14–17. The 75-lb block is released from rest 5 ft above the plate. Determine the compression of each spring when the block momentarily comes to rest after striking the plate. Neglect the mass of the plate. The springs are initially unstretched.

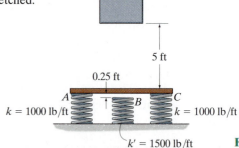

F14–17

F14–18. The 4-kg collar C has a velocity of $v_A = 2$ m/s when it is at A. If the guide rod is smooth, determine the speed of the collar when it is at B. The spring has an unstretched length of $l_0 = 0.2$ m.

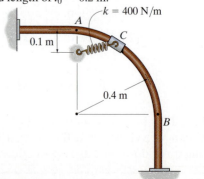

F14–18

PROBLEMS

*14–72. Solve Prob. 14–12 using the conservation of energy equation.

•14–73. Solve Prob. 14–7 using the conservation of energy equation.

14–74. Solve Prob. 14–8 using the conservation of energy equation.

14–75. Solve Prob. 14–18 using the conservation of energy equation.

*14–76. Solve Prob. 14–22 using the conservation of energy equation.

•14–77. Each of the two elastic rubber bands of the slingshot has an unstretched length of 200 mm. If they are pulled back to the position shown and released from rest, determine the speed of the 25-g pellet just after the rubber bands become unstretched. Neglect the mass of the rubber bands. Each rubber band has a stiffness of $k = 50$ N/m.

14.78. Each of the two elastic rubber bands of the slingshot has an unstretched length of 200 mm. If they are pulled back to the position shown and released from rest, determine the maximum height the 25-g pellet will reach if it is fired vertically upward. Neglect the mass of the rubber bands and the change in elevation of the pellet while it is constrained by the rubber bands. Each rubber band has a stiffness $k = 50$ N/m.

14–79. Block A has a weight of 1.5 lb and slides in the smooth horizontal slot. If the block is drawn back to $s = 1.5$ ft and released from rest, determine its speed at the instant $s = 0$. Each of the two springs has a stiffness of $k = 150$ lb/ft and an unstretched length of 0.5 ft.

*14–80. The 2-lb block A slides in the smooth horizontal slot. When $s = 0$ the block is given an initial velocity of 60 ft/s to the right. Determine the maximum horizontal displacement s of the block. Each of the two springs has a stiffness of $k = 150$ lb/ft and an unstretched length of 0.5 ft.

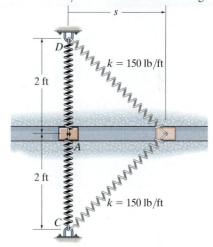

Probs. 14–79/80

•14–81. The 30-lb block A is placed on top of two nested springs B and C and then pushed down to the position shown. If it is then released, determine the maximum height h to which it will rise.

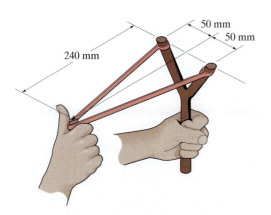

Probs. 14–77/78

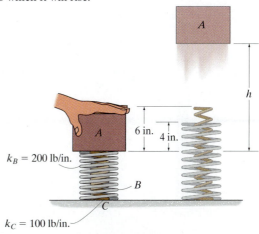

Prob. 14–81

14–82. The spring is unstretched when $s = 1\text{m}$ and the 15-kg block is released from rest at this position. Determine the speed of the block when $s = 3\text{m}$. The spring remains horizontal during the motion, and the contact surfaces between the block and the inclined plane are smooth.

*14–84.** The 5-kg collar slides along the smooth vertical rod. If the collar is nudged from rest at A, determine its speed when it passes point B. The spring has an unstretched length of 200 mm.

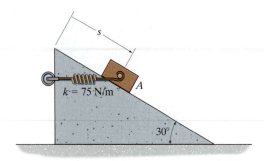

Prob. 14–82

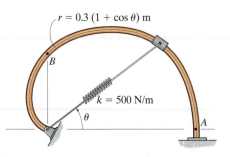

Prob. 14–84

14–83. The vertical guide is smooth and the 5-kg collar is released from rest at A. Determine the speed of the collar when it is at position C. The spring has an unstretched length of 300 mm.

•**14–85.** The cylinder has a mass of 20 kg and is released from rest when $h = 0$. Determine its speed when $h = 3$ m. The springs each have an unstretched length of 2 m.

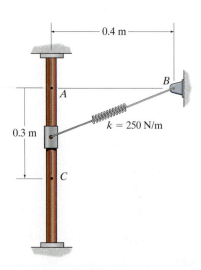

Prob. 14–83

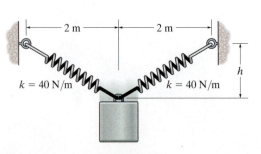

Prob. 14–85

14–86. Tarzan has a mass of 100 kg and from rest swings from the cliff by rigidly holding on to the tree vine, which is 10 m measured from the supporting limb A to his center of mass. Determine his speed just after the vine strikes the lower limb at B. Also, with what force must he hold on to the vine just before and just after the vine contacts the limb at B?

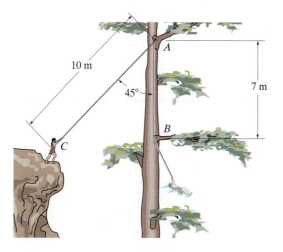

Prob. 14–86

14–87. The roller-coaster car has a mass of 800 kg, including its passenger, and starts from the top of the hill A with a speed $v_A = 3$ m/s. Determine the minimum height h of the hill so that the car travels around both inside loops without leaving the track. Neglect friction, the mass of the wheels, and the size of the car. What is the normal reaction on the car when the car is at B and at C?

*14–88. The roller-coaster car has a mass of 800 kg, including its passenger. If it is released from rest at the top of the hill A, determine the minimum height h of the hill so that the car travels around both inside loops without leaving the track. Neglect friction, the mass of the wheels, and the size of the car. What is the normal reaction on the car when the car is at B and at C?

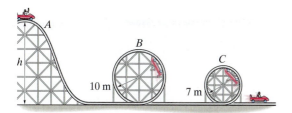

Probs. 14–87/88

•14–89. The roller coaster and its passenger have a total mass m. Determine the smallest velocity it must have when it enters the loop at A so that it can complete the loop and not leave the track. Also, determine the normal force the tracks exert on the car when it comes around to the bottom at C. The radius of curvature of the tracks at B is ρ_B, and at C it is ρ_C. Neglect the size of the car. Points A and C are at the same elevation.

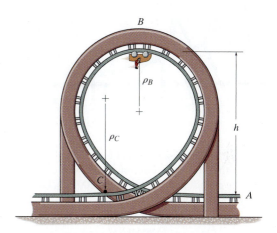

Prob. 14–89

14–90. The 0.5-lb ball is shot from the spring device. The spring has a stiffness $k = 10$ lb/in. and the four cords C and plate P keep the spring compressed 2 in. when no load is on the plate. The plate is pushed back 3 in. from its initial position. If it is then released from rest, determine the speed of the ball when it reaches a position $s = 30$ in. on the smooth inclined plane.

14–91. The 0.5-lb ball is shot from the spring device shown. Determine the smallest stiffness k which is required to shoot the ball a maximum distance $s = 30$ in. up the plane after the spring is pushed back 3 in. and the ball is released from rest. The four cords C and plate P keep the spring compressed 2 in. when no load is on the plate.

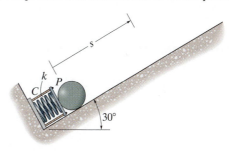

Probs. 14–90/91

14

*14–92. The roller coaster car having a mass m is released from rest at point A. If the track is to be designed so that the car does not leave it at B, determine the required height h. Also, find the speed of the car when it reaches point C. Neglect friction.

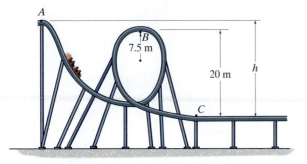

Prob. 14–92

•14–93. When the 50-kg cylinder is released from rest, the spring is subjected to a tension of 60 N. Determine the speed of the cylinder after it has fallen 200 mm. How far has it fallen when it momentarily stops?

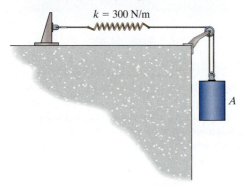

Prob. 14–93

14–94. A pan of negligible mass is attached to two identical springs of stiffness $k = 250$ N/m. If a 10-kg box is dropped from a height of 0.5 m above the pan, determine the maximum vertical displacement d. Initially each spring has a tension of 50 N.

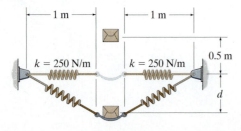

Prob. 14–94

14–95. The man on the bicycle attempts to coast around the ellipsoidal loop without falling off the track. Determine the speed he must maintain at A just before entering the loop in order to perform the stunt. The bicycle and man have a total mass of 85 kg and a center of mass at G. Neglect the mass of the wheels.

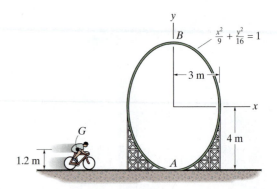

Prob. 14–95

*14–96. The 65-kg skier starts from rest at A. Determine his speed at B and the distance s where he lands at C. Neglect friction.

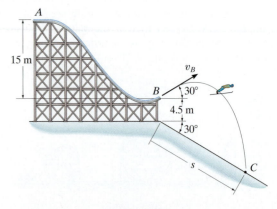

Prob. 14–96

•**14–97.**   The 75-kg man bungee jumps off the bridge at A with an initial downward speed of 1.5 m/s. Determine the required unstretched length of the elastic cord to which he is attached in order that he stops momentarily just above the surface of the water. The stiffness of the elastic cord is $k = 3$ kN/m. Neglect the size of the man.

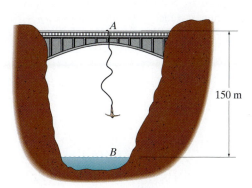

Prob. 14–97

14–98.   The 10-kg block A is released from rest and slides down the smooth plane. Determine the compression x of the spring when the block momentarily stops.

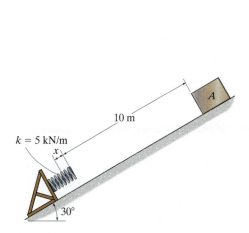

Prob. 14–98

14–99.   The 20-lb smooth collar is attached to the spring that has an unstretched length of 4 ft. If it is released from rest at position A, determine its speed when it reaches point B.

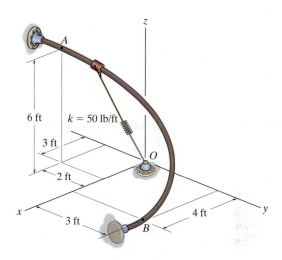

Prob. 14–99

***14–100.**   The 2-kg collar is released from rest at A and travels along the smooth vertical guide. Determine the speed of the collar when it reaches position B. Also, find the normal force exerted on the collar at this position. The spring has an unstretched length of 200 mm.

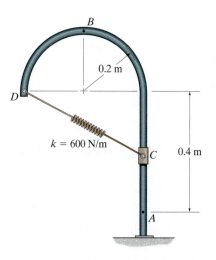

Prob. 14–100

•14–101. A quarter-circular tube AB of mean radius r contains a smooth chain that has a mass per unit length of m_0. If the chain is released from rest from the position shown, determine its speed when it emerges completely from the tube.

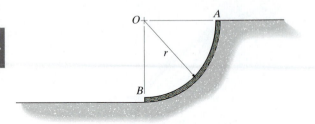

Prob. 14–101

14–102. The ball of mass m is given a speed of $v_A = \sqrt{3gr}$ at position A. When it reaches B, the cord hits the small peg P, after which the ball describes a smaller circular path. Determine the position x of P so that the ball will just be able to reach point C.

14–103. The ball of mass m is given a speed of $v_A = \sqrt{5gr}$ at position A. When it reaches B, the cord hits the peg P, after which the ball describes a smaller circular path. If $x = \frac{2}{3}r$, determine the speed of the ball and the tension in the cord when it is at the highest point C.

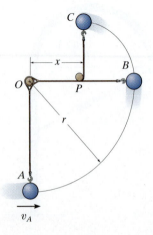

Probs. 14–102/103

***14–104.** If the mass of the earth is M_e, show that the gravitational potential energy of a body of mass m located a distance r from the center of the earth is $V_g = -GM_em/r$. Recall that the gravitational force acting between the earth and the body is $F = G(M_em/r^2)$, Eq. 13–1. For the calculation, locate the datum an "infinite" distance from the earth. Also, prove that $\mathbf{F}$ is a conservative force.

•14–105. A 60-kg satellite travels in free flight along an elliptical orbit such that at A, where $r_A = 20$ Mm, it has a speed $v_A = 40$ Mm/h. What is the speed of the satellite when it reaches point B, where $r_B = 80$ Mm? *Hint:* See Prob. 14–104, where $M_e = 5.976(10^{24})$ kg and $G = 66.73(10^{-12})$ m³/(kg · s²).

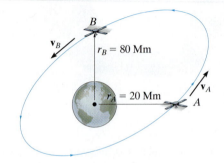

Prob. 14–105

14–106. The double-spring bumper is used to stop the 1500-lb steel billet in the rolling mill. Determine the maximum displacement of the plate A if the billet strikes the plate with a speed of 8 ft/s. Neglect the mass of the springs, rollers and the plates A and B. Take $k_1 = 3000$ lb/ft, $k_2 = 4500$ lb/ft.

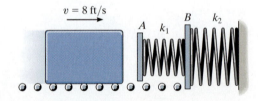

Prob. 14–106

CONCEPTUAL PROBLEMS

P14–1. The roller coaster is momentarily at rest at *A*. Determine the approximate normal force it exerts on the track at *B*. Also determine its approximate acceleration at this point. Use numerical data, and take scaled measurements from the photo with a known height at *A*.

P14–1

P14–2. As the large ring rotates, the operator can apply a breaking mechanism that binds the cars to the ring, which then allows the cars to rotate with the ring. Assuming the passengers are not belted into the cars, determine the smallest speed of the ring (cars) so that no passenger will fall out. When should the operator release the brake so that the cars can achieve their greatest speed as they slide freely on the ring? Estimate the greatest normal force of the seat on a passenger when this speed is reached. Use numerical values to explain your answer.

P14–2

P14–3. The boy pulls the water balloon launcher back, stretching each of the four elastic cords. Estimate the maximum height and the maximum range of the water ballon if it is released from the position shown. Use numerical values and any necessary measurements from the photo. Assume the unstretched length and stiffness of each cord is known.

P14–3

P14–4. The girl is momentarily at rest in the position shown. If the unstretched length and stiffness of each of the two elastic cords is known, determine approximately how far the girl descends before she again becomes momentarily at rest. Use numerical values and take any necessary measurements from the photo.

P14–4

14

CHAPTER REVIEW

Work of a Force

A force does work when it undergoes a displacement along its line of action. If the force varies with the displacement, then the work is $U = \int F \cos \theta \, ds$.

Graphically, this represents the area under the F–s diagram.

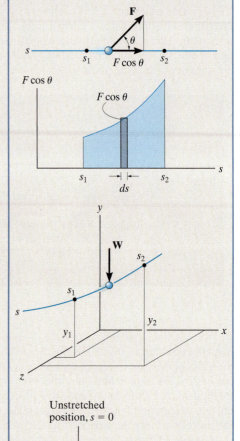

If the force is constant, then for a displacement Δs in the direction of the force, $U = F_c \, \Delta s$. A typical example of this case is the work of a weight, $U = -W \, \Delta y$. Here, Δy is the vertical displacement.

The work done by a spring force, $F = ks$, depends upon the stretch or compression s of the spring.

$$U = \tfrac{1}{2}ks_2^2 - \tfrac{1}{2}ks_1^2$$

The Principle of Work and Energy

If the equation of motion in the tangential direction, $\Sigma F_t = ma_t$, is combined with the kinematic equation, $a_t \, ds = v \, dv$, we obtain the principle of work and energy. This equation states that the initial kinetic energy T, plus the work done ΣU_{1-2} is equal to the final kinetic energy.

$$T_1 + \Sigma U_{1\text{-}2} = T_2$$

14

The principle of work and energy is useful for solving problems that involve force, velocity, and displacement. For application, the free-body diagram of the particle should be drawn in order to identify the forces that do work.

Power and Efficiency

Power is the time rate of doing work. For application, the force $\mathbf{F}$ creating the power and its velocity $\mathbf{v}$ must be specified.

Efficiency represents the ratio of power output to power input. Due to frictional losses, it is always less than one.

$$P = \frac{dU}{dt}$$

$$P = \mathbf{F} \cdot \mathbf{v}$$

$$\epsilon = \frac{\text{power output}}{\text{power input}}$$

Conservation of Energy

A conservative force does work that is independent of its path. Two examples are the weight of a particle and the spring force.

Friction is a nonconservative force since the work depends upon the length of the path. The longer the path, the more work done.

The work done by a conservative force depends upon its position relative to a datum. When this work is referenced from a datum, it is called potential energy. For a weight, it is $V_g = \pm Wy$, and for a spring it is $V_e = +\frac{1}{2}kx^2$.

Mechanical energy consists of kinetic energy T and gravitational and elastic potential energies V. According to the conservation of energy, this sum is constant and has the same value at any position on the path. If only gravitational and spring forces cause motion of the particle, then the conservation-of-energy equation can be used to solve problems involving these conservative forces, displacement, and velocity.

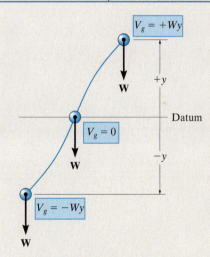

Gravitational potential energy

Elastic potential energy

$$T_1 + V_1 = T_2 + V_2$$

Impulse and momentum principles are required to predict the motion of this golf ball.

Kinetics of a Particle: Impulse and Momentum

15

CHAPTER OBJECTIVES

- To develop the principle of linear impulse and momentum for a particle and apply it to solve problems that involve force, velocity, and time.

- To study the conservation of linear momentum for particles.

- To analyze the mechanics of impact.

- To introduce the concept of angular impulse and momentum.

- To solve problems involving steady fluid streams and propulsion with variable mass.

15.1 Principle of Linear Impulse and Momentum

In this section we will integrate the equation of motion with respect to time and thereby obtain the principle of impulse and momentum. The resulting equation will be useful for solving problems involving force, velocity, and time.

Using kinematics, the equation of motion for a particle of mass m can be written as

$$\Sigma \mathbf{F} = m\mathbf{a} = m\frac{d\mathbf{v}}{dt} \tag{15–1}$$

where $\mathbf{a}$ and $\mathbf{v}$ are both measured from an inertial frame of reference. Rearranging the terms and integrating between the limits $\mathbf{v} = \mathbf{v}_1$ at $t = t_1$ and $\mathbf{v} = \mathbf{v}_2$ at $t = t_2$, we have

$$\Sigma \int_{t_1}^{t_2} \mathbf{F}\,dt = m \int_{\mathbf{v}_1}^{\mathbf{v}_2} d\mathbf{v}$$

or

$$\Sigma \int_{t_1}^{t_2} \mathbf{F} dt = m\mathbf{v}_2 - m\mathbf{v}_1 \qquad (15\text{–}2)$$

This equation is referred to as the *principle of linear impulse and momentum*. From the derivation it can be seen that it is simply a time integration of the equation of motion. It provides a *direct means* of obtaining the particle's final velocity $\mathbf{v}_2$ after a specified time period when the particle's initial velocity is known and the forces acting on the particle are either constant or can be expressed as functions of time. By comparison, if $\mathbf{v}_2$ was determined using the equation of motion, a two-step process would be necessary; i.e., apply $\Sigma \mathbf{F} = m\mathbf{a}$ to obtain $\mathbf{a}$, then integrate $\mathbf{a} = d\mathbf{v}/dt$ to obtain $\mathbf{v}_2$.

Linear Momentum.   Each of the two vectors of the form $\mathbf{L} = m\mathbf{v}$ in Eq. 15–2 is referred to as the particle's linear momentum. Since m is a positive scalar, the linear-momentum vector has the same direction as $\mathbf{v}$, and its magnitude mv has units of mass–velocity, e.g., $\text{kg} \cdot \text{m/s}$, or $\text{slug} \cdot \text{ft/s}$.

Linear Impulse.   The integral $\mathbf{I} = \int \mathbf{F} \, dt$ in Eq. 15–2 is referred to as the *linear impulse*. This term is a vector quantity which measures the effect of a force during the time the force acts. Since time is a positive scalar, the impulse acts in the same direction as the force, and its magnitude has units of force–time, e.g., $\text{N} \cdot \text{s}$ or $\text{lb} \cdot \text{s}$.*

If the force is expressed as a function of time, the impulse can be determined by direct evaluation of the integral. In particular, if the force is constant in both magnitude and direction, the resulting impulse becomes

$$\mathbf{I} = \int_{t_1}^{t_2} \mathbf{F}_c dt = \mathbf{F}_c(t_2 - t_1).$$

Graphically the magnitude of the impulse can be represented by the shaded area under the curve of force versus time, Fig. 15–1. A constant force creates the shaded rectangular area shown in Fig. 15–2.

The impulse tool is used to remove the dent in the trailer fender. To do so its end is first screwed into a hole drilled in the fender, then the weight is gripped and jerked upwards, striking the stop ring. The impulse developed is transferred along the shaft of the tool and pulls suddenly on the dent.

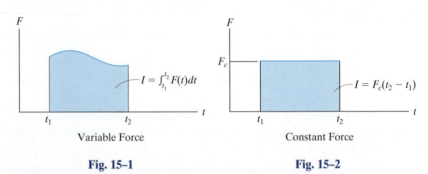

Variable Force

Constant Force

Fig. 15–1

Fig. 15–2

*Although the units for impulse and momentum are defined differently, it can be shown that Eq. 15–2 is dimensionally homogeneous.

Principle of Linear Impulse and Momentum.   For problem solving, Eq. 15–2 will be rewritten in the form

$$m\mathbf{v}_1 + \Sigma \int_{t_1}^{t_2} \mathbf{F}\, dt = m\mathbf{v}_2 \qquad (15\text{–}3)$$

which states that the initial momentum of the particle at time t_1 plus the sum of all the impulses applied to the particle from t_1 to t_2 is equivalent to the final momentum of the particle at time t_2. These three terms are illustrated graphically on the *impulse and momentum diagrams* shown in Fig. 15–3. The two *momentum diagrams* are simply outlined shapes of the particle which indicate the direction and magnitude of the particle's initial and final momenta, $m\mathbf{v}_1$ and $m\mathbf{v}_2$. Similar to the free-body diagram, the *impulse diagram* is an outlined shape of the particle showing all the impulses that act on the particle when it is located at some intermediate point along its path.

   If each of the vectors in Eq. 15–3 is resolved into its x, y, z components, we can write the following three scalar equations of linear impulse and momentum.

$$m(v_x)_1 + \Sigma \int_{t_1}^{t_2} F_x\, dt = m(v_x)_2$$

$$m(v_y)_1 + \Sigma \int_{t_1}^{t_2} F_y\, dt = m(v_y)_2 \qquad (15\text{–}4)$$

$$m(v_z)_1 + \Sigma \int_{t_1}^{t_2} F_z\, dt = m(v_z)_2$$

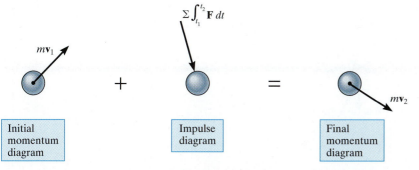

Fig. 15–3

15

Procedure for Analysis

The principle of linear impulse and momentum is used to solve problems involving *force, time,* and *velocity,* since these terms are involved in the formulation. For application it is suggested that the following procedure be used.*

Free-Body Diagram.

- Establish the *x, y, z* inertial frame of reference and draw the particle's free-body diagram in order to account for all the forces that produce impulses on the particle.

- The direction and sense of the particle's initial and final velocities should be established.

- If a vector is unknown, assume that the sense of its components is in the direction of the positive inertial coordinate(s).

- As an alternative procedure, draw the impulse and momentum diagrams for the particle as discussed in reference to Fig. 15–3.

Principle of Impulse and Momentum.

- In accordance with the established coordinate system, apply the principle of linear impulse and momentum, $m\mathbf{v}_1 + \Sigma \int_{t_1}^{t_2} \mathbf{F}\, dt = m\mathbf{v}_2$. If motion occurs in the *x–y* plane, the two scalar component equations can be formulated by either resolving the vector components of $\mathbf{F}$ from the free-body diagram, or by using the data on the impulse and momentum diagrams.

- Realize that every force acting on the particle's free-body diagram will create an impulse, even though some of these forces will do no work.

- Forces that are functions of time must be integrated to obtain the impulse. Graphically, the impulse is equal to the area under the force–time curve.

As the wheels of the pitching machine rotate, they apply frictional impulses to the ball, thereby giving it a linear momentum. These impulses are shown on the impulse diagram. Here both the frictional and normal impulses vary with time. By comparison, the weight impulse is constant and is very small since the time Δt the ball is in contact with the wheels is very small.

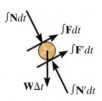

*This procedure will be followed when developing the proofs and theory in the text.

EXAMPLE 15.1

The 100-kg stone shown in Fig. 15–4a is originally at rest on the smooth horizontal surface. If a towing force of 200 N, acting at an angle of 45°, is applied to the stone for 10 s, determine the final velocity and the normal force which the surface exerts on the stone during this time interval.

(a)

SOLUTION

This problem can be solved using the principle of impulse and momentum since it involves force, velocity, and time.

Free-Body Diagram.   See Fig. 15–4b. Since all the forces acting are *constant*, the impulses are simply the product of the force magnitude and 10 s $[\mathbf{I} = \mathbf{F}_c(t_2 - t_1)]$. Note the alternative procedure of drawing the stone's impulse and momentum diagrams, Fig. 15–4c.

Principle of Impulse and Momentum.   Applying Eqs. 15–4 yields

$(\xrightarrow{+})$

$$m(v_x)_1 + \Sigma \int_{t_1}^{t_2} F_x \, dt = m(v_x)_2$$

$$0 + 200 \text{ N} \cos 45°(10 \text{ s}) = (100 \text{ kg})v_2$$

$$v_2 = 14.1 \text{ m/s} \qquad Ans.$$

$(+\uparrow)$

$$m(v_y)_1 + \Sigma \int_{t_1}^{t_2} F_y \, dt = m(v_y)_2$$

$$0 + N_C(10 \text{ s}) - 981 \text{ N}(10 \text{ s}) + 200 \text{ N} \sin 45°(10 \text{ s}) = 0$$

$$N_C = 840 \text{ N} \qquad Ans.$$

(b)

NOTE: Since no motion occurs in the *y* direction, direct application of the equilibrium equation $\Sigma F_y = 0$ gives the same result for N_C.

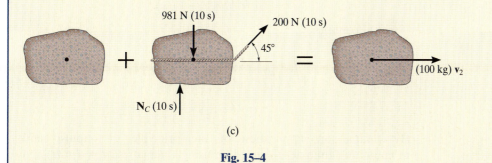

(c)

Fig. 15–4

EXAMPLE 15.2

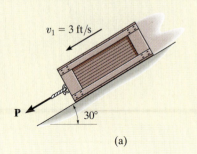

$v_1 = 3$ ft/s

P

30°

(a)

15

The 50-lb crate shown in Fig. 15–5a is acted upon by a force having a variable magnitude $P = (20t)$ lb, where t is in seconds. Determine the crate's velocity 2 s after **P** has been applied. The initial velocity is $v_1 = 3$ ft/s down the plane, and the coefficient of kinetic friction between the crate and the plane is $\mu_k = 0.3$.

SOLUTION

Free-Body Diagram. See Fig. 15–5b. Since the magnitude of force $P = 20t$ varies with time, the impulse it creates must be determined by integrating over the 2-s time interval.

Principle of Impulse and Momentum. Applying Eqs. 15–4 in the x direction, we have

$$(+\swarrow) \qquad m(v_x)_1 + \Sigma \int_{t_1}^{t_2} F_x\, dt = m(v_x)_2$$

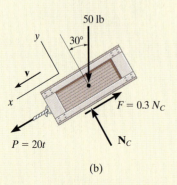

50 lb

y

30°

v

x

$F = 0.3 N_C$

$P = 20t$

N_C

(b)

Fig. 15–5

$$\frac{50 \text{ lb}}{32.2 \text{ ft/s}^2}(3 \text{ ft/s}) + \int_0^{2\text{ s}} 20t\, dt - 0.3N_C(2 \text{ s}) + (50 \text{ lb}) \sin 30°(2 \text{ s}) = \frac{50 \text{ lb}}{32.2 \text{ ft/s}^2}v_2$$

$$4.658 + 40 - 0.6N_C + 50 = 1.553v_2$$

The equation of equilibrium can be applied in the y direction. Why?

$$+\nwarrow\Sigma F_y = 0; \qquad N_C - 50 \cos 30° \text{ lb} = 0$$

Solving,

$$N_C = 43.30 \text{ lb}$$
$$v_2 = 44.2 \text{ ft/s}\swarrow \qquad\qquad Ans.$$

NOTE: We can also solve this problem using the equation of motion. From Fig. 15–5b,

$$+\swarrow\Sigma F_x = ma_x; \quad 20t - 0.3(43.30) + 50 \sin 30° = \frac{50}{32.2}a$$
$$a = 12.88t + 7.734$$

Using kinematics

$$+\swarrow dv = a\, dt; \qquad \int_{3 \text{ ft/s}}^{v} dv = \int_0^{2\text{ s}} (12.88t + 7.734)dt$$

$$v = 44.2 \text{ ft/s} \qquad\qquad Ans.$$

By comparison, application of the principle of impulse and momentum eliminates the need for using kinematics ($a = dv/dt$) and thereby yields an easier method for solution.

EXAMPLE 15.3

Blocks A and B shown in Fig. 15–6a have a mass of 3 kg and 5 kg, respectively. If the system is released from rest, determine the velocity of block B in 6 s. Neglect the mass of the pulleys and cord.

SOLUTION

Free-Body Diagram. See Fig. 15–6b. Since the weight of each block is constant, the cord tensions will also be constant. Furthermore, since the mass of pulley D is neglected, the cord tension $T_A = 2T_B$. Note that the blocks are both assumed to be moving downward in the positive coordinate directions, s_A and s_B.

Principle of Impulse and Momentum.

Block A:

$(+\downarrow)$
$$m(v_A)_1 + \Sigma \int_{t_1}^{t_2} F_y \, dt = m(v_A)_2$$

$$0 - 2T_B(6 \text{ s}) + 3(9.81) \text{ N}(6 \text{ s}) = (3 \text{ kg})(v_A)_2 \qquad (1)$$

Block B:

$(+\downarrow)$
$$m(v_B)_1 + \Sigma \int_{t_1}^{t_2} F_y \, dt = m(v_B)_2$$

$$0 + 5(9.81) \text{ N}(6 \text{ s}) - T_B(6 \text{ s}) = (5 \text{ kg})(v_B)_2 \qquad (2)$$

Kinematics. Since the blocks are subjected to dependent motion, the velocity of A can be related to that of B by using the kinematic analysis discussed in Sec. 12.9. A horizontal datum is established through the fixed point at C, Fig. 15–6a, and the position coordinates, s_A and s_B, are related to the constant total length l of the vertical segments of the cord by the equation

$$2s_A + s_B = l$$

Taking the time derivative yields

$$2v_A = -v_B \qquad (3)$$

As indicated by the negative sign, when B moves downward A moves upward. Substituting this result into Eq. 1 and solving Eqs. 1 and 2 yields

$$(v_B)_2 = 35.8 \text{ m/s} \downarrow \qquad \textit{Ans.}$$

$$T_B = 19.2 \text{ N}$$

NOTE: Realize that the *positive* (downward) direction for $\mathbf{v}_A$ and $\mathbf{v}_B$ is *consistent* in Figs. 15–6a and 15–6b and in Eqs. 1 to 3. This is important since we are seeking a simultaneous solution of equations.

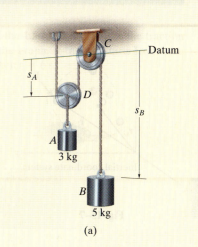

(a)

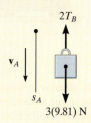

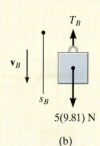

(b)

Fig. 15–6

15

•15–21. The 40-kg slider block is moving to the right with a speed of 1.5 m/s when it is acted upon by the forces $\mathbf{F}_1$ and $\mathbf{F}_2$. If these loadings vary in the manner shown on the graph, determine the speed of the block at $t = 6$ s. Neglect friction and the mass of the pulleys and cords.

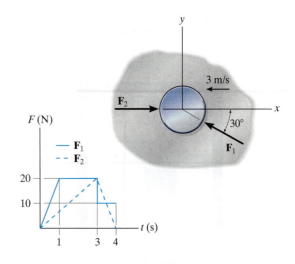

Prob. 15–21

15–22. At the instant the cable fails, the 200-lb crate is traveling up the plane with a speed of 15 ft/s. Determine the speed of the crate 2 s afterward. The coefficient of kinetic friction between the crate and the plane is $\mu_k = 0.20$.

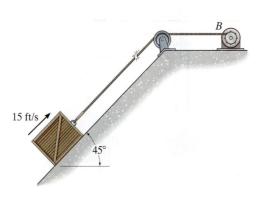

Prob. 15–22

15–23. Forces $\mathbf{F}_1$ and $\mathbf{F}_2$ vary as shown by the graph. The 5-kg smooth disk is traveling to the left with a speed of 3 m/s when $t = 0$. Determine the magnitude and direction of the disk's velocity when $t = 4$ s.

Prob. 15–23

***15–24.** A 0.5-kg particle is acted upon by the force $\mathbf{F} = \{2t^2\mathbf{i} - (3t + 3)\mathbf{j} + (10 - t^2)\mathbf{k}\}$ N, where t is in seconds. If the particle has an initial velocity of $\mathbf{v}_0 = \{5\mathbf{i} + 10\mathbf{j} + 20\mathbf{k}\}$ m/s, determine the magnitude of the velocity of the particle when $t = 3$ s.

•15–25. The train consists of a 30-Mg engine E, and cars A, B, and C, which have a mass of 15 Mg, 10 Mg, and 8 Mg, respectively. If the tracks provide a traction force of $F = 30$ kN on the engine wheels, determine the speed of the train when $t = 30$ s, starting from rest. Also, find the horizontal coupling force at D between the engine E and car A. Neglect rolling resistance.

Prob. 15–25

15–26.   The motor M pulls on the cable with a force of $\mathbf{F}$, which has a magnitude that varies as shown on the graph. If the 20-kg crate is originally resting on the floor such that the cable tension is zero at the instant the motor is turned on, determine the speed of the crate when $t = 6$ s. *Hint:* First determine the time needed to begin lifting the crate.

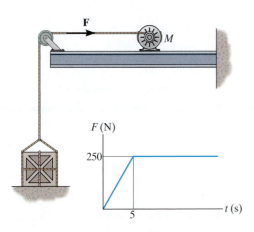

Prob. 15–26

15–27.   The winch delivers a horizontal towing force $\mathbf{F}$ to its cable at A which varies as shown in the graph. Determine the speed of the 70-kg bucket when $t = 18$ s. Originally the bucket is moving upward at $v_1 = 3$ m/s.

*15–28.   The winch delivers a horizontal towing force $\mathbf{F}$ to its cable at A which varies as shown in the graph. Determine the speed of the 80-kg bucket when $t = 24$ s. Originally the bucket is released from rest.

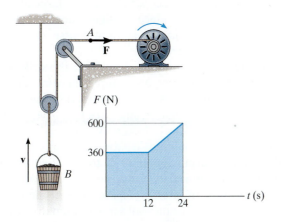

Probs. 15–27/28

•15–29.   The 0.1-lb golf ball is struck by the club and then travels along the trajectory shown. Determine the average impulsive force the club imparts on the ball if the club maintains contact with the ball for 0.5 ms.

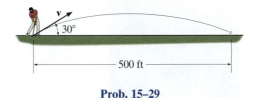

Prob. 15–29

15–30.   The 0.15-kg baseball has a speed of $v = 30$ m/s just before it is struck by the bat. It then travels along the trajectory shown before the outfielder catches it. Determine the magnitude of the average impulsive force imparted to the ball if it is in contact with the bat for 0.75 ms.

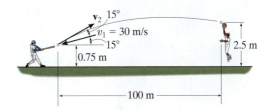

Prob. 15–30

15–31.   The 50-kg block is hoisted up the incline using the cable and motor arrangement shown. The coefficient of kinetic friction between the block and the surface is $\mu_k = 0.4$. If the block is initially moving up the plane at $v_0 = 2$ m/s, and at this instant ($t = 0$) the motor develops a tension in the cord of $T = (300 + 120\sqrt{t})$ N, where t is in seconds, determine the velocity of the block when $t = 2$ s.

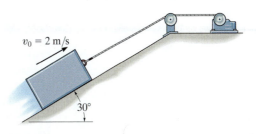

Prob. 15–31

15

15

The hammer in the top photo applies an impulsive force to the stake. During this extremely short time of contact the weight of the stake can be considered nonimpulsive, and provided the stake is driven into soft ground, the impulse of the ground acting on the stake can also be considered nonimpulsive. By contrast, if the stake is used in a concrete chipper to break concrete, then two impulsive forces act on the stake: one at its top due to the chipper and the other on its bottom due to the rigidity of the concrete.

15.3   Conservation of Linear Momentum for a System of Particles

When the sum of the *external impulses* acting on a system of particles is *zero*, Eq. 15–6 reduces to a simplified form, namely,

$$\Sigma m_i(\mathbf{v}_i)_1 = \Sigma m_i(\mathbf{v}_i)_2 \qquad (15\text{–}8)$$

This equation is referred to as the *conservation of linear momentum*. It states that the total linear momentum for a system of particles remains constant during the time period t_1 to t_2. Substituting $m\mathbf{v}_G = \Sigma m_i\mathbf{v}_i$ into Eq. 15–8, we can also write

$$(\mathbf{v}_G)_1 = (\mathbf{v}_G)_2 \qquad (15\text{–}9)$$

which indicates that the velocity $\mathbf{v}_G$ of the mass center for the system of particles does not change if no external impulses are applied to the system.

The conservation of linear momentum is often applied when particles collide or interact. For application, a careful study of the free-body diagram for the *entire* system of particles should be made in order to identify the forces which create either external or internal impulses and thereby determine in what direction(s) linear momentum is conserved. As stated earlier, the *internal impulses* for the system will always cancel out, since they occur in equal but opposite collinear pairs. If the time period over which the motion is studied is *very short*, some of the external impulses may also be neglected or considered approximately equal to zero. The forces causing these negligible impulses are called *nonimpulsive forces*. By comparison, forces which are very large and act for a very short period of time produce a significant change in momentum and are called *impulsive forces*. They, of course, cannot be neglected in the impulse–momentum analysis.

Impulsive forces normally occur due to an explosion or the striking of one body against another, whereas nonimpulsive forces may include the weight of a body, the force imparted by a slightly deformed spring having a relatively small stiffness, or for that matter, any force that is very small compared to other larger (impulsive) forces. When making this distinction between impulsive and nonimpulsive forces, it is important to realize that this only applies during the time t_1 to t_2. To illustrate, consider the effect of striking a tennis ball with a racket as shown in the photo. During the *very short* time of interaction, the force of the racket on the ball is impulsive since it changes the ball's momentum drastically. By comparison, the ball's weight will have a negligible effect on the change in momentum, and therefore it is nonimpulsive. Consequently, it can be neglected from an impulse–momentum analysis during this time. If an impulse–momentum analysis is considered during the much longer time of flight after the racket–ball interaction, then the impulse of the ball's weight is important since it, along with air resistance, causes the change in the momentum of the ball.

Procedure for Analysis

Generally, the principle of linear impulse and momentum or the conservation of linear momentum is applied to a *system of particles* in order to determine the final velocities of the particles *just after* the time period considered. By applying this principle to the entire system, the internal impulses acting within the system, which may be unknown, are *eliminated* from the analysis. For application it is suggested that the following procedure be used.

Free-Body Diagram.

- Establish the *x, y, z* inertial frame of reference and draw the free-body diagram for each particle of the system in order to identify the internal and external forces.

- The conservation of linear momentum applies to the system in a direction which either has no external forces or the forces can be considered nonimpulsive.

- Establish the direction and sense of the particles' initial and final velocities. If the sense is unknown, assume it is along a positive inertial coordinate axis.

- As an alternative procedure, draw the impulse and momentum diagrams for each particle of the system.

Momentum Equations.

- Apply the principle of linear impulse and momentum or the conservation of linear momentum in the appropriate directions.

- If it is necessary to determine the *internal impulse* $\int F \, dt$ acting on only one particle of a system, then the particle must be *isolated* (free-body diagram), and the principle of linear impulse and momentum must be applied *to this particle*.

- After the impulse is calculated, and provided the time Δt for which the impulse acts is known, then the *average impulsive force* F_{avg} can be determined from $F_{avg} = \int F \, dt / \Delta t$.

EXAMPLE 15.4

The 15-Mg boxcar A is coasting at 1.5 m/s on the horizontal track when it encounters a 12-Mg tank car B coasting at 0.75 m/s toward it as shown in Fig. 15–8a. If the cars collide and couple together, determine (a) the speed of both cars just after the coupling, and (b) the average force between them if the coupling takes place in 0.8 s.

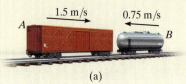

(a)

SOLUTION

Part (a) Free-Body Diagram.* Here we have considered *both* cars as a single system, Fig. 15–8b. By inspection, momentum is conserved in the x direction since the coupling force $\mathbf{F}$ is *internal* to the system and will therefore cancel out. It is assumed both cars, when coupled, move at $\mathbf{v}_2$ in the positive x direction.

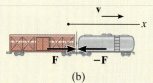

(b)

Conservation of Linear Momentum.

$$(\stackrel{+}{\rightarrow}) \qquad m_A(v_A)_1 + m_B(v_B)_1 = (m_A + m_B)v_2$$

$$(15\,000 \text{ kg})(1.5 \text{ m/s}) - 12\,000 \text{ kg}(0.75 \text{ m/s}) = (27\,000 \text{ kg})v_2$$

$$v_2 = 0.5 \text{ m/s} \rightarrow \qquad\qquad Ans.$$

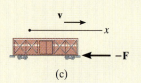

(c)

Fig. 15–8

Part (b). The average (impulsive) coupling force, $\mathbf{F}_{avg}$, can be determined by applying the principle of linear momentum to *either one* of the cars.

Free-Body Diagram. As shown in Fig. 15–8c, by isolating the boxcar the coupling force is *external* to the car.

Principle of Impulse and Momentum. Since $\int F \, dt = F_{avg} \, \Delta t = F_{avg}(0.8 \text{ s})$, we have

$$(\stackrel{+}{\rightarrow}) \qquad m_A(v_A)_1 + \Sigma \int F \, dt = m_A v_2$$

$$(15\,000 \text{ kg})(1.5 \text{ m/s}) - F_{avg}(0.8 \text{ s}) = (15\,000 \text{ kg})(0.5 \text{ m/s})$$

$$F_{avg} = 18.8 \text{ kN} \qquad\qquad Ans.$$

NOTE: Solution was possible here since the boxcar's final velocity was obtained in Part (a). Try solving for F_{avg} by applying the principle of impulse and momentum to the tank car.

*Only horizontal forces are shown on the free-body diagram.

EXAMPLE 15.5

The 1200-lb cannon shown in Fig. 15–9a fires an 8-lb projectile with a muzzle velocity of 1500 ft/s relative to the ground. If firing takes place in 0.03 s, determine (a) the recoil velocity of the cannon just after firing, and (b) the average impulsive force acting on the projectile. The cannon support is fixed to the ground, and the horizontal recoil of the cannon is absorbed by two springs.

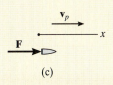

(a)

SOLUTION

Part (a) Free-Body Diagram.* As shown in Fig. 15–9b, we will consider the projectile and cannon as a single system, since the impulsive forces, **F**, between the cannon and projectile are *internal* to the system and will therefore cancel from the analysis. Furthermore, during the time $\Delta t = 0.03$ s, the two recoil springs which are attached to the support each exert a *nonimpulsive force* **F**$_s$ on the cannon. This is because Δt is very short, so that during this time the cannon only moves through a very small distance s. Consequently, $F_s = ks \approx 0$, where k is the spring's stiffness. Hence it can be concluded that momentum for the system is conserved in the *horizontal direction*.

Conservation of Linear Momentum.

$$(\xrightarrow{+}) \qquad m_c(v_c)_1 + m_p(v_p)_1 = -m_c(v_c)_2 + m_p(v_p)_2$$

$$0 + 0 = -\left(\frac{1200 \text{ lb}}{32.2 \text{ ft/s}^2}\right)(v_c)_2 + \left(\frac{8 \text{ lb}}{32.2 \text{ ft/s}^2}\right)(1500 \text{ ft/s})$$

$$(v_c)_2 = 10 \text{ ft/s} \leftarrow \qquad\qquad Ans.$$

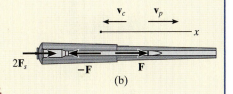

(b)

Part (b). The average impulsive force exerted by the cannon on the projectile can be determined by applying the principle of linear impulse and momentum to the projectile (or to the cannon). Why?

Principle of Impulse and Momentum. From Fig. 15–9c, with $\int F \, dt = F_{\text{avg}} \Delta t = F_{\text{avg}}(0.03)$, we have

$$(\xrightarrow{+}) \qquad m(v_p)_1 + \Sigma \int F \, dt = m(v_p)_2$$

$$0 + F_{\text{avg}}(0.03 \text{ s}) = \left(\frac{8 \text{ lb}}{32.2 \text{ ft/s}^2}\right)(1500 \text{ ft/s})$$

$$F_{\text{avg}} = 12.4(10^3) \text{ lb} = 12.4 \text{ kip} \qquad\qquad Ans.$$

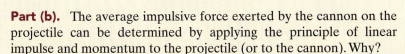

(c)

Fig. 15–9

NOTE: If the cannon is firmly fixed to its support (no springs), the reactive force of the support on the cannon must be considered as an external impulse to the system, since the support would allow no movement of the cannon.

*Only horizontal forces are shown on the free-body diagram.

EXAMPLE | 15.6

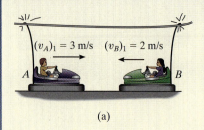

$(v_A)_1 = 3$ m/s $(v_B)_1 = 2$ m/s

(a)

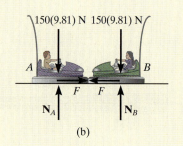

150(9.81) N 150(9.81) N

(b)

Fig. 15–10

The bumper cars A and B in Fig. 15–10a each have a mass of 150 kg and are coasting with the velocities shown before they freely collide head on. If no energy is lost during the collision, determine their velocities after collision.

SOLUTION

Free-Body Diagram.   The cars will be considered as a single system. The free-body diagram is shown in Fig. 15–10b.

Conservation of Momentum.

$$(\xrightarrow{+}) \qquad m_A(v_A)_1 + m_B(v_B)_1 = m_A(v_A)_2 + m_B(v_B)_2$$

$$(150 \text{ kg})(3 \text{ m/s}) + (150 \text{ kg})(-2 \text{ m/s}) = (150 \text{ kg})(v_A)_2 + (150 \text{ kg})(v_B)_2$$

$$(v_A)_2 = 1 - (v_B)_2 \qquad (1)$$

Conservation of Energy.   Since no energy is lost, the conservation of energy theorem gives

$$T_1 + V_1 = T_2 + V_2$$

$$\frac{1}{2}m_A(v_A)_1^2 + \frac{1}{2}m_B(v_B)_1^2 + 0 = \frac{1}{2}m_A(v_A)_2^2 + \frac{1}{2}m_B(v_B)_2^2 + 0$$

$$\frac{1}{2}(150 \text{ kg})(3 \text{ m/s})^2 + \frac{1}{2}(150 \text{ kg})(2 \text{ m/s})^2 + 0 = \frac{1}{2}(150 \text{ kg})(v_A)_2^2$$

$$+ \frac{1}{2}(150 \text{ kg})(v_B)_2{}^2 + 0$$

$$(v_A)_2^2 + (v_B)_2^2 = 13 \qquad (2)$$

Substituting Eq. (1) into (2) and simplifying, we get

$$(v_B)_2^2 - (v_B)_2 - 6 = 0$$

Solving for the two roots,

$$(v_B)_2 = 3 \text{ m/s} \qquad \text{and} \qquad (v_B)_2 = -2 \text{ m/s}$$

Since $(v_B)_2 = -2$ m/s refers to the velocity of B just *before* collision, then the velocity of B just after the collision must be

$$(v_B)_2 = 3 \text{ m/s} \rightarrow \qquad\qquad Ans.$$

Substituting this result into Eq. (1), we obtain

$$(v_A)_2 = 1 - 3 \text{ m/s} = -2 \text{ m/s} = 2 \text{ m/s} \leftarrow \qquad Ans.$$

EXAMPLE | 15.7

An 800-kg rigid pile shown in Fig. 15–11a is driven into the ground using a 300-kg hammer. The hammer falls from rest at a height $y_0 = 0.5$ m and strikes the top of the pile. Determine the impulse which the pile exerts on the hammer if the pile is surrounded entirely by loose sand so that after striking, the hammer does *not* rebound off the pile.

SOLUTION

Conservation of Energy. The velocity at which the hammer strikes the pile can be determined using the conservation of energy equation applied to the hammer. With the datum at the top of the pile, Fig. 15–11a, we have

$$T_0 + V_0 = T_1 + V_1$$

$$\frac{1}{2}m_H(v_H)_0^2 + W_H y_0 = \frac{1}{2}m_H(v_H)_1^2 + W_H y_1$$

$$0 + 300(9.81) \text{ N}(0.5 \text{ m}) = \frac{1}{2}(300 \text{ kg})(v_H)_1^2 + 0$$

$$(v_H)_1 = 3.132 \text{ m/s}$$

Free-Body Diagram. From the physical aspects of the problem, the free-body diagram of the hammer and pile, Fig. 15–11b, indicates that during the *short time* from *just before* to *just after* the collision, the weights of the hammer and pile and the resistance force $\mathbf{F}_s$ of the sand are all *nonimpulsive*. The impulsive force $\mathbf{R}$ is internal to the system and therefore cancels. Consequently, momentum is conserved in the vertical direction during this short time.

Conservation of Momentum. Since the hammer does not rebound off the pile just after collision, then $(v_H)_2 = (v_P)_2 = v_2$.

$$(+\downarrow) \qquad m_H(v_H)_1 + m_P(v_P)_1 = m_H v_2 + m_P v_2$$

$$(300 \text{ kg})(3.132 \text{ m/s}) + 0 = (300 \text{ kg})v_2 + (800 \text{ kg})v_2$$

$$v_2 = 0.8542 \text{ m/s}$$

Principle of Impulse and Momentum. The impulse which the pile imparts to the hammer can now be determined since $\mathbf{v}_2$ is known. From the free-body diagram for the hammer, Fig. 15–11c, we have

$$(+\downarrow) \qquad m_H(v_H)_1 + \Sigma \int_{t_1}^{t_2} F_y \, dt = m_H v_2$$

$$(300 \text{ kg})(3.132 \text{ m/s}) - \int R \, dt = (300 \text{ kg})(0.8542 \text{ m/s})$$

$$\int R \, dt = 683 \text{ N} \cdot \text{s} \qquad \qquad \textit{Ans.}$$

NOTE: The equal but opposite impulse acts on the pile. Try finding this impulse by applying the principle of impulse and momentum to the pile.

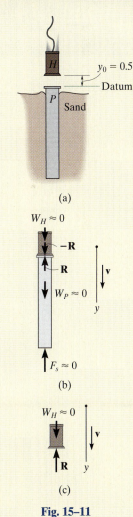

Fig. 15–11

EXAMPLE 15.8

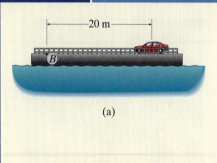

|← 20 m →|

(a)

Fig. 15–12

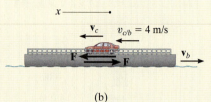

x ─────────•

$\mathbf{v}_c$ $v_{c/b} = 4$ m/s

F ←   → F $\mathbf{v}_b$

(b)

The 1.5-Mg car in Fig. 15–12a moves on the 10-Mg barge to the left with a constant speed of 4 m/s, measured relative to the barge. Neglecting water resistance, determine the velocity of the barge and the displacement of the barge when the car reaches point B. Initially, the car and the barge are at rest relative to the water.

SOLUTION

Free-Body Diagram. If the car and the barge are considered as a single system, the traction force between the car and the barge becomes internal to the system, and so linear momentum will be conserved along the x axis, Fig. 15–12b.

Conservation of Momentum. When writing the conservation of momentum equation, it is important that the velocities be measured from the same inertial coordinate system, assumed here to be fixed. We will also assume that as the car goes to the left the barge goes to the right, as shown in Fig. 15–12b.

Applying the conservation of linear momentum to the car and barge system,

$(\xleftarrow{+})$
$$0 + 0 = m_c v_c - m_b v_b$$

$$0 = (1.5(10^3)\ \text{kg})v_c - (10(10^3)\ \text{kg})v_b$$

$$1.5v_c - 10v_b = 0 \tag{1}$$

Kinematics. Since the velocity of the car relative to the barge is known, then the velocity of the car and the barge can also be related using the relative velocity equation.

$(\xleftarrow{+})$
$$\mathbf{v}_c = \mathbf{v}_b + \mathbf{v}_{c/b}$$

$$v_c = -v_b + 4\ \text{m/s} \tag{2}$$

Solving Eqs. (1) and (2),

$$v_b = 0.5217\ \text{m/s} = 0.522\ \text{m/s} \rightarrow \qquad Ans.$$
$$v_c = 3.478\ \text{m/s} \leftarrow$$

The car travels $s_{c/b} = 20$ m on the barge at a constant relative speed of 4 m/s. Thus, the time for the car to reach point B is

$$s_{c/b} = v_{c/b}\, t$$
$$20\ \text{m} = (4\ \text{m/s})\, t$$
$$t = 5\ \text{s}$$

The displacement of the barge is therefore

$(\xrightarrow{+})\qquad s_b = v_b t = 0.5217\ \text{m/s}\,(5\ \text{s}) = 2.61\ \text{m} \rightarrow \qquad Ans.$

FUNDAMENTAL PROBLEMS

F15–7. The freight cars A and B have a mass of 20 Mg and 15 Mg, respectively. Determine the velocity of A after collision if the cars collide and rebound, such that B moves to the right with a speed of 2 m/s. If A and B are in contact for 0.5 s, find the average impulsive force which acts between them.

3 m/s
A

1.5 m/s
B

F15–7

F15–8. The cart and package have a mass of 20 kg and 5 kg, respectively. If the cart has a smooth surface and it is initially at rest, while the velocity of the package is as shown, determine the final common velocity of the cart and package after the impact.

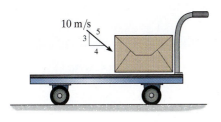

10 m/s
3
5
4

F15–8

F15–9. The 5-kg block A has an initial speed of 5 m/s as it slides down the smooth ramp, after which it collides with the stationary block B of mass 8 kg. If the two blocks couple together after collision, determine their common velocity immediately after collision.

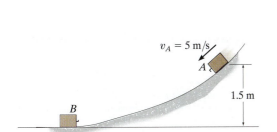

$v_A = 5$ m/s
A
B
1.5 m

F15–9

F15–10. The spring is fixed to block A and block B is pressed against the spring. If the spring is compressed $s = 200$ mm and then the blocks are released, determine their velocity at the instant block B loses contact with the spring. The masses of blocks A and B are 10 kg and 15 kg, respectively.

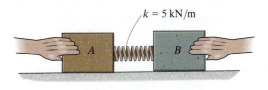

$k = 5$ kN/m
A B

F15–10

F15–11. Blocks A and B have a mass of 15 kg and 10 kg, respectively. If A is stationary and B has a velocity of 15 m/s just before collision, and the blocks couple together after impact, determine the maximum compression of the spring.

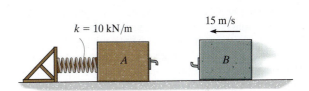

$k = 10$ kN/m
A
15 m/s
B

F15–11

F15–12. The cannon and support without a projectile have a mass of 250 kg. If a 20-kg projectile is fired from the cannon with a velocity of 400 m/s, measured *relative* to the cannon, determine the speed of the projectile as it leaves the barrel of the cannon. Neglect rolling resistance.

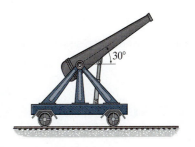

30°

F15–12

15

PROBLEMS

*15–32. The 10-lb cannon ball is fired horizontally by a 500-lb cannon as shown. If the muzzle velocity of the ball is 2000 ft/s, measured relative to the ground, determine the recoil velocity of the cannon just after firing. If the cannon rests on a smooth support and is to be stopped after it has recoiled a distance of 6 in., determine the required stiffness k of the two identical springs, each of which is originally unstretched.

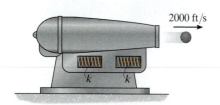

Prob. 15–32

15–33. A railroad car having a mass of 15 Mg is coasting at 1.5 m/s on a horizontal track. At the same time another car having a mass of 12 Mg is coasting at 0.75 m/s in the opposite direction. If the cars meet and couple together, determine the speed of both cars just after the coupling. Find the difference between the total kinetic energy before and after coupling has occurred, and explain qualitatively what happened to this energy.

15–34. The car A has a weight of 4500 lb and is traveling to the right at 3 ft/s. Meanwhile a 3000-lb car B is traveling at 6 ft/s to the left. If the cars crash head-on and become entangled, determine their common velocity just after the collision. Assume that the brakes are not applied during collision.

15–35. The two blocks A and B each have a mass of 5 kg and are suspended from parallel cords. A spring, having a stiffness of $k = 60$ N/m, is attached to B and is compressed 0.3 m against A as shown. Determine the maximum angles θ and ϕ of the cords when the blocks are released from rest and the spring becomes unstretched.

*15–36. Block A has a mass of 4 kg and B has a mass of 6 kg. A spring, having a stiffness of $k = 40$ N/m, is attached to B and is compressed 0.3 m against A as shown. Determine the maximum angles θ and ϕ of the cords after the blocks are released from rest and the spring becomes unstretched.

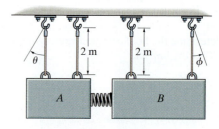

Probs. 15–35/36

•15–37. The winch on the back of the Jeep A is turned on and pulls in the tow rope at 2 m/s measured relative to the Jeep. If both the 1.25-Mg car B and the 2.5-Mg Jeep A are free to roll, determine their velocities at the instant they meet. If the rope is 5 m long, how long will this take?

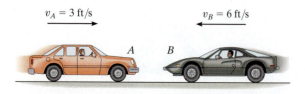

Probs. 15–33/34

Prob. 15–37

15–38. The 40-kg package is thrown with a speed of 4 m/s onto the cart having a mass of 20 kg. If it slides on the smooth surface and strikes the spring, determine the velocity of the cart at the instant the package fully compresses the spring. What is the maximum compression of the spring? Neglect rolling resistance of the cart.

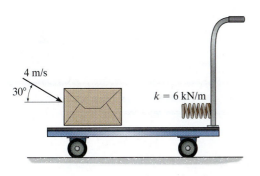

Prob. 15–38

15–39. Two cars A and B have a mass of 2 Mg and 1.5 Mg, respectively. Determine the magnitudes of $\mathbf{v}_A$ and $\mathbf{v}_B$ if the cars collide and stick together while moving with a common speed of 50 km/h in the direction shown.

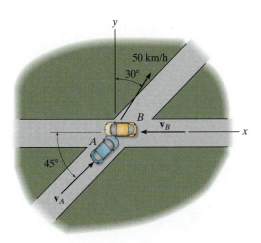

Prob. 15–39

*****15–40.** A 4-kg projectile travels with a horizontal velocity of 600 m/s before it explodes and breaks into two fragments A and B of mass 1.5 kg and 2.5 kg, respectively. If the fragments travel along the parabolic trajectories shown, determine the magnitude of velocity of each fragment just after the explosion and the horizontal distance d_A where segment A strikes the ground at C.

•15–41. A 4-kg projectile travels with a horizontal velocity of 600 m/s before it explodes and breaks into two fragments A and B of mass 1.5 kg and 2.5 kg, respectively. If the fragments travel along the parabolic trajectories shown, determine the magnitude of velocity of each fragment just after the explosion and the horizontal distance d_B where segment B strikes the ground at D.

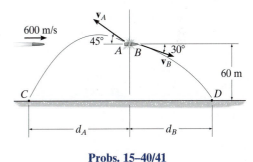

Probs. 15–40/41

15–42. The 75-kg boy leaps off cart A with a horizontal velocity of $v' = 3$ m/s measured relative to the cart. Determine the velocity of cart A just after the jump. If he then lands on cart B with the same velocity that he left cart A, determine the velocity of cart B just after he lands on it. Carts A and B have the same mass of 50 kg and are originally at rest.

Prob. 15–42

15–43. Block A has a mass of 2 kg and slides into an open ended box B with a velocity of 2 m/s. If the box B has a mass of 3 kg and rests on top of a plate P that has a mass of 3 kg, determine the distance the plate moves after it stops sliding on the floor. Also, how long is it after impact before all motion ceases? The coefficient of kinetic friction between the box and the plate is $\mu_k = 0.2$, and between the plate and the floor $\mu'_k = 0.4$. Also, the coefficient of static friction between the plate and the floor is $\mu'_s = 0.5$.

*15–44. Block A has a mass of 2 kg and slides into an open ended box B with a velocity of 2 m/s. If the box B has a mass of 3 kg and rests on top of a plate P that has a mass of 3 kg, determine the distance the plate moves after it stops sliding on the floor. Also, how long is it after impact before all motion ceases? The coefficient of kinetic friction between the box and the plate is $\mu_k = 0.2$, and between the plate and the floor $\mu'_k = 0.1$. Also, the coefficient of static friction between the plate and the floor is $\mu'_s = 0.12$.

Probs. 15–43/44

•15–45. The 20-kg block A is towed up the ramp of the 40-kg cart using the motor M mounted on the side of the cart. If the motor winds in the cable with a constant velocity of 5 m/s, measured relative to the cart, determine how far the cart will move when the block has traveled a distance $s = 2$ m up the ramp. Both the block and cart are at rest when $s = 0$. The coefficient of kinetic friction between the block and the ramp is $\mu_k = 0.2$. Neglect rolling resistance.

15–46. If the 150-lb man fires the 0.2-lb bullet with a horizontal muzzle velocity of 3000 ft/s, measured relative to the 600-lb cart, determine the velocity of the cart just after firing. What is the velocity of the cart when the bullet becomes embedded in the target? During the firing, the man remains at the same position on the cart. Neglect rolling resistance of the cart.

Prob. 15–46

15–47. The free-rolling ramp has a weight of 120 lb. The crate whose weight is 80 lb slides from rest at A, 15 ft down the ramp to B. Determine the ramp's speed when the crate reaches B. Assume that the ramp is smooth, and neglect the mass of the wheels.

*15–48. The free-rolling ramp has a weight of 120 lb. If the 80-lb crate is released from rest at A, determine the distance the ramp moves when the crate slides 15 ft down the ramp to the bottom B.

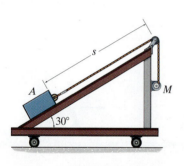

Prob. 15–45

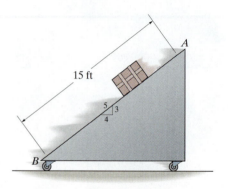

Probs. 15–47/48

•**15–49.** The 5-kg spring-loaded gun rests on the smooth surface. It fires a ball having a mass of 1 kg with a velocity of $v' = 6$ m/s relative to the gun in the direction shown. If the gun is originally at rest, determine the horizontal distance d the ball is from the initial position of the gun at the instant the ball strikes the ground at D. Neglect the size of the gun.

15–50. The 5-kg spring-loaded gun rests on the smooth surface. It fires a ball having a mass of 1 kg with a velocity of $v' = 6$ m/s relative to the gun in the direction shown. If the gun is originally at rest, determine the distance the ball is from the initial position of the gun at the instant the ball reaches its highest elevation C. Neglect the size of the gun.

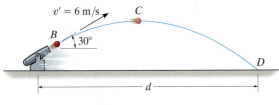

Probs. 15–49/50

*****15–52.** The block of mass m travels at v_1 in the direction θ_1 shown at the top of the smooth slope. Determine its speed v_2 and its direction θ_2 when it reaches the bottom.

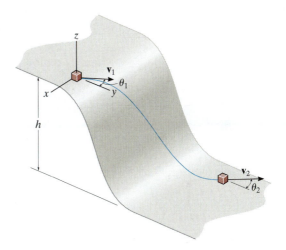

Prob. 15–52

15–51. A man wearing ice skates throws an 8-kg block with an initial velocity of 2 m/s, measured relative to himself, in the direction shown. If he is originally at rest and completes the throw in 1.5 s while keeping his legs rigid, determine the horizontal velocity of the man just after releasing the block. What is the vertical reaction of both his skates on the ice during the throw? The man has a mass of 70 kg. Neglect friction and the motion of his arms.

•**15–53.** The 20-lb cart B is supported on rollers of negligible size. If a 10-lb suitcase A is thrown horizontally onto the cart at 10 ft/s when it is at rest, determine the length of time that A slides relative to B, and the final velocity of A and B. The coefficient of kinetic friction between A and B is $\mu_k = 0.4$.

15–54. The 20-lb cart B is supported on rollers of negligible size. If a 10-lb suitcase A is thrown horizontally onto the cart at 10 ft/s when it is at rest, determine the time t and the distance B moves at the instant A stops relative to B. The coefficient of kinetic friction between A and B is $\mu_k = 0.4$.

Prob. 15–51

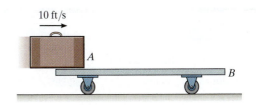

Probs. 15–53/54

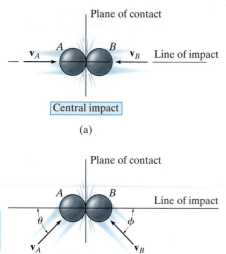

Central impact

(a)

Oblique impact

(b)

Fig. 15–13

15.4   Impact

Impact occurs when two bodies collide with each other during a very *short* period of time, causing relatively large (impulsive) forces to be exerted between the bodies. The striking of a hammer on a nail, or a golf club on a ball, are common examples of impact loadings.

In general, there are two types of impact. *Central impact* occurs when the direction of motion of the mass centers of the two colliding particles is along a line passing through the mass centers of the particles. This line is called the *line of impact*, which is perpendicular to the plane of contact, Fig. 15–13a. When the motion of one or both of the particles make an angle with the line of impact, Fig. 15–13b, the impact is said to be *oblique impact*.

Central Impact. To illustrate the method for analyzing the mechanics of impact, consider the case involving the central impact of the two particles *A* and *B* shown in Fig. 15–14.

- The particles have the initial momenta shown in Fig. 15–14a. Provided $(v_A)_1 > (v_B)_1$, collision will eventually occur.

- During the collision the particles must be thought of as *deformable* or nonrigid. The particles will undergo a *period of deformation* such that they exert an equal but opposite deformation impulse $\int P \, dt$ on each other, Fig. 15–14b.

- Only at the instant of *maximum deformation* will both particles move with a common velocity **v**, since their relative motion is zero, Fig. 15–14c.

- Afterward a *period of restitution* occurs, in which case the particles will either return to their original shape or remain permanently deformed. The equal but opposite *restitution impulse* $\int R \, dt$ pushes the particles apart from one another, Fig. 15–14d. In reality, the physical properties of any two bodies are such that the deformation impulse with *always* be greater than that of restitution, i.e., $\int P \, dt > \int R \, dt$.

- Just after separation the particles will have the final momenta shown in Fig. 15–14e, where $(v_B)_2 > (v_A)_2$.

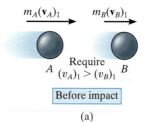

Before impact

(a)

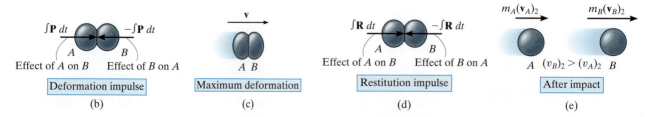

Fig. 15–14

In most problems the initial velocities of the particles will be *known*, and it will be necessary to determine their final velocities $(v_A)_2$ and $(v_B)_2$. In this regard, *momentum* for the *system of particles* is *conserved* since during collision the internal impulses of deformation and restitution *cancel*. Hence, referring to Fig. 15–14a and Fig. 15–14e we require

$(\overset{+}{\rightarrow})$ $\qquad m_A(v_A)_1 + m_B(v_B)_1 = m_A(v_A)_2 + m_B(v_B)_2$ $\qquad$ (15–10)

In order to obtain a second equation necessary to solve for $(v_A)_2$ and $(v_B)_2$, we must apply the principle of impulse and momentum to *each* particle. For example, during the deformation phase for particle A, Figs. 15–14a, 15–14b, and 15–14c, we have

$(\overset{+}{\rightarrow})$ $\qquad m_A(v_A)_1 - \int P \, dt = m_A v$

For the restitution phase, Figs. 15–14c, 15–14d, and 15–14e,

$(\overset{+}{\rightarrow})$ $\qquad m_A v - \int R \, dt = m_A(v_A)_2$

The ratio of the restitution impulse to the deformation impulse is called the *coefficient of restitution*, e. From the above equations, this value for particle A is

$$e = \frac{\int R \, dt}{\int P \, dt} = \frac{v - (v_A)_2}{(v_A)_1 - v}$$

In a similar manner, we can establish e by considering particle B, Fig. 15–14. This yields

$$e = \frac{\int R \, dt}{\int P \, dt} = \frac{(v_B)_2 - v}{v - (v_B)_1}$$

If the unknown v is eliminated from the above two equations, the coefficient of restitution can be expressed in terms of the particles' initial and final velocities as

$(\overset{+}{\rightarrow})$ $\qquad \boxed{e = \dfrac{(v_B)_2 - (v_A)_2}{(v_A)_1 - (v_B)_1}}$ $\qquad$ (15–11)

The quality of a manufactured tennis ball is measured by the height of its bounce, which can be related to its coefficient of restitution. Using the mechanics of oblique impact, engineers can design a separation device to remove substandard tennis balls from a production line.

Provided a value for e is specified, Eqs. 15–10 and 15–11 can be solved simultaneously to obtain $(v_A)_2$ and $(v_B)_2$. In doing so, however, it is important to carefully establish a sign convention for defining the positive direction for both $\mathbf{v}_A$ and $\mathbf{v}_B$ and then use it *consistently* when writing *both* equations. As noted from the application shown, and indicated symbolically by the arrow in parentheses, we have defined the positive direction to the right when referring to the motions of both A and B. Consequently, if a negative value results from the solution of either $(v_A)_2$ or $(v_B)_2$, it indicates motion is to the left.

Coefficient of Restitution.

From Figs. 15–14a and 15–14e, it is seen that Eq. 15–11 states that e is equal to the ratio of the relative velocity of the particles' separation *just after impact*, $(v_B)_2 - (v_A)_2$, to the relative velocity of the particles' approach *just before impact*, $(v_A)_1 - (v_B)_1$. By measuring these relative velocities experimentally, it has been found that e varies appreciably with impact velocity as well as with the size and shape of the colliding bodies. For these reasons the coefficient of restitution is reliable only when used with data which closely approximate the conditions which were known to exist when measurements of it were made. In general e has a value between zero and one, and one should be aware of the physical meaning of these two limits.

Elastic Impact ($e = 1$).

If the collision between the two particles is *perfectly elastic*, the deformation impulse $\left(\int \mathbf{P}\, dt \right)$ is equal and opposite to the restitution impulse $\left(\int \mathbf{R}\, dt \right)$. Although in reality this can never be achieved, $e = 1$ for an elastic collision.

Plastic Impact ($e = 0$).

The impact is said to be *inelastic or plastic* when $e = 0$. In this case there is no restitution impulse $\left(\int \mathbf{R}\, dt = \mathbf{0} \right)$, so that after collision both particles couple or stick *together* and move with a common velocity.

From the above derivation it should be evident that the principle of work and energy cannot be used for the analysis of impact problems since it is not possible to know how the *internal forces* of deformation and restitution vary or displace during the collision. By knowing the particle's velocities before and after collision, however, the energy loss during collision can be calculated on the basis of the difference in the particle's kinetic energy. This energy loss, $\Sigma U_{1-2} = \Sigma T_2 - \Sigma T_1$, occurs because some of the initial kinetic energy of the particle is transformed into thermal energy as well as creating sound and localized deformation of the material when the collision occurs. In particular, if the impact is *perfectly elastic*, no energy is lost in the collision; whereas if the collision is *plastic*, the energy lost during collision is a maximum.

Procedure for Analysis (Central Impact)

In most cases the *final velocities* of two smooth particles are to be determined *just after* they are subjected to direct central impact. Provided the coefficient of restitution, the mass of each particle, and each particle's initial velocity *just before* impact are known, the solution to this problem can be obtained using the following two equations:

• The conservation of momentum applies to the system of particles, $\Sigma m v_1 = \Sigma m v_2$.

• The coefficient of restitution, $e = [(v_B)_2 - (v_A)_2]/[(v_A)_1 - (v_B)_1]$, relates the relative velocities of the particles along the line of impact, just before and just after collision.

When applying these two equations, the sense of an unknown velocity can be assumed. If the solution yields a negative magnitude, the velocity acts in the opposite sense.

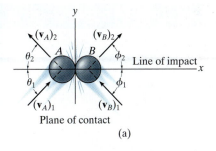

(a)

Plane of contact

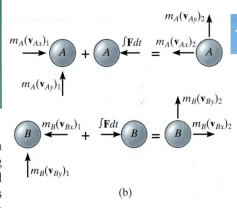

(b)

Fig. 15–15

Oblique Impact.

When oblique impact occurs between two smooth particles, the particles move away from each other with velocities having unknown directions as well as unknown magnitudes. Provided the initial velocities are known, then four unknowns are present in the problem. As shown in Fig. 15–15a, these unknowns may be represented either as $(v_A)_2, (v_B)_2, \theta_2,$ and ϕ_2, or as the x and y components of the final velocities.

Procedure for Analysis (Oblique Impact)

If the y axis is established within the plane of contact and the x axis along the line of impact, the impulsive forces of deformation and restitution act *only in the x direction*, Fig. 15–15b. By resolving the velocity or momentum vectors into components along the x and y axes, Fig. 15–15b, it is then possible to write four independent scalar equations in order to determine $(v_{Ax})_2, (v_{Ay})_2, (v_{Bx})_2,$ and $(v_{By})_2$.

• Momentum of the system is conserved *along the line of impact, x axis*, so that $\Sigma m(v_x)_1 = \Sigma m(v_x)_2$.

• The coefficient of restitution, $e = [(v_{Bx})_2 - (v_{Ax})_2]/[(v_{Ax})_1 - (v_{Bx})_1]$, relates the relative-velocity *components* of the particles *along the line of impact (x axis)*.

• If these two equations are solved simultaneously, we obtain $(v_{Ax})_2$ and $(v_{Bx})_2$.

• Momentum of particle A is conserved along the y axis, perpendicular to the line of impact, since no impulse acts on particle A in this direction. As a result $m_A(v_{Ay})_1 = m_A(v_{Ay})_2$ or $(v_{Ay})_1 = (v_{Ay})_2$

• Momentum of particle B is conserved along the y axis, perpendicular to the line of impact, since no impulse acts on particle B in this direction. Consequently $(v_{By})_1 = (v_{By})_2$.

Application of these four equations is illustrated in Example 15.11.

EXAMPLE 15.9

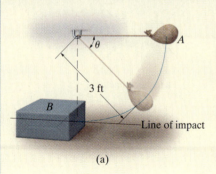

(a)

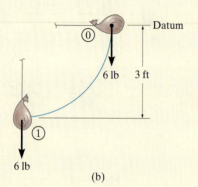

(b)

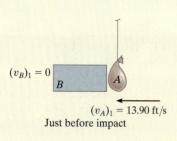

(vB)₁ = 0

(vA)₁ = 13.90 ft/s

Just before impact

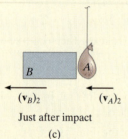

(vB)₂   (vA)₂

Just after impact

(c)

Fig. 15–16

The bag A, having a weight of 6 lb, is released from rest at the position $\theta = 0°$, as shown in Fig. 15–16a. After falling to $\theta = 90°$, it strikes an 18-lb box B. If the coefficient of restitution between the bag and box is $e = 0.5$, determine the velocities of the bag and box just after impact. What is the loss of energy during collision?

SOLUTION

This problem involves central impact. Why? Before analyzing the mechanics of the impact, however, it is first necessary to obtain the velocity of the bag *just before* it strikes the box.

Conservation of Energy. With the datum at $\theta = 0°$, Fig. 15–16b, we have

$$T_0 + V_0 = T_1 + V_1$$

$$0 + 0 = \frac{1}{2}\left(\frac{6\text{ lb}}{32.2\text{ ft/s}^2}\right)(v_A)_1^2 - 6\text{ lb}(3\text{ ft}); \qquad (v_A)_1 = 13.90\text{ ft/s}$$

Conservation of Momentum. After impact we will assume A and B travel to the left. Applying the conservation of momentum to the system, Fig. 15–16c, we have

$$(\overset{+}{\leftarrow}) \qquad m_B(v_B)_1 + m_A(v_A)_1 = m_B(v_B)_2 + m_A(v_A)_2$$

$$0 + \left(\frac{6\text{ lb}}{32.2\text{ ft/s}^2}\right)(13.90\text{ ft/s}) = \left(\frac{18\text{ lb}}{32.2\text{ ft/s}^2}\right)(v_B)_2 + \left(\frac{6\text{ lb}}{32.2\text{ ft/s}^2}\right)(v_A)_2$$

$$(v_A)_2 = 13.90 - 3(v_B)_2 \qquad (1)$$

Coefficient of Restitution. Realizing that for separation to occur after collision $(v_B)_2 > (v_A)_2$, Fig. 15–16c, we have

$$(\overset{+}{\leftarrow}) \qquad e = \frac{(v_B)_2 - (v_A)_2}{(v_A)_1 - (v_B)_1}; \qquad 0.5 = \frac{(v_B)_2 - (v_A)_2}{13.90\text{ ft/s} - 0}$$

$$(v_A)_2 = (v_B)_2 - 6.950 \qquad (2)$$

Solving Eqs. 1 and 2 simultaneously yields

$$(v_A)_2 = -1.74\text{ ft/s} = 1.74\text{ ft/s} \rightarrow \quad \text{and} \quad (v_B)_2 = 5.21\text{ ft/s} \leftarrow \quad Ans.$$

Loss of Energy. Applying the principle of work and energy to the bag and box just before and just after collision, we have

$$\Sigma U_{1-2} = T_2 - T_1;$$

$$\Sigma U_{1-2} = \left[\frac{1}{2}\left(\frac{18\text{ lb}}{32.2\text{ ft/s}^2}\right)(5.21\text{ ft/s})^2 + \frac{1}{2}\left(\frac{6\text{ lb}}{32.2\text{ ft/s}^2}\right)(1.74\text{ ft/s})^2\right]$$

$$- \left[\frac{1}{2}\left(\frac{6\text{ lb}}{32.2\text{ ft/s}^2}\right)(13.9\text{ ft/s})^2\right]$$

$$\Sigma U_{1-2} = -10.1\text{ ft} \cdot \text{lb} \qquad Ans.$$

NOTE: The energy loss occurs due to inelastic deformation during the collision.

EXAMPLE 15.10

Ball *B* shown in Fig. 15–17a has a mass of 1.5 kg and is suspended from the ceiling by a 1-m-long elastic cord. If the cord is *stretched* downward 0.25 m and the ball is released from rest, determine how far the cord stretches after the ball rebounds from the ceiling. The stiffness of the cord is $k = 800$ N/m, and the coefficient of restitution between the ball and ceiling is $e = 0.8$. The ball makes a central impact with the ceiling.

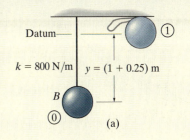

(a)

SOLUTION
First we must obtain the velocity of the ball *just before* it strikes the ceiling using energy methods, then consider the impulse and momentum between the ball and ceiling, and finally again use energy methods to determine the stretch in the cord.

Conservation of Energy. With the datum located as shown in Fig. 15–17a, realizing that initially $y = y_0 = (1 + 0.25)$ m $= 1.25$ m, we have

$$T_0 + V_0 = T_1 + V_1$$

$$\tfrac{1}{2}m(v_B)_0^2 - W_B y_0 + \tfrac{1}{2}ks^2 = \tfrac{1}{2}m(v_B)_1^2 + 0$$

$$0 - 1.5(9.81)\text{N}(1.25 \text{ m}) + \tfrac{1}{2}(800 \text{ N/m})(0.25 \text{ m})^2 = \tfrac{1}{2}(1.5 \text{ kg})(v_B)_1^2$$

$$(v_B)_1 = 2.968 \text{ m/s} \uparrow$$

The interaction of the ball with the ceiling will now be considered using the principles of impact.* Since an unknown portion of the mass of the ceiling is involved in the impact, the conservation of momentum for the ball–ceiling system will not be written. The "velocity" of this portion of ceiling is zero since it (or the earth) are assumed to remain at rest both before and after impact.

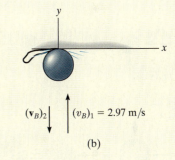

(b)

Coefficient of Restitution. Fig. 15–17b.

$$(+\uparrow) \quad e = \frac{(v_B)_2 - (v_A)_2}{(v_A)_1 - (v_B)_1}; \qquad 0.8 = \frac{(v_B)_2 - 0}{0 - 2.968 \text{ m/s}}$$

$$(v_B)_2 = -2.374 \text{ m/s} = 2.374 \text{ m/s} \downarrow$$

Conservation of Energy. The maximum stretch s_3 in the cord can be determined by again applying the conservation of energy equation to the ball just after collision. Assuming that $y = y_3 = (1 + s_3)$ m, Fig. 15–17c, then

$$T_2 + V_2 = T_3 + V_3$$

$$\tfrac{1}{2}m(v_B)_2^2 + 0 = \tfrac{1}{2}m(v_B)_3^2 - W_B y_3 + \tfrac{1}{2}ks_3^2$$

$$\tfrac{1}{2}(1.5 \text{ kg})(2.37 \text{ m/s})^2 = 0 - 9.81(1.5) \text{ N}(1 \text{ m} + s_3) + \tfrac{1}{2}(800 \text{ N/m})s_3^2$$

$$400s_3^2 - 14.715s_3 - 18.94 = 0$$

Solving this quadratic equation for the positive root yields

$$s_3 = 0.237 \text{ m} = 237 \text{ mm} \qquad \textit{Ans.}$$

* The weight of the ball is considered a nonimpulsive force.

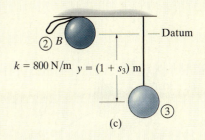

(c)

Fig. 15–17

EXAMPLE | **15.11**

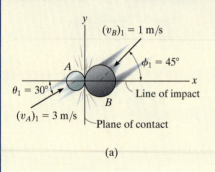

(a)

Two smooth disks A and B, having a mass of 1 kg and 2 kg, respectively, collide with the velocities shown in Fig. 15–18a. If the coefficient of restitution for the disks is $e = 0.75$, determine the x and y components of the final velocity of each disk just after collision.

SOLUTION

This problem involves *oblique impact*. Why? In order to solve it, we have established the x and y axes along the line of impact and the plane of contact, respectively, Fig. 15–18a.

Resolving each of the initial velocities into x and y components, we have

$$(v_{Ax})_1 = 3 \cos 30° = 2.598 \text{ m/s} \quad (v_{Ay})_1 = 3 \sin 30° = 1.50 \text{ m/s}$$

$$(v_{Bx})_1 = -1 \cos 45° = -0.7071 \text{ m/s} \quad (v_{By})_1 = -1 \sin 45° = -0.7071 \text{ m/s}$$

The four unknown velocity components after collision are *assumed to act in the positive directions*, Fig. 15–18b. Since the impact occurs in the x direction (line of impact), the conservation of momentum for *both* disks can be applied in this direction. Why?

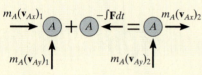

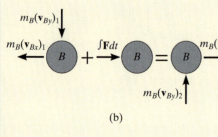

(b)

Conservation of "x" Momentum. In reference to the momentum diagrams, we have

$$(\xrightarrow{+}) \qquad m_A(v_{Ax})_1 + m_B(v_{Bx})_1 = m_A(v_{Ax})_2 + m_B(v_{Bx})_2$$

$$1 \text{ kg}(2.598 \text{ m/s}) + 2 \text{ kg}(-0.707 \text{ m/s}) = 1 \text{ kg}(v_{Ax})_2 + 2 \text{ kg}(v_{Bx})_2$$

$$(v_{Ax})_2 + 2(v_{Bx})_2 = 1.184 \qquad (1)$$

Coefficient of Restitution (x).

$$(\xrightarrow{+}) \qquad e = \frac{(v_{Bx})_2 - (v_{Ax})_2}{(v_{Ax})_1 - (v_{Bx})_1}; \quad 0.75 = \frac{(v_{Bx})_2 - (v_{Ax})_2}{2.598 \text{ m/s} - (-0.7071 \text{ m/s})}$$

$$(v_{Bx})_2 - (v_{Ax})_2 = 2.479 \qquad (2)$$

Solving Eqs. 1 and 2 for $(v_{Ax})_2$ and $(v_{Bx})_2$ yields

$$(v_{Ax})_2 = -1.26 \text{ m/s} = 1.26 \text{ m/s} \leftarrow \qquad (v_{Bx})_2 = 1.22 \text{ m/s} \rightarrow \qquad Ans.$$

Conservation of "y" Momentum. The momentum of *each disk* is *conserved* in the y direction (plane of contact), since the disks are smooth and therefore *no* external impulse acts in this direction. From Fig. 15–18b,

$$(+\uparrow) \; m_A(v_{Ay})_1 = m_A(v_{Ay})_2; \; (v_{Ay})_2 = 1.50 \text{ m/s} \uparrow \qquad Ans.$$

$$(+\uparrow) \; m_B(v_{By})_1 = m_B(v_{By})_2; \; (v_{By})_2 = -0.707 \text{ m/s} = 0.707 \text{ m/s} \downarrow \; Ans.$$

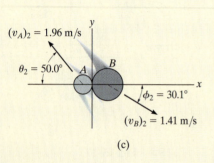

(c)

Fig. 15–18

NOTE: Show that when the velocity components are summed vectorially, one obtains the results shown in Fig. 15–18c.

FUNDAMENTAL PROBLEMS

F15–13. Determine the coefficient of restitution e between ball A and ball B. The velocities of A and B before and after the collision are shown.

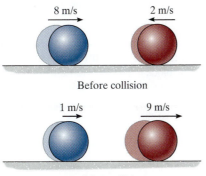

8 m/s   2 m/s

Before collision

1 m/s   9 m/s

After collision

F15–13

F15–14. The 15-Mg tank car A and 25-Mg freight car B travel towards each other with the velocities shown. If the coefficient of restitution between the bumpers is $e = 0.6$, determine the velocity of each car just after the collision.

5 m/s   7 m/s
A B

F15–14

F15–15. The 30-lb package A has a speed of 5 ft/s when it enters the smooth ramp. As it slides down the ramp, it strikes the 80-lb package B which is initially at rest. If the coefficient of restitution between A and B is $e = 0.6$, determine the velocity of B just after the impact.

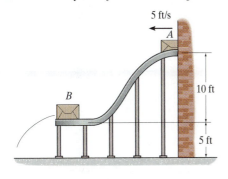

5 ft/s
A

10 ft

B

5 ft

F15–15

F15–16. Blocks A and B weigh 5 lb and 10 lb, respectively. After striking block B, A slides 2 in. to the right, and B slides 3 in. to the right. If the coefficient of kinetic friction between the blocks and the surface is $\mu_k = 0.2$, determine the coefficient of restitution between the blocks. Block B is originally at rest.

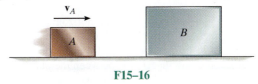

v_A

A B

F15–16

F15–17. The ball strikes the smooth wall with a velocity of $(v_b)_1 = 20$ m/s. If the coefficient of restitution between the ball and the wall is $e = 0.75$, determine the velocity of the ball just after the impact.

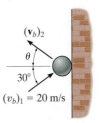

$(v_b)_2$

θ

30°

$(v_b)_1 = 20$ m/s

F15–17

F15–18. Disk A weighs 2 lb and slides on the smooth horizontal plane with a velocity of 3 ft/s. Disk B weighs 11 lb and is initially at rest. If after impact A has a velocity of 1 ft/s, parallel to the positive x axis, determine the speed of disk B after impact.

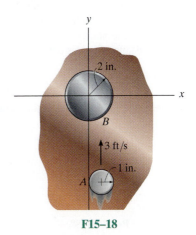

y

2 in.

B

x

3 ft/s

1 in.

A

F15–18

PROBLEMS

15–55. A 1-lb ball A is traveling horizontally at $20\ \text{ft/s}$ when it strikes a 10-lb block B that is at rest. If the coefficient of restitution between A and B is $e = 0.6$, and the coefficient of kinetic friction between the plane and the block is $\mu_k = 0.4$, determine the time for the block B to stop sliding.

***15–56.** A 1-lb ball A is traveling horizontally at $20\ \text{ft/s}$ when it strikes a 10-lb block B that is at rest. If the coefficient of restitution between A and B is $e = 0.6$, and the coefficient of kinetic friction between the plane and the block is $\mu_k = 0.4$, determine the distance block B slides on the plane before it stops sliding.

•15–57. The three balls each have a mass m. If A has a speed v just before a direct collision with B, determine the speed of C after collision. The coefficient of restitution between each ball is e. Neglect the size of each ball.

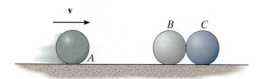

Prob. 15–57

15–58. The 15-lb suitcase A is released from rest at C. After it slides down the smooth ramp, it strikes the 10-lb suitcase B, which is originally at rest. If the coefficient of restitution between the suitcases is $e = 0.3$ and the coefficient of kinetic friction between the floor DE and each suitcase is $\mu_k = 0.4$, determine (a) the velocity of A just before impact, (b) the velocities of A and B just after impact, and (c) the distance B slides before coming to rest.

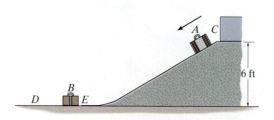

Prob. 15–58

15–59. The 2-kg ball is thrown at the suspended 20-kg block with a velocity of $4\ \text{m/s}$. If the coefficient of restitution between the ball and the block is $e = 0.8$, determine the maximum height h to which the block will swing before it momentarily stops.

***15–60.** The 2-kg ball is thrown at the suspended 20-kg block with a velocity of $4\ \text{m/s}$. If the time of impact between the ball and the block is $0.005\ \text{s}$, determine the average normal force exerted on the block during this time. Take $e = 0.8$.

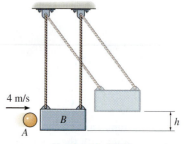

Probs. 15–59/60

•15–61. The slider block B is confined to move within the smooth slot. It is connected to two springs, each of which has a stiffness of $k = 30\ \text{N/m}$. They are originally stretched $0.5\ \text{m}$ when $s = 0$ as shown. Determine the maximum distance, s_{max}, block B moves after it is hit by block A which is originally traveling at $(v_A)_1 = 8\ \text{m/s}$. Take $e = 0.4$ and the mass of each block to be $1.5\ \text{kg}$.

15–62. In Prob. 15–61 determine the average net force between blocks A and B during impact if the impact occurs in $0.005\ \text{s}$.

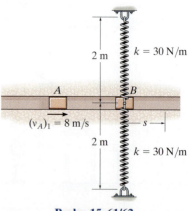

Probs. 15–61/62

15–63. The pile P has a mass of 800 kg and is being driven into *loose sand* using the 300-kg hammer C which is dropped a distance of 0.5 m from the top of the pile. Determine the initial speed of the pile just after it is struck by the hammer. The coefficient of restitution between the hammer and the pile is $e = 0.1$. Neglect the impulses due to the weights of the pile and hammer and the impulse due to the sand during the impact.

*15–64. The pile P has a mass of 800 kg and is being driven into *loose sand* using the 300-kg hammer C which is dropped a distance of 0.5 m from the top of the pile. Determine the distance the pile is driven into the sand after one blow if the sand offers a frictional resistance against the pile of 18 kN. The coefficient of restitution between the hammer and the pile is $e = 0.1$. Neglect the impulses due to the weights of the pile and hammer and the impulse due to the sand during the impact.

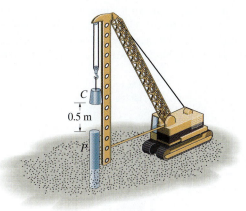

Probs. 15–63/64

•15–65. The girl throws the ball with a horizontal velocity of $v_1 = 8$ ft/s. If the coefficient of restitution between the ball and the ground is $e = 0.8$, determine (a) the velocity of the ball just after it rebounds from the ground and (b) the maximum height to which the ball rises after the first bounce.

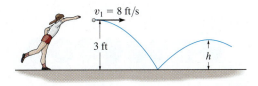

$v_1 = 8$ ft/s

3 ft

h

Prob. 15–65

15–66. During an impact test, the 2000-lb weight is released from rest when $\theta = 60°$. It swings downwards and strikes the concrete blocks, rebounds and swings back up to $\theta = 15°$ before it momentarily stops. Determine the coefficient of restitution between the weight and the blocks. Also, find the impulse transferred between the weight and blocks during impact. Assume that the blocks do not move after impact.

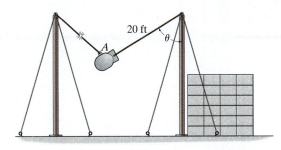

20 ft

θ

A

Prob. 15–66

15–67. The 100-lb crate A is released from rest onto the smooth ramp. After it slides down the ramp it strikes the 200-lb crate B that rests against the spring of stiffness $k = 600$ lb/ft. If the coefficient of restitution between the crates is $e = 0.5$, determine their velocities just after impact. Also, what is the spring's maximum compression? The spring is originally unstretched.

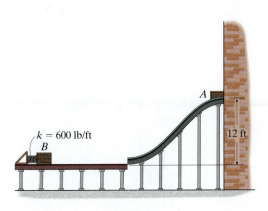

$k = 600$ lb/ft

B

A

12 ft

Prob. 15–67

15

*15–68.   A ball has a mass m and is dropped onto a surface from a height h. If the coefficient of restitution is e between the ball and the surface, determine the time needed for the ball to stop bouncing.

•15–69.   To test the manufactured properties of 2-lb steel balls, each ball is released from rest as shown and strikes the 45° smooth inclined surface. If the coefficient of restitution is to be $e = 0.8$, determine the distance s to where the ball strikes the horizontal plane at A. At what speed does the ball strike point A?

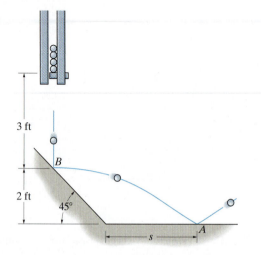

Prob. 15–69

15–70.   Two identical balls A and B of mass m are suspended from cords of length $L/2$ and L, respectively. Ball A is released from rest when $\phi = 90°$ and swings down to $\phi = 0°$, where it strikes B. Determine the speed of each ball just after impact and the maximum angle θ through which B will swing. The coefficient of restitution between the balls is e.

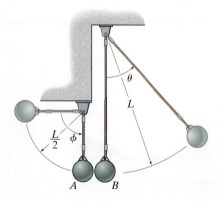

Prob. 15–70

15–71.   The 5-Mg truck and 2-Mg car are traveling with the free-rolling velocities shown just before they collide. After the collision, the car moves with a velocity of 15 km/h to the right *relative* to the truck. Determine the coefficient of restitution between the truck and car and the loss of energy due to the collision.

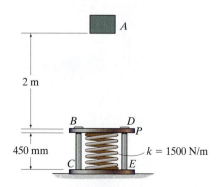

Prob. 15–71

*15–72.   A 10-kg block A is released from rest 2 m above the 5-kg plate P, which can slide freely along the smooth vertical guides BC and DE. Determine the velocity of the block and plate just after impact. The coefficient of restitution between the block and the plate is $e = 0.75$. Also, find the maximum compression of the spring due to impact. The spring has an unstretched length of 600 mm.

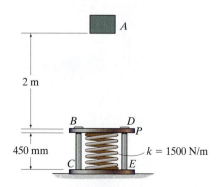

Prob. 15–72

•15–73.   A row of n similar spheres, each of mass m, are placed next to each other as shown. If sphere 1 has a velocity of v_1, determine the velocity of the nth sphere just after being struck by the adjacent $(n - 1)$th sphere. The coefficient of restitution between the spheres is e.

Prob. 15–73

15–74. The three balls each have a mass of *m*. If *A* is released from rest at θ, determine the angle ϕ to which *C* rises after collision. The coefficient of restitution between each ball is *e*.

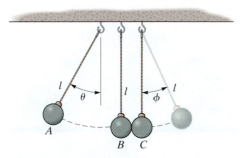

Prob. 15–74

15–75. The cue ball *A* is given an initial velocity $(v_A)_1 = 5$ m/s. If it makes a direct collision with ball *B* ($e = 0.8$), determine the velocity of *B* and the angle θ just after it rebounds from the cushion at *C* ($e' = 0.6$). Each ball has a mass of 0.4 kg. Neglect the size of each ball.

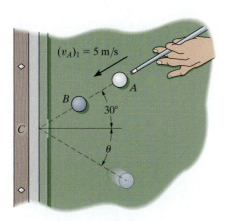

Prob. 15–75

*****15–76.** The girl throws the 0.5-kg ball toward the wall with an initial velocity $v_A = 10$ m/s. Determine (a) the velocity at which it strikes the wall at *B*, (b) the velocity at which it rebounds from the wall if the coefficient of restitution $e = 0.5$, and (c) the distance *s* from the wall to where it strikes the ground at *C*.

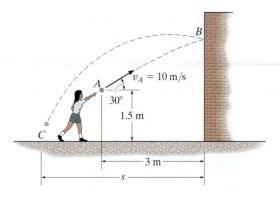

Prob. 15–76

•15–77. A 300-g ball is kicked with a velocity of $v_A = 25$ m/s at point *A* as shown. If the coefficient of restitution between the ball and the field is $e = 0.4$, determine the magnitude and direction θ of the velocity of the rebounding ball at *B*.

Prob. 15–77

15–78. Using a slingshot, the boy fires the 0.2-lb marble at the concrete wall, striking it at *B*. If the coefficient of restitution between the marble and the wall is $e = 0.5$, determine the speed of the marble after it rebounds from the wall.

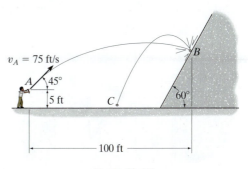

Prob. 15–78

15

15–79. The 2-kg ball is thrown so that it travels horizontally at 10 m/s when it strikes the 6-kg block as it is traveling down the inclined plane at 1 m/s. If the coefficient of restitution between the ball and the block is $e = 0.6$, determine the speeds of the ball and the block just after the impact. Also, what distance does B slide up the plane before it momentarily stops? The coefficient of kinetic friction between the block and the plane is $\mu_k = 0.4$.

*15–80. The 2-kg ball is thrown so that it travels horizontally at 10 m/s when it strikes the 6-kg block as it travels down the smooth inclined plane at 1 m/s. If the coefficient of restitution between the ball and the block is $e = 0.6$, and the impact occurs in 0.006 s, determine the average impulsive force between the ball and block.

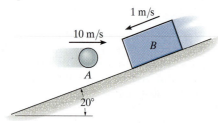

Probs. 15–79/80

•15–81. Two cars A and B each have a weight of 4000 lb and collide on the icy pavement of an intersection. The direction of motion of each car after collision is measured from snow tracks as shown. If the driver in car A states that he was going 44 ft/s (30 mi/h) just before collision and that after collision he applied the brakes so that his car skidded 10 ft before stopping, determine the approximate speed of car B just before the collision. Assume that the coefficient of kinetic friction between the car wheels and the pavement is $\mu_k = 0.15$. *Note:* The line of impact has not been defined; however, this information is not needed for the solution.

Prob. 15–81

15–82. The pool ball A travels with a velocity of 10 m/s just before it strikes ball B, which is at rest. If the masses of A and B are each 200 g, and the coefficient of restitution between them is $e = 0.8$, determine the velocity of both balls just after impact.

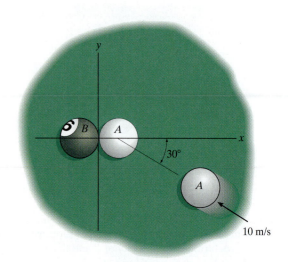

Prob. 15–82

15–83. Two coins A and B have the initial velocities shown just before they collide at point O. If they have weights of $W_A = 13.2(10^{-3})$ lb and $W_B = 6.60(10^{-3})$ lb and the surface upon which they slide is smooth, determine their speeds just after impact. The coefficient of restitution is $e = 0.65$.

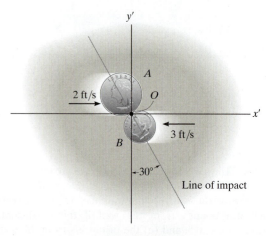

Prob. 15–83

*15–84. Two disks A and B weigh 2 lb and 5 lb, respectively. If they are sliding on the smooth horizontal plane with the velocities shown, determine their velocities just after impact. The coefficient of restitution between the disks is $e = 0.6$.

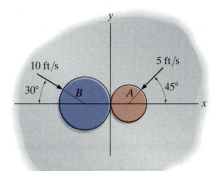

Prob. 15–84

•15–85. Disks A and B have a mass of 15 kg and 10 kg, respectively. If they are sliding on a smooth horizontal plane with the velocities shown, determine their speeds just after impact. The coefficient of restitution between them is $e = 0.8$.

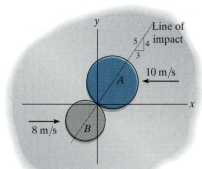

Prob. 15–85

15–86. Disks A and B have a mass of 6 kg and 4 kg, respectively. If they are sliding on the smooth horizontal plane with the velocities shown, determine their speeds just after impact. The coefficient of restitution between the disks is $e = 0.6$.

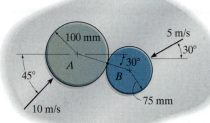

Prob. 15–86

15–87. Disks A and B weigh 8 lb and 2 lb, respectively. If they are sliding on the smooth horizontal plane with the velocities shown, determine their speeds just after impact. The coefficient of restitution between them is $e = 0.5$.

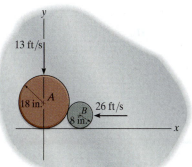

Prob. 15–87

*15–88. Ball A strikes ball B with an initial velocity of $(\mathbf{v}_A)_1$ as shown. If both balls have the same mass and the collision is perfectly elastic, determine the angle θ after collision. Ball B is originally at rest. Neglect the size of each ball.

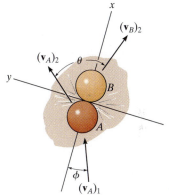

Prob. 15–88

•15–89. Two disks A and B each have a weight of 2 lb and the initial velocities shown just before they collide. If the coefficient of restitution is $e = 0.5$, determine their speeds just after impact.

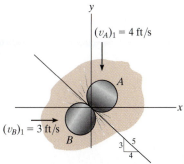

Prob. 15–89

15.5   Angular Momentum

The *angular momentum* of a particle about point O is defined as the "moment" of the particle's linear momentum about O. Since this concept is analogous to finding the moment of a force about a point, the angular momentum, $\mathbf{H}_O$, is sometimes referred to as the *moment of momentum*.

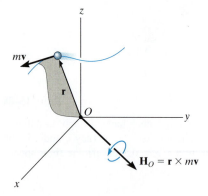

Fig. 15–19

Scalar Formulation.   If a particle moves along a curve lying in the *x*–*y* plane, Fig. 15–19, the angular momentum at any instant can be determined about point O (actually the z axis) by using a scalar formulation. The *magnitude* of $\mathbf{H}_O$ is

$$(H_O)_z = (d)(mv) \tag{15–12}$$

Here d is the moment arm or perpendicular distance from O to the line of action of mv. Common units for $(H_O)_z$ are $\text{kg} \cdot \text{m}^2/\text{s}$ or $\text{slug} \cdot \text{ft}^2/\text{s}$. The *direction* of $\mathbf{H}_O$ is defined by the right-hand rule. As shown, the curl of the fingers of the right hand indicates the sense of rotation of mv about O, so that in this case the thumb (or $\mathbf{H}_O$) is directed perpendicular to the *x*–*y* plane along the $+z$ axis.

Vector Formulation.   If the particle moves along a space curve, Fig. 15–20, the vector cross product can be used to determine the *angular momentum* about O. In this case

$$\mathbf{H}_O = \mathbf{r} \times m\mathbf{v} \tag{15–13}$$

Here $\mathbf{r}$ denotes a position vector drawn from point O to the particle. As shown in the figure, $\mathbf{H}_O$ is *perpendicular* to the shaded plane containing $\mathbf{r}$ and $m\mathbf{v}$.

In order to evaluate the cross product, $\mathbf{r}$ and $m\mathbf{v}$ should be expressed in terms of their Cartesian components, so that the angular momentum can be determined by evaluating the determinant:

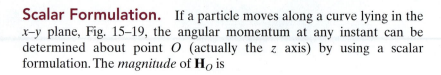

Fig. 15–20

$$\mathbf{H}_O = \begin{vmatrix} \mathbf{i} & \mathbf{j} & \mathbf{k} \\ r_x & r_y & r_z \\ mv_x & mv_y & mv_z \end{vmatrix} \tag{15–14}$$

15.6   Relation Between Moment of a Force and Angular Momentum

The moments about point O of all the forces acting on the particle in Fig. 15–21a can be related to the particle's angular momentum by applying the equation of motion. If the mass of the particle is constant, we may write

$$\Sigma \mathbf{F} = m\dot{\mathbf{v}}$$

The moments of the forces about point O can be obtained by performing a cross-product multiplication of each side of this equation by the position vector $\mathbf{r}$, which is measured from the x, y, z inertial frame of reference. We have

$$\Sigma \mathbf{M}_O = \mathbf{r} \times \Sigma \mathbf{F} = \mathbf{r} \times m\dot{\mathbf{v}}$$

From Appendix B, the derivative of $\mathbf{r} \times m\mathbf{v}$ can be written as

$$\dot{\mathbf{H}}_O = \frac{d}{dt}(\mathbf{r} \times m\mathbf{v}) = \dot{\mathbf{r}} \times m\mathbf{v} + \mathbf{r} \times m\dot{\mathbf{v}}$$

The first term on the right side, $\dot{\mathbf{r}} \times m\mathbf{v} = m(\dot{\mathbf{r}} \times \dot{\mathbf{r}}) = \mathbf{0}$, since the cross product of a vector with itself is zero. Hence, the above equation becomes

$$\boxed{\Sigma \mathbf{M}_O = \dot{\mathbf{H}}_O} \qquad (15\text{–}15)$$

which states that *the resultant moment about point O of all the forces acting on the particle is equal to the time rate of change of the particle's angular momentum about point O.* This result is similar to Eq. 15–1, i.e.,

$$\boxed{\Sigma \mathbf{F} = \dot{\mathbf{L}}} \qquad (15\text{–}16)$$

Here $\mathbf{L} = m\mathbf{v}$, so that *the resultant force acting on the particle is equal to the time rate of change of the particle's linear momentum.*

From the derivations, it is seen that Eqs. 15–15 and 15–16 are actually another way of stating Newton's second law of motion. In other sections of this book it will be shown that these equations have many practical applications when extended and applied to problems involving either a system of particles or a rigid body.

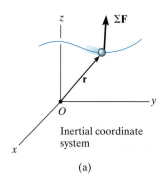

Inertial coordinate system

(a)

Fig. 15–21

15

System of Particles. An equation having the same form as Eq. 15–15 may be derived for the system of particles shown in Fig. 15–21b. The forces acting on the arbitrary *i*th particle of the system consist of a resultant *external force* $\mathbf{F}_i$ and a resultant *internal force* $\mathbf{f}_i$. Expressing the moments of these forces about point O, using the form of Eq. 15–15, we have

$$(\mathbf{r}_i \times \mathbf{F}_i) + (\mathbf{r}_i \times \mathbf{f}_i) = (\dot{\mathbf{H}}_i)_O$$

Here $(\dot{\mathbf{H}}_i)_O$ is the time rate of change in the angular momentum of the *i*th particle about O. Similar equations can be written for each of the other particles of the system. When the results are summed vectorially, the result is

$$\Sigma(\mathbf{r}_i \times \mathbf{F}_i) + \Sigma(\mathbf{r}_i \times \mathbf{f}_i) = \Sigma(\dot{\mathbf{H}}_i)_O$$

The second term is zero since the internal forces occur in equal but opposite collinear pairs, and hence the moment of each pair about point O is zero. Dropping the index notation, the above equation can be written in a simplified form as

$$\Sigma\mathbf{M}_O = \dot{\mathbf{H}}_O \tag{15–17}$$

which states that *the sum of the moments about point O of all the external forces acting on a system of particles is equal to the time rate of change of the total angular momentum of the system about point O.* Although O has been chosen here as the origin of coordinates, it actually can represent any *fixed point* in the inertial frame of reference.

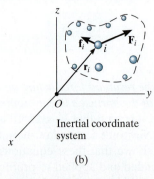

Inertial coordinate system

(b)

Fig. 15–21 (cont.)

EXAMPLE 15.12

The box shown in Fig. 15–22a has a mass m and travels down the smooth circular ramp such that when it is at the angle θ it has a speed v. Determine its angular momentum about point O at this instant and the rate of increase in its speed, i.e., a_t.

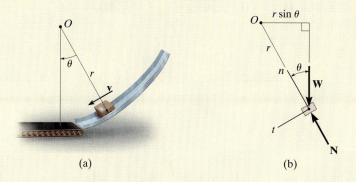

(a)   (b)

Fig. 15–22

SOLUTION

Since **v** is tangent to the path, applying Eq. 15–12 the angular momentum is

$$H_O = rmv \quad \circlearrowleft \qquad\qquad Ans.$$

The rate of increase in its speed (dv/dt) can be found by applying Eq. 15–15. From the free-body diagram of the box, Fig. 15–22b, it can be seen that only the weight $W = mg$ contributes a moment about point O. We have

$$\circlearrowleft + \Sigma M_O = \dot{H}_O; \qquad mg(r \sin \theta) = \frac{d}{dt}(rmv)$$

Since r and m are constant,

$$mgr \sin \theta = rm\frac{dv}{dt}$$

$$\frac{dv}{dt} = g \sin \theta \qquad\qquad Ans.$$

NOTE: This same result can, of course, be obtained from the equation of motion applied in the tangential direction, Fig. 15–22b, i.e.,

$$+\swarrow \Sigma F_t = ma_t; \qquad mg \sin \theta = m\left(\frac{dv}{dt}\right)$$

$$\frac{dv}{dt} = g \sin \theta \qquad\qquad Ans.$$

15.7   Principle of Angular Impulse and Momentum

Principle of Angular Impulse and Momentum.   If Eq. 15–15 is rewritten in the form $\Sigma \mathbf{M}_O \, dt = d\mathbf{H}_O$ and integrated, assuming that at time $t = t_1$, $\mathbf{H}_O = (\mathbf{H}_O)_1$ and at time $t = t_2$, $\mathbf{H}_O = (\mathbf{H}_O)_2$, we have

$$\Sigma \int_{t_1}^{t_2} \mathbf{M}_O \, dt = (\mathbf{H}_O)_2 - (\mathbf{H}_O)_1$$

or

$$(\mathbf{H}_O)_1 + \Sigma \int_{t_1}^{t_2} \mathbf{M}_O \, dt = (\mathbf{H}_O)_2 \qquad (15\text{–}18)$$

This equation is referred to as the *principle of angular impulse and momentum*. The initial and final angular momenta $(\mathbf{H}_O)_1$ and $(\mathbf{H}_O)_2$ are defined as the moment of the linear momentum of the particle $(\mathbf{H}_O = \mathbf{r} \times m\mathbf{v})$ at the instants t_1 and t_2, respectively. The second term on the left side, $\Sigma \int \mathbf{M}_O \, dt$, is called the *angular impulse*. It is determined by integrating, with respect to time, the moments of all the forces acting on the particle over the time period t_1 to t_2. Since the moment of a force about point O is $\mathbf{M}_O = \mathbf{r} \times \mathbf{F}$, the angular impulse may be expressed in vector form as

$$\text{angular impulse} = \int_{t_1}^{t_2} \mathbf{M}_O \, dt = \int_{t_1}^{t_2} (\mathbf{r} \times \mathbf{F}) \, dt \qquad (15\text{–}19)$$

Here $\mathbf{r}$ is a position vector which extends from point O to any point on the line of action of $\mathbf{F}$.

In a similar manner, using Eq. 15–18, the principle of angular impulse and momentum for a system of particles may be written as

$$\Sigma(\mathbf{H}_O)_1 + \Sigma \int_{t_1}^{t_2} \mathbf{M}_O \, dt = \Sigma(\mathbf{H}_O)_2 \qquad (15\text{–}20)$$

Here the first and third terms represent the angular momenta of all the particles $[\Sigma \mathbf{H}_O = \Sigma(\mathbf{r}_i \times m\mathbf{v}_i)]$ at the instants t_1 and t_2. The second term is the sum of the angular impulses given to all the particles from t_1 to t_2. Recall that these impulses are created only by the moments of the external forces acting on the system where, for the ith particle, $\mathbf{M}_O = \mathbf{r}_i \times \mathbf{F}_i$.

Vector Formulation. Using impulse and momentum principles, it is therefore possible to write two equations which define the particle's motion, namely, Eqs. 15–3 and Eqs. 15–18, restated as

$$
\begin{aligned}
m\mathbf{v}_1 + \Sigma \int_{t_1}^{t_2} \mathbf{F}\, dt &= m\mathbf{v}_2 \\
(\mathbf{H}_O)_1 + \Sigma \int_{t_1}^{t_2} \mathbf{M}_O\, dt &= (\mathbf{H}_O)_2
\end{aligned}
\tag{15–21}
$$

Scalar Formulation. In general, the above equations can be expressed in x, y, z component form, yielding a total of six scalar equations. If the particle is confined to move in the x–y plane, three scalar equations can be written to express the motion, namely,

$$
\begin{aligned}
m(v_x)_1 + \Sigma \int_{t_1}^{t_2} F_x\, dt &= m(v_x)_2 \\
m(v_y)_1 + \Sigma \int_{t_1}^{t_2} F_y\, dt &= m(v_y)_2 \\
(H_O)_1 + \Sigma \int_{t_1}^{t_2} M_O\, dt &= (H_O)_2
\end{aligned}
\tag{15–22}
$$

The first two of these equations represent the principle of linear impulse and momentum in the x and y directions, which has been discussed in Sec. 15.1, and the third equation represents the principle of angular impulse and momentum about the z axis.

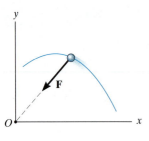

Fig. 15–23

Conservation of Angular Momentum. When the angular impulses acting on a particle are all zero during the time t_1 to t_2, Eq. 15–18 reduces to the following simplified form:

$$(\mathbf{H}_O)_1 = (\mathbf{H}_O)_2 \qquad (15\text{–}23)$$

This equation is known as the *conservation of angular momentum*. It states that from t_1 to t_2 the particle's angular momentum remains constant. Obviously, if no external impulse is applied to the particle, both linear and angular momentum will be conserved. In some cases, however, the particle's angular momentum will be conserved and linear momentum may not. An example of this occurs when the particle is subjected *only* to a *central force* (see Sec. 13.7). As shown in Fig. 15–23, the impulsive central force $\mathbf{F}$ is always directed toward point O as the particle moves along the path. Hence, the angular impulse (moment) created by $\mathbf{F}$ about the z axis is always zero, and therefore angular momentum of the particle is conserved about this axis.

From Eq. 15–20, we can also write the conservation of angular momentum for a system of particles as

$$\Sigma(\mathbf{H}_O)_1 = \Sigma(\mathbf{H}_O)_2 \qquad (15\text{–}24)$$

In this case the summation must include the angular momenta of all particles in the system.

Provided air resistance is neglected, the passengers on this amusement-park ride are subjected to a conservation of angular momentum about the axis of rotation. As shown on the free-body diagram, the line of action of the normal force $\mathbf{N}$ of the seat on the passenger passes through the axis, and the passenger's weight $\mathbf{W}$ is parallel to it. Thus, no angular impulse acts around the z axis.

Procedure for Analysis

When applying the principles of angular impulse and momentum, or the conservation of angular momentum, it is suggested that the following procedure be used.

Free-Body Diagram.

- Draw the particle's free-body diagram in order to determine any axis about which angular momentum may be conserved. For this to occur, the moments of all the forces (or impulses) must either be parallel or pass through the axis so as to create zero moment throughout the time period t_1 to t_2.

- The direction and sense of the particle's initial and final velocities should also be established.

- An alternative procedure would be to draw the impulse and momentum diagrams for the particle.

Momentum Equations.

- Apply the principle of angular impulse and momentum, $(\mathbf{H}_O)_1 + \Sigma \int_{t_1}^{t_2} \mathbf{M}_O dt = (\mathbf{H}_O)_2$, or if appropriate, the conservation of angular momentum, $(\mathbf{H}_O)_1 = (\mathbf{H}_O)_2$.

EXAMPLE 15.13

The 1.5-Mg car travels along the circular road as shown in Fig. 15–24a. If the traction force of the wheels on the road is $F = (150t^2)$ N, where t is in seconds, determine the speed of the car when $t = 5$ s. The car initially travels with a speed of 5 m/s. Negect the size of the car.

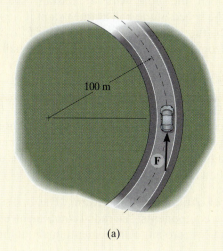

(a)

SOLUTION

Free-Body Diagram. The free-body diagram of the car is shown in Fig. 15–24b. If we apply the principle of angular impulse and momentum about the z axis, then the angular impulse created by the weight, normal force, and radial frictional force will be eliminated since they act parallel to the axis or pass through it.

Principle of Angular Impulse and Momentum.

$$(H_z)_1 + \Sigma \int_{t_1}^{t_2} M_z \, dt = (H_z)_2$$

$$r m_c (v_c)_1 + \int_{t_1}^{t_2} rF \, dt = r m_c (v_c)_2$$

$$(100 \text{ m})(1500 \text{ kg})(5 \text{ m/s}) + \int_0^{5\text{ s}} (100 \text{ m})[(150t^2) \text{ N}] \, dt$$

$$= (100 \text{ m})(1500 \text{ kg})(v_c)_2$$

$$750(10^3) + 5000t^3 \Big|_0^{5\text{ s}} = 150(10^3)(v_c)_2$$

$$(v_c)_2 = 9.17 \text{ m/s} \qquad \qquad Ans.$$

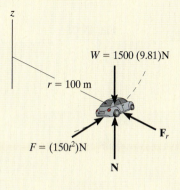

(b)

Fig. 15–24

EXAMPLE | 15.14

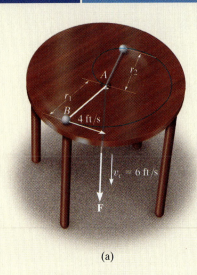

(a)

15

The 0.8-lb ball B, shown in Fig. 15–25a, is attached to a cord which passes through a hole at A in a smooth table. When the ball is $r_1 = 1.75$ ft from the hole, it is rotating around in a circle such that its speed is $v_1 = 4$ ft/s. By applying the force $\mathbf{F}$ the cord is pulled downward through the hole with a constant speed $v_c = 6$ ft/s. Determine (a) the speed of the ball at the instant it is $r_2 = 0.6$ ft from the hole, and (b) the amount of work done by $\mathbf{F}$ in shortening the radial distance from r_1 to r_2. Neglect the size of the ball.

SOLUTION

Part (a) Free-Body Diagram. As the ball moves from r_1 to r_2, Fig. 15–25b, the cord force $\mathbf{F}$ on the ball always passes through the z axis, and the weight and $\mathbf{N}_B$ are parallel to it. Hence the moments, or angular impulses created by these forces, are all *zero* about this axis. Therefore, angular momentum is conserved about the z axis.

Conservation of Angular Momentum. The ball's velocity $\mathbf{v}_2$ is resolved into two components. The radial component, 6 ft/s, is known; however, it produces zero angular momentum about the z axis. Thus,

$$\mathbf{H}_1 = \mathbf{H}_2$$

$$r_1 m_B v_1 = r_2 m_B v_2'$$

$$1.75 \text{ ft}\left(\frac{0.8 \text{ lb}}{32.2 \text{ ft/s}^2}\right)4 \text{ ft/s} = 0.6 \text{ ft}\left(\frac{0.8 \text{ lb}}{32.2 \text{ ft/s}^2}\right)v_2'$$

$$v_2' = 11.67 \text{ ft/s}$$

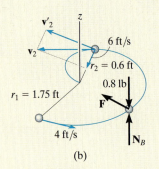

(b)

Fig. 15–25

The speed of the ball is thus

$$v_2 = \sqrt{(11.67 \text{ ft/s})^2 + (6 \text{ ft/s})^2}$$

$$= 13.1 \text{ ft/s}$$

Part (b). The only force that does work on the ball is $\mathbf{F}$. (The normal force and weight do not move vertically.) The initial and final kinetic energies of the ball can be determined so that from the principle of work and energy we have

$$T_1 + \Sigma U_{1-2} = T_2$$

$$\frac{1}{2}\left(\frac{0.8 \text{ lb}}{32.2 \text{ ft/s}^2}\right)(4 \text{ ft/s})^2 + U_F = \frac{1}{2}\left(\frac{0.8 \text{ lb}}{32.2 \text{ ft/s}^2}\right)(13.1 \text{ ft/s})^2$$

$$U_F = 1.94 \text{ ft} \cdot \text{lb} \qquad \qquad Ans.$$

NOTE: The force F is not constant because the normal component of acceleration, $a_n = v^2/r$, changes as r changes.

EXAMPLE 15.15

The 2-kg disk shown in Fig. 15–26a rests on a smooth horizontal surface and is attached to an elastic cord that has a stiffness $k_c = 20$ N/m and is initially unstretched. If the disk is given a velocity $(v_D)_1 = 1.5$ m/s, perpendicular to the cord, determine the rate at which the cord is being stretched and the speed of the disk at the instant the cord is stretched 0.2 m.

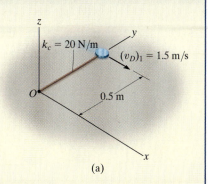

(a)

SOLUTION

Free-Body Diagram. After the disk has been launched, it slides along the path shown in Fig. 15–26b. By inspection, angular momentum about point O (or the z axis) is *conserved*, since none of the forces produce an angular impulse about this axis. Also, when the distance is 0.7 m, only the transverse component $(v_D')_2$ produces angular momentum of the disk about O.

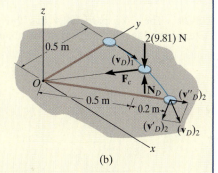

(b)

Fig. 15–26

Conservation of Angular Momentum. The component $(v_D')_2$ can be obtained by applying the conservation of angular momentum about O (the z axis).

$$(H_O)_1 = (H_O)_2$$

$$r_1 m_D (v_D)_1 = r_2 m_D (v_D')_2$$

$$0.5 \text{ m } (2 \text{ kg})(1.5 \text{ m/s}) = 0.7 \text{ m} (2 \text{ kg})(v_D')_2$$

$$(v_D')_2 = 1.071 \text{ m/s}$$

Conservation of Energy. The speed of the disk can be obtained by applying the conservation of energy equation at the point where the disk was launched and at the point where the cord is stretched 0.2 m.

$$T_1 + V_1 = T_2 + V_2$$

$$\tfrac{1}{2} m_D (v_D)_1^2 + \tfrac{1}{2} k x_1^2 = \tfrac{1}{2} m_D (v_D)_2^2 + \tfrac{1}{2} k x_2^2$$

$$\tfrac{1}{2}(2 \text{ kg})(1.5 \text{ m/s})^2 + 0 = \tfrac{1}{2}(2 \text{ kg})(v_D)_2^2 + \tfrac{1}{2}(20 \text{ N/m})(0.2 \text{ m})^2$$

$$(v_D)_2 = 1.360 \text{ m/s} = 1.36 \text{ m/s} \qquad \textit{Ans.}$$

Having determined $(v_D)_2$ and its component $(v_D')_2$, the rate of stretch of the cord, or radial component, $(v_D'')_2$ is determined from the Pythagorean theorem,

$$(v_D'')_2 = \sqrt{(v_D)_2^2 - (v_D')_2^2}$$

$$= \sqrt{(1.360 \text{ m/s})^2 - (1.071 \text{ m/s})^2}$$

$$= 0.838 \text{ m/s} \qquad \textit{Ans.}$$

FUNDAMENTAL PROBLEMS

F15–19. The 2-kg particle A has the velocity shown. Determine its angular momentum $\mathbf{H}_O$ about point O.

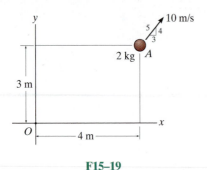

F15–19

F15–20. The 2-kg particle A has the velocity shown. Determine its angular momentum $\mathbf{H}_P$ about point P.

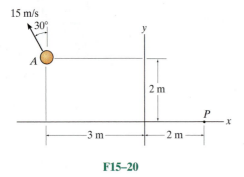

F15–20

F15–21. Initially the 5-kg block is rotating with a constant speed of 2 m/s around the circular path centered at O on the smooth horizontal plane. If a constant tangential force $F = 5$ N is applied to the block, determine its speed when $t = 3$ s. Neglect the size of the block.

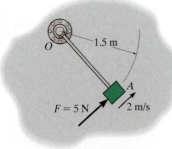

F15–21

F15–22. The 5-kg block is rotating around the circular path centered at O on the smooth horizontal plane when it is subjected to the force $F = (10t)$ N, where t is in seconds. If the block starts from rest, determine its speed when $t = 4$ s. Neglect the size of the block. The force maintains the same constant angle tangent to the path.

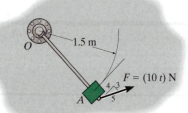

F15–22

F15–23. The 2-kg sphere is attached to the light rigid rod, which rotates in the *horizontal plane* centered at O. If the system is subjected to a couple moment $M = (0.9t^2)$ N · m, where t is in seconds, determine the speed of the sphere at the instant $t = 5$ s starting from rest.

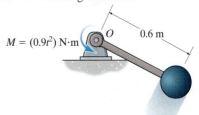

F15–23

F15–24. Two identical 10-kg spheres are attached to the light rigid rod, which rotates in the *horizontal plane* centered at O. If the spheres are subjected to tangential forces of $P = 10$ N, and the rod is subjected to a couple moment $M = (8t)$ N · m, where t is in seconds, determine the speed of the spheres at the instant $t = 4$ s. The system starts from rest. Neglect the size of the spheres.

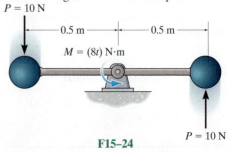

F15–24

PROBLEMS

15–90. The spheres A and B each weighing 4 lb, are welded to the light rods that are rigidly connected to a shaft as shown. If the shaft is subjected to a couple moment of $M = (4t^2 + 2)$ lb · ft, where t is in seconds, determine the speed of A and B when $t = 3$ s. The system starts from rest. Neglect the size of the spheres.

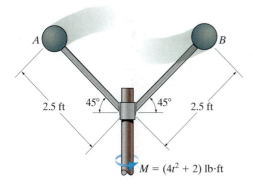

Prob. 15–90

15–91. If the rod of negligible mass is subjected to a couple moment of $M = (30t^2)$ N · m and the engine of the car supplies a traction force of $F = (15t)$ N to the wheels, where t is in seconds, determine the speed of the car at the instant $t = 5$ s. The car starts from rest. The total mass of the car and rider is 150 kg. Neglect the size of the car.

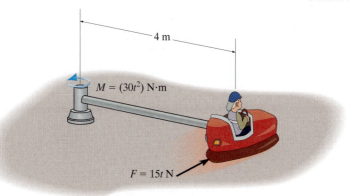

Prob. 15–91

15–92. The 10-lb block rests on a surface for which $\mu_k = 0.5$. It is acted upon by a radial force of 2 lb and a horizontal force of 7 lb, always directed at 30° from the tangent to the path as shown. If the block is initially moving in a circular path with a speed $v_1 = 2$ ft/s at the instant the forces are applied, determine the time required before the tension in cord AB becomes 20 lb. Neglect the size of the block for the calculation.

15–93. The 10-lb block is originally at rest on the smooth surface. It is acted upon by a radial force of 2 lb and a horizontal force of 7 lb, always directed at 30° from the tangent to the path as shown. Determine the time required to break the cord, which requires a tension $T = 30$ lb. What is the speed of the block when this occurs? Neglect the size of the block for the calculation.

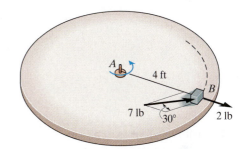

Probs. 15–92/93

15–94. The projectile having a mass of 3 kg is fired from a cannon with a muzzle velocity of $v_0 = 500$ m/s. Determine the projectile's angular momentum about point O at the instant it is at the maximum height of its trajectory.

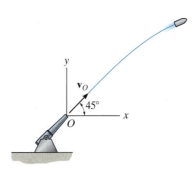

Prob. 15–94

15–95. The 3-lb ball located at *A* is released from rest and travels down the curved path. If the ball exerts a normal force of 5 lb on the path when it reaches point *B*, determine the angular momentum of the ball about the center of curvature, point *O*. *Hint:* Neglect the size of the ball. The radius of curvature at point *B* must first be determined.

•15–97. The two spheres each have a mass of 3 kg and are attached to the rod of negligible mass. If a torque $M = (6e^{0.2t})$ N · m, where *t* is in seconds, is applied to the rod as shown, determine the speed of each of the spheres in 2 s, starting from rest.

15–98. The two spheres each have a mass of 3 kg and are attached to the rod of negligible mass. Determine the time the torque $M = (8t)$ N · m, where *t* is in seconds, must be applied to the rod so that each sphere attains a speed of 3 m/s starting from rest.

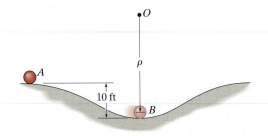

Prob. 15–95

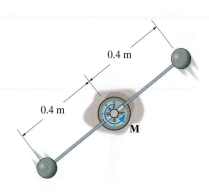

Probs. 15–97/98

*****15–96.** The ball *B* has a mass of 10 kg and is attached to the end of a rod whose mass can be neglected. If the shaft is subjected to a torque $M = (2t^2 + 4)$ N · m, where *t* is in seconds, determine the speed of the ball when *t* = 2 s. The ball has a speed *v* = 2 m/s when *t* = 0.

15–99. An amusement park ride consists of a car which is attached to the cable *OA*. The car rotates in a horizontal circular path and is brought to a speed $v_1 = 4$ ft/s when *r* = 12 ft. The cable is then pulled in at the constant rate of 0.5 ft/s. Determine the speed of the car in 3 s.

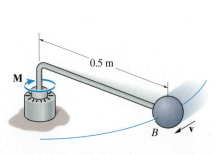

Prob. 15–96

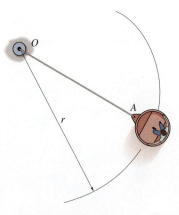

Prob. 15–99

*15–100. An earth satellite of mass 700 kg is launched into a free-flight trajectory about the earth with an initial speed of $v_A = 10$ km/s when the distance from the center of the earth is $r_A = 15$ Mm. If the launch angle at this position is $\phi_A = 70°$, determine the speed v_B of the satellite and its closest distance r_B from the center of the earth. The earth has a mass $M_e = 5.976(10^{24})$ kg. *Hint:* Under these conditions, the satellite is subjected only to the earth's gravitational force, $F = GM_e m_s/r^2$, Eq. 13–1. For part of the solution, use the conservation of energy.

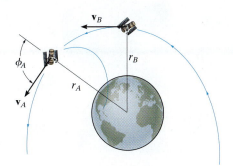

Prob. 15–100

•15–101. The 2-kg ball rotates around a 0.5-m-diameter circular path with a constant speed. If the cord length is shortened from $l = 1$ m to $l' = 0.5$ m, by pulling the cord through the tube, determine the new diameter of the path d'. Also, what is the tension in the cord in each case?

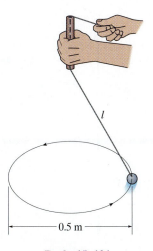

Prob. 15–101

15–102. A gymnast having a mass of 80 kg holds the two rings with his arms down in the position shown as he swings downward. His center of mass is located at point G_1. When he is at the lowest position of his swing, his velocity is $(v_G)_1 = 5$ m/s. At this position he *suddenly* lets his arms come up, shifting his center of mass to position G_2. Determine his new velocity in the upswing and the angle θ to which he swings before momentarily coming to rest. Treat his body as a particle.

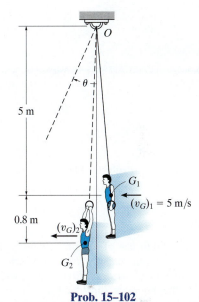

Prob. 15–102

15–103. The four 5-lb spheres are rigidly attached to the crossbar frame having a negligible weight. If a couple moment $M = (0.5t + 0.8)$ lb·ft, where t is in seconds, is applied as shown, determine the speed of each of the spheres in 4 seconds starting from rest. Neglect the size of the spheres.

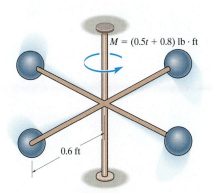

Prob. 15–103

*15–104. At the instant $r = 1.5$ m, the 5-kg disk is given a speed of $v = 5$ m/s, perpendicular to the elastic cord. Determine the speed of the disk and the rate of shortening of the elastic cord at the instant $r = 1.2$ m. The disk slides on the smooth horizontal plane. Neglect its size. The cord has an unstretched length of 0.5 m.

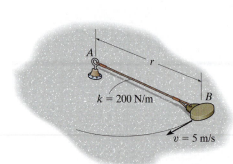

Prob. 15–104

15–106. A small ball bearing of mass m is given a velocity of v_0 at A parallel to the horizontal rim of a smooth bowl. Determine the magnitude of the velocity $\mathbf{v}$ of the ball when it has fallen through a vertical distance h to reach point B. Angle θ is measured from $\mathbf{v}$ to the horizontal at point B.

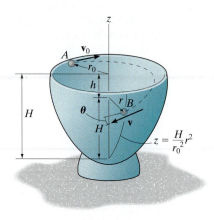

Prob. 15–106

•15–105. The 150-lb car of an amusement park ride is connected to a rotating telescopic boom. When $r = 15$ ft, the car is moving on a horizontal circular path with a speed of 30 ft/s. If the boom is shortened at a rate of 3 ft/s, determine the speed of the car when $r = 10$ ft. Also, find the work done by the axial force $\mathbf{F}$ along the boom. Neglect the size of the car and the mass of the boom.

Prob. 15–105

15–107. When the 2-kg bob is given a horizontal speed of 1.5 m/s, it begins to rotate around the horizontal circular path A. If the force $\mathbf{F}$ on the cord is increased, the bob rises and then rotates around the horizontal circular path B. Determine the speed of the bob around path B. Also, find the work done by force $\mathbf{F}$.

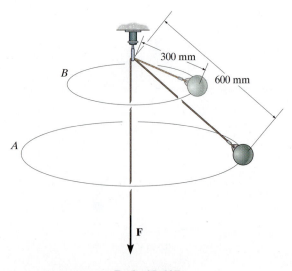

Prob. 15–107

15.8   Steady Flow of a Fluid Stream

Up to this point we have restricted our study of impulse and momentum principles to a system of particles contained within a *closed volume*. In this section, however, we will apply the principle of impulse and momentum to the steady mass flow of fluid particles entering into and then out of a *control volume*. This volume is defined as a region in space where fluid particles can flow into or out of a region. The size and shape of the control volume is frequently made to coincide with the solid boundaries and openings of a pipe, turbine, or pump. Provided the flow of the fluid into the control volume is equal to the flow out, then the flow can be classified as *steady flow*.

Principle of Impulse and Momentum. Consider the steady flow of a fluid stream in Fig. 15–27a that passes through a pipe. The region within the pipe and its openings will be taken as the control volume. As shown, the fluid flows into and out of the control volume with velocities $\mathbf{v}_A$ and $\mathbf{v}_B$, respectively. The change in the direction of the fluid flow within the control volume is caused by an impulse produced by the resultant external force exerted on the control surface by the wall of the pipe. This resultant force can be determined by applying the principle of impulse and momentum to the control volume.

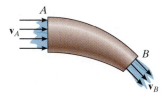

(a)

Fig. 15–27

The conveyor belt must supply frictional forces to the gravel that falls upon it in order to change the momentum of the gravel stream, so that it begins to travel along the belt.

The air on one side of this fan is essentially at rest, and as it passes through the blades its momentum is increased. To change the momentum of the air flow in this manner, the blades must exert a horizontal thrust on the air stream. As the blades turn faster, the equal but opposite thrust of the air on the blades could overcome the rolling resistance of the wheels on the ground and begin to move the frame of the fan.

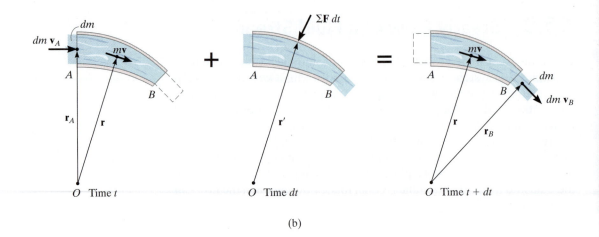

(b)

15

As indicated in Fig. 15–27b, a small amount of fluid having a mass dm is about to enter the control volume through opening A with a velocity of $\mathbf{v}_A$ at time t. Since the flow is considered steady, at time $t + dt$, the same amount of fluid will leave the control volume through opening B with a velocity $\mathbf{v}_B$. The momenta of the fluid entering and leaving the control volume are therefore $dm\,\mathbf{v}_A$ and $dm\,\mathbf{v}_B$, respectively. Also, during the time dt, the momentum of the fluid mass within the control volume remains constant and is denoted as $m\mathbf{v}$. As shown on the center diagram, the resultant external force exerted on the control volume produces the impulse $\Sigma\mathbf{F}\,dt$. If we apply the principle of linear impulse and momentum, we have

$$dm\,\mathbf{v}_A + m\mathbf{v} + \Sigma\mathbf{F}\,dt = dm\,\mathbf{v}_B + m\mathbf{v}$$

If $\mathbf{r}, \mathbf{r}_A, \mathbf{r}_B$ are position vectors measured from point O to the geometric centers of the control volume and the openings at A and B, Fig. 15–27b, then the principle of angular impulse and momentum about O becomes

$$\mathbf{r}_A \times dm\,\mathbf{v}_A + \mathbf{r} \times m\mathbf{v} + \mathbf{r}' \times \Sigma\mathbf{F}\,dt = \mathbf{r} \times m\mathbf{v} + \mathbf{r}_B \times dm\,\mathbf{v}_B$$

Dividing both sides of the above two equations by dt and simplifying, we get

$$\Sigma\mathbf{F} = \frac{dm}{dt}(\mathbf{v}_B - \mathbf{v}_A) \tag{15–25}$$

$$\Sigma\mathbf{M}_O = \frac{dm}{dt}(\mathbf{r}_B \times \mathbf{v}_B - \mathbf{r}_A \times \mathbf{v}_A) \tag{15–26}$$

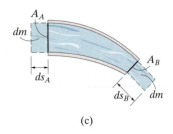

(c)

Fig. 15–27 (cont.)

The term dm/dt is called the *mass flow*. It indicates the constant amount of fluid which flows either into or out of the control volume per unit of time. If the cross-sectional areas and densities of the fluid at the entrance A are A_A, ρ_A and at exit B, A_B, ρ_B, Fig. 15–27c, then for an incompressible fluid, the *continuity of mass* requires $dm = \rho dV = \rho_A(ds_A A_A) = \rho_B(ds_B A_B)$. Hence, during the time dt, since $v_A = ds_A/dt$ and $v_B = ds_B/dt$, we have $dm/dt = \rho_A v_A A_A = \rho_B v_B A_B$ or in general,

$$\frac{dm}{dt} = \rho v A = \rho Q \qquad (15\text{–}27)$$

The term $Q = vA$ measures the volume of fluid flow per unit of time and is referred to as the *discharge* or the *volumetric flow*.

15

Procedure for Analysis

Problems involving steady flow can be solved using the following procedure.

Kinematic Diagram.

- Identify the control volume. If it is *moving*, a *kinematic diagram* may be helpful for determining the entrance and exit velocities of the fluid flowing into and out of its openings since a *relative-motion analysis* of velocity will be involved.

- The measurement of velocities v_A and v_B must be made by an observer fixed in an inertial frame of reference.

- Once the velocity of the fluid flowing into the control volume is determined, the mass flow is calculated using Eq. 15–27.

Free-Body Diagram.

- Draw the free-body diagram of the control volume in order to establish the forces $\Sigma \mathbf{F}$ that act on it. These forces will include the support reactions, the weight of all solid parts and the fluid contained within the control volume, and the static gauge pressure forces of the fluid on the entrance and exit sections.* The gauge pressure is the pressure measured above atmospheric pressure, and so if an opening is exposed to the atmosphere, the gauge pressure there will be zero.

Equations of Steady Flow.

- Apply the equations of steady flow, Eqs. 15–25 and 15–26, using the appropriate components of velocity and force shown on the kinematic and free-body diagrams.

* In the SI system, pressure is measured using the pascal (Pa), where $1\text{Pa} = 1 \text{ N/m}^2$.

EXAMPLE 15.16

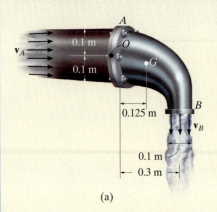

(a)

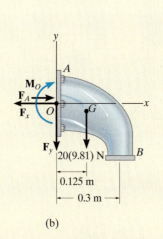

(b)

Fig. 15–28

Determine the components of reaction which the fixed pipe joint at A exerts on the elbow in Fig. 15–28a, if water flowing through the pipe is subjected to a static gauge pressure of 100 kPa at A. The discharge at B is $Q_B = 0.2$ m³/s. Water has a density $\rho_w = 1000$ kg/m³, and the water-filled elbow has a mass of 20 kg and center of mass at G.

SOLUTION

We will consider the control volume to be the outer surface of the elbow. Using a fixed inertial coordinate system, the velocity of flow at A and B and the mass flow rate can be obtained from Eq. 15–27. Since the density of water is constant, $Q_B = Q_A = Q$. Hence,

$$\frac{dm}{dt} = \rho_w Q = (1000 \text{ kg/m}^3)(0.2 \text{ m}^3/\text{s}) = 200 \text{ kg/s}$$

$$v_B = \frac{Q}{A_B} = \frac{0.2 \text{ m}^3/\text{s}}{\pi(0.05 \text{ m})^2} = 25.46 \text{ m/s} \downarrow$$

$$v_A = \frac{Q}{A_A} = \frac{0.2 \text{ m}^3/\text{s}}{\pi(0.1 \text{ m})^2} = 6.37 \text{ m/s} \rightarrow$$

Free-Body Diagram. As shown on the free-body diagram of the control volume (elbow) Fig. 15–28b, the *fixed* connection at A exerts a resultant couple moment $\mathbf{M}_O$ and force components $\mathbf{F}_x$ and $\mathbf{F}_y$ on the elbow. Due to the static pressure of water in the pipe, the pressure force acting on the open control surface at A is $F_A = p_A A_A$. Since 1 kPa = 1000 N/m²,

$$F_A = p_A A_A = [100(10^3) \text{ N/m}^2][\pi(0.1 \text{ m})^2] = 3141.6 \text{ N}$$

There is no static pressure acting at B, since the water is discharged at atmospheric pressure; i.e., the pressure measured by a gauge at B is equal to zero, $p_B = 0$.

Equations of Steady Flow.

$$\xrightarrow{+} \Sigma F_x = \frac{dm}{dt}(v_{Bx} - v_{Ax}); \quad -F_x + 3141.6 \text{ N} = 200 \text{ kg/s}(0 - 6.37 \text{ m/s})$$

$$F_x = 4.41 \text{ kN} \qquad \qquad Ans.$$

$$+\uparrow \Sigma F_y = \frac{dm}{dt}(v_{By} - v_{Ay}); \quad -F_y - 20(9.81) \text{ N} = 200 \text{ kg/s}(-25.46 \text{ m/s} - 0)$$

$$F_y = 4.90 \text{ kN} \qquad \qquad Ans.$$

If moments are summed about point O, Fig. 15–28b, then $\mathbf{F}_x$, $\mathbf{F}_y$, and the static pressure $\mathbf{F}_A$ are eliminated, as well as the moment of momentum of the water entering at A, Fig. 15–28a. Hence,

$$\zeta + \Sigma M_O = \frac{dm}{dt}(d_{OB}v_B - d_{OA}v_A)$$

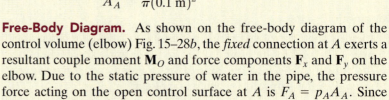

$$M_O + 20(9.81) \text{ N} (0.125 \text{ m}) = 200 \text{ kg/s}[(0.3 \text{ m})(25.46 \text{ m/s}) - 0]$$

$$M_O = 1.50 \text{ kN} \cdot \text{m} \qquad \qquad Ans.$$

EXAMPLE 15.17

A 2-in.-diameter water jet having a velocity of 25 ft/s impinges upon a single moving blade, Fig. 15–29a. If the blade moves with a constant velocity of 5 ft/s away from the jet, determine the horizontal and vertical components of force which the blade is exerting on the water. What power does the water generate on the blade? Water has a specific weight of $\gamma_w = 62.4$ lb/ft^3.

SOLUTION

Kinematic Diagram.   Here the control volume will be the stream of water on the blade. From a fixed inertial coordinate system, Fig. 15–29b, the rate at which water enters the control volume at A is

$$\mathbf{v}_A = \{25\mathbf{i}\} \text{ ft/s}$$

The *relative-flow velocity* within the control volume is $\mathbf{v}_{w/cv} = \mathbf{v}_w - \mathbf{v}_{cv} = 25\mathbf{i} - 5\mathbf{i} = \{20\mathbf{i}\}$ ft/s. Since the control volume is moving with a velocity of $\mathbf{v}_{cv} = \{5\mathbf{i}\}$ ft/s, the velocity of flow at B measured from the fixed x, y axes is the vector sum, shown in Fig. 15–29b. Here,

$$\mathbf{v}_B = \mathbf{v}_{cv} + \mathbf{v}_{w/cv}$$
$$= \{5\mathbf{i} + 20\mathbf{j}\} \text{ ft/s}$$

Thus, the mass flow of water *onto* the control volume that undergoes a momentum change is

$$\frac{dm}{dt} = \rho_w(v_{w/cv})A_A = \left(\frac{62.4}{32.2}\right)(20)\left[\pi\left(\frac{1}{12}\right)^2\right] = 0.8456 \text{ slug/s}$$

Free-Body Diagram.   The free-body diagram of the control volume is shown in Fig. 15–29c. The weight of the water will be neglected in the calculation, since this force will be small compared to the reactive components $\mathbf{F}_x$ and $\mathbf{F}_y$.

Equations of Steady Flow.

$$\Sigma\mathbf{F} = \frac{dm}{dt}(\mathbf{v}_B - \mathbf{v}_A)$$
$$-F_x\mathbf{i} + F_y\mathbf{j} = 0.8456(5\mathbf{i} + 20\mathbf{j} - 25\mathbf{i})$$

Equating the respective $\mathbf{i}$ and $\mathbf{j}$ components gives

$$F_x = 0.8456(20) = 16.9 \text{ lb} \leftarrow \qquad Ans.$$
$$F_y = 0.8456(20) = 16.9 \text{ lb} \uparrow \qquad Ans.$$

The water exerts equal but opposite forces on the blade.
Since the water force which causes the blade to move forward horizontally with a velocity of 5 ft/s is $F_x = 16.9$ lb, then from Eq. 14–10 the power is

$$P = \mathbf{F} \cdot \mathbf{v}; \qquad P = \frac{16.9 \text{ lb}(5 \text{ ft/s})}{550 \text{ hp}/(\text{ft} \cdot \text{lb/s})} = 0.154 \text{ hp}$$

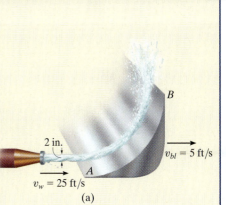

(a)

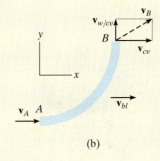

(b)

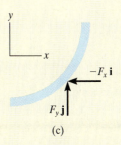

(c)

Fig. 15–29

15

*15.9   Propulsion with Variable Mass

A Control Volume That Loses Mass.

Consider a device such as a rocket which at an instant of time has a mass m and is moving forward with a velocity $\mathbf{v}$, Fig. 15–30a. At this same instant the amount of mass m_e is expelled from the device with a mass flow velocity $\mathbf{v}_e$. For the analysis, the control volume will include *both the mass m of the device and the expelled mass m_e*. The impulse and momentum diagrams for the control volume are shown in Fig. 15–30b. During the time dt, its velocity is increased from $\mathbf{v}$ to $\mathbf{v} + d\mathbf{v}$ since an amount of mass dm_e has been ejected and thereby gained in the exhaust. This increase in forward velocity, however, does not change the velocity $\mathbf{v}_e$ of the expelled mass, as seen by a fixed observer, since this mass moves with a constant velocity once it has been ejected. The impulses are created by $\Sigma \mathbf{F}_{cv}$, which represents the resultant of all the external forces, such as drag or weight, that *act on the control volume* in the direction of motion. This force resultant *does not include* the force which causes the control volume to move forward, since this force (called a *thrust*) is *internal to the control volume*; that is, the thrust acts with equal magnitude but opposite direction on the mass m of the device and the expelled exhaust mass m_e.* Applying the principle of impulse and momentum to the control volume, Fig. 15–30b, we have

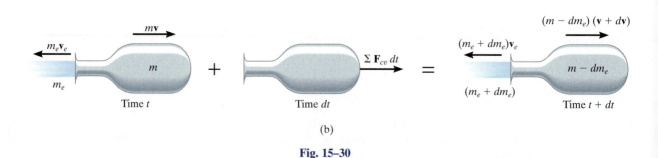

Control Volume

$$(\xrightarrow{+})\quad mv - m_e v_e + \Sigma F_{cv}\, dt = (m - dm_e)(v + dv) - (m_e + dm_e)v_e$$

or

$$\Sigma F_{cv}\, dt = -v\, dm_e + m\, dv - dm_e\, dv - v_e\, dm_e$$

(a)

$m\mathbf{v}$

$m_e\mathbf{v}_e$

m_e

m

Time t

$+$

$\Sigma \mathbf{F}_{cv}\, dt$

Time dt

$=$

$(m - dm_e)\,(\mathbf{v} + d\mathbf{v})$

$(m_e + dm_e)\mathbf{v}_e$

$(m_e + dm_e)$

$m - dm_e$

Time $t + dt$

(b)

Fig. 15–30

---

*$\Sigma \mathbf{F}$ represents the external resultant force *acting on the control volume*, which is different from $\mathbf{F}$, the resultant force acting only on the device.

Without loss of accuracy, the third term on the right side may be neglected since it is a "second-order" differential. Dividing by dt gives

$$\Sigma F_{cv} = m\frac{dv}{dt} - (v + v_e)\frac{dm_e}{dt}$$

The velocity of the device as seen by an observer moving with the particles of the ejected mass is $v_{D/e} = (v + v_e)$, and so the final result can be written as

$$\Sigma F_{cv} = m\frac{dv}{dt} - v_{D/e}\frac{dm_e}{dt} \qquad (15\text{--}28)$$

Here the term dm_e/dt represents the rate at which mass is being ejected.

To illustrate an application of Eq. 15–28, consider the rocket shown in Fig. 15–31, which has a weight **W** and is moving upward against an atmospheric drag force $\mathbf{F}_D$. The control volume to be considered consists of the mass of the rocket and the mass of ejected gas m_e. Applying Eq. 15–28 gives

$(+\uparrow)$ $\qquad -F_D - W = \dfrac{W}{g}\dfrac{dv}{dt} - v_{D/e}\dfrac{dm_e}{dt}$

The last term of this equation represents the *thrust* **T** which the engine exhaust exerts on the rocket, Fig. 15–31. Recognizing that $dv/dt = a$, we can therefore write

$(+\uparrow)$ $\qquad T - F_D - W = \dfrac{W}{g}a$

If a free-body diagram of the rocket is drawn, it becomes obvious that this equation represents an application of $\Sigma \mathbf{F} = m\mathbf{a}$ for the rocket.

A Control Volume That Gains Mass.

A device such as a scoop or a shovel may gain mass as it moves forward. For example, the device shown in Fig. 15–32a has a mass m and moves forward with a velocity **v**. At this instant, the device is collecting a particle stream of mass m_i. The flow velocity $\mathbf{v}_i$ of this injected mass is constant and independent of the velocity **v** such that $v > v_i$. The control volume to be considered here includes both the mass of the device and the mass of the injected particles.

Control Volume

Fig. 15–31

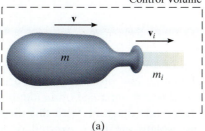

Control Volume

(a)

Fig. 15–32

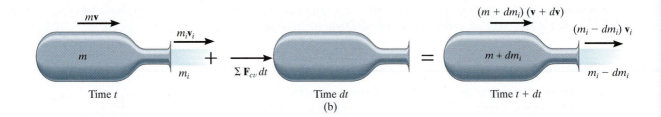

Time t Time dt Time $t + dt$

(b)

15

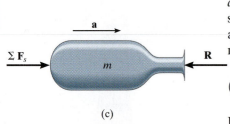

(c)

Fig. 15–32 (cont.)

The impulse and momentum diagrams are shown in Fig. 15–32b. Along with an increase in mass dm_i gained by the device, there is an assumed increase in velocity $d\mathbf{v}$ during the time interval dt. This increase is caused by the impulse created by $\Sigma \mathbf{F}_{cv}$, the resultant of all the external forces *acting on the control volume* in the direction of motion. The force summation does not include the retarding force of the injected mass acting on the device. Why? Applying the principle of impulse and momentum to the control volume, we have

$$(\xrightarrow{+}) \quad mv + m_i v_i + \Sigma F_{cv}\, dt = (m + dm_i)(v + dv) + (m_i - dm_i)v_i$$

Using the same procedure as in the previous case, we may write this equation as

$$\Sigma F_{cv} = m\frac{dv}{dt} + (v - v_i)\frac{dm_i}{dt}$$

Since the velocity of the device as seen by an observer moving with the particles of the injected mass is $v_{D/i} = (v - v_i)$, the final result can be written as

$$\boxed{\Sigma F_{cv} = m\frac{dv}{dt} + v_{D/i}\frac{dm_i}{dt}} \qquad (15\text{–}29)$$

where dm_i/dt is the rate of mass injected into the device. The last term in this equation represents the magnitude of force $\mathbf{R}$, which the injected mass *exerts on the device*, Fig. 15–32c. Since $dv/dt = a$, Eq. 15–29 becomes

$$\Sigma F_{cv} - R = ma$$

This is the application of $\Sigma \mathbf{F} = m\mathbf{a}$.

As in the case of steady flow, problems which are solved using Eqs. 15–28 and 15–29 should be accompanied by an identified control volume and the necessary free-body diagram. With this diagram one can then determine ΣF_{cv} and isolate the force exerted on the device by the particle stream.

The scraper box behind this tractor represents a device that gains mass. If the tractor maintains a constant velocity v, then $dv/dt = 0$ and, because the soil is originally at rest, $v_{D/i} = v$. Applying Eq. 15–29, the horizontal towing force on the scraper box is then $T = 0 + v(dm/dt)$, where dm/dt is the rate of soil accumulated in the box.

EXAMPLE | 15.18

The initial combined mass of a rocket and its fuel is m_0. A total mass m_f of fuel is consumed at a constant rate of $dm_e/dt = c$ and expelled at a constant speed of u relative to the rocket. Determine the maximum velocity of the rocket, i.e., at the instant the fuel runs out. Neglect the change in the rocket's weight with altitude and the drag resistance of the air. The rocket is fired vertically from rest.

SOLUTION

Since the rocket loses mass as it moves upward, Eq. 15–28 can be used for the solution. The only *external force* acting on the *control volume* consisting of the rocket and a portion of the expelled mass is the weight **W**, Fig. 15–33. Hence,

$$+\uparrow \Sigma F_{cv} = m\frac{dv}{dt} - v_{D/e}\frac{dm_e}{dt}; \qquad -W = m\frac{dv}{dt} - uc \qquad (1)$$

The rocket's velocity is obtained by integrating this equation.

At any given instant t during the flight, the mass of the rocket can be expressed as $m = m_0 - (dm_e/dt)t = m_0 - ct$. Since $W = mg$, Eq. 1 becomes

$$-(m_0 - ct)g = (m_0 - ct)\frac{dv}{dt} - uc$$

Separating the variables and integrating, realizing that $v = 0$ at $t = 0$, we have

$$\int_0^v dv = \int_0^t \left(\frac{uc}{m_0 - ct} - g\right) dt$$

$$v = -u\ln(m_0 - ct) - gt \bigg|_0^t = u\ln\left(\frac{m_0}{m_0 - ct}\right) - gt \qquad (2)$$

Note that liftoff requires the first term on the right to be greater than the second during the initial phase of motion. The time t' needed to consume all the fuel is

$$m_f = \left(\frac{dm_e}{dt}\right)t' = ct'$$

Hence,

$$t' = m_f/c$$

Substituting into Eq. 2 yields

$$v_{max} = u\ln\left(\frac{m_0}{m_0 - m_f}\right) - \frac{gm_f}{c} \qquad \qquad Ans.$$

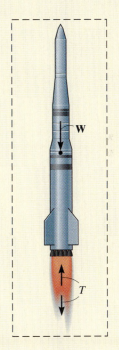

Fig. 15–33

15

EXAMPLE 15.19

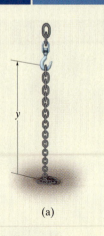

(a)

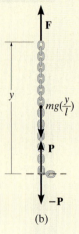

(b)

(c)

Fig. 15–34

A chain of length l, Fig. 15–34a, has a mass m. Determine the magnitude of force **F** required to (a) raise the chain with a constant speed v_c, starting from rest when $y = 0$; and (b) lower the chain with a constant speed v_c, starting from rest when $y = l$.

SOLUTION

Part (a). As the chain is raised, all the suspended links are given a sudden downward impulse by each added link which is lifted off the ground. Thus, the *suspended portion* of the chain may be considered as a device which is *gaining mass*. The control volume to be considered is the length of chain y which is suspended by **F** at any instant, including the next link which is about to be added but is still at rest, Fig. 15–34b. The forces acting on the control volume *exclude* the internal forces **P** and $-$**P**, which act between the added link and the suspended portion of the chain. Hence, $\Sigma F_{cv} = F - mg(y/l)$.

To apply Eq. 15–29, it is also necessary to find the rate at which mass is being added to the system. The velocity $\mathbf{v}_c$ of the chain is equivalent to $\mathbf{v}_{D/i}$. Why? Since v_c is constant, $dv_c/dt = 0$ and $dy/dt = v_c$. Integrating, using the initial condition that $y = 0$ when $t = 0$, gives $y = v_c t$. Thus, the mass of the control volume at any instant is $m_{cv} = m(y/l) = m(v_c t/l)$, and therefore the *rate* at which mass is *added* to the suspended chain is

$$\frac{dm_i}{dt} = m\left(\frac{v_c}{l}\right)$$

Applying Eq. 15–29 using this data, we have

$$+\uparrow \Sigma F_{cv} = m\frac{dv_c}{dt} + v_{D/i}\frac{dm_i}{dt}$$

$$F - mg\left(\frac{y}{l}\right) = 0 + v_c m\left(\frac{v_c}{l}\right)$$

Hence,

$$F = (m/l)(gy + v_c^2) \qquad \textit{Ans.}$$

Part (b). When the chain is being lowered, the links which are expelled (given zero velocity) *do not* impart an impulse to the *remaining* suspended links. Why? Thus, the control volume in Part (a) will not be considered. Instead, the equation of motion will be used to obtain the solution. At time t the portion of chain still off the floor is y. The free-body diagram for a suspended portion of the chain is shown in Fig. 15–34c. Thus,

$$+\uparrow \Sigma F = ma; \qquad F - mg\left(\frac{y}{l}\right) = 0$$

$$F = mg\left(\frac{y}{l}\right) \qquad \textit{Ans.}$$

PROBLEMS

*15–108.   A scoop in front of the tractor collects snow at a rate of 200 kg/s. Determine the resultant traction force **T** that must be developed on all the wheels as it moves forward on level ground at a constant speed of 5 km/h. The tractor has a mass of 5 Mg.

•15–109.   A four-engine commercial jumbo jet is cruising at a constant speed of 800 km/h in level flight when all four engines are in operation. Each of the engines is capable of discharging combustion gases with a velocity of 775 m/s relative to the plane. If during a test two of the engines, one on each side of the plane, are shut off, determine the new cruising speed of the jet. Assume that air resistance (drag) is proportional to the square of the speed, that is, $F_D = cv^2$, where c is a constant to be determined. Neglect the loss of mass due to fuel consumption.

Prob. 15–109

15–110.   The jet dragster when empty has a mass of 1.25 Mg and carries 250 kg of solid propellant fuel. Its engine is capable of burning the fuel at a constant rate of 50 kg/s, while ejecting it at 1500 m/s relative to the dragster. Determine the maximum speed attained by the dragster starting from rest. Assume air resistance is $F_D = (10v^2)$ N, where v is the dragster's velocity in m/s. Neglect rolling resistance.

Prob. 15–110

15–111.   The 150-lb fireman is holding a hose which has a nozzle diameter of 1 in. and hose diameter of 2 in. If the velocity of the water at discharge is 60 ft/s, determine the resultant normal and frictional force acting on the man's feet at the ground. Neglect the weight of the hose and the water within it. $\gamma_w = 62.4 \ \text{lb/ft}^3$.

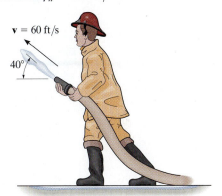

Prob. 15–111

*15–112.   When operating, the air-jet fan discharges air with a speed of $v_B = 20$ m/s into a slipstream having a diameter of 0.5 m. If air has a density of 1.22 kg/m³, determine the horizontal and vertical components of reaction at C and the vertical reaction at each of the two wheels, D, when the fan is in operation. The fan and motor have a mass of 20 kg and a center of mass at G. Neglect the weight of the frame. Due to symmetry, both of the wheels support an equal load. Assume the air entering the fan at A is essentially at rest.

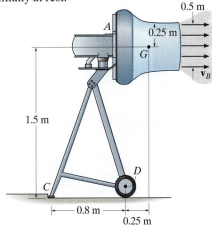

Prob. 15–112

15

•15–113. The blade divides the jet of water having a diameter of 3 in. If one-fourth of the water flows downward while the other three-fourths flows upwards, and the total flow is $Q = 0.5$ ft^3/s, determine the horizontal and vertical components of force exerted on the blade by the jet, $\gamma_w = 62.4$ lb/ft^3.

15–115. The fire boat discharges two streams of seawater, each at a flow of 0.25 m^3/s and with a nozzle velocity of 50 m/s. Determine the tension developed in the anchor chain, needed to secure the boat. The density of seawater is $\rho_{sw} = 1020$ kg/m^3.

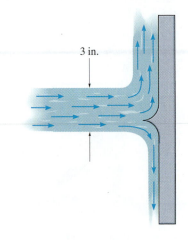

3 in.

Prob. 15–113

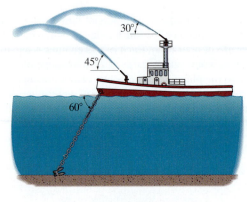

30°

45°

60°

Prob. 15–115

15–114. The toy sprinkler for children consists of a 0.2-kg cap and a hose that has a mass per length of 30 g/m. Determine the required rate of flow of water through the 5-mm-diameter tube so that the sprinkler will lift 1.5 m from the ground and hover from this position. Neglect the weight of the water in the tube. $\rho_w = 1$ Mg/m^3.

*15–116. A speedboat is powered by the jet drive shown. Seawater is drawn into the pump housing at the rate of 20 ft^3/s through a 6-in.-diameter intake A. An impeller accelerates the water flow and forces it out horizontally through a 4-in.- diameter nozzle B. Determine the horizontal and vertical components of thrust exerted on the speedboat. The specific weight of seawater is $\gamma_{sw} = 64.3$ lb/ft^3.

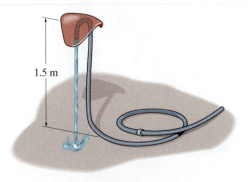

1.5 m

Prob. 15–114

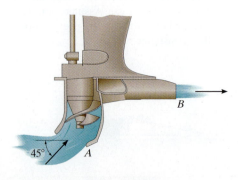

B

45°   A

Prob. 15–116

15–126.
scoops u
train. If t
12 ft/s, d
shoveling

15–127.
forward
measured
opposite
as shown
determin
required
collection
$\rho_w = 1$ M

•15–117. The fan blows air at 6000 ft³/min. If the fan has a weight of 30 lb and a center of gravity at G, determine the smallest diameter d of its base so that it will not tip over. The specific weight of air is $\gamma = 0.076$ lb/ft³.

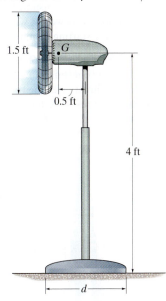

Prob. 15–117

15–118. The elbow for a 5-in-diameter buried pipe is subjected to a static pressure of 10 lb/in². The speed of the water passing through it is $v = 8$ ft/s. Assuming the pipe connections at A and B do not offer any vertical force resistance on the elbow, determine the resultant vertical force $\mathbf{F}$ that the soil must then exert on the elbow in order to hold it in equilibrium. Neglect the weight of the elbow and the water within it. $\gamma_w = 62.4$ lb/ft³.

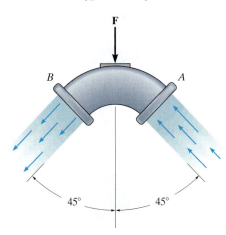

Prob. 15–118

15–119. The hemispherical bowl of mass m is held in equilibrium by the vertical jet of water discharged through a nozzle of diameter d. If the discharge of the water through the nozzle is Q, determine the height h at which the bowl is suspended. The water density is ρ_w. Neglect the weight of the water jet.

Prob. 15–119

*15–120. The chute is used to divert the flow of water, $Q = 0.6$ m³/s. If the water has a cross-sectional area of 0.05 m², determine the force components at the pin D and roller C necessary for equilibrium. Neglect the weight of the chute and weight of the water on the chute. $\rho_w = 1$ Mg/m³.

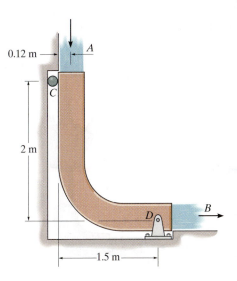

Prob. 15–120

*15–128.
the rate
determin
of the bel
to pull the

15

•15–131. The 12-Mg jet airplane has a constant speed of 950 km/h when it is flying along a horizontal straight line. Air enters the intake scoops S at the rate of 50 m³/s. If the engine burns fuel at the rate of 0.4 kg/s and the gas (air and fuel) is exhausted relative to the plane with a speed of 450 m/s, determine the resultant drag force exerted on the plane by air resistance. Assume that air has a constant density of 1.22 kg/m³. *Hint:* Since mass both enters and exits the plane, Eqs. 15–28 and 15–29 must be combined to yield

$$\Sigma F_s = m\frac{dv}{dt} - v_{D/e}\frac{dm_e}{dt} + v_{D/i}\frac{dm_i}{dt}.$$

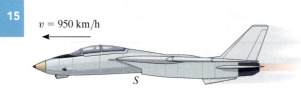

$v = 950$ km/h

S

Prob. 15–131

*15–132. The cart has a mass M and is filled with water that has a mass m_0. If a pump ejects the water through a nozzle having a cross-sectional area A at a constant rate of v_0 relative to the cart, determine the velocity of the cart as a function of time. What is the maximum speed of the cart assuming all the water can be pumped out? The frictional resistance to forward motion is F. The density of the water is ρ.

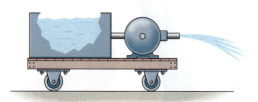

Prob. 15–132

•15–133. The truck has a mass of 50 Mg when empty. When it is unloading 5 m³ of sand at a constant rate of 0.8 m³/s, the sand flows out the back at a speed of 7 m/s, measured relative to the truck, in the direction shown. If the truck is free to roll, determine its initial acceleration just as the sand begins to fall out. Neglect the mass of the wheels and any frictional resistance to motion. The density of sand is $\rho_s = 1520$ kg/m³.

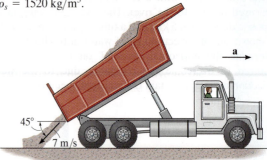

a

45°

7 m/s

Prob. 15–133

15–134. The truck has a mass m_0 and is used to tow the smooth chain having a total length l and a mass per unit of length m'. If the chain is originally piled up, determine the tractive force **F** that must be supplied by the rear wheels of the truck necessary to maintain a constant speed v while the chain is being drawn out.

v

F

Prob. 15–134

15–135. The chain has a total length $L < d$ and a mass per unit length of m'. If a portion h of the chain is suspended over the table and released, determine the velocity of its end A as a function of its position y. Neglect friction.

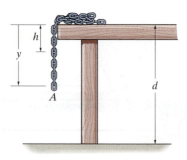

h

y

A

d

Prob. 15–135

•15–1
and B
carries
vertica
reactio
total v
with a
water
respec
flanges
$\gamma_w = $

15–122
water
$v_A = $
vertica
necess:
the we
pipe. T
the dia

15

*15–136. A commercial jet aircraft has a mass of 150 Mg and is cruising at a constant speed of 850 km/h in level flight ($\theta = 0°$). If each of the two engines draws in air at a rate of 1000 kg/s and ejects it with a velocity of 900 m/s, relative to the aircraft, determine the maximum angle of inclination θ at which the aircraft can fly with a constant speed of 750 km/h. Assume that air resistance (drag) is proportional to the square of the speed, that is, $F_D = cv^2$, where c is a constant to be determined. The engines are operating with the same power in both cases. Neglect the amount of fuel consumed.

Prob. 15–136

•15–137. A coil of heavy open chain is used to reduce the stopping distance of a sled that has a mass M and travels at a speed of v_0. Determine the required mass per unit length of the chain needed to slow down the sled to $(1/2)v_0$ within a distance $x = s$ if the sled is hooked to the chain at $x = 0$. Neglect friction between the chain and the ground.

15–138. The car is used to scoop up water that is lying in a trough at the tracks. Determine the force needed to pull the car forward at constant velocity **v** for each of the three cases. The scoop has a cross-sectional area A and the density of water is ρ_w.

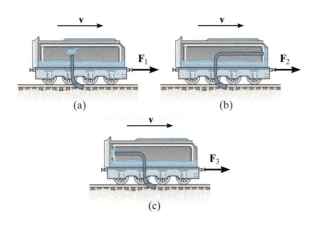

(a)

(b)

(c)

Prob. 15–138

15–139. A rocket has an empty weight of 500 lb and carries 300 lb of fuel. If the fuel is burned at the rate of 1.5 lb/s and ejected with a velocity of 4400 ft/s relative to the rocket, determine the maximum speed attained by the rocket starting from rest. Neglect the effect of gravitation on the rocket.

*15–140. Determine the magnitude of force **F** as a function of time, which must be applied to the end of the cord at A to raise the hook H with a constant speed $v = 0.4$ m/s. Initially the chain is at rest on the ground. Neglect the mass of the cord and the hook. The chain has a mass of 2 kg/m.

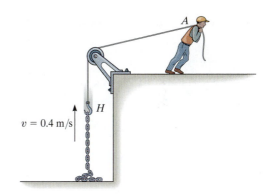

$v = 0.4$ m/s

H

Prob. 15–140

•15–141. The earthmover initially carries 10 m³ of sand having a density of 1520 kg/m³. The sand is unloaded horizontally through a 2.5-m² dumping port P at a rate of 900 kg/s measured relative to the port. If the earthmover maintains a constant resultant tractive force $F = 4$ kN at its front wheels to provide forward motion, determine its acceleration when half the sand is dumped. When empty, the earthmover has a mass of 30 Mg. Neglect any resistance to forward motion and the mass of the wheels. The rear wheels are free to roll.

F

P

Prob. 15–141

15–142. The earthmover initially carries $10 \, \text{m}^3$ of sand having a density of $1520 \, \text{kg/m}^3$. The sand is unloaded horizontally through a 2.5-m^2 dumping port P at a rate of $900 \, \text{kg/s}$ measured relative to the port. Determine the resultant tractive force $\mathbf{F}$ at its front wheels if the acceleration of the earthmover is $0.1 \, \text{m/s}^2$ when half the sand is dumped. When empty, the earthmover has a mass of 30 Mg. Neglect any resistance to forward motion and the mass of the wheels. The rear wheels are free to roll.

Prob. 15–142

15–143. The jet is traveling at a speed of 500 mi/h, 30° with the horizontal. If the fuel is being spent at $3 \, \text{lb/s}$, and the engine takes in air at $400 \, \text{lb/s}$, whereas the exhaust gas (air and fuel) has a relative speed of $32\,800 \, \text{ft/s}$, determine the acceleration of the plane at this instant. The drag resistance of the air is $F_D = (0.7v^2) \, \text{lb}$, where the speed is measured in ft/s. The jet has a weight of 15 000 lb. *Hint:* See Prob. 15–131.

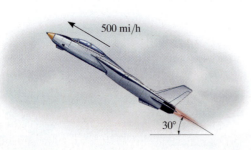

Prob. 15–143

***15–144.** The rocket has an initial mass m_0, including the fuel. For practical reasons desired for the crew, it is required that it maintain a constant upward acceleration a_0. If the fuel is expelled from the rocket at a relative speed $v_{e/r}$ determine the rate at which the fuel should be consumed to maintain the motion. Neglect air resistance, and assume that the gravitational acceleration is constant.

Prob. 15–144

•15–145. If the chain is lowered at a constant speed, determine the normal reaction exerted on the floor as a function of time. The chain has a weight of 5 lb/ft and a total length of 20 ft.

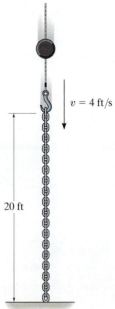

Prob. 15–145

CONCEPTUAL PROBLEMS

P15–1. The baseball travels to the left when it is struck by the bat. If the ball then moves horizontally to the right, determine which measurements you could make in order to determine the net impulse given to the ball. Use numerical values to give an example of how this can be done.

P15–1

P15–2. The steel wrecking "ball" is suspended from the boom using an old rubber tire A. The crane operator lifts the ball then allows it to drop freely to break up the concrete. Explain, using appropriate numerical data, why it is a good idea to use the rubber tire for this work.

P15–2

P15–3. The train engine on the left, A, is at rest, and the one on the right, B, is coasting to the left. If the engines are identical, use numerical values to show how to determine the maximum compression in each of the spring bumpers that are mounted in the front of the engines. Each engine is free to roll.

P15–3

P15–4. Three train cars each have the same mass and are rolling freely when they strike the fixed bumper. Legs AB and BC on the bumper are pin-connected at their ends and the angle BAC is 30° and BCA is 60°. Compare the average impulse in each leg needed to stop the motion if the cars have no bumper and if the cars have a spring bumper. Use appropriate numerical values to explain your answer.

P15–4

CHAPTER REVIEW

Impulse

An impulse is defined as the product of force and time. Graphically it represents the area under the F–t diagram. If the force is constant, then the impulse becomes $I = F_c(t_2 - t_1)$.

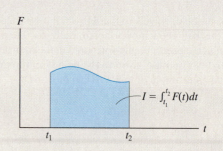

Principle of Impulse and Momentum

When the equation of motion, $\Sigma \mathbf{F} = m\mathbf{a}$, and the kinematic equation, $a = dv/dt$, are combined, we obtain the principle of impulse and momentum. This is a vector equation that can be resolved into rectangular components and used to solve problems that involve force, velocity, and time. For application, the free-body diagram should be drawn in order to account for all the impulses that act on the particle.

$$m\mathbf{v}_1 + \Sigma \int_{t_1}^{t_2} \mathbf{F}\, dt = m\mathbf{v}_2$$

Conservation of Linear Momentum

If the principle of impulse and momentum is applied to a *system of particles*, then the collisions between the particles produce internal impulses that are equal, opposite, and collinear, and therefore cancel from the equation. Furthermore, if an external impulse is small, that is, the force is small and the time is short, then the impulse can be classified as nonimpulsive and can be neglected. Consequently, momentum for the system of particles is conserved.

$$\Sigma m_i(\mathbf{v}_i)_1 = \Sigma m_i(\mathbf{v}_i)_2$$

The conservation-of-momentum equation is useful for finding the final velocity of a particle when internal impulses are exerted between two particles and the initial velocities of the particles is known. If the internal impulse is to be determined, then one of the particles is isolated and the principle of impulse and momentum is applied to this particle.

Impact

When two particles A and B have a direct impact, the internal impulse between them is equal, opposite, and collinear. Consequently the conservation of momentum for this system applies along the line of impact.

$$m_A(v_A)_1 + m_B(v_B)_1 = m_A(v_A)_2 + m_B(v_B)_2$$

If the final velocities are unknown, a second equation is needed for solution. We must use the coefficient of restitution, e. This experimentally determined coefficient depends upon the physical properties of the colliding particles. It can be expressed as the ratio of their relative velocity after collision to their relative velocity before collision. If the collision is elastic, no energy is lost and $e = 1$. For a plastic collision $e = 0$.

$$e = \frac{(v_B)_2 - (v_A)_2}{(v_A)_1 - (v_B)_1}$$

If the impact is oblique, then the conservation of momentum for the system and the coefficient-of-restitution equation apply along the line of impact. Also, conservation of momentum for each particle applies perpendicular to this line (plane of impact) because no impulse acts on the particles in this direction.

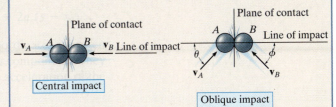

Central impact

Oblique impact

15

Principle of Angular Impulse and Momentum

The moment of the linear momentum about an axis (z) is called the angular momentum.

$$(H_O)_z = (d)(mv)$$

The principle of angular impulse and momentum is often used to eliminate unknown impulses by summing the moments about an axis through which the lines of action of these impulses produce no moment. For this reason, a free-body diagram should accompany the solution.

$$(\mathbf{H}_O)_1 + \Sigma \int_{t_1}^{t_2} \mathbf{M}_O \, dt = (\mathbf{H}_O)_2$$

Steady Fluid Streams

Impulse-and-momentum methods are often used to determine the forces that a device exerts on the mass flow of a fluid—liquid or gas. To do so, a free-body diagram of the fluid mass in contact with the device is drawn in order to identify these forces. Also, the velocity of the fluid as it flows into and out of a control volume for the device is calculated. The equations of steady flow involve summing the forces and the moments to determine these reactions.

$$\Sigma \mathbf{F} = \frac{dm}{dt}(\mathbf{v}_B - \mathbf{v}_A)$$

$$\Sigma \mathbf{M}_O = \frac{dm}{dt}(\mathbf{r}_B \times \mathbf{v}_B - \mathbf{r}_A \times \mathbf{v}_A)$$

Propulsion with Variable Mass

Some devices, such as a rocket, lose mass as they are propelled forward. Others gain mass, such as a shovel. We can account for this mass loss or gain by applying the principle of impulse and momentum to a control volume for the device. From this equation, the force exerted on the device by the mass flow can then be determined.

$$\Sigma F_{cv} = m\frac{dv}{dt} - v_{D/e}\frac{dm_e}{dt}$$

Loses Mass

$$\Sigma F_{cv} = m\frac{dv}{dt} + v_{D/i}\frac{dm_i}{dt}$$

Gains Mass

Relative Motion. If the origin of a *translating* coordinate system is established at particle A, then for particle B,

$$\mathbf{r}_B = \mathbf{r}_A + \mathbf{r}_{B/A}$$
$$\mathbf{v}_B = \mathbf{v}_A + \mathbf{v}_{B/A}$$
$$\mathbf{a}_B = \mathbf{a}_A + \mathbf{a}_{B/A}$$

Here the relative motion is measured by an observer fixed in the translating coordinate system.

Kinetics. Problems in kinetics involve the analysis of forces which cause the motion. When applying the equations of kinetics, it is absolutely necessary that measurements of the motion be made from an *inertial coordinate system,* i.e., one that does not rotate and is either fixed or translates with constant velocity. If a problem requires *simultaneous solution* of the equations of kinetics and kinematics, then it is important that the coordinate systems selected for writing each of the equations define the *positive directions* of the axes in the *same* manner.

Equations of Motion. These equations are used to solve for the particle's acceleration or the forces causing the motion. If they are used to determine a particle's position, velocity, or time of motion, then kinematics will also have to be considered to complete the solution. Before applying the equations of motion, always draw a free-body diagram to identify all the forces acting on the particle. Also, establish the direction of the particle's acceleration or its components. (A kinetic diagram may accompany the solution in order to graphically account for the $m\mathbf{a}$ vector.)

$$\Sigma F_x = ma_x \qquad \Sigma F_n = ma_n \qquad \Sigma F_r = ma_r$$
$$\Sigma F_y = ma_y \qquad \Sigma F_t = ma_t \qquad \Sigma F_\theta = ma_\theta$$
$$\Sigma F_z = ma_z \qquad \Sigma F_b = 0 \qquad \Sigma F_z = ma_z$$

Work and Energy. The equation of work and energy represents an integrated form of the tangential equation of motion, $\Sigma F_t = ma_t$, combined with kinematics ($a_t \, ds = v \, dv$). *It is used to solve problems involving force, velocity, and displacement.* Before applying this equation, *always draw a free-body diagram* in order to identify the forces which do work on the particle.

$$T_1 + \Sigma U_{1-2} = T_2$$

where

$$T = \tfrac{1}{2}mv^2 \qquad \text{(kinetic energy)}$$
$$U_F = \int_{s_1}^{s_2} F \cos \theta \, ds \qquad \text{(work of a variable force)}$$
$$U_{F_c} = F_c \cos \theta (s_2 - s_1) \qquad \text{(work of a constant force)}$$
$$U_W = -W \, \Delta y \qquad \text{(work of a weight)}$$
$$U_s = -(\tfrac{1}{2}ks_2^2 - \tfrac{1}{2}ks_1^2) \qquad \text{(work of an elastic spring)}$$

If the forces acting on the particle are *conservative forces,* i.e., those that *do not* cause a dissipation of energy, such as friction, then apply the conservation of energy equation. This equation is easier to use than the equation of work and energy since it applies at only *two points* on the path and *does not* require calculation of the work done by a force as the particle moves along the path.

$$T_1 + V_1 = T_2 + V_2$$

where $V = V_g + V_e$ and

$$V_g = Wy \quad \text{(gravitational potential energy)}$$
$$V_e = \tfrac{1}{2}ks^2 \quad \text{(elastic potential energy)}$$

If the *power* developed by a force is to be determined, use

$$P = \frac{dU}{dt} = \mathbf{F} \cdot \mathbf{v}$$

where $\mathbf{v}$ is the velocity of the particle acted upon by the force $\mathbf{F}$.

Impulse and Momentum. The equation of *linear impulse and momentum* is an integrated form of the equation of motion, $\Sigma\mathbf{F} = m\mathbf{a}$, combined with kinematics ($\mathbf{a} = d\mathbf{v}/dt$). *It is used to solve problems involving force, velocity, and time.* Before applying this equation, one should *always draw the free-body diagram,* in order to identify all the forces that cause impulses on the particle. From the diagram the impulsive and nonimpulsive forces should be identified. Recall that the nonimpulsive forces can be neglected in the analysis during the time of impact. Also, establish the direction of the particle's velocity just before and just after the impulses are applied. As an alternative procedure, impulse and momentum diagrams may accompany the solution in order to graphically account for the terms in the equation.

$$m\mathbf{v}_1 + \Sigma \int_{t_1}^{t_2} \mathbf{F}\, dt = m\mathbf{v}_2$$

If several particles are involved in the problem, consider applying the *conservation of momentum* to the system in order to eliminate the internal impulses from the analysis. This can be done in a specified direction, provided no external impulses act on the particles in that direction.

$$\Sigma m\mathbf{v}_1 = \Sigma m\mathbf{v}_2$$

If the problem involves impact and the coefficient of restitution e is given, then apply the following equation.

$$e = \frac{(v_B)_2 - (v_A)_2}{(v_A)_1 - (v_B)_1} \quad \text{(along line of impact)}$$

Remember that during impact the principle of work and energy cannot be used, since the particles deform and therefore the work due to the internal forces will be unknown. The principle of work and energy can be used, however, to determine the energy loss during the collision once the particle's initial and final velocities are determined.

The *principle of angular impulse and momentum* and the *conservation of angular momentum* can be applied about an axis in order to *eliminate* some of the unknown impulses acting on the particle during the period when its motion is studied. Investigation of the particle's free-body diagram (or the impulse diagram) will aid in choosing the axis for application.

$$(\mathbf{H}_O)_1 + \Sigma \int_{t_1}^{t_2} \mathbf{M}_O \, dt = (\mathbf{H}_O)_2$$

$$(\mathbf{H}_O)_1 = (\mathbf{H}_O)_2$$

The following problems provide an opportunity for applying the above concepts. They are presented in *random order* so that practice may be gained in identifying the various types of problems and developing the skills necessary for their solution.

REVIEW PROBLEMS

R1–1. The ball is thrown horizontally with a speed of 8 m/s. Find the equation of the path, $y = f(x)$, and then determine the ball's velocity and the normal and tangential components of acceleration when $t = 0.25$ s.

R1–2. Cartons having a mass of 5 kg are required to move along the assembly line with a constant speed of 8 m/s. Determine the smallest radius of curvature, ρ, for the conveyor so the cartons do not slip. The coefficients of static and kinetic friction between a carton and the conveyor are $\mu_s = 0.7$ and $\mu_k = 0.5$, respectively.

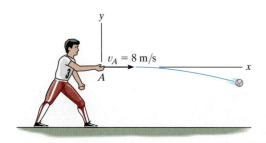

Prob. R1–1

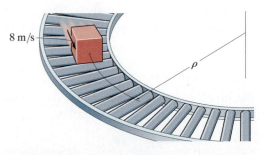

Prob. R1–2

R1–3. A small metal particle travels downward through a fluid medium while being subjected to the attraction of a magnetic field such that its position is $s = (15t^3 - 3t)$ mm, where t is in seconds. Determine (a) the particle's displacement from $t = 2$ s to $t = 4$ s, and (b) the velocity and acceleration of the particle when $t = 5$ s.

*R1–4. The flight path of a jet aircraft as it takes off is defined by the parametric equations $x = 1.25t^2$ and $y = 0.03t^3$, where t is the time after take-off, measured in seconds, and x and y are given in meters. If the plane starts to level off at $t = 40$ s, determine at this instant (a) the horizontal distance it is from the airport, (b) its altitude, (c) its speed, and (d) the magnitude of its acceleration.

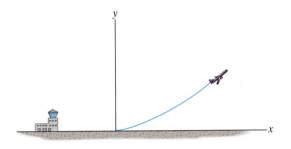

Prob. R1–4

R1–5. The boy jumps off the flat car at A with a velocity of $v' = 4$ ft/s relative to the car as shown. If he lands on the second flat car B, determine the final speed of both cars after the motion. Each car has a weight of 80 lb. The boy's weight is 60 lb. Both cars are originally at rest. Neglect the mass of the car's wheels.

Prob. R1–5

R1-6. The man A has a weight of 175 lb and jumps from rest at a height $h = 8$ ft onto a platform P that has a weight of 60 lb. The platform is mounted on a spring, which has a stiffness $k = 200$ lb/ft. Determine (a) the velocities of A and P just after impact and (b) the maximum compression imparted to the spring by the impact. Assume the coefficient of restitution between the man and the platform is $e = 0.6$, and the man holds himself rigid during the motion.

R1–7. The man A has a weight of 100 lb and jumps from rest onto the platform P that has a weight of 60 lb. The platform is mounted on a spring, which has a stiffness $k = 200$ lb/ft. If the coefficient of restitution between the man and the platform is $e = 0.6$, and the man holds himself rigid during the motion, determine the required height h of the jump if the maximum compression of the spring is 2 ft.

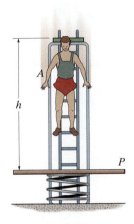

Probs. R1–6/7

*R1–8. The baggage truck A has a mass of 800 kg and is used to pull each of the 300-kg cars. Determine the tension in the couplings at B and C if the tractive force $\mathbf{F}$ on the truck is $F = 480$ N. What is the speed of the truck when $t = 2$ s, starting from rest? The car wheels are free to roll. Neglect the mass of the wheels.

R1–9. The baggage truck A has a mass of 800 kg and is used to pull each of the 300-kg cars. If the tractive force $\mathbf{F}$ on the truck is $F = 480$ N, determine the acceleration of the truck. What is the acceleration of the truck if the coupling at C suddenly fails? The car wheels are free to roll. Neglect the mass of the wheels.

Probs. R1–8/9

R1–10. A car travels at 80 ft/s when the brakes are suddenly applied, causing a constant deceleration of 10 ft/s². Determine the time required to stop the car and the distance traveled before stopping.

R1–11. Determine the speed of block B if the end of the cable at C is pulled downward with a speed of 10 ft/s. What is the relative velocity of the block with respect to C?

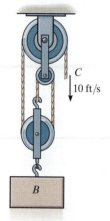

Prob. R1–11

*R1–12.** The skier starts fom rest at A and travels down the ramp. If friction and air resistance can be neglected, determine his speed v_B when he reaches B. Also, compute the distance s to where he strikes the ground at C, if he makes the jump traveling horizontally at B. Neglect the skier's size. He has a mass of 70 kg.

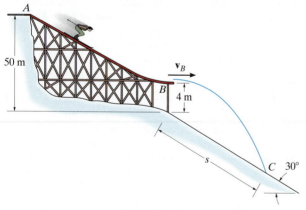

Prob. R1–12

R1–13. The position of a particle is defined by $\mathbf{r} = \{5(\cos 2t)\mathbf{i} + 4(\sin 2t)\mathbf{j}\}$ m, where t is in seconds and the arguments for the sine and cosine are given in radians. Determine the magnitudes of the velocity and acceleration of the particle when $t = 1$ s. Also, prove that the path of the particle is elliptical.

R1–14. The 5-lb cylinder falls past A with a speed $v_A = 10$ ft/s onto the platform. Determine the maximum displacement of the platform, caused by the collision. The spring has an unstretched length of 1.75 ft and is originally kept in compression by the 1-ft-long cables attached to the platform. Neglect the mass of the platform and spring and any energy lost during the collision.

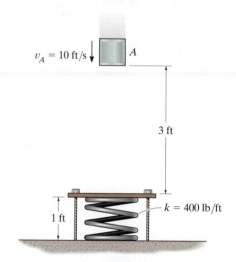

Prob. R1–14

R1–15. The block has a mass of 50 kg and rests on the surface of the cart having a mass of 75 kg. If the spring which is attached to the cart and not the block is compressed 0.2 m and the system is released from rest, determine the speed of the block after the spring becomes undeformed. Neglect the mass of the cart's wheels and the spring in the calculation. Also neglect friction. Take $k = 300$ N/m.

*R1–16.** The block has a mass of 50 kg and rests on the surface of the cart having a mass of 75 kg. If the spring which is attached to the cart and not the block is compressed 0.2 m and the system is released from rest, determine the speed of the block with respect to the cart after the spring becomes undeformed. Neglect the mass of the cart's wheels and the spring in the calculation. Also neglect friction. Take $k = 300$ N/m.

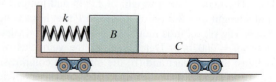

Probs. R1–15/16

R1–17. A ball is launched from point A at an angle of 30°. Determine the maximum and minimum speed v_A it can have so that it lands in the container.

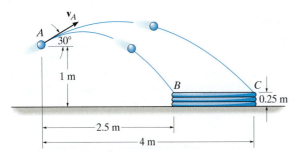

Prob. R1–17

R1–18. At the instant shown, cars A and B travel at speeds of 55 mi/h and 40 mi/h, respectively. If B is increasing its speed by 1200 mi/h², while A maintains its constant speed, determine the velocity and acceleration of B with respect to A. Car B moves along a curve having a radius of curvature of 0.5 mi.

R1–19. At the instant shown, cars A and B travel at speeds of 55 mi/h and 40 mi/h, respectively. If B is decreasing its speed at 1500 mi/h² while A is increasing its speed at 800 mi/h², determine the acceleration of B with respect to A. Car B moves along a curve having a radius of curvature of 0.75 mi.

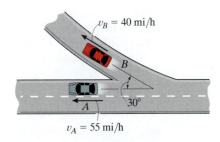

Probs. R1–18/19

*R1–20. Four inelastic cables C are attached to a plate P and hold the 1-ft-long spring 0.25 ft in compression when *no weight* is on the plate. There is also an undeformed spring nested within this compressed spring. If the block, having a weight of 10 lb, is moving downward at $v = 4$ ft/s, when it is 2 ft above the plate, determine the maximum compression in each spring after it strikes the plate. Neglect the mass of the plate and springs and any energy lost in the collision.

R1–21. Four inelastic cables C are attached to plate P and hold the 1-ft-long spring 0.25 ft in compression when *no weight* is on the plate. There is also a 0.5-ft-long undeformed spring nested within this compressed spring. Determine the speed v of the 10-lb block when it is 2 ft above the plate, so that after it strikes the plate, it compresses the nested spring, having a stiffness of 50 lb/in., an amount of 0.20 ft. Neglect the mass of the plate and springs and any energy lost in the collision.

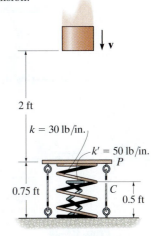

Probs. R1–20/21

R1–22. The 2-kg spool S fits loosely on the rotating inclined rod for which the coefficient of static friction is $\mu_s = 0.2$. If the spool is located 0.25 m from A, determine the minimum constant speed the spool can have so that it does not slip down the rod.

R1–23. The 2-kg spool S fits loosely on the rotating inclined rod for which the coefficient of static friction is $\mu_s = 0.2$. If the spool is located 0.25 m from A, determine the maximum constant speed the spool can have so that it does not slip up the rod.

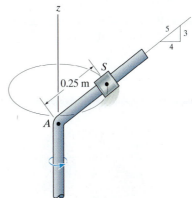

Probs. R1–22/23

*R1–24. The winding drum D draws in the cable at an accelerated rate of 5 m/s^2. Determine the cable tension if the suspended crate has a mass of 800 kg.

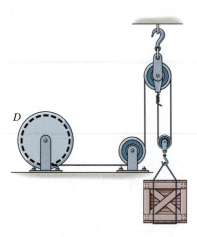

Prob. R1–24

R1–25. The bottle rests at a distance of 3 ft from the center of the horizontal platform. If the coefficient of static friction between the bottle and the platform is $\mu_s = 0.3$, determine the maximum speed that the bottle can attain before slipping. Assume the angular motion of the platform is slowly increasing.

R1–26. Work Prob. R1–25 assuming that the platform starts rotating from rest so that the speed of the bottle is increased at 2 ft/s^2.

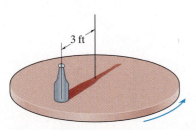

Probs. R1–25/26

R1–27. The 150-lb man lies against the cushion for which the coefficient of static friction is $\mu_s = 0.5$. Determine the resultant normal and frictional forces the cushion exerts on him if, due to rotation about the z axis, he has a constant speed $v = 20$ ft/s. Neglect the size of the man. Take $\theta = 60°$.

*R1–28. The 150-lb man lies against the cushion for which the coefficient of static friction is $\mu_s = 0.5$. If he rotates about the z axis with a constant speed $v = 30$ ft/s, determine the smallest angle θ of the cushion at which he will begin to slip up the cushion.

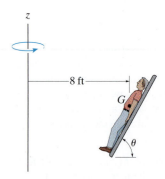

Probs. R1–27/28

R1–29. The motor pulls on the cable at A with a force $F = (30 + t^2)$ lb, where t is in seconds. If the 34-lb crate is originally at rest on the ground when $t = 0$, determine its speed when $t = 4$ s. Neglect the mass of the cable and pulleys. *Hint:* First find the time needed to begin lifting the crate.

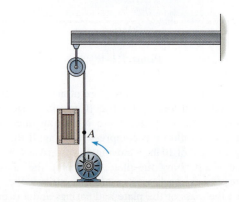

Prob. R1–29

R1–30. The motor pulls on the cable at A with a force $F = (e^{2t})$ lb, where t is in seconds. If the 34-lb crate is originally at rest on the ground when $t = 0$, determine the crate's velocity when $t = 2$ s. Neglect the mass of the cable and pulleys. *Hint:* First find the time needed to begin lifting the crate.

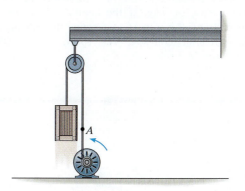

Prob. R1–30

R1–31. The collar has a mass of 2 kg and travels along the smooth *horizontal* rod defined by the equiangular spiral $r = (e^\theta)$ m, where θ is in radians. Determine the tangential force $\mathbf{F}$ and the normal force $\mathbf{N}$ acting on the collar when $\theta = 45°$, if force $\mathbf{F}$ maintains a constant angular motion $\dot{\theta} = 2$ rad/s.

*R1–32. The collar has a mass of 2 kg and travels along the smooth *horizontal* rod defined by the equiangular spiral $r = (e^\theta)$ m, where θ is in radians. Determine the tangential force $\mathbf{F}$ and the normal force $\mathbf{N}$ acting on the collar when $\theta = 90°$, if force $\mathbf{F}$ maintains a constant angular motion $\dot{\theta} = 2$ rad/s.

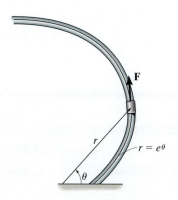

Probs. R1–31/32

R1–33. The acceleration of a particle along a straight line is defined by $a = (2t - 9)$ m/s², where t is in seconds. When $t = 0$, $s = 1$ m and $v = 10$ m/s. When $t = 9$ s, determine (a) the particle's position, (b) the total distance traveled, and (c) the velocity. Assume the positive direction is to the right.

R1–34. The 400-kg mine car is hoisted up the incline using the cable and motor M. For a short time, the force in the cable is $F = (3200t^2)$ N, where t is in seconds. If the car has an initial velocity $v_1 = 2$ m/s when $t = 0$, determine its velocity when $t = 2$ s.

R1–35. The 400-kg mine car is hoisted up the incline using the cable and motor M. For a short time, the force in the cable is $F = (3200t^2)$ N, where t is in seconds. If the car has an initial velocity $v_1 = 2$ m/s at $s = 0$ and $t = 0$, determine the distance it moves up the plane when $t = 2$ s.

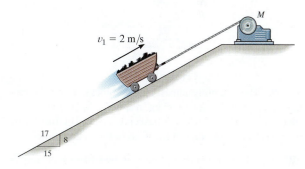

Probs. R1–34/35

*R1–36. The rocket sled has a mass of 4 Mg and travels along the smooth horizontal track such that it maintains a constant power output of 450 kW. Neglect the loss of fuel mass and air resistance, and determine how far the sled must travel to reach a speed of $v = 60$ m/s starting from rest.

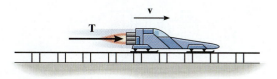

Prob. R1–36

R1–37. The collar has a mass of 20 kg and can slide freely on the smooth rod. The attached springs are undeformed when $d = 0.5$ m. Determine the speed of the collar after the applied force $F = 100$ N causes it to be displaced so that $d = 0.3$ m. When $d = 0.5$ m the collar is at rest.

R1–38. The collar has a mass of 20 kg and can slide freely on the smooth rod. The attached springs are both compressed 0.4 m when $d = 0.5$ m. Determine the speed of the collar after the applied force $F = 100$ N causes it to be displaced so that $d = 0.3$ m. When $d = 0.5$ m the collar is at rest.

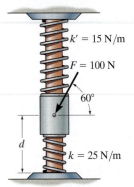

k' = 15 N/m

F = 100 N

60°

d

k = 25 N/m

Probs. R1–37/38

R1–39. The assembly consists of two blocks A and B which have masses of 20 kg and 30 kg, respectively. Determine the speed of each block when B descends 1.5 m. The blocks are released from rest. Neglect the mass of the pulleys and cords.

*R1–40. The assembly consists of two blocks A and B, which have masses of 20 kg and 30 kg, respectively. Determine the distance B must descend in order for A to achieve a speed of 3 m/s starting from rest.

B

A

Probs. R1–39/40

R1–41. Block A, having a mass m, is released from rest, falls a distance h and strikes the plate B having a mass $2m$. If the coefficient of restitution between A and B is e, determine the velocity of the plate just after collision. The spring has a stiffness k.

R1–42. Block A, having a mass of 2 kg, is released from rest, falls a distance $h = 0.5$ m, and strikes the plate B having a mass of 3 kg. If the coefficient of restitution between A and B is $e = 0.6$, determine the velocity of the block just after collision. The spring has a stiffness $k = 30$ N/m.

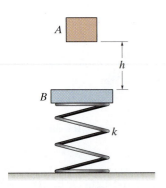

A

h

B

k

Probs. R1–41/42

R1–43. The cylindrical plug has a weight of 2 lb and it is free to move within the confines of the smooth pipe. The spring has a stiffness $k = 14$ lb/ft and when no motion occurs the distance $d = 0.5$ ft. Determine the force of the spring on the plug when the plug is at rest with respect to the pipe. The plug travels in a circle with a constant speed of 15 ft/s, which is caused by the rotation of the pipe about the vertical axis. Neglect the size of the plug.

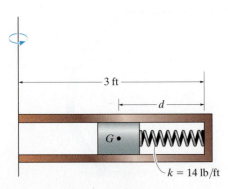

3 ft

d

G•

k = 14 lb/ft

Prob. R1–43

*R1–44. A 20-g bullet is fired horizontally into the 300-g block which rests on the smooth surface. After the bullet becomes embedded into the block, the block moves to the right 0.3 m before momentarily coming to rest. Determine the speed $(v_B)_1$ of the bullet. The spring has a stiffness $k = 200$ N/m and is originally unstretched.

R1–45. The 20-g bullet is fired horizontally at $(v_B)_1 = 1200$ m/s into the 300-g block which rests on the smooth surface. Determine the distance the block moves to the right before momentarily coming to rest. The spring has a stiffness $k = 200$ N/m and is originally unstretched.

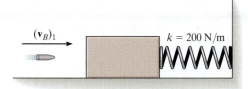

Probs. R1–44/45

R1–46. A particle of mass m is fired at an angle θ_0 with a velocity v_0 in a liquid that develops a drag resistance $F = -kv$, where k is a constant. Determine the maximum or terminal speed reached by the particle.

R1–47. A projectile of mass m is fired into a liquid at an angle θ_0 with an initial velocity v_0 as shown. If the liquid develops a friction or drag resistance on the projectile which is proportional to its velocity, i.e., $F = -kv$, where k is a constant, determine the x and y components of its position at any instant. Also, what is the maximum distance x_{max} that it travels?

*R1–48. The position of particles A and B are $\mathbf{r}_A = \{3t\mathbf{i} + 9t(2 - t)\mathbf{j}\}$ m and $\mathbf{r}_B = \{3(t^2 - 2t + 2)\mathbf{i} + 3(t - 2)\mathbf{j}\}$ m, respectively, where t is in seconds. Determine the point where the particles collide and their speeds just before the collision. How long does it take before the collision occurs?

R1–49. Determine the speed of the automobile if it has the acceleration shown and is traveling on a road which has a radius of curvature of $\rho = 50$ m. Also, what is the automobile's rate of increase in speed?

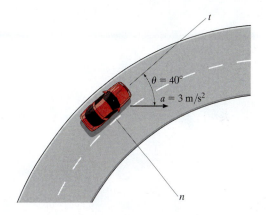

Prob. R1–49

R1–50. The spring has a stiffness $k = 3$ lb/ft and an unstretched length of 2 ft. If it is attached to the 5-lb smooth collar and the collar is released from rest at A, determine the speed of the collar just before it strikes the end of the rod at B. Neglect the size of the collar.

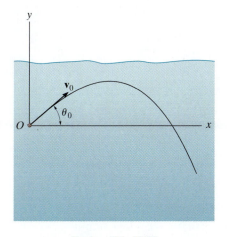

Probs. R1–46/47

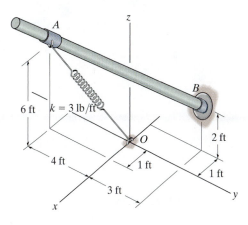

Prob. R1–50

The blades of each wind turbine rotate about a fixed axis with variable angular motion.

Planar Kinematics of a Rigid Body

16

CHAPTER OBJECTIVES

- To classify the various types of rigid-body planar motion.
- To investigate rigid-body translation and angular motion about a fixed axis.
- To study planar motion using an absolute motion analysis.
- To provide a relative motion analysis of velocity and acceleration using a translating frame of reference.
- To show how to find the instantaneous center of zero velocity and determine the velocity of a point on a body using this method.
- To provide a relative-motion analysis of velocity and acceleration using a rotating frame of reference.

16.1   Planar Rigid-Body Motion

In this chapter, the planar kinematics of a rigid body will be discussed. This study is important for the design of gears, cams, and mechanisms used for many mechanical operations. Once the kinematics is thoroughly understood, then we can apply the equations of motion, which relate the forces on the body to the body's motion.

The *planar motion* of a body occurs when all the particles of a rigid body move along paths which are equidistant from a fixed plane. There are three types of rigid body planar motion, in order of increasing complexity, they are

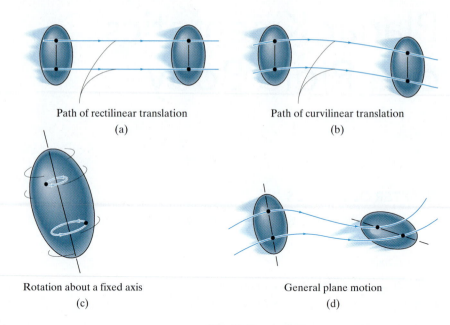

Path of rectilinear translation

(a)

Path of curvilinear translation

(b)

Rotation about a fixed axis

(c)

General plane motion

(d)

Fig. 16–1

- *Translation.* This type of motion occurs when a line in the body remains parallel to its original orientation throughout the motion. When the paths of motion for any two points on the body are parallel lines, the motion is called *rectilinear translation*, Fig. 16–1a. If the paths of motion are along curved lines which are equidistant, the motion is called *curvilinear translation*, Fig. 16–1b.

- *Rotation about a fixed axis.* When a rigid body rotates about a fixed axis, all the particles of the body, except those which lie on the axis of rotation, move along circular paths, Fig. 16–1c.

- *General plane motion.* When a body is subjected to general plane motion, it undergoes a combination of translation *and* rotation, Fig. 16–1d. The translation occurs within a reference plane, and the rotation occurs about an axis perpendicular to the reference plane.

In the following sections we will consider each of these motions in detail. Examples of bodies undergoing these motions are shown in Fig. 16–2.

General plane motion

Curvilinear translation

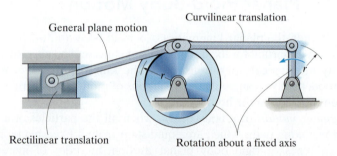

Rectilinear translation

Rotation about a fixed axis

Fig. 16–2

16.2 Translation

Consider a rigid body which is subjected to either rectilinear or curvilinear translation in the *x*–*y* plane, Fig. 16–3.

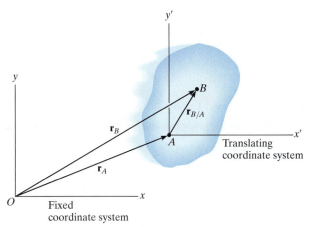

Fig. 16–3

Position. The locations of points *A* and *B* on the body are defined with respect to fixed *x, y* reference frame using *position vectors* $\mathbf{r}_A$ and $\mathbf{r}_B$. The translating x', y' coordinate system is *fixed in the body* and has its origin at *A*, hereafter referred to as the *base point*. The position of *B* with respect to *A* is denoted by the *relative-position vector* $\mathbf{r}_{B/A}$ ("**r** of *B* with respect to *A*"). By vector addition,

$$\mathbf{r}_B = \mathbf{r}_A + \mathbf{r}_{B/A}$$

Velocity. A relation between the instantaneous velocities of *A* and *B* is obtained by taking the time derivative of this equation, which yields $\mathbf{v}_B = \mathbf{v}_A + d\mathbf{r}_{B/A}/dt$. Here $\mathbf{v}_A$ and $\mathbf{v}_B$ denote *absolute velocities* since these vectors are measured with respect to the *x, y* axes. The term $d\mathbf{r}_{B/A}/dt = \mathbf{0}$, since the *magnitude* of $\mathbf{r}_{B/A}$ is *constant* by definition of a rigid body, and because the body is translating the *direction* of $\mathbf{r}_{B/A}$ is also *constant*. Therefore,

$$\mathbf{v}_B = \mathbf{v}_A$$

Acceleration. Taking the time derivative of the velocity equation yields a similar relationship between the instantaneous accelerations of *A* and *B*:

$$\mathbf{a}_B = \mathbf{a}_A$$

The above two equations indicate that *all points in a rigid body subjected to either rectilinear or curvilinear translation move with the same velocity and acceleration*. As a result, the kinematics of particle motion, discussed in Chapter 12, can also be used to specify the kinematics of points located in a translating rigid body.

Riders on this amusement ride are subjected to curvilinear translation, since the vehicle moves in a circular path yet it always remains in an upright position.

16.3   Rotation about a Fixed Axis

When a body rotates about a fixed axis, any point P located in the body travels along a *circular path*. To study this motion it is first necessary to discuss the angular motion of the body about the axis.

Angular Motion.   Since a point is without dimension, it cannot have angular motion. *Only lines or bodies undergo angular motion.* For example, consider the body shown in Fig. 16–4a and the angular motion of a radial line r located within the shaded plane.

Angular Position.   At the instant shown, the *angular position* of r is defined by the angle θ, measured from a *fixed* reference line to r.

Angular Displacement.   The change in the angular position, which can be measured as a differential $d\theta$, is called the *angular displacement.** This vector has a *magnitude* of $d\theta$, measured in degrees, radians, or revolutions, where 1 rev = 2π rad. Since motion is about a *fixed axis*, the direction of $d\theta$ is *always* along this axis. Specifically, the *direction* is determined by the right-hand rule; that is, the fingers of the right hand are curled with the sense of rotation, so that in this case the thumb, or $d\theta$, points upward, Fig. 16–4a. In two dimensions, as shown by the top view of the shaded plane, Fig. 16–4b, both θ and $d\theta$ are counterclockwise, and so the thumb points outward from the page.

Angular Velocity.   The time rate of change in the angular position is called the *angular velocity* $\boldsymbol{\omega}$ (omega). Since $d\theta$ occurs during an instant of time dt, then,

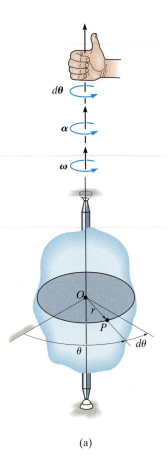

(a)

$$(\zeta+) \qquad \boxed{\omega = \frac{d\theta}{dt}} \qquad (16\text{–}1)$$

This vector has a *magnitude* which is often measured in rad/s. It is expressed here in scalar form since its *direction* is also along the axis of rotation, Fig. 16–4a. When indicating the angular motion in the shaded plane, Fig. 16–4b, we can refer to the sense of rotation as clockwise or counterclockwise. Here we have *arbitrarily* chosen counterclockwise rotations as *positive* and indicated this by the curl shown in parentheses next to Eq. 16–1 Realize, however, that the directional sense of $\boldsymbol{\omega}$ is actually outward from the page.

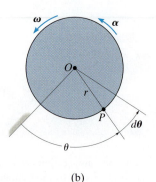

(b)

Fig. 16–4

*It is shown in Sec. 20.1 that finite rotations or finite angular displacements are *not* vector quantities, although differential rotations $d\theta$ are vectors.

Angular Acceleration. The *angular acceleration* $\boldsymbol{\alpha}$ (alpha) measures the time rate of change of the angular velocity. The *magnitude* of this vector is

$(\circlearrowleft+)$
$$\alpha = \frac{d\omega}{dt}$$
(16–2)

Using Eq. 16–1, it is also possible to express α as

$(\circlearrowleft+)$
$$\alpha = \frac{d^2\theta}{dt^2}$$
(16–3)

The line of action of $\boldsymbol{\alpha}$ is the same as that for $\boldsymbol{\omega}$, Fig. 16–4a; however, its sense of *direction* depends on whether $\boldsymbol{\omega}$ is increasing or decreasing. If $\boldsymbol{\omega}$ is decreasing, then $\boldsymbol{\alpha}$ is called an *angular deceleration* and therefore has a sense of direction which is opposite to $\boldsymbol{\omega}$.

By eliminating dt from Eqs. 16–1 and 16–2, we obtain a differential relation between the angular acceleration, angular velocity, and angular displacement, namely,

$(\circlearrowleft+)$
$$\alpha\, d\theta = \omega\, d\omega$$
(16–4)

The similarity between the differential relations for angular motion and those developed for rectilinear motion of a particle ($v = ds/dt$, $a = dv/dt$, and $a\, ds = v\, dv$) should be apparent.

Constant Angular Acceleration.   If the angular acceleration of the body is *constant*, $\boldsymbol{\alpha} = \boldsymbol{\alpha}_c$, then Eqs. 16–1, 16–2, and 16–4, when integrated, yield a set of formulas which relate the body's angular velocity, angular position, and time. These equations are similar to Eqs. 12–4 to 12–6 used for rectilinear motion. The results are

$(\circlearrowleft+)$
$(\circlearrowleft+)$
$(\circlearrowleft+)$
$$\omega = \omega_0 + \alpha_c t \tag{16–5}$$
$$\theta = \theta_0 + \omega_0 t + \tfrac{1}{2}\alpha_c t^2 \tag{16–6}$$
$$\omega^2 = \omega_0^2 + 2\alpha_c(\theta - \theta_0) \tag{16–7}$$
Constant Angular Acceleration

Here θ_0 and ω_0 are the initial values of the body's angular position and angular velocity, respectively.

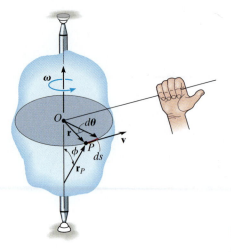

(c)

Fig. 16–4 (cont.)

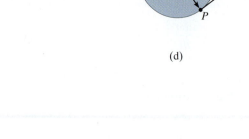

(d)

Motion of Point *P*.

As the rigid body in Fig. 16–4c rotates, point *P* travels along a *circular path* of radius *r* with center at point *O*. This path is contained within the shaded plane shown in top view, Fig. 16–4d.

Position and Displacement. The position of *P* is defined by the position vector **r**, which extends from *O* to *P*. If the body rotates $d\theta$ then *P* will displace $ds = r\,d\theta$.

Velocity. The velocity of *P* has a magnitude which can be found by dividing $ds = r\,d\theta$ by dt so that

$$v = \omega r \tag{16–8}$$

As shown in Figs. 16–4c and 16–4d, the *direction* of **v** is *tangent* to the circular path.

Both the magnitude and direction of **v** can also be accounted for by using the cross product of $\boldsymbol{\omega}$ and $\mathbf{r}_P$ (see Appendix B). Here, $\mathbf{r}_P$ is directed from *any point* on the axis of rotation to point *P*, Fig. 16–4c. We have

$$\mathbf{v} = \boldsymbol{\omega} \times \mathbf{r}_P \tag{16–9}$$

The order of the vectors in this formulation is important, since the cross product is not commutative, i.e., $\boldsymbol{\omega} \times \mathbf{r}_P \neq \mathbf{r}_P \times \boldsymbol{\omega}$. Notice in Fig. 16–4c how the correct direction of **v** is established by the right-hand rule. The fingers of the right hand are curled from $\boldsymbol{\omega}$ toward $\mathbf{r}_P$ ($\boldsymbol{\omega}$ "cross" $\mathbf{r}_P$). The thumb indicates the correct direction of **v**, which is tangent to the path in the direction of motion. From Eq. B–8, the magnitude of **v** in Eq. 16–9 is $v = \omega r_P \sin \phi$, and since $r = r_P \sin \phi$, Fig. 16–4c, then $v = \omega r$, which agrees with Eq. 16–8. As a special case, the position vector **r** can be chosen for $\mathbf{r}_P$. Here **r** lies in the plane of motion and again the velocity of point *P* is

$$\mathbf{v} = \boldsymbol{\omega} \times \mathbf{r} \tag{16–10}$$

Acceleration.   The acceleration of P can be expressed in terms of its normal and tangential components. Since $a_t = dv/dt$ and $a_n = v^2/\rho$, where $\rho = r$, $v = \omega r$, and $\alpha = d\omega/dt$, we have

$$a_t = \alpha r \qquad (16\text{–}11)$$

$$a_n = \omega^2 r \qquad (16\text{–}12)$$

The *tangential component of acceleration*, Figs. 16–4e and 16–4f, represents the time rate of change in the velocity's magnitude. If the speed of P is increasing, then $\mathbf{a}_t$ acts in the same direction as $\mathbf{v}$; if the speed is decreasing, $\mathbf{a}_t$ acts in the opposite direction of $\mathbf{v}$; and finally, if the speed is constant, $\mathbf{a}_t$ is zero.

The *normal component of acceleration* represents the time rate of change in the velocity's direction. The *direction* of $\mathbf{a}_n$ is always toward O, the center of the circular path, Figs. 16–4e and 16–4f.

Like the velocity, the acceleration of point P can be expressed in terms of the vector cross product. Taking the time derivative of Eq. 16–9 we have

$$\mathbf{a} = \frac{d\mathbf{v}}{dt} = \frac{d\boldsymbol{\omega}}{dt} \times \mathbf{r}_P + \boldsymbol{\omega} \times \frac{d\mathbf{r}_P}{dt}$$

Recalling that $\boldsymbol{\alpha} = d\boldsymbol{\omega}/dt$, and using Eq. 16–9 ($d\mathbf{r}_P/dt = \mathbf{v} = \boldsymbol{\omega} \times \mathbf{r}_P$), yields

$$\mathbf{a} = \boldsymbol{\alpha} \times \mathbf{r}_P + \boldsymbol{\omega} \times (\boldsymbol{\omega} \times \mathbf{r}_P) \qquad (16\text{–}13)$$

From the definition of the cross product, the first term on the right has a magnitude $a_t = \alpha r_P \sin \phi = \alpha r$, and by the right-hand rule, $\boldsymbol{\alpha} \times \mathbf{r}_P$ is in the direction of $\mathbf{a}_t$, Fig. 16–4e. Likewise, the second term has a magnitude $a_n = \omega^2 r_P \sin \phi = \omega^2 r$, and applying the right-hand rule twice, first to determine the result $\mathbf{v}_P = \boldsymbol{\omega} \times \mathbf{r}_P$ then $\boldsymbol{\omega} \times \mathbf{v}_P$, it can be seen that this result is in the same direction as $\mathbf{a}_n$, shown in Fig. 16–4e. Noting that this is also the *same* direction as $-\mathbf{r}$, which lies in the plane of motion, we can express $\mathbf{a}_n$ in a much simpler form as $\mathbf{a}_n = -\omega^2 \mathbf{r}$. Hence, Eq. 16–13 can be identified by its two components as

$$\begin{aligned} \mathbf{a} &= \mathbf{a}_t + \mathbf{a}_n \\ &= \boldsymbol{\alpha} \times \mathbf{r} - \omega^2 \mathbf{r} \end{aligned} \qquad (16\text{–}14)$$

Since $\mathbf{a}_t$ and $\mathbf{a}_n$ are perpendicular to one another, if needed the magnitude of acceleration can be determined from the Pythagorean theorem; namely, $a = \sqrt{a_n^2 + a_t^2}$, Fig. 16–4f.

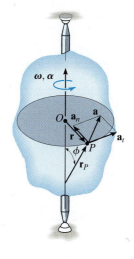

(e)

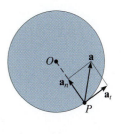

(f)

16

The gears used in the operation of a crane all rotate about fixed axes. Engineers must be able to relate their angular motions in order to properly design this gear system.

Important Points

- A body can undergo two types of translation. During rectilinear translation all points follow parallel straight-line paths, and during curvilinear translation the points follow curved paths that are the same shape and are equidistant from one another.

- All the points on a translating body move with the same velocity and acceleration.

- Points located on a body that rotates about a fixed axis follow circular paths.

- The relation $\alpha\, d\theta = \omega\, d\omega$ is derived from $\alpha = d\omega/dt$ and $\omega = d\theta/dt$ by eliminating dt.

- Once angular motions ω and α are known, the velocity and acceleration of any point on the body can be determined.

- The velocity always acts tangent to the path of motion.

- The acceleration has two components. The tangential acceleration measures the rate of change in the magnitude of the velocity and can be determined from $a_t = \alpha r$. The normal acceleration measures the rate of change in the direction of the velocity and can be determined from $a_n = \omega^2 r$.

Procedure for Analysis

The velocity and acceleration of a point located on a rigid body that is rotating about a fixed axis can be determined using the following procedure.

Angular Motion.

- Establish the positive sense of rotation about the axis of rotation and show it alongside each kinematic equation as it is applied.
- If a relation is known between any *two* of the four variables α, ω, θ, and t, then a third variable can be obtained by using one of the following kinematic equations which relates all three variables.

$$\omega = \frac{d\theta}{dt} \qquad \alpha = \frac{d\omega}{dt} \qquad \alpha \, d\theta = \omega \, d\omega$$

- If the body's angular acceleration is *constant*, then the following equations can be used:

$$\omega = \omega_0 + \alpha_c t$$

$$\theta = \theta_0 + \omega_0 t + \tfrac{1}{2}\alpha_c t^2$$

$$\omega^2 = \omega_0^2 + 2\alpha_c(\theta - \theta_0)$$

- Once the solution is obtained, the sense of θ, ω, and α is determined from the algebraic signs of their numerical quantities.

Motion of Point P.

- In most cases the velocity of P and its two components of acceleration can be determined from the scalar equations

$$v = \omega r$$

$$a_t = \alpha r$$

$$a_n = \omega^2 r$$

- If the geometry of the problem is difficult to visualize, the following vector equations should be used:

$$\mathbf{v} = \boldsymbol{\omega} \times \mathbf{r}_P = \boldsymbol{\omega} \times \mathbf{r}$$

$$\mathbf{a}_t = \boldsymbol{\alpha} \times \mathbf{r}_P = \boldsymbol{\alpha} \times \mathbf{r}$$

$$\mathbf{a}_n = \boldsymbol{\omega} \times (\boldsymbol{\omega} \times \mathbf{r}_P) = -\omega^2 \mathbf{r}$$

- Here $\mathbf{r}_P$ is directed from any point on the axis of rotation to point P, whereas $\mathbf{r}$ lies in the plane of motion of P. Either of these vectors, along with $\boldsymbol{\omega}$ and $\boldsymbol{\alpha}$, should be expressed in terms of its $\mathbf{i}$, $\mathbf{j}$, $\mathbf{k}$ components, and, if necessary, the cross products determined using a determinant expansion (see Eq. B–12).

16

EXAMPLE 16.1

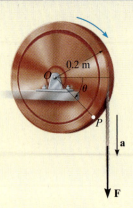

Fig. 16–5

A cord is wrapped around a wheel in Fig. 16–5, which is initially at rest when $\theta = 0$. If a force is applied to the cord and gives it an acceleration $a = (4t)$ m/s^2, where t is in seconds, determine, as a function of time, (a) the angular velocity of the wheel, and (b) the angular position of line OP in radians.

SOLUTION

Part (a). The wheel is subjected to rotation about a fixed axis passing through point O. Thus, point P on the wheel has motion about a circular path, and the acceleration of this point has *both* tangential and normal components. The tangential component is $(a_P)_t = (4t)$ m/s^2, since the cord is wrapped around the wheel and moves *tangent* to it. Hence the angular acceleration of the wheel is

$$(\zeta +) \qquad (a_P)_t = \alpha r$$

$$(4t) \text{ m/s}^2 = \alpha(0.2 \text{ m})$$

$$\alpha = (20t) \text{ rad/s}^2 \, \zeta$$

Using this result, the wheel's angular velocity ω can now be determined from $\alpha = d\omega/dt$, since this equation relates α, t, and ω. Integrating, with the initial condition that $\omega = 0$ when $t = 0$, yields

$$(\zeta +) \qquad \alpha = \frac{d\omega}{dt} = (20t) \text{ rad/s}^2$$

$$\int_0^\omega d\omega = \int_0^t 20t \, dt$$

$$\omega = 10t^2 \text{ rad/s} \, \zeta \qquad\qquad Ans.$$

Part (b). Using this result, the angular position θ of OP can be found from $\omega = d\theta/dt$, since this equation relates θ, ω, and t. Integrating, with the initial condition $\theta = 0$ when $t = 0$, we have

$$(\zeta +) \qquad \frac{d\theta}{dt} = \omega = (10t^2) \text{ rad/s}$$

$$\int_0^\theta d\theta = \int_0^t 10t^2 \, dt$$

$$\theta = 3.33t^3 \text{ rad} \qquad\qquad Ans.$$

NOTE: We cannot use the equation of constant angular acceleration, since α is a function of time.

EXAMPLE 16.2

The motor shown in the photo is used to turn a wheel and attached blower contained within the housing. The details of the design are shown in Fig. 16–6a. If the pulley A connected to the motor begins to rotate from rest with a constant angular acceleration of $\alpha_A = 2 \text{ rad/s}^2$, determine the magnitudes of the velocity and acceleration of point P on the wheel, after the pulley has turned two revolutions. Assume the transmission belt does not slip on the pulley and wheel.

SOLUTION

Angular Motion. First we will convert the two revolutions to radians. Since there are 2π rad in one revolution, then

$$\theta_A = 2 \text{ rev} \left(\frac{2\pi \text{ rad}}{1 \text{ rev}} \right) = 12.57 \text{ rad}$$

Since α_A is constant, the angular velocity of pulley A is therefore

$(\circlearrowleft +)$
$$\omega^2 = \omega_0^2 + 2\alpha_c(\theta - \theta_0)$$
$$\omega_A^2 = 0 + 2(2 \text{ rad/s}^2)(12.57 \text{ rad} - 0)$$
$$\omega_A = 7.090 \text{ rad/s}$$

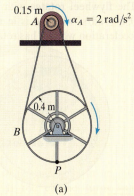

The belt has the same speed and tangential component of acceleration as it passes over the pulley and wheel. Thus,

$$v = \omega_A r_A = \omega_B r_B; \quad 7.090 \text{ rad/s} (0.15 \text{ m}) = \omega_B(0.4 \text{ m})$$
$$\omega_B = 2.659 \text{ rad/s}$$
$$a_t = \alpha_A r_A = \alpha_B r_B; \quad 2 \text{ rad/s}^2 (0.15 \text{ m}) = \alpha_B(0.4 \text{ m})$$
$$\alpha_B = 0.750 \text{ rad/s}^2$$

(a)

Motion of P. As shown on the kinematic diagram in Fig. 16–6b, we have

$$v_P = \omega_B r_B = 2.659 \text{ rad/s} (0.4 \text{ m}) = 1.06 \text{ m/s} \qquad Ans.$$
$$(a_P)_t = \alpha_B r_B = 0.750 \text{ rad/s}^2 (0.4 \text{ m}) = 0.3 \text{ m/s}^2$$
$$(a_P)_n = \omega_B^2 r_B = (2.659 \text{ rad/s})^2(0.4 \text{ m}) = 2.827 \text{ m/s}^2$$

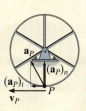

Thus

$$a_P = \sqrt{(0.3 \text{ m/s}^2)^2 + (2.827 \text{ m/s}^2)^2} = 2.84 \text{ m/s}^2 \qquad Ans.$$

(b)

Fig. 16–6

EXAMPLE | **16.3**

The end of rod R shown in Fig. 16–7 maintains contact with the cam by means of a spring. If the cam rotates about an axis passing through point O with an angular acceleration α and angular velocity ω, determine the velocity and acceleration of the rod when the cam is in the arbitrary position θ.

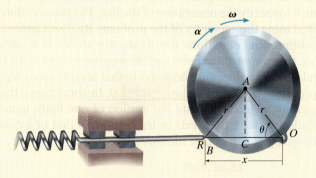

Fig. 16–7

SOLUTION

Position Coordinate Equation. Coordinates θ and x are chosen in order to relate the *rotational motion* of the line segment OA on the cam to the *rectilinear translation* of the rod. These coordinates are measured from the *fixed point O* and can be related to each other using trigonometry. Since $OC = CB = r\cos\theta$, Fig. 16–7, then

$$x = 2r\cos\theta$$

Time Derivatives. Using the chain rule of calculus, we have

$$\frac{dx}{dt} = -2r(\sin\theta)\frac{d\theta}{dt}$$

$$v = -2r\omega\sin\theta \qquad\qquad Ans.$$

$$\frac{dv}{dt} = -2r\left(\frac{d\omega}{dt}\right)\sin\theta - 2r\omega(\cos\theta)\frac{d\theta}{dt}$$

$$a = -2r(\alpha\sin\theta + \omega^2\cos\theta) \qquad\qquad Ans.$$

NOTE: The negative signs indicate that v and a are opposite to the direction of positive x. This seems reasonable when you visualize the motion.

EXAMPLE 16.4

At a given instant, the cylinder of radius r, shown in Fig. 16–8, has an angular velocity $\boldsymbol{\omega}$ and angular acceleration $\boldsymbol{\alpha}$. Determine the velocity and acceleration of its center G if the cylinder rolls without slipping.

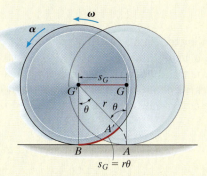

Fig. 16–8

SOLUTION

Position Coordinate Equation. The cylinder undergoes general plane motion since it simultaneously translates and rotates. By inspection, point G moves in a *straight line* to the left, from G to G', as the cylinder rolls, Fig. 16–8. Consequently its new position G' will be specified by the *horizontal* position coordinate s_G, which is measured from G to G'. Also, as the cylinder rolls (without slipping), the arc length $A'B$ on the rim which was in contact with the ground from A to B, is equivalent to s_G. Consequently, the motion requires the radial line GA to rotate θ to the position $G'A'$. Since the arc $A'B = r\theta$, then G travels a distance

$$s_G = r\theta$$

Time Derivatives. Taking successive time derivatives of this equation, realizing that r is constant, $\omega = d\theta/dt$, and $\alpha = d\omega/dt$, gives the necessary relationships:

$$s_G = r\theta$$

$$v_G = r\omega \qquad\qquad\qquad \textit{Ans.}$$

$$a_G = r\alpha \qquad\qquad\qquad \textit{Ans.}$$

NOTE: Remember that these relationships are valid only if the cylinder (disk, wheel, ball, etc.) rolls *without* slipping.

EXAMPLE 16.5

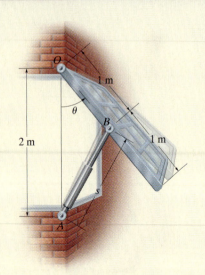

Fig. 16–9

The large window in Fig. 16–9 is opened using a hydraulic cylinder *AB*. If the cylinder extends at a constant rate of 0.5 m/s, determine the angular velocity and angular acceleration of the window at the instant $\theta = 30°$.

SOLUTION

Position Coordinate Equation.   The angular motion of the window can be obtained using the coordinate θ, whereas the extension or motion *along the hydraulic cylinder* is defined using a coordinate s, which measures its length from the fixed point A to the moving point B. These coordinates can be related using the law of cosines, namely,

$$s^2 = (2 \text{ m})^2 + (1 \text{ m})^2 - 2(2 \text{ m})(1 \text{ m}) \cos \theta$$

$$s^2 = 5 - 4 \cos \theta \qquad (1)$$

When $\theta = 30°$,

$$s = 1.239 \text{ m}$$

Time Derivatives.   Taking the time derivatives of Eq. 1, we have

$$2s \frac{ds}{dt} = 0 - 4(-\sin \theta) \frac{d\theta}{dt}$$

$$s(v_s) = 2(\sin \theta)\omega \qquad (2)$$

Since $v_s = 0.5$ m/s, then at $\theta = 30°$,

$$(1.239 \text{ m})(0.5 \text{ m/s}) = 2 \sin 30° \omega$$

$$\omega = 0.6197 \text{ rad/s} = 0.620 \text{ rad/s} \qquad \textit{Ans.}$$

Taking the time derivative of Eq. 2 yields

$$\frac{ds}{dt} v_s + s \frac{dv_s}{dt} = 2(\cos \theta) \frac{d\theta}{dt} \omega + 2(\sin \theta) \frac{d\omega}{dt}$$

$$v_s^2 + sa_s = 2(\cos \theta)\omega^2 + 2(\sin \theta)\alpha$$

Since $a_s = dv_s/dt = 0$, then

$$(0.5 \text{ m/s})^2 + 0 = 2 \cos 30°(0.6197 \text{ rad/s})^2 + 2 \sin 30° \alpha$$

$$\alpha = -0.415 \text{ rad/s}^2 \qquad \textit{Ans.}$$

Because the result is negative, it indicates the window has an angular deceleration.

PROBLEMS

*16–36.   Rod CD presses against AB, giving it an angular velocity. If the angular velocity of AB is maintained at $\omega = 5$ rad/s, determine the required magnitude of the velocity **v** of CD as a function of the angle θ of rod AB.

16–38.   The block moves to the left with a constant velocity $\mathbf{v}_0$. Determine the angular velocity and angular acceleration of the bar as a function of θ.

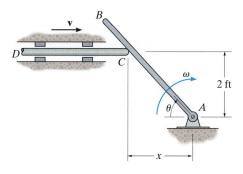

Prob. 16–36

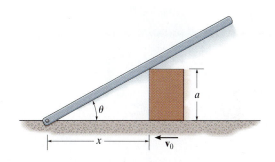

Prob. 16–38

•16–37.   The scaffold S is raised by moving the roller at A toward the pin at B. If A is approaching B with a speed of 1.5 ft/s, determine the speed at which the platform rises as a function of θ. The 4-ft links are pin connected at their midpoint.

16–39.   Determine the velocity and acceleration of platform P as a function of the angle θ of cam C if the cam rotates with a constant angular velocity $\boldsymbol{\omega}$. The pin connection does not cause interference with the motion of P on C. The platform is constrained to move vertically by the smooth vertical guides.

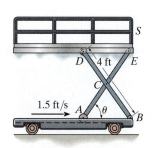

Prob. 16–37

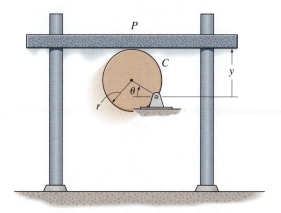

Prob. 16–39

*16–40. Disk *A* rolls without slipping over the surface of the *fixed* cylinder *B*. Determine the angular velocity of *A* if its center *C* has a speed $v_C = 5$ m/s. How many revolutions will *A* rotate about its center just after link *DC* completes one revolution?

16–42. The pins at *A* and *B* are constrained to move in the vertical and horizontal tracks. If the slotted arm is causing *A* to move downward at $\mathbf{v}_A$, determine the velocity of *B* as a function of θ.

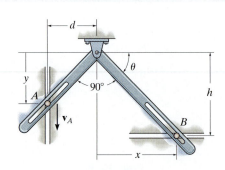

Prob. 16–42

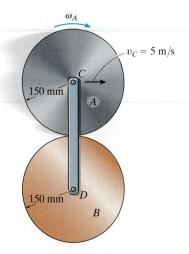

Prob. 16–40

•16–41. Crank *AB* rotates with a constant angular velocity of 5 rad/s. Determine the velocity of block *C* and the angular velocity of link *BC* at the instant $\theta = 30°$.

16–43. End *A* of the bar moves to the left with a constant velocity $\mathbf{v}_A$. Determine the angular velocity $\boldsymbol{\omega}$ and angular acceleration $\boldsymbol{\alpha}$ of the bar as a function of its position *x*.

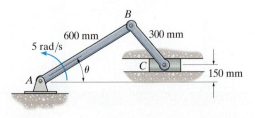

Prob. 16–41

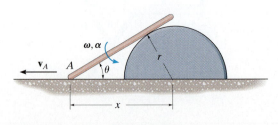

Prob. 16–43

*16–44.   Determine the velocity and acceleration of the plate at the instant $\theta = 30°$, if at this instant the circular cam is rotating about the fixed point O with an angular velocity $\omega = 4$ rad/s and an angular acceleration $\alpha = 2$ rad/s^2.

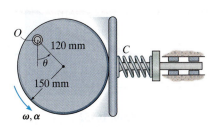

Prob. 16–44

•16–45.   At the instant $\theta = 30°$, crank AB rotates with an angular velocity and angular acceleration of $\omega = 10$ rad/s and $\alpha = 2$ rad/s^2, respectively. Determine the velocity and acceleration of the slider block C at this instant. Take $a = b = 0.3$ m.

16–46.   At the instant $\theta = 30°$, crank AB rotates with an angular velocity and angular acceleration of $\omega = 10$ rad/s and $\alpha = 2$ rad/s^2, respectively. Determine the angular velocity and angular acceleration of the connecting rod BC at this instant. Take $a = 0.3$ m and $b = 0.5$ m.

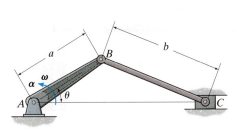

Probs. 16–45/46

16–47.   The bridge girder G of a bascule bridge is raised and lowered using the drive mechanism shown. If the hydraulic cylinder AB shortens at a constant rate of 0.15 m/s, determine the angular velocity of the bridge girder at the instant $\theta = 60°$.

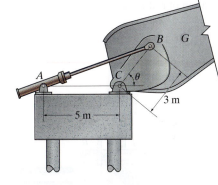

Prob. 16–47

*16–48.   The man pulls on the rope at a constant rate of 0.5 m/s. Determine the angular velocity and angular acceleration of beam AB when $\theta = 60°$. The beam rotates about A. Neglect the thickness of the beam and the size of the pulley.

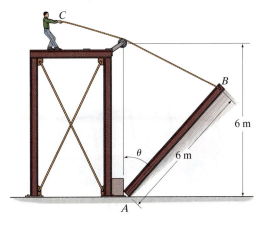

Prob. 16–48

EXAMPLE 16.9

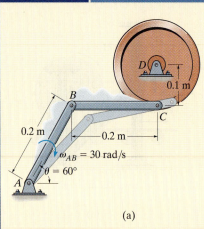

(a)

The bar AB of the linkage shown in Fig. 16–16a has a clockwise angular velocity of 30 rad/s when $\theta = 60°$. Determine the angular velocities of member BC and the wheel at this instant.

SOLUTION (VECTOR ANALYSIS)

Kinematic Diagram. By inspection, the velocities of points B and C are defined by the rotation of link AB and the wheel about their fixed axes. The position vectors and the angular velocity of each member are shown on the kinematic diagram in Fig. 16–16b. To solve, we will write the appropriate kinematic equation for each member.

Velocity Equation. Link AB (rotation about a fixed axis):

$$\mathbf{v}_B = \boldsymbol{\omega}_{AB} \times \mathbf{r}_B$$

$$= (-30\mathbf{k}) \times (0.2 \cos 60°\mathbf{i} + 0.2 \sin 60°\mathbf{j})$$

$$= \{5.20\mathbf{i} - 3.0\mathbf{j}\} \text{ m/s}$$

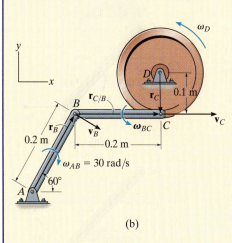

(b)

Fig. 16–16

Link BC (general plane motion):

$$\mathbf{v}_C = \mathbf{v}_B + \boldsymbol{\omega}_{BC} \times \mathbf{r}_{C/B}$$

$$v_C\mathbf{i} = 5.20\mathbf{i} - 3.0\mathbf{j} + (\omega_{BC}\mathbf{k}) \times (0.2\mathbf{i})$$

$$v_C\mathbf{i} = 5.20\mathbf{i} + (0.2\omega_{BC} - 3.0)\mathbf{j}$$

$$v_C = 5.20 \text{ m/s}$$

$$0 = 0.2\omega_{BC} - 3.0$$

$$\omega_{BC} = 15 \text{ rad/s} \circlearrowright \qquad\qquad Ans.$$

Wheel (rotation about a fixed axis):

$$\mathbf{v}_C = \boldsymbol{\omega}_D \times \mathbf{r}_C$$

$$5.20\mathbf{i} = (\omega_D\mathbf{k}) \times (-0.1\mathbf{j})$$

$$5.20 = 0.1\omega_D$$

$$\omega_D = 52.0 \text{ rad/s} \circlearrowright \qquad\qquad Ans.$$

FUNDAMENTAL PROBLEMS

F16–7. If roller A moves to the right with a constant velocity of $v_A = 3$ m/s, determine the angular velocity of the link and the velocity of roller B at the instant $\theta = 30°$.

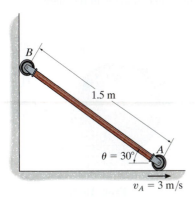

F16–7

F16–8. The wheel rolls without slipping with an angular velocity of $\omega = 10$ rad/s. Determine the magnitude of the velocity of point B at the instant shown.

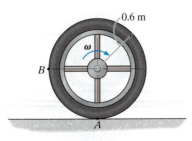

F16–8

F16–9. Determine the angular velocity of the spool. The cable wraps around the inner core, and the spool does not slip on the platform P.

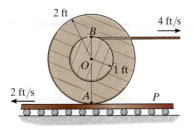

F16–9

F16–10. If crank OA rotates with an angular velocity of $\omega = 12$ rad/s, determine the velocity of piston B and the angular velocity of rod AB at the instant shown.

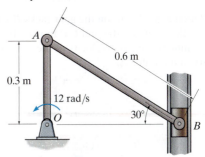

F16–10

F16–11. If rod AB slides along the horizontal slot with a velocity of 60 ft/s, determine the angular velocity of link BC at the instant shown.

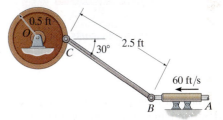

F16–11

F16–12. End A of the link has a velocity of $v_A = 3$ m/s. Determine the velocity of the peg at B at this instant. The peg is constrained to move along the slot.

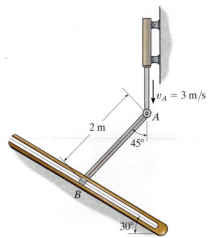

F16–12

16

EXAMPLE 16.14

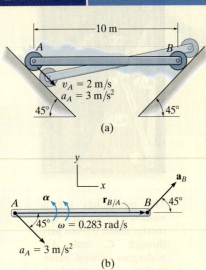

(a)

(b)

The rod AB shown in Fig. 16–27a is confined to move along the inclined planes at A and B. If point A has an acceleration of 3 m/s² and a velocity of 2 m/s, both directed down the plane at the instant the rod is horizontal, determine the angular acceleration of the rod at this instant.

SOLUTION I (VECTOR ANALYSIS)

We will apply the acceleration equation to points A and B on the rod. To do so it is first necessary to determine the angular velocity of the rod. Show that it is $\omega = 0.283$ rad/s $\circlearrowright$ using either the velocity equation or the method of instantaneous centers.

Kinematic Diagram. Since points A and B both move along straight-line paths, they have *no* components of acceleration normal to the paths. There are two unknowns in Fig. 16–27b, namely, a_B and α.

Acceleration Equation.

$$\mathbf{a}_B = \mathbf{a}_A + \boldsymbol{\alpha} \times \mathbf{r}_{B/A} - \omega^2 \mathbf{r}_{B/A}$$

$$a_B \cos 45°\mathbf{i} + a_B \sin 45°\mathbf{j} = 3 \cos 45°\mathbf{i} - 3 \sin 45°\mathbf{j} + (\alpha\mathbf{k}) \times (10\mathbf{i}) - (0.283)^2(10\mathbf{i})$$

Carrying out the cross product and equating the $\mathbf{i}$ and $\mathbf{j}$ components yields

$$a_B \cos 45° = 3 \cos 45° - (0.283)^2(10) \qquad (1)$$

$$a_B \sin 45° = -3 \sin 45° + \alpha(10) \qquad (2)$$

Solving, we have

$$a_B = 1.87 \text{ m/s}^2 \measuredangle 45°$$

$$\alpha = 0.344 \text{ rad/s}^2 \circlearrowright \qquad \qquad Ans.$$

SOLUTION II (SCALAR ANALYSIS)

From the kinematic diagram, showing the relative-acceleration components $(\mathbf{a}_{B/A})_t$ and $(\mathbf{a}_{B/A})_n$, Fig. 16–27c, we have

$$\mathbf{a}_B = \mathbf{a}_A + (\mathbf{a}_{B/A})_t + (\mathbf{a}_{B/A})_n$$

$$\begin{bmatrix} a_B \\ \measuredangle 45° \end{bmatrix} = \begin{bmatrix} 3 \text{ m/s}^2 \\ \measuredangle 45° \end{bmatrix} + \begin{bmatrix} \alpha(10 \text{ m}) \\ \uparrow \end{bmatrix} + \begin{bmatrix} (0.283 \text{ rad/s})^2(10 \text{ m}) \\ \leftarrow \end{bmatrix}$$

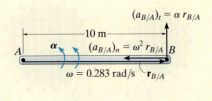

(c)

Fig. 16–27

Equating the x and y components yields Eqs. 1 and 2, and the solution proceeds as before.

EXAMPLE 16.15

At a given instant, the cylinder of radius r, shown in Fig. 16–28a, has an angular velocity $\boldsymbol{\omega}$ and angular acceleration $\boldsymbol{\alpha}$. Determine the velocity and acceleration of its center G and the acceleration of the contact point at A if it rolls without slipping.

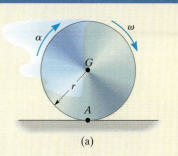

(a)

SOLUTION (VECTOR ANALYSIS)

Velocity Analysis. Since no slipping occurs, at the instant A contacts the ground, $\mathbf{v}_A = \mathbf{0}$. Thus, from the kinematic diagram in Fig. 16–28b we have

$$\mathbf{v}_G = \mathbf{v}_A + \boldsymbol{\omega} \times \mathbf{r}_{G/A}$$
$$v_G\mathbf{i} = \mathbf{0} + (-\omega\mathbf{k}) \times (r\mathbf{j})$$
$$v_G = \omega r \qquad (1) \; Ans.$$

This same result can also be obtained directly by noting that point A represents the instantaneous center of zero velocity.

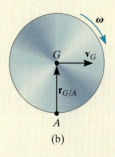

(b)

Kinematic Diagram. Since the motion of G is always along a *straight line*, then its acceleration can be determined by taking the time derivative of its velocity, which gives

$$a_G = \frac{dv_G}{dt} = \frac{d\omega}{dt}r$$
$$a_G = \alpha r \qquad (2) \; Ans.$$

Acceleration Equation. The magnitude and direction of $\mathbf{a}_A$ is unknown, Fig. 16–28c.

$$\mathbf{a}_G = \mathbf{a}_A + \boldsymbol{\alpha} \times \mathbf{r}_{G/A} - \omega^2\mathbf{r}_{G/A}$$
$$\alpha r \, \mathbf{i} = (a_A)_x\mathbf{i} + (a_A)_y\mathbf{j} + (-\alpha\mathbf{k}) \times (r\mathbf{j}) - \omega^2(r\mathbf{j})$$

Evaluating the cross product and equating the $\mathbf{i}$ and $\mathbf{j}$ components yields

$$(a_A)_x = 0 \qquad\qquad Ans.$$
$$(a_A)_y = \omega^2 r \qquad\qquad Ans.$$

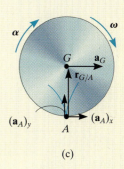

(c)

Fig. 16–28

NOTE: The results, that $v_G = \omega r$ and $a_G = \alpha r$, can be applied to any circular object, such as a ball, cylinder, disk, etc., that rolls *without* slipping. Also, the fact that $a_A = \omega^2 r$ indicates that the instantaneous center of zero velocity, point A, is *not* a point of zero acceleration.

EXAMPLE 16.16

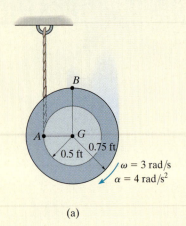

(a)

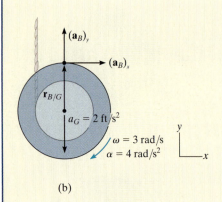

(b)

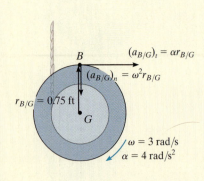

(c)

Fig. 16–29

The spool shown in Fig. 16–29a unravels from the cord, such that at the instant shown it has an angular velocity of 3 rad/s and an angular acceleration of 4 rad/s². Determine the acceleration of point B.

SOLUTION I (VECTOR ANALYSIS)

The spool "appears" to be rolling downward without slipping at point A. Therefore, we can use the results of Example 16.15 to determine the acceleration of point G, i.e.,

$$a_G = \alpha r = (4 \text{ rad/s}^2)(0.5 \text{ ft}) = 2 \text{ ft/s}^2$$

We will apply the acceleration equation to points G and B.

Kinematic Diagram.   Point B moves along a *curved path* having an *unknown* radius of curvature.* Its acceleration will be represented by its unknown x and y components as shown in Fig. 16–29b.

Acceleration Equation.

$$\mathbf{a}_B = \mathbf{a}_G + \boldsymbol{\alpha} \times \mathbf{r}_{B/G} - \omega^2 \mathbf{r}_{B/G}$$

$$(a_B)_x\mathbf{i} + (a_B)_y\mathbf{j} = -2\mathbf{j} + (-4\mathbf{k}) \times (0.75\mathbf{j}) - (3)^2(0.75\mathbf{j})$$

Equating the **i** and **j** terms, the component equations are

$$(a_B)_x = 4(0.75) = 3 \text{ ft/s}^2 \rightarrow \qquad (1)$$

$$(a_B)_y = -2 - 6.75 = -8.75 \text{ ft/s}^2 = 8.75 \text{ ft/s}^2 \downarrow \qquad (2)$$

The magnitude and direction of $\mathbf{a}_B$ are therefore

$$a_B = \sqrt{(3)^2 + (8.75)^2} = 9.25 \text{ ft/s}^2 \qquad \textit{Ans.}$$

$$\theta = \tan^{-1}\frac{8.75}{3} = 71.1° \qquad \textit{Ans.}$$

SOLUTION II (SCALAR ANALYSIS)

This problem may be solved by writing the scalar component equations directly. The kinematic diagram in Fig. 16–29c shows the relative-acceleration components $(\mathbf{a}_{B/G})_t$ and $(\mathbf{a}_{B/G})_n$. Thus,

$$\mathbf{a}_B = \mathbf{a}_G + (\mathbf{a}_{B/G})_t + (\mathbf{a}_{B/G})_n$$

$$\begin{bmatrix} (a_B)_x \\ \rightarrow \end{bmatrix} + \begin{bmatrix} (a_B)_y \\ \uparrow \end{bmatrix}$$

$$= \begin{bmatrix} 2 \text{ ft/s}^2 \\ \downarrow \end{bmatrix} + \begin{bmatrix} 4 \text{ rad/s}^2 (0.75 \text{ ft}) \\ \rightarrow \end{bmatrix} + \begin{bmatrix} (3 \text{ rad/s})^2(0.75 \text{ ft}) \\ \downarrow \end{bmatrix}$$

The x and y components yield Eqs. 1 and 2 above.

*Realize that the path's radius of curvature ρ is not equal to the radius of the spool since the spool is not rotating about point G. Furthermore, ρ is not defined as the distance from A (IC) to B, since the location of the IC depends only on the velocity of a point and not the geometry of its path.

EXAMPLE 16.17

The collar C in Fig. 16–30a moves downward with an acceleration of 1 m/s^2. At the instant shown, it has a speed of 2 m/s which gives links CB and AB an angular velocity $\omega_{AB} = \omega_{CB} = 10 \text{ rad/s}$. (See Example 16.8.) Determine the angular accelerations of CB and AB at this instant.

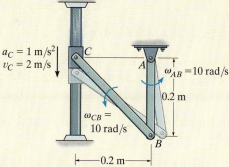

(a)

SOLUTION (VECTOR ANALYSIS)

Kinematic Diagram.   The kinematic diagrams of *both* links AB and CB are shown in Fig. 16–30b. To solve, we will apply the appropriate kinematic equation to each link.

Acceleration Equation.

Link AB (rotation about a fixed axis):

$$\mathbf{a}_B = \boldsymbol{\alpha}_{AB} \times \mathbf{r}_B - \omega_{AB}^2 \mathbf{r}_B$$
$$\mathbf{a}_B = (\alpha_{AB}\mathbf{k}) \times (-0.2\mathbf{j}) - (10)^2(-0.2\mathbf{j})$$
$$\mathbf{a}_B = 0.2\alpha_{AB}\mathbf{i} + 20\mathbf{j}$$

Note that $\mathbf{a}_B$ has n and t components since it moves along a *circular* path.

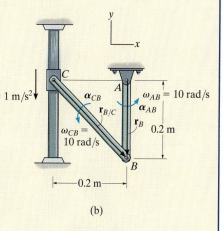

(b)

Fig. 16–30

Link BC (general plane motion): Using the result for $\mathbf{a}_B$ and applying Eq. 16–18, we have

$$\mathbf{a}_B = \mathbf{a}_C + \boldsymbol{\alpha}_{CB} \times \mathbf{r}_{B/C} - \omega_{CB}^2 \mathbf{r}_{B/C}$$
$$0.2\alpha_{AB}\mathbf{i} + 20\mathbf{j} = -1\mathbf{j} + (\alpha_{CB}\mathbf{k}) \times (0.2\mathbf{i} - 0.2\mathbf{j}) - (10)^2(0.2\mathbf{i} - 0.2\mathbf{j})$$
$$0.2\alpha_{AB}\mathbf{i} + 20\mathbf{j} = -1\mathbf{j} + 0.2\alpha_{CB}\mathbf{j} + 0.2\alpha_{CB}\mathbf{i} - 20\mathbf{i} + 20\mathbf{j}$$

Thus,

$$0.2\alpha_{AB} = 0.2\alpha_{CB} - 20$$
$$20 = -1 + 0.2\alpha_{CB} + 20$$

Solving,

$$\alpha_{CB} = 5 \text{ rad/s}^2 \text{↻} \qquad\qquad \textit{Ans.}$$
$$\alpha_{AB} = -95 \text{ rad/s}^2 = 95 \text{ rad/s}^2 \text{↺} \qquad \textit{Ans.}$$

16

EXAMPLE 16.18

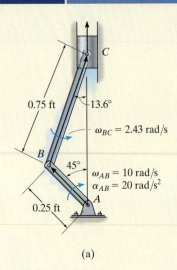

(a)

The crankshaft AB turns with a clockwise angular acceleration of 20 rad/s^2, Fig. 16–31a. Determine the acceleration of the piston at the instant AB is in the position shown. At this instant $\omega_{AB} = 10 \text{ rad/s}$ and $\omega_{BC} = 2.43 \text{ rad/s}$ (See Example 16.13.)

SOLUTION (VECTOR ANALYSIS)

Kinematic Diagram.   The kinematic diagrams for both AB and BC are shown in Fig. 16–31b. Here $\mathbf{a}_C$ is vertical since C moves along a straight-line path.

Acceleration Equation.   Expressing each of the position vectors in Cartesian vector form

$$\mathbf{r}_B = \{-0.25 \sin 45°\mathbf{i} + 0.25 \cos 45°\mathbf{j}\} \text{ ft} = \{-0.177\mathbf{i} + 0.177\mathbf{j}\} \text{ ft}$$

$$\mathbf{r}_{C/B} = \{0.75 \sin 13.6°\mathbf{i} + 0.75 \cos 13.6°\mathbf{j}\} \text{ ft} = \{0.177\mathbf{i} + 0.729\mathbf{j}\} \text{ ft}$$

Crankshaft AB (rotation about a fixed axis):

$$\mathbf{a}_B = \boldsymbol{\alpha}_{AB} \times \mathbf{r}_B - \omega_{AB}^2 \mathbf{r}_B$$

$$= (-20\mathbf{k}) \times (-0.177\mathbf{i} + 0.177\mathbf{j}) - (10)^2(-0.177\mathbf{i} + 0.177\mathbf{j})$$

$$= \{21.21\mathbf{i} - 14.14\mathbf{j}\} \text{ ft/s}^2$$

Connecting Rod BC (general plane motion): Using the result for $\mathbf{a}_B$ and noting that $\mathbf{a}_C$ is in the vertical direction, we have

$$\mathbf{a}_C = \mathbf{a}_B + \boldsymbol{\alpha}_{BC} \times \mathbf{r}_{C/B} - \omega_{BC}^2 \mathbf{r}_{C/B}$$

$$a_C\mathbf{j} = 21.21\mathbf{i} - 14.14\mathbf{j} + (\alpha_{BC}\mathbf{k}) \times (0.177\mathbf{i} + 0.729\mathbf{j}) - (2.43)^2(0.177\mathbf{i} + 0.729\mathbf{j})$$

$$a_C\mathbf{j} = 21.21\mathbf{i} - 14.14\mathbf{j} + 0.177\alpha_{BC}\mathbf{j} - 0.729\alpha_{BC}\mathbf{i} - 1.04\mathbf{i} - 4.30\mathbf{j}$$

$$0 = 20.17 - 0.729\alpha_{BC}$$

$$a_C = 0.177\alpha_{BC} - 18.45$$

Solving yields

$$\alpha_{BC} = 27.7 \text{ rad/s}^2 \circlearrowright$$

$$a_C = -13.5 \text{ ft/s}^2 \qquad \qquad \qquad Ans.$$

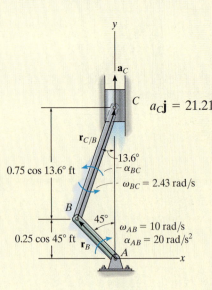

(b)

Fig. 16–31

NOTE: Since the piston is moving upward, the negative sign for a_C indicates that the piston is decelerating, i.e., $\mathbf{a}_C = \{-13.5\mathbf{j}\} \text{ ft/s}^2$. This causes the speed of the piston to decrease until AB becomes vertical, at which time the piston is momentarily at rest.

FUNDAMENTAL PROBLEMS

F16–19. At the instant shown, end A of the rod has the velocity and acceleration shown. Determine the angular acceleration of the rod and acceleration of end B of the rod.

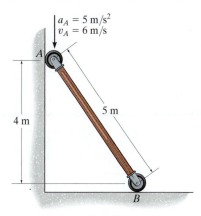

$a_A = 5 \text{ m/s}^2$
$v_A = 6 \text{ m/s}$

4 m

5 m

F16–19

F16–20. The gear rolls on the fixed rack with an angular velocity of $\omega = 12 \text{ rad/s}$ and angular acceleration of $\alpha = 6 \text{ rad/s}^2$. Determine the acceleration of point A.

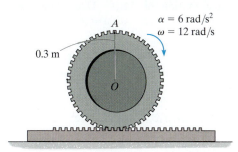

$\alpha = 6 \text{ rad/s}^2$
$\omega = 12 \text{ rad/s}$

0.3 m

F16–20

F16–21. The gear rolls on the fixed rack B. At the instant shown, the center O of the gear moves with a velocity of $v_O = 6 \text{ m/s}$ and acceleration of $a_O = 3 \text{ m/s}^2$. Determine the angular acceleration of the gear and acceleration of point A at this instant.

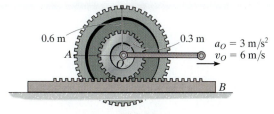

0.6 m   0.3 m $a_O = 3 \text{ m/s}^2$
$v_O = 6 \text{ m/s}$

F16–21

F16–22. At the instant shown, cable AB has a velocity of 3 m/s and acceleration of 1.5 m/s², while the gear rack has a velocity of 1.5 m/s and acceleration of 0.75 m/s². Determine the angular acceleration of the gear at this instant.

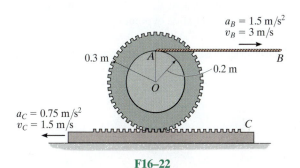

$a_B = 1.5 \text{ m/s}^2$
$v_B = 3 \text{ m/s}$

0.3 m   0.2 m

$a_C = 0.75 \text{ m/s}^2$
$v_C = 1.5 \text{ m/s}$

F16–22

F16–23. At the instant shown, the wheel rotates with an angular velocity of $\omega = 12 \text{ rad/s}$ and an angular acceleration of $\alpha = 6 \text{ rad/s}^2$. Determine the angular acceleration of link BC and the acceleration of piston C at this instant.

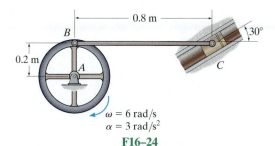

0.3 m   1.2 m

$\alpha = 6 \text{ rad/s}^2$
$\omega = 12 \text{ rad/s}$

F16–23

F16–24. At the instant shown, wheel A rotates with an angular velocity of $\omega = 6 \text{ rad/s}$ and an angular acceleration of $\alpha = 3 \text{ rad/s}^2$. Determine the angular acceleration of link BC and the acceleration of piston C.

0.8 m

0.2 m   30°

$\omega = 6 \text{ rad/s}$
$\alpha = 3 \text{ rad/s}^2$

F16–24

16

PROBLEMS

•16–109. The disk is moving to the left such that it has an angular acceleration $\alpha = 8$ rad/s^2 and angular velocity $\omega = 3$ rad/s at the instant shown. If it does not slip at A, determine the acceleration of point B.

16–110. The disk is moving to the left such that it has an angular acceleration $\alpha = 8$ rad/s^2 and angular velocity $\omega = 3$ rad/s at the instant shown. If it does not slip at A, determine the acceleration of point D.

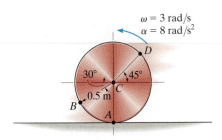

Probs. 16–109/110

16–111. The hoop is cast on the rough surface such that it has an angular velocity $\omega = 4$ rad/s and an angular acceleration $\alpha = 5$ rad/s^2. Also, its center has a velocity $v_O = 5$ m/s and a deceleration $a_O = 2$ m/s^2. Determine the acceleration of point A at this instant.

*16–112. The hoop is cast on the rough surface such that it has an angular velocity $\omega = 4$ rad/s and an angular acceleration $\alpha = 5$ rad/s^2. Also, its center has a velocity of $v_O = 5$ m/s and a deceleration $a_O = 2$ m/s^2. Determine the acceleration of point B at this instant.

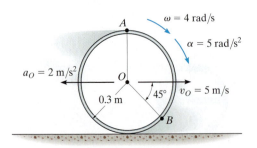

Probs. 16–111/112

•16–113. At the instant shown, the slider block B is traveling to the right with the velocity and acceleration shown. Determine the angular acceleration of the wheel at this instant.

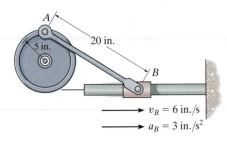

Prob. 16–113

16–114. The ends of bar AB are confined to move along the paths shown. At a given instant, A has a velocity of 8 ft/s and an acceleration of 3 ft/s^2. Determine the angular velocity and angular acceleration of AB at this instant.

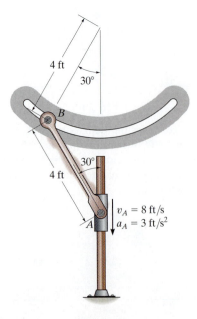

Prob. 16–114

16–115. Rod *AB* has the angular motion shown. Determine the acceleration of the collar *C* at this instant.

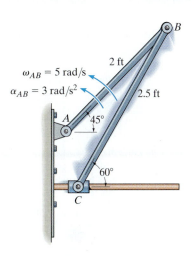

Prob. 16–115

•16–117. The hydraulic cylinder *D* extends with a velocity of $v_B = 4$ ft/s and an acceleration of $a_B = 1.5$ ft/s². Determine the acceleration of *A* at the instant shown.

16–118. The hydraulic cylinder *D* extends with a velocity of $v_B = 4$ ft/s and an acceleration of $a_B = 1.5$ ft/s². Determine the acceleration of *C* at the instant shown.

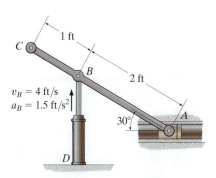

Probs. 16–117/118

*16–116. At the given instant member *AB* has the angular motions shown. Determine the velocity and acceleration of the slider block *C* at this instant.

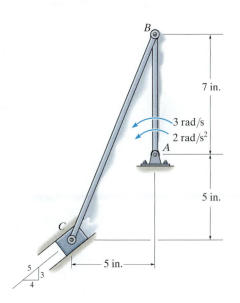

Prob. 16–116

16–119. The slider block moves with a velocity of $v_B = 5$ ft/s and an acceleration of $a_B = 3$ ft/s². Determine the angular acceleration of rod *AB* at the instant shown.

*16–120. The slider block moves with a velocity of $v_B = 5$ ft/s and an acceleration of $a_B = 3$ ft/s². Determine the acceleration of *A* at the instant shown.

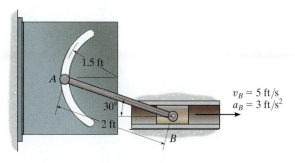

Probs. 16–119/120

•**16–121.** Crank *AB* rotates with an angular velocity of $\omega_{AB} = 6$ rad/s and an angular acceleration of $\alpha_{AB} = 2$ rad/s². Determine the acceleration of *C* and the angular acceleration of *BC* at the instant shown.

16–123. Pulley *A* rotates with the angular velocity and angular acceleration shown. Determine the angular acceleration of pulley *B* at the instant shown.

16–124. Pulley *A* rotates with the angular velocity and angular acceleration shown. Determine the acceleration of block *E* at the instant shown.

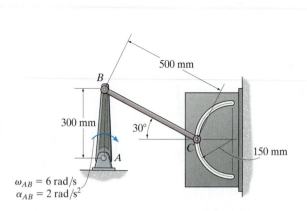

Prob. 16–121

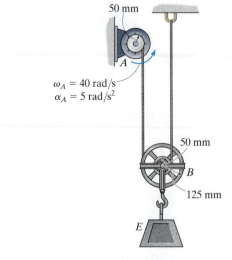

Probs. 16–123/124

16–122. The hydraulic cylinder extends with a velocity of $v_A = 1.5$ m/s and an acceleration of $a_A = 0.5$ m/s². Determine the angular acceleration of link *ABC* and the acceleration of end *C* at the instant shown. Point *B* is pin connected to the slider block.

•**16–125.** The hydraulic cylinder is extending with the velocity and acceleration shown. Determine the angular acceleration of crank *AB* and link *BC* at the instant shown.

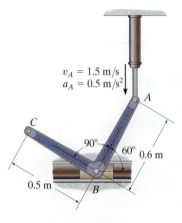

Prob. 16–122

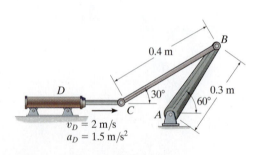

Prob. 16–125

16–126. A cord is wrapped around the inner spool of the gear. If it is pulled with a constant velocity **v**, determine the velocities and accelerations of points A and B. The gear rolls on the fixed gear rack.

*16–128. At a given instant, the gear has the angular motion shown. Determine the accelerations of points A and B on the link and the link's angular acceleration at this instant.

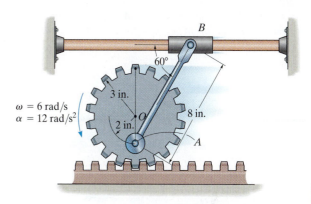

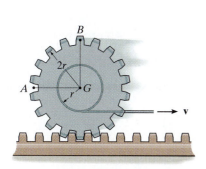

Prob. 16–126

Prob. 16–128

16–127. At a given instant, the gear racks have the velocities and accelerations shown. Determine the acceleration of points A and B.

•16–129. Determine the angular acceleration of link AB if link CD has the angular velocity and angular deceleration shown.

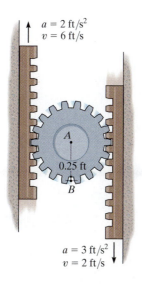

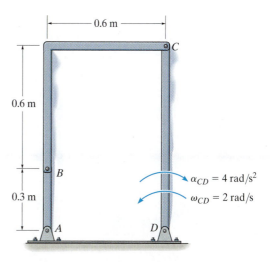

Prob. 16–127

Prob. 16–129

16–130. Gear A is held fixed, and arm DE rotates clockwise with an angular velocity of $\omega_{DE} = 6$ rad/s and an angular acceleration of $\alpha_{DE} = 3$ rad/s^2. Determine the angular acceleration of gear B at the instant shown.

16–131. Gear A rotates counterclockwise with a constant angular velocity of $\omega_A = 10$ rad/s, while arm DE rotates clockwise with an angular velocity of $\omega_{DE} = 6$ rad/s and an angular acceleration of $\alpha_{DE} = 3$ rad/s^2. Determine the angular acceleration of gear B at the instant shown.

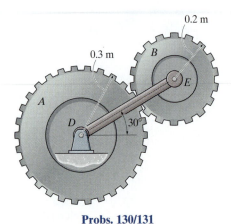

Probs. 130/131

*16–132. If end A of the rod moves with a constant velocity of $v_A = 6$ m/s, determine the angular velocity and angular acceleration of the rod and the acceleration of end B at the instant shown.

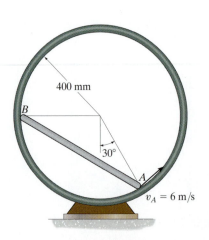

Prob. 16–132

•16–133. The retractable wing-tip float is used on an airplane able to land on water. Determine the angular accelerations α_{CD}, α_{BD}, and α_{AB} at the instant shown if the trunnion C travels along the horizontal rotating screw with an acceleration of $a_C = 0.5$ ft/s^2. In the position shown, $v_C = 0$. Also, points A and E are pin connected to the wing and points A and C are coincident at the instant shown.

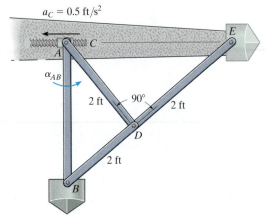

Prob. 16–133

16–134. Determine the angular velocity and the angular acceleration of the plate CD of the stone-crushing mechanism at the instant AB is horizontal. At this instant $\theta = 30°$ and $\phi = 90°$. Driving link AB is turning with a constant angular velocity of $\omega_{AB} = 4$ rad/s.

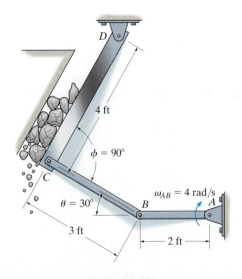

Prob. 16–134

16.8   Relative-Motion Analysis using Rotating Axes

In the previous sections the relative-motion analysis for velocity and acceleration was described using a translating coordinate system. This type of analysis is useful for determining the motion of points on the *same* rigid body, or the motion of points located on several pin-connected bodies. In some problems, however, rigid bodies (mechanisms) are constructed such that *sliding* will occur at their connections. The kinematic analysis for such cases is best performed if the motion is analyzed using a coordinate system which both *translates* and *rotates*. Furthermore, this frame of reference is useful for analyzing the motions of two points on a mechanism which are *not* located in the *same* body and for specifying the kinematics of particle motion when the particle moves along a rotating path.

In the following analysis two equations will be developed which relate the velocity and acceleration of two points, one of which is the origin of a moving frame of reference subjected to both a translation and a rotation in the plane.*

Position.   Consider the two points A and B shown in Fig. 16–32a. Their location is specified by the position vectors $\mathbf{r}_A$ and $\mathbf{r}_B$, which are measured with respect to the fixed X, Y, Z coordinate system. As shown in the figure, the "base point" A represents the origin of the x, y, z coordinate system, which is assumed to be both translating and rotating with respect to the X, Y, Z system. The position of B with respect to A is specified by the relative-position vector $\mathbf{r}_{B/A}$. The components of this vector may be expressed either in terms of unit vectors along the X, Y axes, i.e., $\mathbf{I}$ and $\mathbf{J}$, or by unit vectors along the x, y axes, i.e., $\mathbf{i}$ and $\mathbf{j}$. For the development which follows, $\mathbf{r}_{B/A}$ will be measured with respect to the moving x, y frame of reference. Thus, if B has coordinates (x_B, y_B), Fig. 16–32a, then

$$\mathbf{r}_{B/A} = x_B\mathbf{i} + y_B\mathbf{j}$$

Using vector addition, the three position vectors in Fig. 16–32a are related by the equation

$$\mathbf{r}_B = \mathbf{r}_A + \mathbf{r}_{B/A} \qquad (16\text{–}19)$$

At the instant considered, point A has a velocity $\mathbf{v}_A$ and an acceleration $\mathbf{a}_A$, while the angular velocity and angular acceleration of the x, y axes are Ω (omega) and $\dot{\Omega} = d\Omega/dt$, respectively.

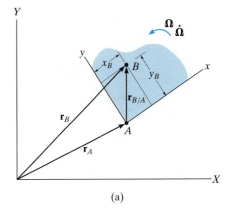

(a)

Fig. 16–32

16

*The more general, three-dimensional motion of the points is developed in Sec. 20.4.

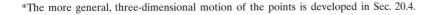

Velocity.   The velocity of point B is determined by taking the time derivative of Eq. 16–19, which yields

$$\mathbf{v}_B = \mathbf{v}_A + \frac{d\mathbf{r}_{B/A}}{dt} \tag{16–20}$$

The last term in this equation is evaluated as follows:

$$\frac{d\mathbf{r}_{B/A}}{dt} = \frac{d}{dt}(x_B\mathbf{i} + y_B\mathbf{j})$$

$$= \frac{dx_B}{dt}\mathbf{i} + x_B\frac{d\mathbf{i}}{dt} + \frac{dy_B}{dt}\mathbf{j} + y_B\frac{d\mathbf{j}}{dt}$$

$$= \left(\frac{dx_B}{dt}\mathbf{i} + \frac{dy_B}{dt}\mathbf{j}\right) + \left(x_B\frac{d\mathbf{i}}{dt} + y_B\frac{d\mathbf{j}}{dt}\right) \tag{16–21}$$

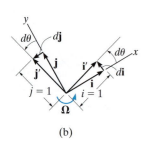

(b)

The two terms in the first set of parentheses represent the components of velocity of point B as measured by an observer attached to the moving x, y, z coordinate system. These terms will be denoted by vector $(\mathbf{v}_{B/A})_{xyz}$. In the second set of parentheses the instantaneous time rate of change of the unit vectors $\mathbf{i}$ and $\mathbf{j}$ is measured by an observer located in the fixed X, Y, Z coordinate system. These changes, $d\mathbf{i}$ and $d\mathbf{j}$, are due *only* to the *rotation $d\theta$* of the x, y, z axes, causing $\mathbf{i}$ to become $\mathbf{i}' = \mathbf{i} + d\mathbf{i}$ and $\mathbf{j}$ to become $\mathbf{j}' = \mathbf{j} + d\mathbf{j}$, Fig. 16–32b. As shown, the *magnitudes* of both $d\mathbf{i}$ and $d\mathbf{j}$ equal $1\,d\theta$, since $i = i' = j = j' = 1$. The *direction* of $d\mathbf{i}$ is defined by $+\mathbf{j}$, since $d\mathbf{i}$ is tangent to the path described by the arrowhead of $\mathbf{i}$ in the limit as $\Delta t \rightarrow dt$. Likewise, $d\mathbf{j}$ acts in the $-\mathbf{i}$ direction, Fig. 16–32b. Hence,

$$\frac{d\mathbf{i}}{dt} = \frac{d\theta}{dt}(\mathbf{j}) = \Omega\mathbf{j} \qquad \frac{d\mathbf{j}}{dt} = \frac{d\theta}{dt}(-\mathbf{i}) = -\Omega\mathbf{i}$$

Viewing the axes in three dimensions, Fig. 16–32c, and noting that $\Omega = \Omega\mathbf{k}$, we can express the above derivatives in terms of the cross product as

$$\frac{d\mathbf{i}}{dt} = \Omega \times \mathbf{i} \qquad \frac{d\mathbf{j}}{dt} = \Omega \times \mathbf{j} \tag{16–22}$$

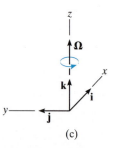

(c)

Fig. 16–32 (cont.)

Substituting these results into Eq. 16–21 and using the distributive property of the vector cross product, we obtain

$$\frac{d\mathbf{r}_{B/A}}{dt} = (\mathbf{v}_{B/A})_{xyz} + \Omega \times (x_B\mathbf{i} + y_B\mathbf{j}) = (\mathbf{v}_{B/A})_{xyz} + \Omega \times \mathbf{r}_{B/A} \tag{16–23}$$

Hence, Eq. 16–20 becomes

$$\mathbf{v}_B = \mathbf{v}_A + \mathbf{\Omega} \times \mathbf{r}_{B/A} + (\mathbf{v}_{B/A})_{xyz} \qquad (16\text{–}24)$$

where

$\mathbf{v}_B$ = velocity of B, measured from the X, Y, Z reference

$\mathbf{v}_A$ = velocity of the origin A of the x, y, z reference, measured from the X, Y, Z reference

$(\mathbf{v}_{B/A})_{xyz}$ = velocity of "B with respect to A," as measured by an observer attached to the rotating x, y, z reference

$\mathbf{\Omega}$ = angular velocity of the x, y, z reference, measured from the X, Y, Z reference

$\mathbf{r}_{B/A}$ = position of B with respect to A

Comparing Eq. 16–24 with Eq. 16–16 ($\mathbf{v}_B = \mathbf{v}_A + \mathbf{\Omega} \times \mathbf{r}_{B/A}$), which is valid for a translating frame of reference, it can be seen that the only difference between these two equations is represented by the term $(\mathbf{v}_{B/A})_{xyz}$.

When applying Eq. 16–24 it is often useful to understand what each of the terms represents. In order of appearance, they are as follows:

$\mathbf{v}_B$ $\begin{cases} \text{absolute velocity of } B \\ \\ \end{cases}$ $\begin{cases} \text{motion of } B \text{ observed} \\ \text{from the } X, Y, Z \text{ frame} \end{cases}$

(equals)

$\mathbf{v}_A$ $\begin{cases} \text{absolute velocity of the} \\ \text{origin of } x, y, z \text{ frame} \end{cases}$

(plus)

$\mathbf{\Omega} \times \mathbf{r}_{B/A}$ $\begin{cases} \text{angular velocity effect caused} \\ \text{by rotation of } x, y, z \text{ frame} \end{cases}$ $\left.\begin{array}{c} \\ \\ \end{array}\right\}$ $\begin{cases} \text{motion of } x, y, z \text{ frame} \\ \text{observed from the} \\ X, Y, Z \text{ frame} \end{cases}$

(plus)

$(\mathbf{v}_{B/A})_{xyz}$ $\begin{cases} \text{velocity of } B \\ \text{with respect to } A \end{cases}$ $\begin{cases} \text{motion of } B \text{ observed} \\ \text{from the } x, y, z \text{ frame} \end{cases}$

Acceleration. The acceleration of B, observed from the X, Y, Z coordinate system, may be expressed in terms of its motion measured with respect to the rotating system of coordinates by taking the time derivative of Eq. 16–24.

$$\frac{d\mathbf{v}_B}{dt} = \frac{d\mathbf{v}_A}{dt} + \frac{d\mathbf{\Omega}}{dt} \times \mathbf{r}_{B/A} + \mathbf{\Omega} \times \frac{d\mathbf{r}_{B/A}}{dt} + \frac{d(\mathbf{v}_{B/A})_{xyz}}{dt}$$

$$\mathbf{a}_B = \mathbf{a}_A + \dot{\mathbf{\Omega}} \times \mathbf{r}_{B/A} + \mathbf{\Omega} \times \frac{d\mathbf{r}_{B/A}}{dt} + \frac{d(\mathbf{v}_{B/A})_{xyz}}{dt} \qquad (16\text{–}25)$$

Here $\dot{\mathbf{\Omega}} = d\mathbf{\Omega}/dt$ is the angular acceleration of the x, y, z coordinate system. Since $\mathbf{\Omega}$ is always perpendicular to the plane of motion, then $\dot{\mathbf{\Omega}}$ measures *only the change in magnitude* of $\mathbf{\Omega}$. The derivative $d\mathbf{r}_{B/A}/dt$ is defined by Eq. 16–23, so that

$$\mathbf{\Omega} \times \frac{d\mathbf{r}_{B/A}}{dt} = \mathbf{\Omega} \times (\mathbf{v}_{B/A})_{xyz} + \mathbf{\Omega} \times (\mathbf{\Omega} \times \mathbf{r}_{B/A}) \qquad (16\text{–}26)$$

Finding the time derivative of $(\mathbf{v}_{B/A})_{xyz} = (v_{B/A})_x \mathbf{i} + (v_{B/A})_y \mathbf{j}$,

$$\frac{d(\mathbf{v}_{B/A})_{xyz}}{dt} = \left[\frac{d(v_{B/A})_x}{dt}\mathbf{i} + \frac{d(v_{B/A})_y}{dt}\mathbf{j}\right] + \left[(v_{B/A})_x \frac{d\mathbf{i}}{dt} + (v_{B/A})_y \frac{d\mathbf{j}}{dt}\right]$$

The two terms in the first set of brackets represent the components of acceleration of point B as measured by an observer attached to the rotating coordinate system. These terms will be denoted by $(\mathbf{a}_{B/A})_{xyz}$. The terms in the second set of brackets can be simplified using Eqs. 16–22.

$$\frac{d(\mathbf{v}_{B/A})_{xyz}}{dt} = (\mathbf{a}_{B/A})_{xyz} + \mathbf{\Omega} \times (\mathbf{v}_{B/A})_{xyz}$$

Substituting this and Eq. 16–26 into Eq. 16–25 and rearranging terms,

$$\mathbf{a}_B = \mathbf{a}_A + \dot{\mathbf{\Omega}} \times \mathbf{r}_{B/A} + \mathbf{\Omega} \times (\mathbf{\Omega} \times \mathbf{r}_{B/A}) + 2\mathbf{\Omega} \times (\mathbf{v}_{B/A})_{xyz} + (\mathbf{a}_{B/A})_{xyz}$$

$$(16\text{–}27)$$

where

$\mathbf{a}_B$ = acceleration of B, measured from the X, Y, Z reference

$\mathbf{a}_A$ = acceleration of the origin A of the x, y, z reference, measured from the X, Y, Z reference

$(\mathbf{a}_{B/A})_{xyz}, (\mathbf{v}_{B/A})_{xyz}$ = acceleration and velocity of B with respect to A, as measured by an observer attached to the *rotating x, y, z* reference

$\dot{\mathbf{\Omega}}, \mathbf{\Omega}$ = angular acceleration and angular velocity of the x, y, z reference, measured from the X, Y, Z reference

$\mathbf{r}_{B/A}$ = position of B with respect to A

If Eq. 16–27 is compared with Eq. 16–18, written in the form $\mathbf{a}_B = \mathbf{a}_A + \dot{\boldsymbol{\Omega}} \times \mathbf{r}_{B/A} + \boldsymbol{\Omega} \times (\boldsymbol{\Omega} \times \mathbf{r}_{B/A})$, which is valid for a translating frame of reference, it can be seen that the difference between these two equations is represented by the terms $2\boldsymbol{\Omega} \times (\mathbf{v}_{B/A})_{xyz}$ and $(\mathbf{a}_{B/A})_{xyz}$. In particular, $2\boldsymbol{\Omega} \times (\mathbf{v}_{B/A})_{xyz}$ is called the *Coriolis acceleration*, named after the French engineer G. C. Coriolis, who was the first to determine it. This term represents the difference in the acceleration of B as measured from nonrotating and rotating x, y, z axes. As indicated by the vector cross product, the Coriolis acceleration will *always* be perpendicular to both $\boldsymbol{\Omega}$ and $(\mathbf{v}_{B/A})_{xyz}$. It is an important component of the acceleration which must be considered whenever rotating reference frames are used. This often occurs, for example, when studying the accelerations and forces which act on rockets, long-range projectiles, or other bodies having motions whose measurements are significantly affected by the rotation of the earth.

The following interpretation of the terms in Eq. 16–27 may be useful when applying this equation to the solution of problems.

$\mathbf{a}_B$ $\left\{\begin{array}{l}\text{absolute acceleration of } B\end{array}\right.$ $\left.\begin{array}{l}\text{motion of } B \text{ observed} \\ \text{from the } X, Y, Z \text{ frame}\end{array}\right\}$

(equals)

$\mathbf{a}_A$ $\left\{\begin{array}{l}\text{absolute acceleration of the} \\ \text{origin of } x, y, z \text{ frame}\end{array}\right.$

(plus)

$\dot{\boldsymbol{\Omega}} \times \mathbf{r}_{B/A}$ $\left\{\begin{array}{l}\text{angular acceleration effect} \\ \text{caused by rotation of } x, y, z \\ \text{frame}\end{array}\right.$ $\left.\begin{array}{l}\text{motion of} \\ x, y, z \text{ frame} \\ \text{observed from} \\ \text{the } X, Y, Z \text{ frame}\end{array}\right.$

(plus)

$\boldsymbol{\Omega} \times (\boldsymbol{\Omega} \times \mathbf{r}_{B/A})$ $\left\{\begin{array}{l}\text{angular velocity effect caused} \\ \text{by rotation of } x, y, z \text{ frame}\end{array}\right.$

(plus)

$2\boldsymbol{\Omega} \times (\mathbf{v}_{B/A})_{xyz}$ $\left\{\begin{array}{l}\text{combined effect of } B \text{ moving} \\ \text{relative to } x, y, z \text{ coordinates} \\ \text{and rotation of } x, y, z \text{ frame}\end{array}\right\}$ interacting motion

(plus)

$(\mathbf{a}_{B/A})_{xyz}$ $\left\{\begin{array}{l}\text{acceleration of } B \text{ with} \\ \text{respect to } A\end{array}\right.$ $\left.\begin{array}{l}\text{motion of } B \text{ observed} \\ \text{from the } x, y, z \text{ frame}\end{array}\right\}$

16

16

Procedure for Analysis

Equations 16–24 and 16–27 can be applied to the solution of problems involving the planar motion of particles or rigid bodies using the following procedure.

Coordinate Axes.

- Choose an appropriate location for the origin and proper orientation of the axes for both fixed X, Y, Z and moving x, y, z reference frames.

- Most often solutions are easily obtained if at the instant considered:

    1. the origins are coincident
    2. the corresponding axes are collinear
    3. the corresponding axes are parallel

- The moving frame should be selected fixed to the body or device along which the relative motion occurs.

Kinematic Equations.

- After defining the origin A of the moving reference and specifying the moving point B, Eqs. 16–24 and 16–27 should be written in symbolic form

$$\mathbf{v}_B = \mathbf{v}_A + \boldsymbol{\Omega} \times \mathbf{r}_{B/A} + (\mathbf{v}_{B/A})_{xyz}$$

$$\mathbf{a}_B = \mathbf{a}_A + \dot{\boldsymbol{\Omega}} \times \mathbf{r}_{B/A} + \boldsymbol{\Omega} \times (\boldsymbol{\Omega} \times \mathbf{r}_{B/A}) + 2\boldsymbol{\Omega} \times (\mathbf{v}_{B/A})_{xyz} + (\mathbf{a}_{B/A})_{xyz}$$

- The Cartesian components of all these vectors may be expressed along either the X, Y, Z axes or the x, y, z axes. The choice is arbitrary provided a consistent set of unit vectors is used.

- Motion of the moving reference is expressed by $\mathbf{v}_A, \mathbf{a}_A, \boldsymbol{\Omega}$, and $\dot{\boldsymbol{\Omega}}$; and motion of B with respect to the moving reference is expressed by $\mathbf{r}_{B/A}, (\mathbf{v}_{B/A})_{xyz}$, and $(\mathbf{a}_{B/A})_{xyz}$.

The rotation of the dumping bin of the truck about point C is operated by the extension of the hydraulic cylinder AB. To determine the rotation of the bin due to this extension, we can use the equations of relative motion and fix the x, y axes to the cylinder so that the relative motion of the cylinder's extension occurs along the y axis.

EXAMPLE 16.19

At the instant $\theta = 60°$, the rod in Fig. 16–33 has an angular velocity of 3 rad/s and an angular acceleration of 2 rad/s². At this same instant, collar C travels outward along the rod such that when $x = 0.2$ m the velocity is 2 m/s and the acceleration is 3 m/s², both measured relative to the rod. Determine the Coriolis acceleration and the velocity and acceleration of the collar at this instant.

SOLUTION

Coordinate Axes. The origin of both coordinate systems is located at point O, Fig. 16–33. Since motion of the collar is reported relative to the rod, the moving x, y, z frame of reference is *attached* to the rod.

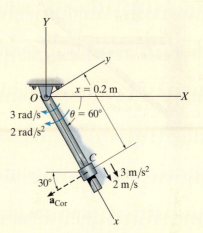

Fig. 16–33

Kinematic Equations.

$$\mathbf{v}_C = \mathbf{v}_O + \boldsymbol{\Omega} \times \mathbf{r}_{C/O} + (\mathbf{v}_{C/O})_{xyz} \qquad (1)$$

$$\mathbf{a}_C = \mathbf{a}_O + \dot{\boldsymbol{\Omega}} \times \mathbf{r}_{C/O} + \boldsymbol{\Omega} \times (\boldsymbol{\Omega} \times \mathbf{r}_{C/O}) + 2\boldsymbol{\Omega} \times (\mathbf{v}_{C/O})_{xyz} + (\mathbf{a}_{C/O})_{xyz} \qquad (2)$$

It will be simpler to express the data in terms of $\mathbf{i}, \mathbf{j}, \mathbf{k}$ component vectors rather than $\mathbf{I}, \mathbf{J}, \mathbf{K}$ components. Hence,

Motion of moving reference	Motion of C with respect to moving reference
$\mathbf{v}_O = 0$	$\mathbf{r}_{C/O} = \{0.2\mathbf{i}\}$ m
$\mathbf{a}_O = 0$	$(\mathbf{v}_{C/O})_{xyz} = \{2\mathbf{i}\}$ m/s
$\boldsymbol{\Omega} = \{-3\mathbf{k}\}$ rad/s	$(\mathbf{a}_{C/O})_{xyz} = \{3\mathbf{i}\}$ m/s²
$\dot{\boldsymbol{\Omega}} = \{-2\mathbf{k}\}$ rad/s²	

The Coriolis acceleration is defined as

$$\mathbf{a}_{Cor} = 2\boldsymbol{\Omega} \times (\mathbf{v}_{C/O})_{xyz} = 2(-3\mathbf{k}) \times (2\mathbf{i}) = \{-12\mathbf{j}\} \text{ m/s}^2 \quad Ans.$$

This vector is shown dashed in Fig. 16–33. If desired, it may be resolved into $\mathbf{I}, \mathbf{J}$ components acting along the X and Y axes, respectively.

The velocity and acceleration of the collar are determined by substituting the data into Eqs. 1 and 2 and evaluating the cross products, which yields

$$\mathbf{v}_C = \mathbf{v}_O + \boldsymbol{\Omega} \times \mathbf{r}_{C/O} + (\mathbf{v}_{C/O})_{xyz}$$

$$= 0 + (-3\mathbf{k}) \times (0.2\mathbf{i}) + 2\mathbf{i}$$

$$= \{2\mathbf{i} - 0.6\mathbf{j}\} \text{ m/s} \qquad\qquad Ans.$$

$$\mathbf{a}_C = \mathbf{a}_O + \dot{\boldsymbol{\Omega}} \times \mathbf{r}_{C/O} + \boldsymbol{\Omega} \times (\boldsymbol{\Omega} \times \mathbf{r}_{C/O}) + 2\boldsymbol{\Omega} \times (\mathbf{v}_{C/O})_{xyz} + (\mathbf{a}_{C/O})_{xyz}$$

$$= 0 + (-2\mathbf{k}) \times (0.2\mathbf{i}) + (-3\mathbf{k}) \times [(-3\mathbf{k}) \times (0.2\mathbf{i})] + 2(-3\mathbf{k}) \times (2\mathbf{i}) + 3\mathbf{i}$$

$$= 0 - 0.4\mathbf{j} - 1.80\mathbf{i} - 12\mathbf{j} + 3\mathbf{i}$$

$$= \{1.20\mathbf{i} - 12.4\mathbf{j}\} \text{ m/s}^2 \qquad\qquad Ans.$$

16

*16–156. A ride in an amusement park consists of a rotating arm AB having a constant angular velocity $\omega_{AB} = 2$ rad/s about point A and a car mounted at the end of the arm which has a constant angular velocity $\boldsymbol{\omega}' = \{-0.5\mathbf{k}\}$ rad/s, measured relative to the arm. At the instant shown, determine the velocity and acceleration of the passenger at C.

•16–157. A ride in an amusement park consists of a rotating arm AB that has an angular acceleration of $\alpha_{AB} = 1$ rad/s^2 when $\omega_{AB} = 2$ rad/s at the instant shown. Also at this instant the car mounted at the end of the arm has an angular acceleration of $\boldsymbol{\alpha}' = \{-0.6\mathbf{k}\}$ rad/s^2 and angular velocity of $\boldsymbol{\omega}' = \{-0.5\mathbf{k}\}$ rad/s, measured relative to the arm. Determine the velocity and acceleration of the passenger C at this instant.

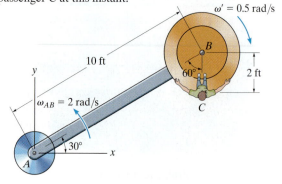

Probs. 16–156/157

16–158. The "quick-return" mechanism consists of a crank AB, slider block B, and slotted link CD. If the crank has the angular motion shown, determine the angular motion of the slotted link at this instant.

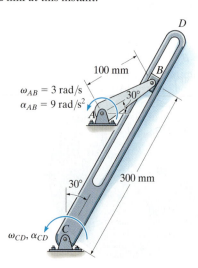

Prob. 16–158

16–159. The quick return mechanism consists of the crank CD and the slotted arm AB. If the crank rotates with the angular velocity and angular acceleration at the instant shown, determine the angular velocity and angular acceleration of AB at this instant.

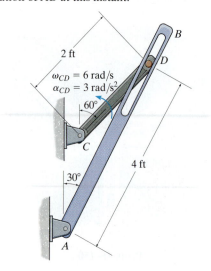

Prob. 16–159

*16–160. The Geneva mechanism is used in a packaging system to convert constant angular motion into intermittent angular motion. The star wheel A makes one sixth of a revolution for each full revolution of the driving wheel B and the attached guide C. To do this, pin P, which is attached to B, slides into one of the radial slots of A, thereby turning wheel A, and then exits the slot. If B has a constant angular velocity of $\omega_B = 4$ rad/s, determine $\boldsymbol{\omega}_A$ and $\boldsymbol{\alpha}_A$ of wheel A at the instant shown.

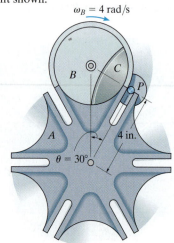

Prob. 16–160

CONCEPTUAL PROBLEMS

P16–1. An electric motor turns the tire at A at a constant angular velocity, and friction then causes the tire to roll without slipping on the inside rim of the Ferris Wheel. Using appropriate numerical values, determine the magnitude of the velocity and acceleration of passengers in one of the baskets. Do passengers in the other baskets experience this same motion? Explain.

P16–1

P16–2. The crank AB turns counterclockwise at a constant rate ω causing the connecting arm CD and rocking beam DE to move. Draw a sketch showing the location of the IC for the connecting arm when $\theta = 0°, 90°, 180°,$ and $270°$. Also, how was the curvature of the head at E determined, and why is it curved in this way?

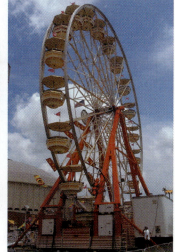

P16–2

P16–3. The bi-fold hangar door is opened by cables that move upward at a constant speed. Determine the position θ of panel BC when the angular velocity of BC is equal but opposite to the angular velocity of AB. Also, what is this angular velocity? Panel BC is pinned at C and has a height which is different from the height of BA. Use appropriate numerical values to explain your result.

P16–3

P16–4. If the tires do not slip on the pavement, determine the points on the tire that have a maximum and minimum speed and the points that have a maximum and minimum acceleration. Use appropriate numerical values for the car's speed and tire size to explain your result.

P16–4

16

CHAPTER REVIEW

Rigid-Body Planar Motion

A rigid body undergoes three types of planar motion: translation, rotation about a fixed axis, and general plane motion.

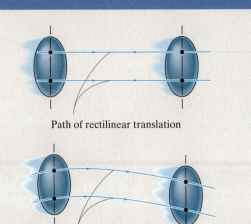

Path of rectilinear translation

Translation

When a body has rectilinear translation, all the particles of the body travel along parallel straight-line paths. If the paths have the same radius of curvature, then curvilinear translation occurs. Provided we know the motion of one of the particles, then the motion of all of the others is also known.

Path of curvilinear translation

Rotation about a Fixed Axis

For this type of motion, all of the particles move along circular paths. Here, all line segments in the body undergo the same angular displacement, angular velocity, and angular acceleration.

Once the angular motion of the body is known, then the velocity of any particle a distance r from the axis can be obtained.

The acceleration of any particle has two components. The tangential component accounts for the change in the magnitude of the velocity, and the normal component accounts for the change in the velocity's direction.

Rotation about a fixed axis

$$\omega = d\theta/dt \qquad\qquad \omega = \omega_0 + \alpha_c t$$

$$\alpha = d\omega/dt \quad \text{or} \quad \theta = \theta_0 + \omega_0 t + \tfrac{1}{2}\alpha_c t^2$$

$$\alpha\,d\theta = \omega\,d\omega \qquad\qquad \omega^2 = \omega_0^2 + 2\alpha_c(\theta - \theta_0)$$

$$\text{Constant } \alpha_c$$

$$v = \omega r \qquad\qquad a_t = \alpha r, \quad a_n = \omega^2 r$$

General Plane Motion

When a body undergoes general plane motion, it simultaneously translates and rotates. There are several methods for analyzing this motion.

Absolute Motion Analysis

If the motion of a point on a body or the angular motion of a line is known, then it may be possible to relate this motion to that of another point or line using an absolute motion analysis. To do so, linear position coordinates s or angular position coordinates θ are established (measured from a fixed point or line). These position coordinates are then related using the geometry of the body. The time derivative of this equation gives the relationship between the velocities and/or the angular velocities. A second time derivative relates the accelerations and/or the angular accelerations.

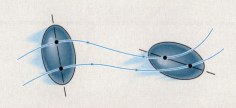

General plane motion

Relative-Motion using Translating Axes

General plane motion can also be analyzed using a relative-motion analysis between two points A and B located on the body. This method considers the motion in parts: first a translation of the selected base point A, then a relative "rotation" of the body about point A, which is measured from a translating axis. Since the relative motion is viewed as circular motion about the base point, point B will have a velocity $\mathbf{v}_{B/A}$ that is tangent to the circle. It also has two components of acceleration, $(\mathbf{a}_{B/A})_t$ and $(\mathbf{a}_{B/A})_n$. It is also important to realize that $\mathbf{a}_A$ and $\mathbf{a}_B$ will have tangential and normal components if these points move along curved paths.

$$\mathbf{v}_B = \mathbf{v}_A + \boldsymbol{\omega} \times \mathbf{r}_{B/A}$$

$$\mathbf{a}_B = \mathbf{a}_A + \boldsymbol{\alpha} \times \mathbf{r}_{B/A} - \omega^2 \mathbf{r}_{B/A}$$

Instantaneous Center of Zero Velocity

If the base point A is selected as having zero velocity, then the relative velocity equation becomes $\mathbf{v}_B = \boldsymbol{\omega} \times \mathbf{r}_{B/A}$. In this case, motion appears as if the body rotates about an instantaneous axis passing through A.

The instantaneous center of rotation (IC) can be established provided the directions of the velocities of any two points on the body are known, or the velocity of a point and the angular velocity are known. Since a radial line r will always be perpendicular to each velocity, then the IC is at the point of intersection of these two radial lines. Its measured location is determined from the geometry of the body. Once it is established, then the velocity of any point P on the body can be determined from $v = \omega r$, where r extends from the IC to point P.

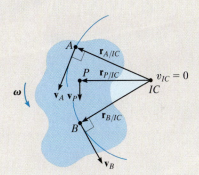

Relative Motion using Rotating Axes

Problems that involve connected members that slide relative to one another or points not located on the same body can be analyzed using a relative-motion analysis referenced from a rotating frame. This gives rise to the term $2\boldsymbol{\Omega} \times (\mathbf{v}_{B/A})_{xyz}$ that is called the Coriolis acceleration.

$$\mathbf{v}_B = \mathbf{v}_A + \boldsymbol{\Omega} \times \mathbf{r}_{B/A} + (\mathbf{v}_{B/A})_{xyz}$$

$$\mathbf{a}_B = \mathbf{a}_A + \dot{\boldsymbol{\Omega}} \times \mathbf{r}_{B/A} + \boldsymbol{\Omega} \times (\boldsymbol{\Omega} \times \mathbf{r}_{B/A}) + 2\boldsymbol{\Omega} \times (\mathbf{v}_{B/A})_{xyz} + (\mathbf{a}_{B/A})_{xyz}$$

The forces acting on this dragster as it begins to accelerate are quite severe and must be accounted for in the design of its structure.

Planar Kinetics of a Rigid Body: Force and Acceleration

17

CHAPTER OBJECTIVES

- To introduce the methods used to determine the mass moment of inertia of a body.
- To develop the planar kinetic equations of motion for a symmetric rigid body.
- To discuss applications of these equations to bodies undergoing translation, rotation about a fixed axis, and general plane motion.

17.1 Mass Moment of Inertia

Since a body has a definite size and shape, an applied nonconcurrent force system can cause the body to both translate and rotate. The translational aspects of the motion were studied in Chapter 13 and are governed by the equation $\mathbf{F} = m\mathbf{a}$. It will be shown in the next section that the rotational aspects, caused by a moment $\mathbf{M}$, are governed by an equation of the form $\mathbf{M} = I\boldsymbol{\alpha}$. The symbol I in this equation is termed the mass moment of inertia. By comparison, the *moment of inertia* is a measure of the resistance of a body to *angular acceleration* ($\mathbf{M} = I\boldsymbol{\alpha}$) in the same way that *mass* is a measure of the body's resistance to *acceleration* ($\mathbf{F} = m\mathbf{a}$).

The flywheel on the engine of this tractor has a large moment of inertia about its axis of rotation. Once it is set into motion, it will be difficult to stop, and this in turn will prevent the engine from stalling and instead will allow it to maintain a constant power.

Fig. 17–1

We define the *moment of inertia* as the integral of the "second moment" about an axis of all the elements of mass dm which compose the body.* For example, the body's moment of inertia about the z axis in Fig. 17–1 is

$$I = \int_m r^2 \, dm \qquad (17\text{--}1)$$

Here the "moment arm" r is the perpendicular distance from the z axis to the arbitrary element dm. Since the formulation involves r, the value of I is different for each axis about which it is computed. In the study of planar kinetics, the axis chosen for analysis generally passes through the body's mass center G and is always perpendicular to the plane of motion. The moment of inertia about this axis will be denoted as I_G. Since r is squared in Eq. 17–1, the mass moment of inertia is always a *positive* quantity. Common units used for its measurement are $\text{kg} \cdot \text{m}^2$ or $\text{slug} \cdot \text{ft}^2$.

If the body consists of material having a variable density, $\rho = \rho\,(x, y, z)$, the elemental mass dm of the body can be expressed in terms of its density and volume as $dm = \rho \, dV$. Substituting dm into Eq. 17–1, the body's moment of inertia is then computed using *volume elements* for integration; i.e.,

$$I = \int_V r^2 \rho \, dV \qquad (17\text{--}2)$$

*Another property of the body, which measures the symmetry of the body's mass with respect to a coordinate system, is the product of inertia. This property applies to the three-dimensional motion of a body and will be discussed in Chapter 21.

In the special case of ρ being a *constant*, this term may be factored out of the integral, and the integration is then purely a function of geometry,

$$I = \rho \int_V r^2 \, dV \qquad (17\text{–}3)$$

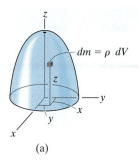

(a)

When the volume element chosen for integration has infinitesimal dimensions in all three directions, Fig. 17–2a, the moment of inertia of the body must be determined using "triple integration." The integration process can, however, be simplified to a *single integration* provided the chosen volume element has a differential size or thickness in only *one* direction. Shell or disk elements are often used for this purpose.

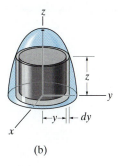

(b)

Procedure for Analysis

To obtain the moment of inertia by integration, we will consider only symmetric bodies having volumes which are generated by revolving a curve about an axis. An example of such a body is shown in Fig. 17–2a. Two types of differential elements can be chosen.

Shell Element.

- If a *shell element* having a height z, radius $r = y$, and thickness dy is chosen for integration, Fig. 17–2b, then the volume is $dV = (2\pi y)(z)dy$.

- This element may be used in Eq. 17–2 or 17–3 for determining the moment of inertia I_z of the body about the z axis, since the *entire element*, due to its "thinness," lies at the *same* perpendicular distance $r = y$ from the z axis (see Example 17.1).

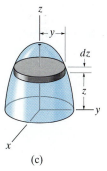

(c)

Fig. 17–2

Disk Element.

- If a disk element having a radius y and a thickness dz is chosen for integration, Fig. 17–2c, then the volume is $dV = (\pi y^2)dz$.

- This element is *finite* in the radial direction, and consequently its parts *do not* all lie at the *same radial distance r* from the z axis. As a result, Eq. 17–2 or 17–3 *cannot* be used to determine I_z directly. Instead, to perform the integration it is first necessary to determine the moment of inertia *of the element* about the z axis and then integrate this result (see Example 17.2).

EXAMPLE | 17.1

Determine the moment of inertia of the cylinder shown in Fig. 17–3a about the z axis. The density of the material, ρ, is constant.

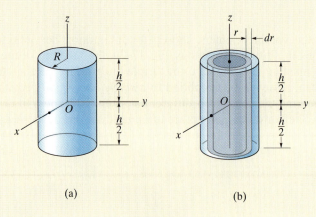

(a)   (b)

Fig. 17–3

SOLUTION

Shell Element.   This problem can be solved using the *shell element* in Fig. 17–3b and a single integration. The volume of the element is $dV = (2\pi r)(h)\, dr$, so that its mass is $dm = \rho dV = \rho(2\pi h r\, dr)$. Since the *entire element* lies at the same distance r from the z axis, the moment of inertia *of the element* is

$$dI_z = r^2 dm = \rho 2\pi h r^3\, dr$$

Integrating over the entire region of the cylinder yields

$$I_z = \int_m r^2\, dm = \rho 2\pi h \int_0^R r^3\, dr = \frac{\rho\pi}{2} R^4 h$$

The mass of the cylinder is

$$m = \int_m dm = \rho 2\pi h \int_0^R r\, dr = \rho\pi h R^2$$

so that

$$I_z = \frac{1}{2} mR^2 \qquad\qquad\qquad \textit{Ans.}$$

EXAMPLE | 17.2

If the density of the material is 5 slug/ft³, determine the moment of inertia of the solid in Fig 17–4a about the y axis.

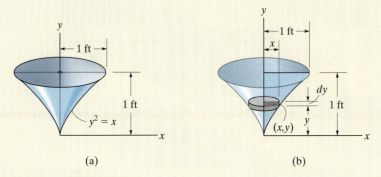

(a)   (b)

Fig. 17–4

SOLUTION

Disk Element.   The moment of inertia will be found using a *disk element*, as shown in Fig. 17–4b. Here the element intersects the curve at the arbitrary point (x, y) and has a mass

$$dm = \rho \, dV = \rho(\pi x^2) \, dy$$

Although all portions of the element are *not* located at the same distance from the y axis, it is still possible to determine the moment of inertia dI_y *of the element* about the y axis. In the preceding example it was shown that the moment of inertia of a cylinder about its longitudinal axis is $I = \frac{1}{2} mR^2$, where m and R are the mass and radius of the cylinder. Since the height is not involved in this formula, the disk itself can be thought of as a cylinder. Thus, for the disk element in Fig. 17–4b, we have

$$dI_y = \tfrac{1}{2}(dm)x^2 = \tfrac{1}{2}[\rho(\pi x^2) \, dy]x^2$$

Substituting $x = y^2$, $\rho = 5$ slug/ft³, and integrating with respect to y, from $y = 0$ to $y = 1$ ft, yields the moment of inertia for the entire solid.

$$I_y = \frac{\pi(5 \text{ slug/ft}^3)}{2} \int_0^{1 \text{ ft}} x^4 \, dy = \frac{\pi(5)}{2} \int_0^{1 \text{ ft}} y^8 \, dy = 0.873 \text{ slug} \cdot \text{ft}^2 \; \textit{Ans.}$$

17

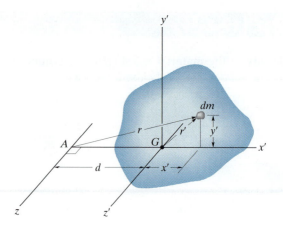

Fig. 17–5

Parallel-Axis Theorem.

If the moment of inertia of the body about an axis passing through the body's mass center is known, then the moment of inertia about any other *parallel axis* can be determined by using the *parallel-axis theorem*. This theorem can be derived by considering the body shown in Fig. 17–5. Here the z' axis passes through the mass center G, whereas the corresponding *parallel z axis* lies at a constant distance d away. Selecting the differential element of mass dm, which is located at point (x', y'), and using the Pythagorean theorem, $r^2 = (d + x')^2 + y'^2$, we can express the moment of inertia of the body about the z axis as

$$I = \int_m r^2\, dm = \int_m [(d + x')^2 + y'^2]\, dm$$

$$= \int_m (x'^2 + y'^2)\, dm + 2d \int_m x'\, dm + d^2 \int_m dm$$

Since $r'^2 = x'^2 + y'^2$, the first integral represents I_G. The second integral equals *zero*, since the z' axis passes through the body's mass center, i.e., $\int x'\,dm = \overline{x}'m = 0$ since $\overline{x}' = 0$. Finally, the third integral represents the total mass m of the body. Hence, the moment of inertia about the z axis can be written as

$$I = I_G + md^2 \qquad (17\text{--}4)$$

where

I_G = moment of inertia about the z' axis passing through the mass center G

m = mass of the body

d = perpendicular distance between the parallel z and z' axes

Radius of Gyration.   Occasionally, the moment of inertia of a body about a specified axis is reported in handbooks using the *radius of gyration, k*. This is a geometrical property which has units of length. When it and the body's mass m are known, the body's moment of inertia is determined from the equation

$$I = mk^2 \quad \text{or} \quad k = \sqrt{\frac{I}{m}} \qquad (17\text{--}5)$$

Note the *similarity* between the definition of k in this formula and r in the equation $dI = r^2\,dm$, which defines the moment of inertia of an elemental mass dm of the body about an axis.

Composite Bodies.   If a body consists of a number of simple shapes such as disks, spheres, and rods, the moment of inertia of the body about any axis can be determined by adding algebraically the moments of inertia of all the composite shapes computed about the axis. Algebraic addition is necessary since a composite part must be considered as a negative quantity if it has already been counted as a piece of another part—for example, a "hole" subtracted from a solid plate. The parallel-axis theorem is needed for the calculations if the center of mass of each composite part does not lie on the axis. For the calculation, then, $I = \Sigma(I_G + md^2)$. Here I_G for each of the composite parts is determined by integration, or for simple shapes, such as rods and disks, it can be found from a table, such as the one given on the inside back cover of this book.

17

EXAMPLE | 17.3

If the plate shown in Fig. 17–6a has a density of 8000 kg/m³ and a thickness of 10 mm, determine its moment of inertia about an axis directed perpendicular to the page and passing through point O.

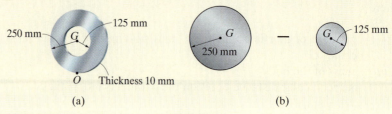

(a)   (b)

Fig. 17–6

SOLUTION

The plate consists of two composite parts, the 250-mm-radius disk *minus* a 125-mm-radius disk, Fig. 17–6b. The moment of inertia about O can be determined by computing the moment of inertia of each of these parts about O and then adding the results *algebraically*. The calculations are performed by using the parallel-axis theorem in conjunction with the data listed in the table on the inside back cover.

Disk. The moment of inertia of a disk about the centroidal axis perpendicular to the plane of the disk is $I_G = \frac{1}{2}mr^2$. The mass center of the disk is located at a distance of 0.25 m from point O. Thus,

$$m_d = \rho_d V_d = 8000 \text{ kg/m}^3 \left[\pi (0.25 \text{ m})^2 (0.01 \text{ m}) \right] = 15.71 \text{ kg}$$

$$(I_d)_O = \frac{1}{2}m_d r_d^2 + m_d d^2$$

$$= \frac{1}{2}(15.71 \text{ kg})(0.25 \text{ m})^2 + (15.71 \text{ kg})(0.25 \text{ m})^2$$

$$= 1.473 \text{ kg} \cdot \text{m}^2$$

Hole. For the 125-mm-radius disk (hole), we have

$$m_h = \rho_h V_h = 8000 \text{ kg/m}^3 \left[\pi (0.125 \text{ m})^2 (0.01 \text{ m}) \right] = 3.927 \text{ kg}$$

$$(I_h)_O = \frac{1}{2}m_h r_h^2 + m_h d^2$$

$$= \frac{1}{2}(3.927 \text{ kg})(0.125 \text{ m})^2 + (3.927 \text{ kg})(0.25 \text{ m})^2$$

$$= 0.276 \text{ kg} \cdot \text{m}^2$$

The moment of inertia of the plate about point O is therefore

$$I_O = (I_d)_O - (I_h)_O$$

$$= 1.473 \text{ kg} \cdot \text{m}^2 - 0.276 \text{ kg} \cdot \text{m}^2$$

$$= 1.20 \text{ kg} \cdot \text{m}^2 \qquad\qquad Ans.$$

EXAMPLE 17.4

The pendulum in Fig. 17–7 is suspended from the pin at O and consists of two thin rods, each having a weight of 10 lb. Determine the moment of inertia of the pendulum about an axis passing through (a) point O, and (b) the mass center G of the pendulum.

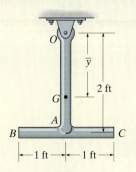

Fig. 17–7

SOLUTION

Part (a). Using the table on the inside back cover, the moment of inertia of rod OA about an axis perpendicular to the page and passing through point O of the rod is $I_O = \frac{1}{3}ml^2$. Hence,

$$(I_{OA})_O = \frac{1}{3}ml^2 = \frac{1}{3}\left(\frac{10 \text{ lb}}{32.2 \text{ ft/s}^2}\right)(2 \text{ ft})^2 = 0.414 \text{ slug} \cdot \text{ft}^2$$

This same value can be obtained using $I_G = \frac{1}{12}ml^2$ and the parallel-axis theorem.

$$(I_{OA})_O = \frac{1}{12}ml^2 + md^2 = \frac{1}{12}\left(\frac{10 \text{ lb}}{32.2 \text{ ft/s}^2}\right)(2 \text{ ft})^2 + \left(\frac{10 \text{ lb}}{32.2 \text{ ft/s}^2}\right)(1 \text{ ft})^2$$

$$= 0.414 \text{ slug} \cdot \text{ft}^2$$

For rod BC we have

$$(I_{BC})_O = \frac{1}{12}ml^2 + md^2 = \frac{1}{12}\left(\frac{10 \text{ lb}}{32.2 \text{ ft/s}^2}\right)(2 \text{ ft})^2 + \left(\frac{10 \text{ lb}}{32.2 \text{ ft/s}^2}\right)(2 \text{ ft})^2$$

$$= 1.346 \text{ slug} \cdot \text{ft}^2$$

The moment of inertia of the pendulum about O is therefore

$$I_O = 0.414 + 1.346 = 1.76 \text{ slug} \cdot \text{ft}^2 \qquad \textit{Ans.}$$

Part (b). The mass center G will be located relative to point O. Assuming this distance to be $\bar{y}$, Fig. 17–7, and using the formula for determining the mass center, we have

$$\bar{y} = \frac{\Sigma \tilde{y}m}{\Sigma m} = \frac{1(10/32.2) + 2(10/32.2)}{(10/32.2) + (10/32.2)} = 1.50 \text{ ft}$$

The moment of inertia I_G may be found in the same manner as I_O, which requires successive applications of the parallel-axis theorem to transfer the moments of inertia of rods OA and BC to G. A more direct solution, however, involves using the result for I_O, i.e.,

$$I_O = I_G + md^2; \quad 1.76 \text{ slug} \cdot \text{ft}^2 = I_G + \left(\frac{20 \text{ lb}}{32.2 \text{ ft/s}^2}\right)(1.50 \text{ ft})^2$$

$$I_G = 0.362 \text{ slug} \cdot \text{ft}^2 \qquad \textit{Ans.}$$

17

PROBLEMS

•17–1. Determine the moment of inertia I_y for the slender rod. The rod's density ρ and cross-sectional area A are constant. Express the result in terms of the rod's total mass m.

17–3. The paraboloid is formed by revolving the shaded area around the x axis. Determine the radius of gyration k_x. The density of the material is $\rho = 5 \text{ Mg/m}^3$.

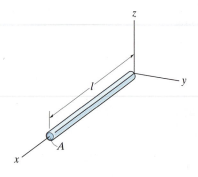

Prob. 17–1

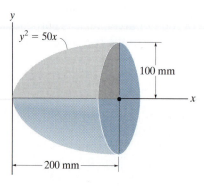

Prob. 17–3

17–2. The right circular cone is formed by revolving the shaded area around the x axis. Determine the moment of inertia I_x and express the result in terms of the total mass m of the cone. The cone has a constant density ρ.

*17–4. The frustum is formed by rotating the shaded area around the x axis. Determine the moment of inertia I_x and express the result in terms of the total mass m of the frustum. The frustum has a constant density ρ.

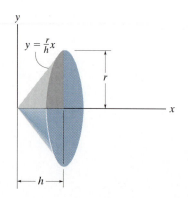

Prob. 17–2

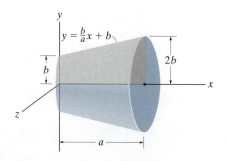

Prob. 17–4

•**17–5.** The paraboloid is formed by revolving the shaded area around the x axis. Determine the moment of inertia about the x axis and express the result in terms of the total mass m of the paraboloid. The material has a constant density ρ.

17–7. Determine the moment of inertia of the homogeneous pyramid of mass m about the z axis. The density of the material is ρ. *Suggestion:* Use a rectangular plate element having a volume of $dV = (2x)(2y)dz$.

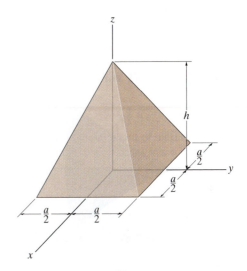

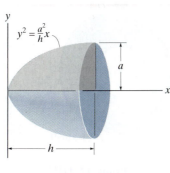

Prob. 17–5

Prob 17–7

17–6. The hemisphere is formed by rotating the shaded area around the y axis. Determine the moment of inertia I_y and express the result in terms of the total mass m of the hemisphere. The material has a constant density ρ.

*****17–8.** Determine the mass moment of inertia I_z of the cone formed by revolving the shaded area around the z axis. The density of the material is ρ. Express the result in terms of the mass m of the cone.

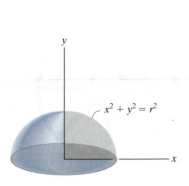

Prob. 17–6

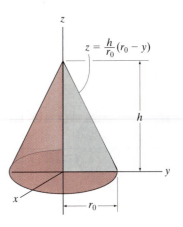

Prob. 17–8

17

*17–36. The forklift travels forward with a constant speed of 9 ft/s. Determine the shortest stopping distance without causing any of the wheels to leave the ground. The forklift has a weight of 2000 lb with center of gravity at G_1, and the load weighs 900 lb with center of gravity at G_2. Neglect the weight of the wheels.

•17–37. If the forklift's rear wheels supply a combined traction force of $F_A = 300$ lb, determine its acceleration and the normal reactions on the pairs of rear wheels and front wheels. The forklift has a weight of 2000 lb, with center of gravity at G_1, and the load weighs 900 lb, with center of gravity at G_2. The front wheels are free to roll. Neglect the weight of the wheels.

17–39. The forklift and operator have a combined weight of 10 000 lb and center of mass at G. If the forklift is used to lift the 2000-lb concrete pipe, determine the maximum vertical acceleration it can give to the pipe so that it does not tip forward on its front wheels.

*17–40. The forklift and operator have a combined weight of 10 000 lb and center of mass at G. If the forklift is used to lift the 2000-lb concrete pipe, determine the normal reactions on each of its four wheels if the pipe is given an upward acceleration of 4 ft/s^2.

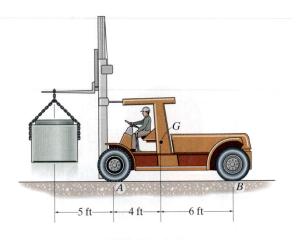

Probs. 17–39/40

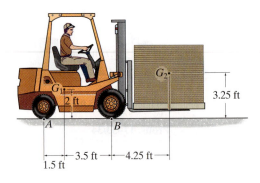

Probs. 17–36/37

17–38. Each uniform box on the stack of four boxes has a weight of 8 lb. The stack is being transported on the dolly, which has a weight of 30 lb. Determine the maximum force **F** which the woman can exert on the handle in the direction shown so that no box on the stack will tip or slip. The coefficient of the static friction at all points of contact is $\mu_s = 0.5$. The dolly wheels are free to roll. Neglect their mass.

•17–41. The car, having a mass of 1.40 Mg and mass center at G_c, pulls a loaded trailer having a mass of 0.8 Mg and mass center at G_t. Determine the normal reactions on both the car's front and rear wheels and the trailer's wheels if the driver applies the car's rear brakes C and causes the car to skid. Take $\mu_C = 0.4$ and assume the hitch at A is a pin or ball-and-socket joint. The wheels at B and D are free to roll. Neglect their mass and the mass of the driver.

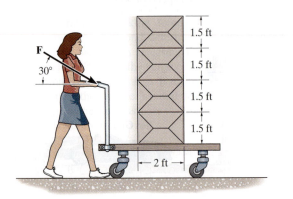

Prob. 17–38

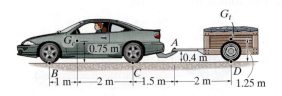

Prob. 17–41

17–42. The uniform crate has a mass of 50 kg and rests on the cart having an inclined surface. Determine the smallest acceleration that will cause the crate either to tip or slip relative to the cart. What is the magnitude of this acceleration? The coefficient of static friction between the crate and the cart is $\mu_s = 0.5$.

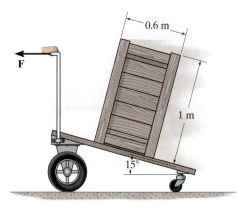

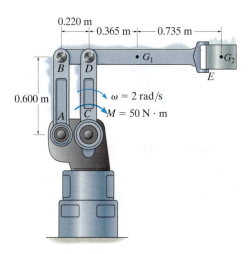

Prob. 17–42

17–43. Arm *BDE* of the industrial robot is activated by applying the torque of $M = 50$ N·m to link *CD*. Determine the reactions at pins *B* and *D* when the links are in the position shown and have an angular velocity of 2 rad/s. Arm *BDE* has a mass of 10 kg with center of mass at G_1. The container held in its grip at *E* has a mass of 12 kg with center of mass at G_2. Neglect the mass of links *AB* and *CD*.

*17–44.** The handcart has a mass of 200 kg and center of mass at *G*. Determine the normal reactions at each of the two wheels at *A* and at *B* if a force of $P = 50$ N is applied to the handle. Neglect the mass of the wheels.

•17–45.** The handcart has a mass of 200 kg and center of mass at *G*. Determine the largest magnitude of force **P** that can be applied to the handle so that the wheels at *A* or *B* continue to maintain contact with the ground. Neglect the mass of the wheels.

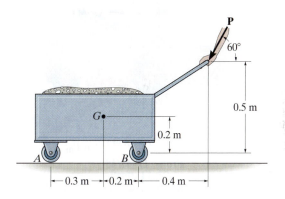

Probs. 17–44/45

17–46. The jet aircraft is propelled by four engines to increase its speed uniformly from rest to 100 m/s in a distance of 500 m. Determine the thrust **T** developed by each engine and the normal reaction on the nose wheel *A*. The aircraft's total mass is 150 Mg and the mass center is at point *G*. Neglect air and rolling resistance and the effect of lift.

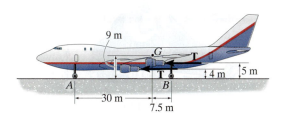

Prob. 17–46

Prob. 17–43

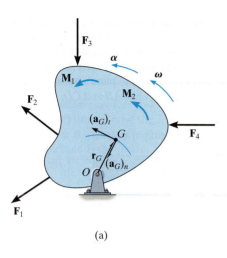

(a)

The free-body and kinetic diagrams for the body are shown in Fig. 17–14b. The two components $m(\mathbf{a}_G)_t$ and $m(\mathbf{a}_G)_n$, shown on the kinetic diagram, are associated with the tangential and normal components of acceleration of the body's mass center. The $I_G\boldsymbol{\alpha}$ vector acts in the same *direction* as $\boldsymbol{\alpha}$ and has a *magnitude* of $I_G\alpha$, where I_G is the body's moment of inertia calculated about an axis which is perpendicular to the page and passes through G. From the derivation given in Sec. 17.2, the equations of motion which apply to the body can be written in the form

$$\Sigma F_n = m(a_G)_n = m\omega^2 r_G$$
$$\Sigma F_t = m(a_G)_t = m\alpha r_G \qquad (17\text{–}14)$$
$$\Sigma M_G = I_G\alpha$$

The moment equation can be replaced by a moment summation about any arbitrary point P on or off the body provided one accounts for the moments $\Sigma(\mathcal{M}_k)_P$ produced by $I_G\boldsymbol{\alpha}$, $m(\mathbf{a}_G)_t$, and $m(\mathbf{a}_G)_n$ about the point. Often it is convenient to sum moments about the pin at O in order to eliminate the *unknown* force $\mathbf{F}_O$. From the kinetic diagram, Fig. 17–14b, this requires

$$\zeta + \Sigma M_O = \Sigma(\mathcal{M}_k)_O; \qquad \Sigma M_O = r_G m(a_G)_t + I_G\alpha \qquad (17\text{–}15)$$

Note that the moment of $m(\mathbf{a}_G)_n$ is not included here since the line of action of this vector passes through O. Substituting $(a_G)_t = r_G\alpha$, we may rewrite the above equation as $\zeta + \Sigma M_O = (I_G + mr_G^2)\alpha$. From the parallel-axis theorem, $I_O = I_G + md^2$, and therefore the term in parentheses represents the *moment of inertia of the body about the fixed axis of rotation passing through O.** Consequently, we can write the three equations of motion for the body as

$$\Sigma F_n = m(a_G)_n = m\omega^2 r_G$$
$$\Sigma F_t = m(a_G)_t = m\alpha r_G \qquad (17\text{–}16)$$
$$\Sigma M_O = I_O\alpha$$

When using these equations, remember that "$I_O\alpha$" accounts for the "moment" of *both* $m(\mathbf{a}_G)_t$ and $I_G\boldsymbol{\alpha}$ about point O, Fig. 17–14b. In other words, $\Sigma M_O = \Sigma(\mathcal{M}_k)_O = I_O\alpha$, as indicated by Eqs. 17–15 and 17–16.

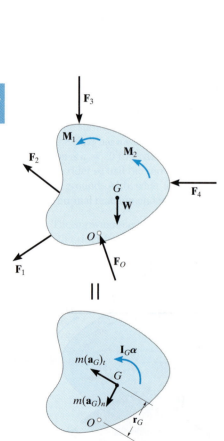

(b)

Fig. 17–14 (cont.)

---

* The result $\Sigma M_O = I_O\alpha$ can also be obtained *directly* from Eq. 17–6 by selecting point P to coincide with O, realizing that $(a_P)_x = (a_P)_y = 0$.

Procedure for Analysis

Kinetic problems which involve the rotation of a body about a fixed axis can be solved using the following procedure.

Free-Body Diagram.

*   Establish the inertial n, t coordinate system and specify the direction and sense of the accelerations $(\mathbf{a}_G)_n$ and $(\mathbf{a}_G)_t$ and the angular acceleration $\boldsymbol{\alpha}$ of the body. Recall that $(\mathbf{a}_G)_t$ must act in a direction which is in accordance with the rotational sense of $\boldsymbol{\alpha}$, whereas $(\mathbf{a}_G)_n$ always acts toward the axis of rotation, point O.

*   Draw the free-body diagram to account for all the external forces and couple moments that act on the body.

*   Determine the moment of inertia I_G or I_O.

*   Identify the unknowns in the problem.

*   If it is decided that the rotational equation of motion $\Sigma M_P = \Sigma(\mathcal{M}_k)_P$ is to be used, i.e., P is a point other than G or O, then consider drawing the kinetic diagram in order to help "visualize" the "moments" developed by the components $m(\mathbf{a}_G)_n$, $m(\mathbf{a}_G)_t$, and $I_G\boldsymbol{\alpha}$ when writing the terms for the moment sum $\Sigma(\mathcal{M}_k)_P$.

Equations of Motion.

*   Apply the three equations of motion in accordance with the established sign convention.

*   If moments are summed about the body's mass center, G, then $\Sigma M_G = I_G\alpha$, since $(m\mathbf{a}_G)_t$ and $(m\mathbf{a}_G)_n$ create no moment about G.

*   If moments are summed about the pin support O on the axis of rotation, then $(m\mathbf{a}_G)_n$ creates no moment about O, and it can be shown that $\Sigma M_O = I_O\alpha$.

Kinematics.

*   Use kinematics if a complete solution cannot be obtained strictly from the equations of motion.

*   If the angular acceleration is variable, use

$$\alpha = \frac{d\omega}{dt} \qquad \alpha\, d\theta = \omega\, d\omega \qquad \omega = \frac{d\theta}{dt}$$

*   If the angular acceleration is constant, use

$$\omega = \omega_0 + \alpha_c t$$
$$\theta = \theta_0 + \omega_0 t + \tfrac{1}{2}\alpha_c t^2$$
$$\omega^2 = \omega_0^2 + 2\alpha_c(\theta - \theta_0)$$

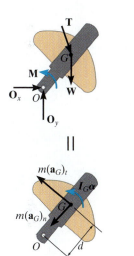

The crank on the oil-pumping rig undergoes rotation about a fixed axis which is caused by a driving torque **M** of the motor. The loadings shown on the free-body diagram cause the effects shown on the kinetic diagram. If moments are summed about the mass center, G, then $\Sigma M_G = I_G\alpha$. However, if moments are summed about point O, noting that $(a_G)_t = \alpha d$, then $\zeta + \Sigma M_O = I_G\alpha + m(a_G)_t d + m(a_G)_n(0) = (I_G + md^2)\alpha = I_O\alpha$.

17

EXAMPLE 17.9

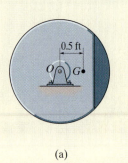

(a)

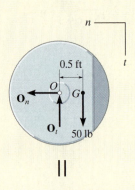

||

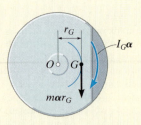

(b)

Fig. 17–15

The unbalanced 50-lb flywheel shown in Fig. 17–15a has a radius of gyration of $k_G = 0.6$ ft about an axis passing through its mass center G. If it is released from rest, determine the horizontal and vertical components of reaction at the pin O.

SOLUTION

Free-Body and Kinetic Diagrams. Since G moves in a circular path, it will have both normal and tangential components of acceleration. Also, since α, which is caused by the flywheel's weight, acts clockwise, the tangential component of acceleration must act downward. Why? Since $\omega = 0$, only $m(a_G)_t = m\alpha r_G$ and $I_G\alpha$ are shown on the kinematic diagram in Fig. 17–15b. Here, the moment of inertia about G is

$$I_G = mk_G^2 = (50 \text{ lb}/32.2 \text{ ft/s}^2)(0.6 \text{ ft})^2 = 0.559 \text{ slug} \cdot \text{ft}^2$$

The three unknowns are O_n, O_t, and α.

Equations of Motion.

$$\xleftarrow{+} \Sigma F_n = m\omega^2 r_G; \qquad\qquad O_n = 0 \qquad\qquad Ans.$$

$$+\downarrow \Sigma F_t = m\alpha r_G; \quad -O_t + 50 \text{ lb} = \left(\frac{50 \text{ lb}}{32.2 \text{ ft/s}^2}\right)(\alpha)(0.5 \text{ ft}) \qquad (1)$$

$$\zeta + \Sigma M_G = I_G\alpha; \quad O_t(0.5 \text{ ft}) = (0.5590 \text{ slug} \cdot \text{ft}^2)\alpha$$

Solving,

$$\alpha = 26.4 \text{ rad/s}^2 \qquad O_t = 29.5 \text{ lb} \qquad\qquad Ans.$$

Moments can also be summed about point O in order to eliminate $\mathbf{O}_n$ and $\mathbf{O}_t$ and thereby obtain a *direct solution* for $\boldsymbol{\alpha}$, Fig. 17–15b. This can be done in one of *two* ways.

$$\zeta + \Sigma M_O = \Sigma(\mathcal{M}_k)_O;$$

$$(50 \text{ lb})(0.5 \text{ ft}) = (0.5590 \text{ slug} \cdot \text{ft}^2)\alpha + \left[\left(\frac{50 \text{ lb}}{32.2 \text{ ft/s}^2}\right)\alpha(0.5 \text{ ft})\right](0.5 \text{ ft})$$

$$50 \text{ lb}(0.5 \text{ ft}) = 0.9472\alpha \qquad (2)$$

If $\Sigma M_O = I_O\alpha$ is applied, then by the parallel-axis theorem the moment of inertia of the flywheel about O is

$$I_O = I_G + mr_G^2 = 0.559 + \left(\frac{50}{32.2}\right)(0.5)^2 = 0.9472 \text{ slug} \cdot \text{ft}^2$$

Hence,

$$\zeta + \Sigma M_O = I_O\alpha; \quad (50 \text{ lb})(0.5 \text{ ft}) = (0.9472 \text{ slug} \cdot \text{ft}^2)\alpha$$

which is the same as Eq. 2. Solving for α and substituting into Eq. 1 yields the answer for O_t obtained previously.

EXAMPLE 17.10

At the instant shown in Fig. 17–16a, the 20-kg slender rod has an angular velocity of $\omega = 5$ rad/s. Determine the angular acceleration and the horizontal and vertical components of reaction of the pin on the rod at this instant.

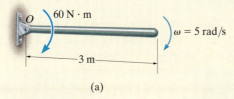

(a)

SOLUTION
Free-Body and Kinetic Diagrams. Fig. 17–16b. As shown on the kinetic diagram, point G moves around a circular path and so it has two components of acceleration. It is important that the tangential component $a_t = \alpha r_G$ act downward since it must be in accordance with the rotational sense of α. The three unknowns are O_n, O_t, and α.

Equation of Motion.

$$\xleftarrow{} \Sigma F_n = m\omega^2 r_G; \qquad O_n = (20\text{ kg})(5\text{ rad/s})^2(1.5\text{ m})$$

$$+\downarrow \Sigma F_t = m\alpha r_G; \qquad -O_t + 20(9.81)\text{N} = (20\text{ kg})(\alpha)(1.5\text{ m})$$

$$\zeta + \Sigma M_G = I_G\alpha; \quad O_t(1.5\text{ m}) + 60\text{ N}\cdot\text{m} = \left[\tfrac{1}{12}(20\text{ kg})(3\text{ m})^2\right]\alpha$$

Solving

$$O_n = 750\text{ N} \qquad O_t = 19.05\text{ N} \qquad \alpha = 5.90\text{ rad/s}^2 \qquad Ans.$$

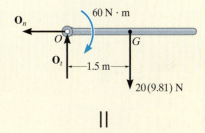

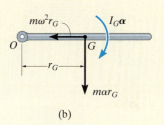

(b)

Fig. 17–16

A more direct solution to this problem would be to sum moments about point O to eliminate $\mathbf{O}_n$ and $\mathbf{O}_t$ and obtain a *direct solution* for α. Here,

$$\zeta + \Sigma M_O = \Sigma(\mathcal{M}_k)_O; \quad 60\text{ N}\cdot\text{m} + 20(9.81)\text{ N}(1.5\text{ m}) =$$

$$\left[\tfrac{1}{12}(20\text{ kg})(3\text{ m})^2\right]\alpha + [20\text{ kg}(\alpha)(1.5\text{ m})](1.5\text{ m})$$

$$\alpha = 5.90\text{ rad/s}^2 \qquad Ans.$$

Also, since $I_O = \tfrac{1}{3}ml^2$ for a slender rod, we can apply

$$\zeta + \Sigma M_O = I_O\alpha; \, 60\text{ N}\cdot\text{m} + 20(9.81)\text{ N}(1.5\text{ m}) = \left[\tfrac{1}{3}(20\text{ kg})(3\text{ m})^2\right]\alpha$$

$$\alpha = 5.90\text{ rad/s}^2 \qquad Ans.$$

NOTE: By comparison, the last equation provides the simplest solution for α and *does not* require use of the kinetic diagram.

17

EXAMPLE 17.11

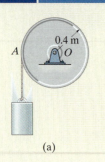

(a)

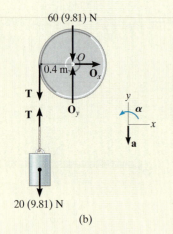

(b)

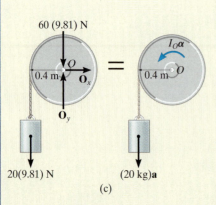

(c)

Fig. 17–17

The drum shown in Fig. 17–17a has a mass of 60 kg and a radius of gyration $k_O = 0.25$ m. A cord of negligible mass is wrapped around the periphery of the drum and attached to a block having a mass of 20 kg. If the block is released, determine the drum's angular acceleration.

SOLUTION I

Free-Body Diagram. Here we will consider the drum and block separately, Fig. 17–17b. Assuming the block accelerates *downward* at **a**, it creates a *counterclockwise* angular acceleration **α** of the drum. The moment of inertia of the drum is

$$I_O = mk_O^2 = (60 \text{ kg})(0.25 \text{ m})^2 = 3.75 \text{ kg} \cdot \text{m}^2$$

There are five unknowns, namely O_x, O_y, T, a, and α.

Equations of Motion. Applying the translational equations of motion $\Sigma F_x = m(a_G)_x$ and $\Sigma F_y = m(a_G)_y$ to the drum is of no consequence to the solution, since these equations involve the unknowns O_x and O_y. Thus, for the drum and block, respectively,

$$\zeta + \Sigma M_O = I_O\alpha; \qquad T(0.4 \text{ m}) = (3.75 \text{ kg} \cdot \text{m}^2)\alpha \qquad (1)$$

$$+\uparrow \Sigma F_y = m(a_G)_y; \quad -20(9.81)\text{N} + T = -(20 \text{ kg})a \qquad (2)$$

Kinematics. Since the point of contact A between the cord and drum has a tangential component of acceleration **a**, Fig. 17–17a, then

$$\zeta + a = \alpha r; \qquad\qquad a = \alpha(0.4 \text{ m}) \qquad (3)$$

Solving the above equations,

$$T = 106 \text{ N} \quad a = 4.52 \text{ m/s}^2$$

$$\alpha = 11.3 \text{ rad/s}^2\,\zeta \qquad\qquad\qquad Ans.$$

SOLUTION II

Free-Body and Kinetic Diagrams. The cable tension T can be eliminated from the analysis by considering the drum and block as a *single system*, Fig. 17–17c. The kinetic diagram is shown since moments will be summed about point O.

Equations of Motion. Using Eq. 3 and applying the moment equation about O to eliminate the unknowns O_x and O_y, we have

$$\zeta + \Sigma M_O = \Sigma(\mathcal{M}_k)_O; \quad [20(9.81)\text{N}] \,(0.4 \text{ m}) =$$

$$(3.75 \text{ kg} \cdot \text{m}^2)\alpha + [20 \text{ kg}(\alpha \, 0.4 \text{ m})](0.4 \text{ m})$$

$$\alpha = 11.3 \text{ rad/s}^2 \qquad\qquad\qquad Ans.$$

NOTE: If the block were *removed* and a force of 20(9.81) N were applied to the cord, show that $\alpha = 20.9$ rad/s^2. This value is larger since the block has an inertia, or resistance to acceleration.

EXAMPLE 17.12

The slender rod shown in Fig. 17–18a has a mass m and length l and is released from rest when $\theta = 0°$. Determine the horizontal and vertical components of force which the pin at A exerts on the rod at the instant $\theta = 90°$.

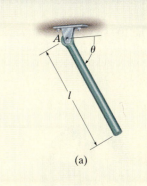

(a)

SOLUTION

Free-Body Diagram. The free-body diagram for the rod in the general position θ is shown in Fig. 17–18b. For convenience, the force components at A are shown acting in the n and t directions. Note that $\boldsymbol{\alpha}$ acts clockwise and so $(\mathbf{a}_G)_t$ acts in the $+ t$ direction.

The moment of inertia of the rod about point A is $I_A = \frac{1}{3}ml^2$.

Equations of Motion. Moments will be summed about A in order to eliminate A_n and A_t.

$$+\nwarrow\Sigma F_n = m\omega^2 r_G; \qquad A_n - mg\sin\theta = m\omega^2(l/2) \qquad (1)$$

$$+\swarrow\Sigma F_t = m\alpha r_G; \qquad A_t + mg\cos\theta = m\alpha(l/2) \qquad (2)$$

$$\zeta + \Sigma M_A = I_A\alpha; \qquad mg\cos\theta(l/2) = \left(\tfrac{1}{3}ml^2\right)\alpha \qquad (3)$$

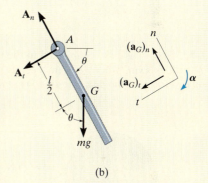

(b)

Fig. 17–18

Kinematics. For a given angle θ there are four unknowns in the above three equations: A_n, A_t, ω, and α. As shown by Eq. 3, α is *not constant;* rather, it depends on the position θ of the rod. The necessary fourth equation is obtained using kinematics, where α and ω can be related to θ by the equation

$$(\zeta+) \qquad\qquad \omega\, d\omega = \alpha\, d\theta \qquad (4)$$

Note that the positive clockwise direction for this equation *agrees* with that of Eq. 3. This is important since we are seeking a simultaneous solution.

In order to solve for ω at $\theta = 90°$, eliminate α from Eqs. 3 and 4, which yields

$$\omega\, d\omega = (1.5g/l)\cos\theta\, d\theta$$

Since $\omega = 0$ at $\theta = 0°$, we have

$$\int_0^\omega \omega\, d\omega = (1.5g/l)\int_{0°}^{90°}\cos\theta\, d\theta$$

$$\omega^2 = 3g/l$$

Substituting this value into Eq. 1 with $\theta = 90°$ and solving Eqs. 1 to 3 yields

$$\alpha = 0$$
$$A_t = 0 \quad A_n = 2.5mg \qquad\qquad Ans.$$

NOTE: If $\Sigma M_A = \Sigma(\mathcal{M}_k)_A$ is used, one must account for the moments of $I_G\boldsymbol{\alpha}$ and $m(\mathbf{a}_G)_t$ about A.

FUNDAMENTAL PROBLEMS

F17–7. The 100-kg wheel has a radius of gyration about its center O of $k_O = 500$ mm. If the wheel starts from rest, determine its angular velocity in $t = 3$ s.

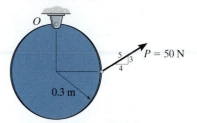

F17–7

F17–8. The 50-kg disk is subjected to the couple moment of $M = (9t)$ N·m, where t is in seconds. Determine the angular velocity of the disk when $t = 4$ s starting from rest.

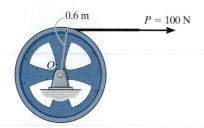

F17–8

F17–9. At the instant shown, the uniform 30-kg slender rod has a counterclockwise angular velocity of $\omega = 6$ rad/s. Determine the tangential and normal components of reaction of pin O on the rod and the angular acceleration of the rod at this instant.

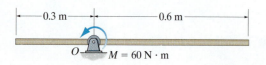

F17–9

F17–10. At the instant shown, the 30-kg disk has a counterclockwise angular velocity of $\omega = 10$ rad/s. Determine the tangential and normal components of reaction of the pin O on the disk and the angular acceleration of the disk at this instant.

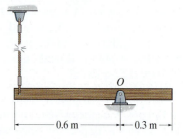

F17–10

F17–11. The uniform slender rod has a mass of 15 kg. Determine the horizontal and vertical components of reaction at the pin O, and the angular acceleration of the rod just after the cord is cut.

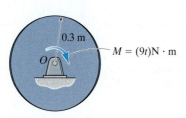

F17–11

F17–12. The uniform 30-kg slender rod is being pulled by the cord that passes over the small smooth peg at A. If the rod has an angular velocity of $\omega = 6$ rad/s at the instant shown, determine the tangential and normal components of reaction at the pin O and the angular acceleration of the rod.

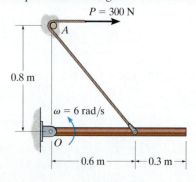

F17–12

PROBLEMS

*17–56. The four fan blades have a total mass of 2 kg and moment of inertia $I_O = 0.18$ kg·m² about an axis passing through the fan's center O. If the fan is subjected to a moment of $M = 3(1 - e^{-0.2t})$ N·m, where t is in seconds, determine its angular velocity when $t = 4$ s starting from rest.

Prob. 17–56

17–58. The single blade PB of the fan has a mass of 2 kg and a moment of inertia $I_G = 0.18$ kg·m² about an axis passing through its center of mass G. If the blade is subjected to an angular acceleration $\alpha = 5$ rad/s², and has an angular velocity $\omega = 6$ rad/s when it is in the vertical position shown, determine the internal normal force N, shear force V, and bending moment M, which the hub exerts on the blade at point P.

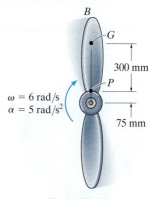

Prob. 17–58

•17–57. Cable is unwound from a spool supported on small rollers at A and B by exerting a force of $T = 300$ N on the cable in the direction shown. Compute the time needed to unravel 5 m of cable from the spool if the spool and cable have a total mass of 600 kg and a centroidal radius of gyration of $k_O = 1.2$ m. For the calculation, neglect the mass of the cable being unwound and the mass of the rollers at A and B. The rollers turn with no friction.

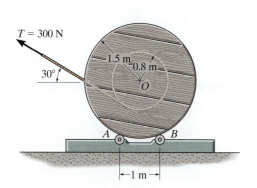

Prob. 17–57

17–59. The uniform spool is supported on small rollers at A and B. Determine the constant force $\mathbf{P}$ that must be applied to the cable in order to unwind 8 m of cable in 4 s starting from rest. Also calculate the normal forces on the spool at A and B during this time. The spool has a mass of 60 kg and a radius of gyration about O of $k_O = 0.65$ m. For the calculation neglect the mass of the cable and the mass of the rollers at A and B.

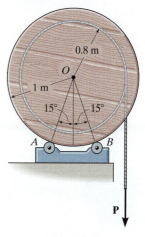

Prob. 17–59

*17–60. A motor supplies a constant torque $M = 2$ N·m to a 50-mm-diameter shaft O connected to the center of the 30-kg flywheel. The resultant bearing friction $\mathbf{F}$, which the bearing exerts on the shaft, acts tangent to the shaft and has a magnitude of 50 N. Determine how long the torque must be applied to the shaft to increase the flywheel's angular velocity from 4 rad/s to 15 rad/s. The flywheel has a radius of gyration $k_O = 0.15$ m about its center O.

•17–61. If the motor in Prob. 17–60 is disengaged from the shaft once the flywheel is rotating at 15 rad/s, so that $M = 0$, determine how long it will take before the resultant bearing frictional force $F = 50$ N stops the flywheel from rotating.

Probs. 17–60/61

17–62. The pendulum consists of a 30-lb sphere and a 10-lb slender rod. Compute the reaction at the pin O just after the cord AB is cut.

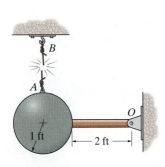

Prob. 17–62

17–63. The 4-kg slender rod is supported horizontally by a spring at A and a cord at B. Determine the angular acceleration of the rod and the acceleration of the rod's mass center at the instant the cord at B is cut. *Hint:* The stiffness of the spring is not needed for the calculation.

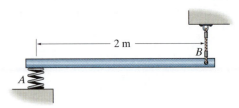

Prob. 17–63

*17–64. The passengers, the gondola, and its swing frame have a total mass of 50 Mg, a mass center at G, and a radius of gyration $k_B = 3.5$ m. Additionally, the 3-Mg steel block at A can be considered as a point of concentrated mass. Determine the horizontal and vertical components of reaction at pin B if the gondola swings freely at $\omega = 1$ rad/s when it reaches its lowest point as shown. Also, what is the gondola's angular acceleration at this instant?

•17–65. The passengers, the gondola, and its swing frame have a total mass of 50 Mg, a mass center at G, and a radius of gyration $k_B = 3.5$ m. Additionally, the 3-Mg steel block at A can be considered as a point of concentrated mass. Determine the angle θ to which the gondola will swing before it stops momentarily, if it has an angular velocity of $\omega = 1$ rad/s at its lowest point.

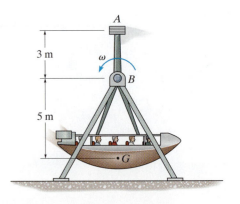

Probs. 17–64/65

17–66. The kinetic diagram representing the general rotational motion of a rigid body about a fixed axis passing through O is shown in the figure. Show that $I_G\alpha$ may be eliminated by moving the vectors $m(\mathbf{a}_G)_t$ and $m(\mathbf{a}_G)_n$ to point P, located a distance $r_{GP} = k_G^2/r_{OG}$ from the center of mass G of the body. Here k_G represents the radius of gyration of the body about an axis passing through G. The point P is called the *center of percussion* of the body.

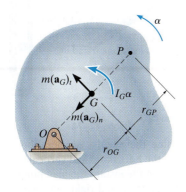

Prob. 17–66

17–67. Determine the position r_P of the center of percussion P of the 10-lb slender bar. (See Prob. 17–66.) What is the horizontal component of force that the pin at A exerts on the bar when it is struck at P with a force of $F = 20$ lb?

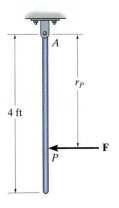

Prob. 17–67

*****17–68.** The 150-kg wheel has a radius of gyration about its center of mass O of $k_O = 250$ mm. If it rotates counterclockwise with an angular velocity of $\omega = 1200$ rev/min at the instant the tensile forces $T_A = 2000$ N and $T_B = 1000$ N are applied to the brake band at A and B, determine the time needed to stop the wheel.

•17–69. The 150-kg wheel has a radius of gyration about its center of mass O of $k_O = 250$ mm. If it rotates counterclockwise with an angular velocity of $\omega = 1200$ rev/min and the tensile force applied to the brake band at A is $T_A = 2000$ N, determine the tensile force $\mathbf{T}_B$ in the band at B so that the wheel stops in 50 revolutions after $\mathbf{T}_A$ and $\mathbf{T}_B$ are applied.

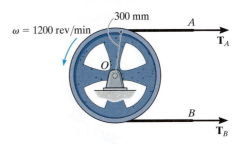

Probs. 17–68/69

17–70. The 100-lb uniform rod is at rest in a vertical position when the cord attached to it at B is subjected to a force of $P = 50$ lb. Determine the rod's initial angular acceleration and the magnitude of the reactive force that pin A exerts on the rod. Neglect the size of the smooth peg at C.

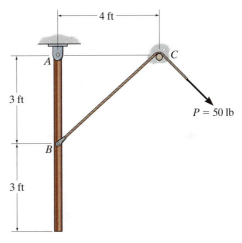

Prob. 17–70

EXAMPLE 17.13

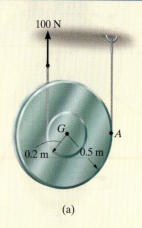

100 N

G

A

0.2 m   0.5 m

(a)

Determine the angular acceleration of the spool in Fig. 17–20a. The spool has a mass of 8 kg and a radius of gyration of $k_G = 0.35$ m. The cords of negligible mass are wrapped around its inner hub and outer rim.

SOLUTION I

Free-Body Diagram. Fig. 17–20b. The 100-N force causes $\mathbf{a}_G$ to act upward. Also, α acts clockwise, since the spool winds around the cord at A.

There are three unknowns T, a_G, and α. The moment of inertia of the spool about its mass center is

$$I_G = mk_G^2 = 8 \text{ kg}(0.35 \text{ m})^2 = 0.980 \text{ kg} \cdot \text{m}^2$$

Equations of Motion.

$$+\uparrow \Sigma F_y = m(a_G)_y; \qquad T + 100 \text{ N} - 78.48 \text{ N} = (8 \text{ kg})a_G \quad (1)$$

$$\zeta + \Sigma M_G = I_G\alpha; \quad 100 \text{ N}(0.2 \text{ m}) - T(0.5 \text{ m}) = (0.980 \text{ kg} \cdot \text{m}^2)\alpha \quad (2)$$

Kinematics. A complete solution is obtained if kinematics is used to relate a_G to α. In this case the spool "rolls without slipping" on the cord at A. Hence, we can use the results of Example 16.4 or 16.15, so that

$$(\zeta +)\, a_G = \alpha r. \qquad\qquad a_G = \alpha\,(0.5 \text{ m}) \quad (3)$$

Solving Eqs. 1 to 3, we have

$$\alpha = 10.3 \text{ rad/s}^2 \qquad\qquad Ans.$$
$$a_G = 5.16 \text{ m/s}^2$$
$$T = 19.8 \text{ N}$$

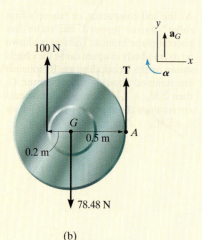

y

$\mathbf{a}_G$

x

T

α

100 N

G

0.5 m

A

0.2 m

78.48 N

(b)

SOLUTION II

Equations of Motion. We can eliminate the unknown T by summing moments about point A. From the free-body and kinetic diagrams Figs. 17–20b and 17–20c, we have

$$\zeta + \Sigma M_A = \Sigma(\mathcal{M}_k)_A; \qquad 100 \text{ N}(0.7 \text{ m}) - 78.48 \text{ N}(0.5 \text{ m})$$
$$= (0.980 \text{ kg} \cdot \text{m}^2)\alpha + [(8 \text{ kg})a_G](0.5 \text{ m})$$

Using Eq. (3),

$$\alpha = 10.3 \text{ rad/s}^2 \qquad\qquad Ans.$$

SOLUTION III

Equations of Motion. The simplest way to solve this problem is to realize that point A is the IC for the spool. Then Eq. 17–19 applies.

$$\zeta + \Sigma M_A = I_A\alpha; \quad (100 \text{ N})(0.7 \text{ m}) - (78.48 \text{ N})(0.5 \text{ m})$$
$$= [0.980 \text{ kg} \cdot \text{m}^2 + (8 \text{ kg})(0.5 \text{ m})^2]\alpha$$
$$\alpha = 10.3 \text{ rad/s}^2$$

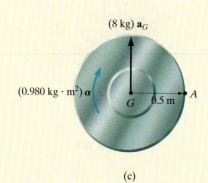

(8 kg) $\mathbf{a}_G$

(0.980 kg · m²) α

G   0.5 m   A

(c)

Fig. 17–20

EXAMPLE 17.14

The 50-lb wheel shown in Fig. 17–21a has a radius of gyration $k_G = 0.70$ ft. If a 35-lb·ft couple moment is applied to the wheel, determine the acceleration of its mass center G. The coefficients of static and kinetic friction between the wheel and the plane at A are $\mu_s = 0.3$ and $\mu_k = 0.25$, respectively.

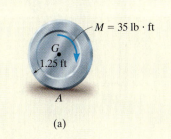

$M = 35$ lb·ft

(a)

SOLUTION

Free-Body Diagram. By inspection of Fig. 17–21b, it is seen that the couple moment causes the wheel to have a clockwise angular acceleration of α. As a result, the acceleration of the mass center, $\mathbf{a}_G$, is directed to the right. The moment of inertia is

$$I_G = mk_G^2 = \frac{50 \text{ lb}}{32.2 \text{ ft/s}^2}(0.70 \text{ ft})^2 = 0.7609 \text{ slug}\cdot\text{ft}^2$$

The unknowns are N_A, F_A, a_G, and α.

Equations of Motion.

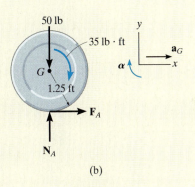

(b)

Fig. 17–21

$$\xrightarrow{+}\ \Sigma F_x = m(a_G)_x; \qquad F_A = \left(\frac{50 \text{ lb}}{32.2 \text{ ft/s}^2}\right)a_G \qquad (1)$$

$$+\uparrow\Sigma F_y = m(a_G)_y; \qquad N_A - 50 \text{ lb} = 0 \qquad (2)$$

$$\zeta+\Sigma M_G = I_G\alpha; \qquad 35 \text{ lb}\cdot\text{ft} - 1.25 \text{ ft}(F_A) = (0.7609 \text{ slug}\cdot\text{ft}^2)\alpha \quad (3)$$

A fourth equation is needed for a complete solution.

Kinematics (No Slipping). If this assumption is made, then

$$(\zeta+) \qquad\qquad a_G = (1.25 \text{ ft})\alpha \qquad (4)$$

Solving Eqs. 1 to 4,

$$N_A = 50.0 \text{ lb} \qquad\qquad F_A = 21.3 \text{ lb}$$
$$\alpha = 11.0 \text{ rad/s}^2 \qquad a_G = 13.7 \text{ ft/s}^2$$

This solution requires that no slipping occurs, i.e., $F_A \leq \mu_s N_A$. However, since 21.3 lb $> 0.3(50$ lb$) = 15$ lb, the wheel slips as it rolls.

(Slipping). Equation 4 is not valid, and so $F_A = \mu_k N_A$, or

$$F_A = 0.25 N_A \qquad (5)$$

Solving Eqs. 1 to 3 and 5 yields

$$N_A = 50.0 \text{ lb} \qquad F_A = 12.5 \text{ lb}$$
$$\alpha = 25.5 \text{ rad/s}^2$$
$$a_G = 8.05 \text{ ft/s}^2 \rightarrow \qquad\qquad \textit{Ans.}$$

17

EXAMPLE 17.15

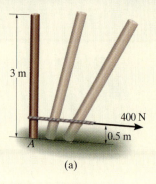

3 m

400 N

0.5 m

A

(a)

The uniform slender pole shown in Fig. 17–22a has a mass of 100 kg. If the coefficients of static and kinetic friction between the end of the pole and the surface are $\mu_s = 0.3$ and $\mu_k = 0.25$, respectively, determine the pole's angular acceleration at the instant the 400-N horizontal force is applied. The pole is originally at rest.

SOLUTION

Free-Body Diagram. Figure 17–22b. The path of motion of the mass center G will be along an unknown curved path having a radius of curvature ρ, which is initially on a vertical line. However, there is no normal or y component of acceleration since the pole is originally at rest, i.e., $\mathbf{v}_G = \mathbf{0}$, so that $(a_G)_y = v_G^2/\rho = 0$. We will assume the mass center accelerates to the right and that the pole has a clockwise angular acceleration of α. The unknowns are N_A, F_A, a_G, and α.

Equation of Motion.

$$\stackrel{+}{\rightarrow} \Sigma F_x = m(a_G)_x; \qquad 400\text{ N} - F_A = (100\text{ kg})a_G \qquad (1)$$

$$+\uparrow \Sigma F_y = m(a_G)_y; \qquad N_A - 981\text{ N} = 0 \qquad (2)$$

$$\zeta + \Sigma M_G = I_G\alpha; \quad F_A(1.5\text{ m}) - (400\text{ N})(1\text{ m}) = ([\tfrac{1}{12}(100\text{ kg})(3\text{ m})^2)\alpha \;(3)$$

A fourth equation is needed for a complete solution.

Kinematics (No Slipping). With this assumption, point A acts as a "pivot" so that α is clockwise, then a_G is directed to the right.

$$a_G = \alpha r_{AG}; \qquad a_G = (1.5\text{ m})\,\alpha \qquad (4)$$

Solving Eqs. 1 to 4 yields

$$N_A = 981\text{ N} \qquad F_A = 300\text{ N}$$
$$a_G = 1\text{ m/s}^2 \qquad \alpha = 0.667\text{ rad/s}^2$$

The assumption of no slipping requires $F_A \leq \mu_s N_A$. However, $300\text{ N} > 0.3(981\text{ N}) = 294\text{ N}$ and so the pole slips at A.

(Slipping). For this case Eq. 4 does *not* apply. Instead the frictional equation $F_A = \mu_k N_A$ must be used. Hence,

$$F_A = 0.25N_A \qquad (5)$$

Solving Eqs. 1 to 3 and 5 simultaneously yields

$$N_A = 981\text{ N} \qquad F_A = 245\text{ N} \qquad a_G = 1.55\text{ m/s}^2$$
$$\alpha = -0.428\text{ rad/s}^2 = 0.428\text{ rad/s}^2 \,\zeta \qquad \qquad Ans.$$

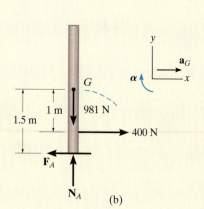

y

$\mathbf{a}_G$

α

x

G

1 m

1.5 m

981 N

400 N

F_A

N_A

(b)

Fig. 17–22

EXAMPLE | 17.16

The uniform 50-kg bar in Fig. 17–23a is held in the equilibrium position by cords AC and BD. Determine the tension in BD and the angular acceleration of the bar immediately after AC is cut.

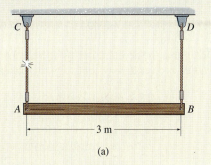

(a)

SOLUTION

Free-Body Diagram. Fig. 17–23b. There are four unknowns, T_B, $(a_G)_x$, $(a_G)_y$, and α.

Equations of Motion.

$$\xrightarrow{+} \Sigma F_x = m(a_G)_x; \qquad 0 = (50 \text{ kg } a_G)_x$$

$$(a_G)_x = 0$$

$$+\uparrow \Sigma F_y = m(a_G)_y; \quad T_B - 50(9.81)\text{N} = -(50 \text{ kg } a_G)_y \qquad (1)$$

$$\zeta + \Sigma M_G = I_G\alpha; \qquad T_B(1.5 \text{ m}) = \left[\frac{1}{12}(50 \text{ kg})(3 \text{ m})^2\right]\alpha \qquad (2)$$

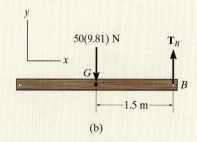

(b)

Kinematics. Since the bar is at rest just after the cable is cut, then its angular velocity and the velocity of point B at this instant are equal to zero. Thus $(a_B)_n = v_B^2/\rho_{BD} = 0$. Therefore, $\mathbf{a}_B$ only has a tangential component, which is directed along the x axis, Fig. 17–23c. Applying the relative acceleration equation to points G and B,

$$\mathbf{a}_G = \mathbf{a}_B + \boldsymbol{\alpha} \times \mathbf{r}_{G/B} - \omega^2\mathbf{r}_{G/B}$$

$$-(a_G)_y\mathbf{j} = a_B\mathbf{i} + (\alpha\mathbf{k}) \times (-1.5\mathbf{i}) - \mathbf{0}$$

$$-(a_G)_y\mathbf{j} = a_B\mathbf{i} - 1.5\alpha\mathbf{j}$$

Equating the $\mathbf{i}$ and $\mathbf{j}$ components of both sides of this equation,

$$0 = a_B$$

$$(a_G)_y = 1.5\alpha \qquad (3)$$

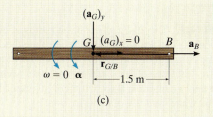

(c)

Fig. 17–23

Solving Eqs. (1) through (3) yields

$$\alpha = 4.905 \text{ rad/s}^2 \qquad\qquad \textit{Ans.}$$

$$T_B = 123 \text{ N} \qquad\qquad \textit{Ans.}$$

$$(a_G)_y = 7.36 \text{ m/s}^2$$

17

FUNDAMENTAL PROBLEMS

F17–13. The uniform 60-kg slender bar is initially at rest on a smooth horizontal plane when the forces are applied. Determine the acceleration of the bar's mass center and the angular acceleration of the bar at this instant.

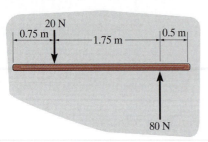

F17–13

F17–14. The 100-kg cylinder rolls without slipping on the horizontal plane. Determine the acceleration of its mass center and its angular acceleration.

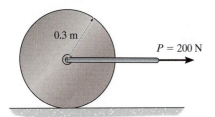

F17–14

F17–15. The 20-kg wheel has a radius of gyration about its center O of $k_O = 300$ mm. When the wheel is subjected to the couple moment, it slips as it rolls. Determine the angular acceleration of the wheel and the acceleration of the wheel's center O. The coefficient of kinetic friction between the wheel and the plane is $\mu_k = 0.5$.

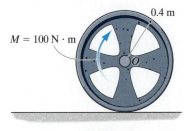

F17–15

F17–16. The 20-kg sphere rolls down the inclined plane without slipping. Determine the angular acceleration of the sphere and the acceleration of its mass center.

F17–16

F17–17. The 200-kg spool has a radius of gyration about its mass center of $k_G = 300$ mm. If the couple moment is applied to the spool and the coefficient of kinetic friction between the spool and the ground is $\mu_k = 0.2$, determine the angular acceleration of the spool, the acceleration of G and the tension in the cable.

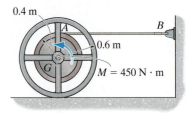

F17–17

F17–18. The 12-kg slender rod is pinned to a small roller A that slides freely along the slot. If the rod is released from rest at $\theta = 0°$, determine the angular acceleration of the rod and the acceleration of the roller immediately after the release.

F17–18

PROBLEMS

17–91. If a disk *rolls without slipping* on a horizontal surface, show that when moments are summed about the instantaneous center of zero velocity, IC, it is possible to use the moment equation $\Sigma M_{IC} = I_{IC}\alpha$, where I_{IC} represents the moment of inertia of the disk calculated about the instantaneous axis of zero velocity.

*17–92. The 10-kg semicircular disk is rotating at $\omega = 4$ rad/s at the instant $\theta = 60°$. Determine the normal and frictional forces it exerts on the ground at A at this instant. Assume the disk does not slip as it rolls.

•17–93. The semicircular disk having a mass of 10 kg is rotating at $\omega = 4$ rad/s at the instant $\theta = 60°$. If the coefficient of static friction at A is $\mu_s = 0.5$, determine if the disk slips at this instant.

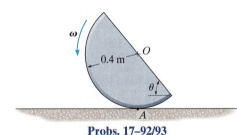

Probs. 17–92/93

17–94. The uniform 50-lb board is suspended from cords at C and D. If these cords are subjected to constant forces of 30 lb and 45 lb, respectively, determine the initial acceleration of the board's center and the board's angular acceleration. Assume the board is a thin plate. Neglect the mass of the pulleys at E and F.

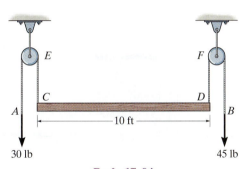

30 lb

45 lb

Prob. 17–94

17–95. The rocket consists of the main section A having a mass of 10 Mg and a center of mass at G_A. The two identical booster rockets B and C each have a mass of 2 Mg with centers of mass at G_B and G_C, respectively. At the instant shown, the rocket is traveling vertically and is at an altitude where the acceleration due to gravity is $g = 8.75$ m/s². If the booster rockets B and C suddenly supply a thrust of $T_B = 30$ kN and $T_C = 20$ kN, respectively, determine the angular acceleration of the rocket. The radius of gyration of A about G_A is $k_A = 2$ m and the radii of gyration of B and C about G_B and G_C are $k_B = k_C = 0.75$ m.

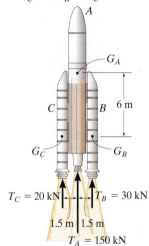

Prob. 17–95

*17–96. The 75-kg wheel has a radius of gyration about the z axis of $k_z = 150$ mm. If the belt of negligible mass is subjected to a force of $P = 150$ N, determine the acceleration of the mass center and the angular acceleration of the wheel. The surface is smooth and the wheel is free to slide.

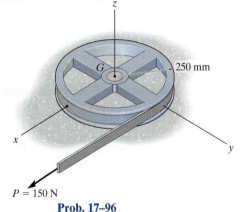

Prob. 17–96

17

17–111. The 15-lb cylinder is initially at rest on a 5-lb plate. If a couple moment $M = 40$ lb · ft is applied to the cylinder, determine the angular acceleration of the cylinder and the time needed for the end B of the plate to travel 3 ft to the right and strike the wall. Assume the cylinder does not slip on the plate, and neglect the mass of the rollers under the plate.

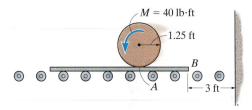

Prob. 17–111

17–114. The 20-kg disk A is attached to the 10-kg block B using the cable and pulley system shown. If the disk rolls without slipping, determine its angular acceleration and the acceleration of the block when they are released. Also, what is the tension in the cable? Neglect the mass of the pulleys.

17–115. Determine the minimum coefficient of static friction between the disk and the surface in Prob. 17–114 so that the disk will roll without slipping. Neglect the mass of the pulleys.

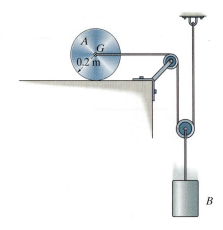

Probs. 17–114/115

*17–112.** The assembly consists of an 8-kg disk and a 10-kg bar which is pin connected to the disk. If the system is released from rest, determine the angular acceleration of the disk. The coefficients of static and kinetic friction between the disk and the inclined plane are $\mu_s = 0.6$ and $\mu_k = 0.4$, respectively. Neglect friction at B.

•**17–113.** Solve Prob. 17–112 if the bar is removed. The coefficients of static and kinetic friction between the disk and inclined plane are $\mu_s = 0.15$ and $\mu_k = 0.1$, respectively.

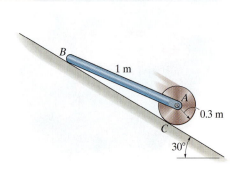

Probs. 17–112/113

*17–116.** The 20-kg square plate is pinned to the 5-kg smooth collar. Determine the initial angular acceleration of the plate when $P = 100$ N is applied to the collar. The plate is originally at rest.

•**17–117.** The 20-kg square plate is pinned to the 5-kg smooth collar. Determine the initial acceleration of the collar when $P = 100$ N is applied to the collar. The plate is originally at rest.

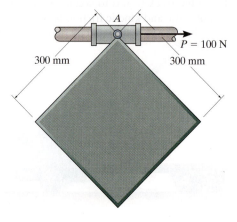

Probs. 17–116/117

17–118. The spool has a mass of 100 kg and a radius of gyration of $k_G = 200$ mm about its center of mass G. If a vertical force of $P = 200$ N is applied to the cable, determine the acceleration of G and the angular acceleration of the spool. The coefficients of static and kinetic friction between the rail and the spool are $\mu_s = 0.3$ and $\mu_k = 0.25$, respectively.

17–119. The spool has a mass of 100 kg and a radius of gyration of $k_G = 200$ mm about its center of mass G. If a vertical force of $P = 500$ N is applied to the cable, determine the acceleration of G and the angular acceleration of the spool. The coefficients of static and kinetic friction between the rail and the spool are $\mu_s = 0.2$ and $\mu_k = 0.15$, respectively.

•17–121. The 75-kg wheel has a radius of gyration about its mass center of $k_G = 375$ mm. If it is subjected to a torque of $M = 100$ N·m, determine its angular acceleration. The coefficients of static and kinetic friction between the wheel and the ground are $\mu_s = 0.2$ and $\mu_k = 0.15$, respectively.

17–122. The 75-kg wheel has a radius of gyration about its mass center of $k_G = 375$ mm. If it is subjected to a torque of $M = 150$ N·m, determine its angular acceleration. The coefficients of static and kinetic friction between the wheel and the ground are $\mu_s = 0.2$ and $\mu_k = 0.15$, respectively.

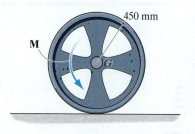

Probs. 17–121/122

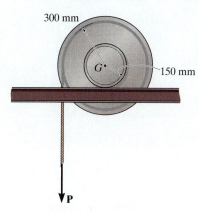

Probs. 17–118/119

*17–120. If the truck accelerates at a constant rate of 6 m/s^2, starting from rest, determine the initial angular acceleration of the 20-kg ladder. The ladder can be considered as a uniform slender rod. The support at B is smooth.

17–123. The 500-kg concrete culvert has a mean radius of 0.5 m. If the truck has an acceleration of 3 m/s^2, determine the culvert's angular acceleration. Assume that the culvert does not slip on the truck bed, and neglect its thickness.

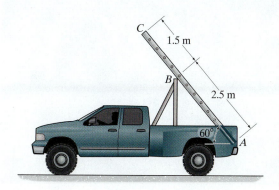

Prob. 17–120

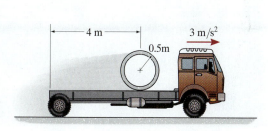

Prob. 17–123

CONCEPTUAL PROBLEMS

P17–1. The truck is used to pull the heavy container. To be most effective at providing traction to the rear wheels at *A*, is it best to keep the container where it is or place it at the front of the trailer? Use appropriate numerical values to explain your answer.

P17–1

P17–2. The tractor is about to tow the plane to the right. Is it possible for the driver to cause the front wheel of the plane to lift off the ground as he accelerates the tractor? Draw the free-body and kinetic diagrams and explain algebraically (letters) if and how this might be possible.

P17–2

P17–3. How can you tell the driver is accelerating this SUV? To explain your answer, draw the free-body and kinetic diagrams. Here power is supplied to the rear wheels. Would the photo look the same if power were supplied to the front wheels? Will the accelerations be the same? Use appropriate numerical values to explain your answers.

P17–3

P17–4. Here is something you should not try at home, at least not without wearing a helmet! Draw the free-body and kinetic diagrams and show what the rider must do to maintain this position. Use appropriate numerical values to explain your answer.

P17–4

CHAPTER REVIEW

Moment of Inertia

The moment of inertia is a measure of the resistance of a body to a change in its angular velocity. It is defined by $I = \int r^2 dm$ and will be different for each axis about which it is computed.

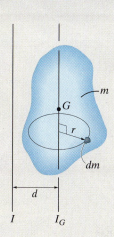

Many bodies are composed of simple shapes. If this is the case, then tabular values of I can be used, such as the ones given on the inside back cover of this book. To obtain the moment of inertia of a composite body about any specified axis, the moment of inertia of each part is determined about the axis and the results are added together. Doing this often requires use of the parallel-axis theorem.

$$I = I_G + md^2$$

Planar Equations of Motion

The equations of motion define the translational, and rotational motion of a rigid body. In order to account for all of the terms in these equations, a free-body diagram should always accompany their application, and for some problems, it may also be convenient to draw the kinetic diagram which shows $m\mathbf{a}_G$ and $I_G\boldsymbol{\alpha}$.

$$\Sigma F_x = m(a_G)_x$$
$$\Sigma F_y = m(a_G)_y$$
$$\Sigma M_G = 0$$
Rectilinear translation

$$\Sigma F_n = m(a_G)_n$$
$$\Sigma F_t = m(a_G)_t$$
$$\Sigma M_G = 0$$
Curvilinear translation

$$\Sigma F_n = m(a_G)_n = m\omega^2 r_G$$

$$\Sigma F_t = m(a_G)_t = m\alpha r_G$$

$$\Sigma M_G = I_G\alpha \text{ or } \Sigma M_O = I_O\alpha$$
Rotation About a Fixed Axis

$$\Sigma F_x = m(a_G)_x$$
$$\Sigma F_x = m(a_G)_x$$
$$\Sigma M_G = I_G\alpha \text{ or } \Sigma M_P = \Sigma(\mathcal{M}_k)_P$$

General Plane Motion

17

The principle of work and energy plays an important role in the motion of the draw works used to lift pipe on this drilling rig.

Planar Kinetics of a Rigid Body: Work and Energy

18

CHAPTER OBJECTIVES

- To develop formulations for the kinetic energy of a body, and define the various ways a force and couple do work.

- To apply the principle of work and energy to solve rigid–body planar kinetic problems that involve force, velocity, and displacement.

- To show how the conservation of energy can be used to solve rigid–body planar kinetic problems.

18.1 Kinetic Energy

In this chapter we will apply work and energy methods to solve planar motion problems involving force, velocity, and displacement. But first it will be necessary to develop a means of obtaining the body's kinetic energy when the body is subjected to translation, rotation about a fixed axis, or general plane motion.

To do this we will consider the rigid body shown in Fig. 18–1, which is represented here by a *slab* moving in the inertial x–y reference plane. An arbitrary ith particle of the body, having a mass dm, is located a distance r from the arbitrary point P. If at the *instant* shown the particle has a velocity $\mathbf{v}_i$, then the particle's kinetic energy is $T_i = \frac{1}{2} dm\, v_i^2$.

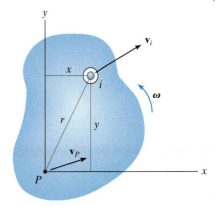

Fig. 18–1

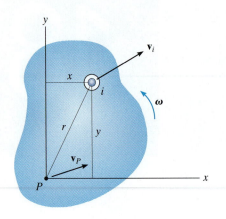

Fig. 18–1

The kinetic energy of the entire body is determined by writing similar expressions for each particle of the body and integrating the results, i.e.,

$$T = \frac{1}{2} \int_m dm \, v_i^2$$

This equation may also be expressed in terms of the velocity of point P. If the body has an angular velocity $\boldsymbol{\omega}$, then from Fig. 18–1 we have

$$\begin{aligned} \mathbf{v}_i &= \mathbf{v}_P + \mathbf{v}_{i/P} \\ &= (v_P)_x \mathbf{i} + (v_P)_y \mathbf{j} + \omega \mathbf{k} \times (x\mathbf{i} + y\mathbf{j}) \\ &= [(v_P)_x - \omega y]\mathbf{i} + [(v_P)_y + \omega x]\mathbf{j} \end{aligned}$$

The square of the magnitude of $\mathbf{v}_i$ is thus

$$\begin{aligned} \mathbf{v}_i \cdot \mathbf{v}_i = v_i^2 &= [(v_P)_x - \omega y]^2 + [(v_P)_y + \omega x]^2 \\ &= (v_P)_x^2 - 2(v_P)_x \omega y + \omega^2 y^2 + (v_P)_y^2 + 2(v_P)_y \omega x + \omega^2 x^2 \\ &= v_P^2 - 2(v_P)_x \omega y + 2(v_P)_y \omega x + \omega^2 r^2 \end{aligned}$$

Substituting this into the equation of kinetic energy yields

$$T = \frac{1}{2}\left(\int_m dm\right)v_P^2 - (v_P)_x \omega \left(\int_m y \, dm\right) + (v_P)_y \omega \left(\int_m x \, dm\right) + \frac{1}{2}\omega^2 \left(\int_m r^2 \, dm\right)$$

The first integral on the right represents the entire mass m of the body. Since $\overline{y}m = \int y \, dm$ and $\overline{x}m = \int x \, dm$, the second and third integrals locate the body's center of mass G with respect to P. The last integral represents the body's moment of inertia I_P, computed about the z axis passing through point P. Thus,

$$T = \tfrac{1}{2}mv_P^2 - (v_P)_x \omega \overline{y}m + (v_P)_y \omega \overline{x}m + \tfrac{1}{2}I_P\omega^2 \qquad (18\text{–}1)$$

As a special case, if point P coincides with the mass center G of the body, then $\overline{y} = \overline{x} = 0$, and therefore

$$T = \tfrac{1}{2}mv_G^2 + \tfrac{1}{2}I_G\omega^2 \qquad (18\text{–}2)$$

Both terms on the right side are *always positive*, since v_G and ω are squared. The first term represents the translational kinetic energy, referenced from the mass center, and the second term represents the body's rotational kinetic energy about the mass center.

Translation.   When a rigid body of mass m is subjected to either rectilinear or curvilinear *translation*, Fig. 18–2, the kinetic energy due to rotation is zero, since $\boldsymbol{\omega} = \mathbf{0}$. The kinetic energy of the body is therefore

$$T = \tfrac{1}{2}mv_G^2 \qquad (18\text{–}3)$$

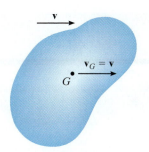

Translation

Fig. 18–2

Rotation About a Fixed Axis.   When a rigid body *rotates about a fixed axis* passing through point O, Fig. 18–3, the body has both *translational* and *rotational* kinetic energy so that

$$T = \tfrac{1}{2}mv_G^2 + \tfrac{1}{2}I_G\omega^2 \qquad (18\text{–}4)$$

The body's kinetic energy may also be formulated for this case by noting that $v_G = r_G\omega$, so that $T = \tfrac{1}{2}(I_G + mr_G^2)\omega^2$. By the parallel–axis theorem, the terms inside the parentheses represent the moment of inertia I_O of the body about an axis perpendicular to the plane of motion and passing through point O. Hence,*

$$T = \tfrac{1}{2}I_O\omega^2 \qquad (18\text{–}5)$$

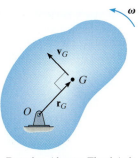

Rotation About a Fixed Axis

Fig. 18–3

From the derivation, this equation will give the same result as Eq. 18–4, since it accounts for *both* the translational and rotational kinetic energies of the body.

General Plane Motion.   When a rigid body is subjected to general plane motion, Fig. 18–4, it has an angular velocity $\boldsymbol{\omega}$ and its mass center has a velocity $\mathbf{v}_G$. Therefore, the kinetic energy is

$$T = \tfrac{1}{2}mv_G^2 + \tfrac{1}{2}I_G\omega^2 \qquad (18\text{–}6)$$

This equation can also be expressed in terms of the body's motion about its instantaneous center of zero velocity i.e.,

$$T = \tfrac{1}{2}I_{IC}\omega^2 \qquad (18\text{–}7)$$

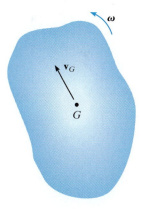

General Plane Motion

Fig. 18–4

where I_{IC} is the moment of inertia of the body about its instantaneous center. The proof is similar to that of Eq. 18–5. (See Prob. 18–1.)

---

*The similarity between this derivation and that of $\Sigma M_O = I_O\alpha$, Eq. 17–16, should be noted. Also the same result can be obtained directly from Eq. 18–1 by selecting point P at O, realizing that $v_O = 0$.

18

The total kinetic energy of this soil compactor consists of the kinetic energy of the body or frame of the machine due to its translation, and the translational and rotational kinetic energies of the roller and the wheels due to their general plane motion. Here we exclude the additional kinetic energy developed by the moving parts of the engine and drive train.

System of Bodies. Because energy is a scalar quantity, the total kinetic energy for a system of *connected* rigid bodies is the sum of the kinetic energies of all its moving parts. Depending on the type of motion, the kinetic energy of *each body* is found by applying Eq. 18–2 or the alternative forms mentioned above.

18.2 The Work of a Force

Several types of forces are often encountered in planar kinetics problems involving a rigid body. The work of each of these forces has been presented in Sec. 14.1 and is listed below as a summary.

Work of a Variable Force. If an external force $\mathbf{F}$ acts on a body, the work done by the force when the body moves along the path s, Fig. 18–5, is

$$U_F = \int \mathbf{F} \cdot d\mathbf{r} = \int_s F \cos \theta \, ds \qquad (18\text{–}8)$$

Here θ is the angle between the "tails" of the force and the differential displacement. The integration must account for the variation of the force's direction and magnitude.

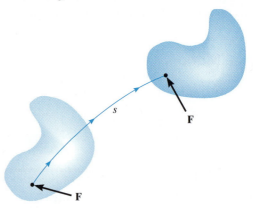

Fig. 18–5

Work of a Constant Force. If an external force $\mathbf{F}_c$ acts on a body, Fig. 18–6, and maintains a constant magnitude F_c and constant direction θ, while the body undergoes a translation s, then the above equation can be integrated, so that the work becomes

$$U_{F_c} = (F_c \cos \theta)s \qquad (18\text{–}9)$$

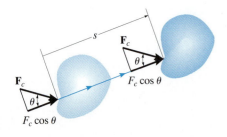

Fig. 18–6

Work of a Weight.

The weight of a body does work only when the body's center of mass G undergoes a *vertical displacement* Δy. If this displacement is *upward*, Fig. 18–7, the work is negative, since the weight is opposite to the displacement.

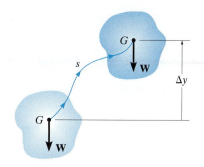

$$U_W = -W\,\Delta y \qquad (18\text{–}10)$$

Likewise, if the displacement is *downward* $(-\Delta y)$ the work becomes *positive*. In both cases the elevation change is considered to be small so that $\mathbf{W}$, which is caused by gravitation, is constant.

Fig. 18–7

Work of a Spring Force.

If a linear elastic spring is attached to a body, the spring force $F_s = ks$ *acting on the body* does work when the spring either stretches or compresses from s_1 to a *further* position s_2. In both cases the work will be *negative* since the *displacement of the body* is in the opposite direction to the force, Fig. 18–8. The work is

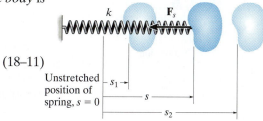

$$U_s = -\left(\tfrac{1}{2}ks_2^2 - \tfrac{1}{2}ks_1^2\right) \qquad (18\text{–}11)$$

Unstretched position of spring, $s = 0$

where $|s_2| > |s_1|$.

Fig. 18–8

Forces That Do No Work.

There are some external forces that do no work when the body is displaced. These forces act either at *fixed points* on the body, or they have a direction *perpendicular to their displacement*. Examples include the reactions at a pin support about which a body rotates, the normal reaction acting on a body that moves along a fixed surface, and the weight of a body when the center of gravity of the body moves in a *horizontal plane*, Fig. 18–9. A frictional force $\mathbf{F}_f$ acting on a round body as it *rolls without slipping* over a rough surface also does no work.* This is because, during any *instant of time dt*, $\mathbf{F}_f$ acts at a point on the body which has *zero velocity* (instantaneous center, IC) and so the work done by the force on the point is zero. In other words, the point is not displaced in the direction of the force during this instant. Since $\mathbf{F}_f$ contacts successive points for only an instant, the work of $\mathbf{F}_f$ will be zero.

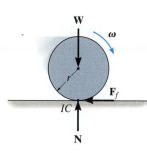

*The work done by a frictional force *when the body slips* is discussed in Sec. 14.3.

Fig. 18–9

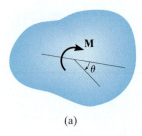

(a)

18.3   The Work of a Couple Moment

Consider the body in Fig. 18–10a, which is subjected to a couple moment $M = Fr$. If the body undergoes a differential displacement, then the work done by the couple forces can be found by considering the displacement as the sum of a separate translation plus rotation. When the body *translates*, the work of each force is produced only by the *component of displacement* along the line of action of the forces ds_t, Fig. 18–10b. Clearly the "positive" work of one force *cancels* the "negative" work of the other. When the body undergoes a differential rotation $d\theta$ about the arbitrary point O, Fig. 18–10c, then each force undergoes a displacement $ds_\theta = (r/2) \, d\theta$ in the direction of the force. Hence, the total work done is

$$dU_M = F\left(\frac{r}{2} d\theta\right) + F\left(\frac{r}{2} d\theta\right) = (Fr) \, d\theta$$

$$= M \, d\theta$$

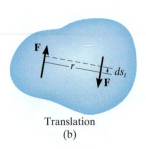

Translation
(b)

The work is *positive* when $\mathbf{M}$ and $d\boldsymbol{\theta}$ have the *same sense of direction* and *negative* if these vectors are in the *opposite sense*.

When the body rotates in the plane through a finite angle θ measured in radians, from θ_1 to θ_2, the work of a couple moment is therefore

$$U_M = \int_{\theta_1}^{\theta_2} M \, d\theta \qquad (18\text{–}12)$$

If the couple moment $\mathbf{M}$ has a *constant magnitude*, then

$$U_M = M(\theta_2 - \theta_1) \qquad (18\text{–}13)$$

Rotation
(c)

Fig. 18–10

EXAMPLE 18.1

The bar shown in Fig. 18–11a has a mass of 10 kg and is subjected to a couple moment of $M = 50\ \text{N} \cdot \text{m}$ and a force of $P = 80\ \text{N}$, which is always applied perpendicular to the end of the bar. Also, the spring has an unstretched length of 0.5 m and remains in the vertical position due to the roller guide at B. Determine the total work done by all the forces acting on the bar when it has rotated downward from $\theta = 0°$ to $\theta = 90°$.

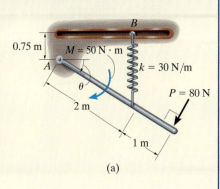

(a)

SOLUTION
First the free-body diagram of the bar is drawn in order to account for all the forces that act on it, Fig. 18–11b.

Weight W. Since the weight $10(9.81)\ \text{N} = 98.1\ \text{N}$ is displaced downward 1.5 m, the work is

$$U_W = 98.1\ \text{N}(1.5\ \text{m}) = 147.2\ \text{J}$$

Why is the work positive?

Couple Moment M. The couple moment rotates through an angle of $\theta = \pi/2$ rad. Hence,

$$U_M = 50\ \text{N} \cdot \text{m}(\pi/2) = 78.5\ \text{J}$$

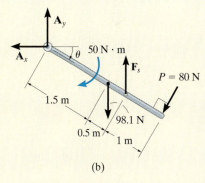

(b)

Fig. 18–11

Spring Force F_s. When $\theta = 0°$ the spring is stretched $(0.75\ \text{m} - 0.5\ \text{m}) = 0.25\ \text{m}$, and when $\theta = 90°$, the stretch is $(2\ \text{m} + 0.75\ \text{m}) - 0.5\ \text{m} = 2.25\ \text{m}$. Thus,

$$U_s = -\left[\tfrac{1}{2}(30\ \text{N/m})(2.25\ \text{m})^2 - \tfrac{1}{2}(30\ \text{N/m})(0.25\ \text{m})^2\right] = -75.0\ \text{J}$$

By inspection the spring does negative work on the bar since F_s acts in the opposite direction to displacement. This checks with the result.

Force P. As the bar moves downward, the force is displaced through a distance of $(\pi/2)(3\ \text{m}) = 4.712\ \text{m}$. The work is positive. Why?

$$U_P = 80\ \text{N}(4.712\ \text{m}) = 377.0\ \text{J}$$

Pin Reactions. Forces $\mathbf{A}_x$ and $\mathbf{A}_y$ do no work since they are not displaced.

Total Work. The work of all the forces when the bar is displaced is thus

$$U = 147.2\ \text{J} + 78.5\ \text{J} - 75.0\ \text{J} + 377.0\ \text{J} = 528\ \text{J} \qquad Ans.$$

18

18.4   Principle of Work and Energy

By applying the principle of work and energy developed in Sec. 14.2 to each of the particles of a rigid body and adding the results algebraically, since energy is a scalar, the principle of work and energy for a rigid body becomes

$$T_1 + \Sigma U_{1-2} = T_2 \qquad (18-14)$$

This equation states that the body's initial translational *and* rotational kinetic energy, plus the work done by all the external forces and couple moments acting on the body as the body moves from its initial to its final position, is equal to the body's final translational *and* rotational kinetic energy. Note that the work of the body's *internal forces* does not have to be considered. These forces occur in equal but opposite collinear pairs, so that when the body moves, the work of one force cancels that of its counterpart. Furthermore, since the body is rigid, *no relative movement* between these forces occurs, so that no internal work is done.

When several rigid bodies are pin connected, connected by inextensible cables, or in mesh with one another, Eq. 18–14 can be applied to the *entire system* of connected bodies. In all these cases the internal forces, which hold the various members together, do no work and hence are eliminated from the analysis.

The work of the torque or moment developed by the driving gears on the motors is transformed into kinetic energy of rotation of the drum.

Procedure for Analysis

The principle of work and energy is used to solve kinetic problems that involve *velocity, force*, and *displacement*, since these terms are involved in the formulation. For application, it is suggested that the following procedure be used.

Kinetic Energy (Kinematic Diagrams).

- The kinetic energy of a body is made up of two parts. Kinetic energy of translation is referenced to the velocity of the mass center, $T = \frac{1}{2}mv_G^2$, and kinetic energy of rotation is determined using the moment of inertia of the body about the mass center, $T = \frac{1}{2}I_G\omega^2$. In the special case of rotation about a fixed axis (or rotation about the *IC*), these two kinetic energies are combined and can be expressed as $T = \frac{1}{2}I_O\omega^2$, where I_O is the moment of inertia about the axis of rotation.

- *Kinematic diagrams* for velocity may be useful for determining v_G and ω or for establishing a *relationship* between v_G and ω.*

Work (Free–Body Diagram).

- Draw a free–body diagram of the body when it is located at an intermediate point along the path in order to account for all the forces and couple moments which do work on the body as it moves along the path.

- A force does work when it moves through a displacement in the direction of the force.

- Forces that are functions of displacement must be integrated to obtain the work. Graphically, the work is equal to the area under the force–displacement curve.

- The work of a weight is the product of its magnitude and the vertical displacement, $U_W = Wy$. It is positive when the weight moves downwards.

- The work of a spring is of the form $U_s = \frac{1}{2}ks^2$, where k is the spring stiffness and s is the stretch or compression of the spring.

- The work of a couple is the product of the couple moment and the angle in radians through which it rotates, $U_M = M\theta$.

- Since *algebraic addition* of the work terms is required, it is important that the proper sign of each term be specified. Specifically, work is *positive* when the force (couple moment) is in the *same direction* as its displacement (rotation); otherwise, it is negative.

Principle of Work and Energy.

- Apply the principle of work and energy, $T_1 + \Sigma U_{1-2} = T_2$. Since this is a scalar equation, it can be used to solve for only one unknown when it is applied to a single rigid body.

*A brief review of Secs. 16.5 to 16.7 may prove helpful when solving problems, since computations for kinetic energy require a kinematic analysis of velocity.

EXAMPLE 18.2

The 30–kg disk shown in Fig. 18–12a is pin supported at its center. Determine the number of revolutions it must make to attain an angular velocity of 20 rad/s starting from rest. It is acted upon by a constant force $F = 10$ N, which is applied to a cord wrapped around its periphery, and a constant couple moment $M = 5$ N·m. Neglect the mass of the cord in the calculation.

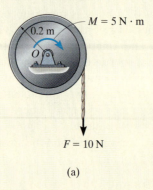

$M = 5$ N·m

0.2 m

O

$F = 10$ N

(a)

SOLUTION

Kinetic Energy.   Since the disk rotates about a fixed axis, and it is initially at rest, then

$$T_1 = 0$$

$$T_2 = \tfrac{1}{2}I_O\omega_2^2 = \tfrac{1}{2}\left[\tfrac{1}{2}(30\text{ kg})(0.2\text{ m})^2\right](20\text{ rad/s})^2 = 120\text{ J}$$

Work (Free–Body Diagram).   As shown in Fig. 18–12b, the pin reactions $\mathbf{O}_x$ and $\mathbf{O}_y$ and the weight (294.3 N) do no work, since they are not displaced. The *couple moment*, having a constant magnitude, does positive work $U_M = M\theta$ as the disk *rotates* through a clockwise angle of θ rad, and the *constant force* $\mathbf{F}$ does positive work $U_{Fc} = Fs$ as the cord moves downward $s = \theta r = \theta(0.2\text{ m})$.

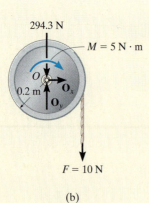

294.3 N

$M = 5$ N·m

O

0.2 m $\mathbf{O}_x$

$\mathbf{O}_y$

$F = 10$ N

(b)

Fig. 18–12

Principle of Work and Energy.

$$\{T_1\} + \{\Sigma U_{1-2}\} = \{T_2\}$$

$$\{T_1\} + \{M\theta + Fs\} = \{T_2\}$$

$$\{0\} + \{(5\text{ N·m})\theta + (10\text{ N})\theta(0.2\text{ m})\} = \{120\text{ J}\}$$

$$\theta = 17.14\text{ rad} = 17.14\text{ rad}\left(\frac{1\text{ rev}}{2\pi\text{ rad}}\right) = 2.73\text{ rev} \qquad Ans.$$

EXAMPLE 18.3

The wheel shown in Fig. 18–13a weighs 40 lb and has a radius of gyration $k_G = 0.6$ ft about its mass center G. If it is subjected to a clockwise couple moment of 15 lb·ft and rolls from rest without slipping, determine its angular velocity after its center G moves 0.5 ft. The spring has a stiffness $k = 10$ lb/ft and is initially unstretched when the couple moment is applied.

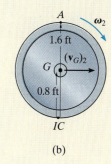

(a)

SOLUTION

Kinetic Energy (Kinematic Diagram). Since the wheel is initially at rest,

$$T_1 = 0$$

The kinematic diagram of the wheel when it is in the final position is shown in Fig. 18–13b. The final kinetic energy is determined from

$$T_2 = \tfrac{1}{2}I_{IC}\omega_2^2$$

$$= \frac{1}{2}\left[\frac{40\ \text{lb}}{32.2\ \text{ft/s}^2}(0.6\ \text{ft})^2 + \left(\frac{40\ \text{lb}}{32.2\ \text{ft/s}^2}\right)(0.8\ \text{ft})^2\right]\omega_2^2$$

$$T_2 = 0.6211\ \omega_2^2$$

Work (Free–Body Diagram). As shown in Fig. 18–13c, only the spring force $\mathbf{F}_s$ and the couple moment do work. The normal force does not move along its line of action and the frictional force does *no* work, since the wheel does not slip as it rolls.

The work of $\mathbf{F}_s$ is found using $U_s = -\tfrac{1}{2}ks^2$. Here the work is negative since $\mathbf{F}_s$ is in the opposite direction to displacement. Since the wheel does not slip when the center G moves 0.5 ft, then the wheel rotates $\theta = s_G/r_{G/IC} = 0.5\ \text{ft}/0.8\ \text{ft} = 0.625$ rad, Fig. 18–13b. Hence, the spring stretches $s = \theta r_{A/IC} = (0.625\ \text{rad})(1.6\ \text{ft}) = 1$ ft.

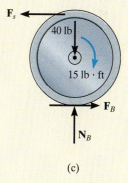

(c)

Fig. 18–13

Principle of Work and Energy.

$$\{T_1\} + \{\Sigma U_{1-2}\} = \{T_2\}$$

$$\{T_1\} + \left\{M\theta - \tfrac{1}{2}ks^2\right\} = \{T_2\}$$

$$\{0\} + \left\{15\ \text{lb·ft}(0.625\ \text{rad}) - \frac{1}{2}(10\ \text{lb/ft})(1\ \text{ft})^2\right\} = \{0.6211\ \omega_2^2\ \text{ft·lb}\}$$

$$\omega_2 = 2.65\ \text{rad/s} \circlearrowright \qquad\qquad Ans.$$

18

EXAMPLE 18.4

The 700-kg pipe is equally suspended from the two tines of the fork lift shown in the photo. It is undergoing a swinging motion such that when $\theta = 30°$ it is momentarily at rest. Determine the normal and frictional forces acting on each tine which are needed to support the pipe at the instant $\theta = 0°$. Measurements of the pipe and the suspender are shown in Fig. 18–14a. Neglect the mass of the suspender and the thickness of the pipe.

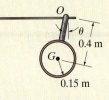

(a)

Fig. 18–14

SOLUTION

We must use the equations of motion to find the forces on the tines since these forces do no work. Before doing this, however, we will apply the principle of work and energy to determine the angular velocity of the pipe when $\theta = 0°$.

Kinetic Energy (Kinematic Diagram). Since the pipe is originally at rest, then

$$T_1 = 0$$

The final kinetic energy may be computed with reference to either the fixed point O or the center of mass G. For the calculation we will consider the pipe to be a thin ring so that $I_G = mr^2$. If point G is considered, we have

$$T_2 = \tfrac{1}{2}m(v_G)_2^2 + \tfrac{1}{2}I_G\omega_2^2$$
$$= \tfrac{1}{2}(700 \text{ kg})[(0.4 \text{ m})\omega_2]^2 + \tfrac{1}{2}[700 \text{ kg}(0.15 \text{ m})^2]\omega_2^2$$
$$= 63.875\omega_2^2$$

If point O is considered then the parallel-axis theorem must be used to determine I_O. Hence,

$$T_2 = \tfrac{1}{2}I_O\omega_2^2 = \tfrac{1}{2}[700 \text{ kg}(0.15 \text{ m})^2 + 700 \text{ kg}(0.4 \text{ m})^2]\omega_2^2$$
$$= 63.875\omega_2^2$$

Work (Free-Body Diagram). Fig. 18–14*b*. The normal and frictional forces on the tines do no work since they do not move as the pipe swings. The weight does positive work since the weight moves downward through a vertical distance $\Delta y = 0.4 \text{ m} - 0.4 \cos 30° \text{ m} = 0.05359 \text{ m}$.

Principle of Work and Energy.

$$\{T_1\} + \{\Sigma U_{1-2}\} = \{T_2\}$$

$$\{0\} + \{700(9.81) \text{ N}(0.05359 \text{ m})\} = \{63.875\omega_2^2\}$$

$$\omega_2 = 2.400 \text{ rad/s}$$

Equations of Motion. Referring to the free-body and kinetic diagrams shown in Fig. 18–14*c*, and using the result for ω_2, we have

$$\xleftarrow{+} \Sigma F_t = m(a_G)_t; \quad F_T = (700 \text{ kg})(a_G)_t$$

$$+\uparrow \Sigma F_n = m(a_G)_n; \quad N_T - 700(9.81) \text{ N} = (700 \text{ kg})(2.400 \text{ rad/s})^2(0.4 \text{ m})$$

$$\zeta + \Sigma M_O = I_O \alpha; \quad 0 = [(700 \text{ kg})(0.15 \text{ m})^2 + (700 \text{ kg})(0.4 \text{ m})^2]\alpha$$

Since $(a_G)_t = (0.4 \text{ m})\alpha$, then

$$\alpha = 0, \quad (a_G)_t = 0$$

$$F_T = 0$$

$$N_T = 8.480 \text{ kN}$$

There are two tines used to support the load, therefore

$$F'_T = 0 \qquad\qquad\qquad Ans.$$

$$N'_T = \frac{8.480 \text{ kN}}{2} = 4.24 \text{ kN} \qquad\qquad Ans.$$

NOTE: Due to the swinging motion the tines are subjected to a *greater* normal force than would be the case if the load were static, in which case $N'_T = 700(9.81) \text{ N}/2 = 3.43 \text{ kN}$.

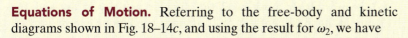

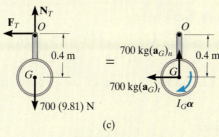

(c)

Fig. 18–14

EXAMPLE | 18.5

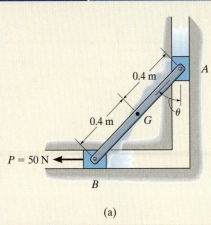

(a)

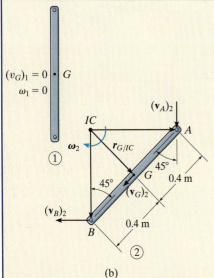

(b)

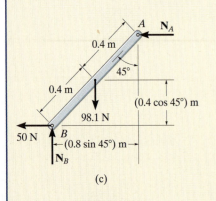

(c)

Fig. 18–15

The 10–kg rod shown in Fig. 18–15a is constrained so that its ends move along the grooved slots. The rod is initially at rest when $\theta = 0°$. If the slider block at B is acted upon by a horizontal force $P = 50$ N, determine the angular velocity of the rod at the instant $\theta = 45°$. Neglect friction and the mass of blocks A and B.

SOLUTION

Why can the principle of work and energy be used to solve this problem?

Kinetic Energy (Kinematic Diagrams). Two kinematic diagrams of the rod, when it is in the initial position 1 and final position 2, are shown in Fig. 18–15b. When the rod is in position 1, $T_1 = 0$ since $(\mathbf{v}_G)_1 = \boldsymbol{\omega}_1 = \mathbf{0}$. In position 2 the angular velocity is $\boldsymbol{\omega}_2$ and the velocity of the mass center is $(\mathbf{v}_G)_2$. Hence, the kinetic energy is

$$T_2 = \tfrac{1}{2}m(v_G)_2^2 + \tfrac{1}{2}I_G\omega_2^2$$
$$= \tfrac{1}{2}(10 \text{ kg})(v_G)_2^2 + \tfrac{1}{2}\Big[\tfrac{1}{12}(10 \text{ kg})(0.8 \text{ m})^2\Big]\omega_2^2$$
$$= 5(v_G)_2^2 + 0.2667(\omega_2)^2$$

The two unknowns $(v_G)_2$ and ω_2 can be related from the instantaneous center of zero velocity for the rod. Fig. 18–15b. It is seen that as A moves downward with a velocity $(\mathbf{v}_A)_2$, B moves horizontally to the left with a velocity $(\mathbf{v}_B)_2$. Knowing these directions, the IC is located as shown in the figure. Hence,

$$(v_G)_2 = r_{G/IC}\omega_2 = (0.4 \tan 45° \text{ m})\omega_2$$
$$= 0.4\omega_2$$

Therefore,

$$T_2 = 0.8\omega_2^2 + 0.2667\omega_2^2 = 1.0667\omega_2^2$$

Of course, we can also determine this result using $T_2 = \tfrac{1}{2}I_{IC}\omega_2^2$.

Work (Free–Body Diagram). Fig. 18–15c. The normal forces $\mathbf{N}_A$ and $\mathbf{N}_B$ do no work as the rod is displaced. Why? The 98.1-N weight is displaced a vertical distance of $\Delta y = (0.4 - 0.4 \cos 45°)$ m; whereas the 50-N force moves a horizontal distance of $s = (0.8 \sin 45°)$ m. Both of these forces do positive work. Why?

Principle of Work and Energy.

$$\{T_1\} + \{\Sigma U_{1-2}\} = \{T_2\}$$
$$\{T_1\} + \{W \, \Delta y + Ps\} = \{T_2\}$$
$$\{0\} + \{98.1 \text{ N}(0.4 \text{ m} - 0.4 \cos 45° \text{ m}) + 50 \text{ N}(0.8 \sin 45° \text{ m})\}$$
$$= \{1.0667\omega_2^2 \text{ J}\}$$

Solving for ω_2 gives

$$\omega_2 = 6.11 \text{ rad/s} \qquad\qquad Ans.$$

FUNDAMENTAL PROBLEMS

F18–1. The 80-kg wheel has a radius of gyration about its mass center O of $k_O = 400$ mm. Determine its angular velocity after it has rotated 20 revolutions starting from rest.

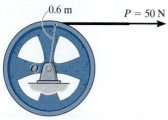

F18–1

F18–2. The uniform 50-lb slender rod is subjected to a couple moment of $M = 100$ lb · ft. If the rod is at rest when $\theta = 0°$, determine its angular velocity when $\theta = 90°$.

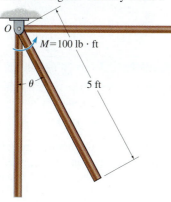

F18–2

F18–3. The uniform 50-kg slender rod is at rest in the position shown when $P = 600$ N is applied. Determine the angular velocity of the rod when the rod reaches the vertical position.

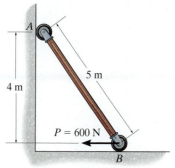

F18–3

F18–4. The 50-kg wheel is subjected to a force of 50 N. If the wheel starts from rest and rolls without slipping, determine its angular velocity after it has rotated 10 revolutions. The radius of gyration of the wheel about its mass center O is $k_O = 0.3$ m.

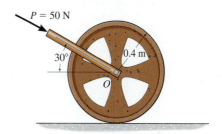

F18–4

F18–5. If the uniform 30-kg slender rod starts from rest at the position shown, determine its angular velocity after it has rotated 4 revolutions. The forces remain perpendicular to the rod.

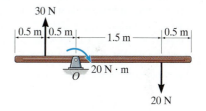

F18–5

F18–6. The 20-kg wheel has a radius of gyration about its center O of $k_O = 300$ mm. When it is subjected to a couple moment of $M = 50$ N · m, it rolls without slipping. Determine the angular velocity of the wheel after its center O has traveled through a distance of $s_O = 20$ m, starting from rest.

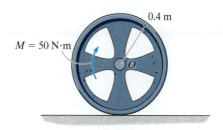

F18–6

18

PROBLEMS

•**18–1.** At a given instant the body of mass m has an angular velocity $\boldsymbol{\omega}$ and its mass center has a velocity $\mathbf{v}_G$. Show that its kinetic energy can be represented as $T = \frac{1}{2}I_{IC}\omega^2$, where I_{IC} is the moment of inertia of the body computed about the instantaneous axis of zero velocity, located a distance $r_{G/IC}$ from the mass center as shown.

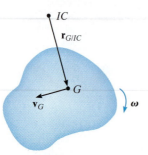

Prob. 18–1

18–2. The double pulley consists of two parts that are attached to one another. It has a weight of 50 lb and a radius of gyration about its center of $k_O = 0.6$ ft. If it rotates with an angular velocity of 20 rad/s clockwise, determine the kinetic energy of the system. Assume that neither cable slips on the pulley.

Prob. 18–2

18–3. A force of $P = 20$ N is applied to the cable, which causes the 175-kg reel to turn without slipping on the two rollers A and B of the dispenser. Determine the angular velocity of the reel after it has rotated two revolutions starting from rest. Neglect the mass of the cable. Each roller can be considered as an 18-kg cylinder, having a radius of 0.1 m. The radius of gyration of the reel about its center axis is $k_G = 0.42$ m.

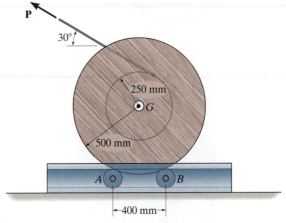

Prob. 18–3

***18–4.** The spool of cable, originally at rest, has a mass of 200 kg and a radius of gyration of $k_G = 325$ mm. If the spool rests on two small rollers A and B and a constant horizontal force of $P = 400$ N is applied to the end of the cable, determine the angular velocity of the spool when 8 m of cable has been unwound. Neglect friction and the mass of the rollers and unwound cable.

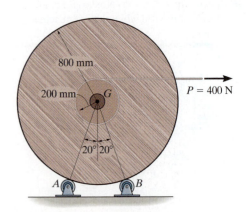

Prob. 18–4

•18–5.   The pendulum of the Charpy impact machine has a mass of 50 kg and a radius of gyration of $k_A = 1.75$ m. If it is released from rest when $\theta = 0°$, determine its angular velocity just before it strikes the specimen $S, \theta = 90°$.

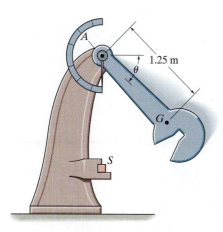

Prob. 18–5

18–6.   The two tugboats each exert a constant force **F** on the ship. These forces are always directed perpendicular to the ship's centerline. If the ship has a mass m and a radius of gyration about its center of mass G of k_G, determine the angular velocity of the ship after it turns 90°. The ship is originally at rest.

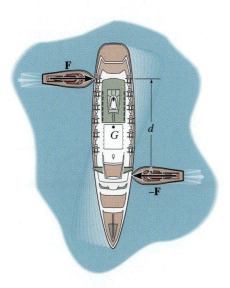

Prob. 18–6

18–7.   The drum has a mass of 50 kg and a radius of gyration about the pin at O of $k_O = 0.23$ m. Starting from rest, the suspended 15-kg block B is allowed to fall 3 m without applying the brake ACD. Determine the speed of the block at this instant. If the coefficient of kinetic friction at the brake pad C is $\mu_k = 0.5$, determine the force **P** that must be applied at the brake handle which will then stop the block after it descends *another* 3 m. Neglect the thickness of the handle.

*18–8.   The drum has a mass of 50 kg and a radius of gyration about the pin at O of $k_O = 0.23$ m. If the 15-kg block is moving downward at 3 m/s, and a force of $P = 100$ N is applied to the brake arm, determine how far the block descends from the instant the brake is applied until it stops. Neglect the thickness of the handle. The coefficient of kinetic friction at the brake pad is $\mu_k = 0.5$.

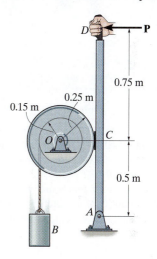

Probs. 18–7/8

•18–9.   The spool has a weight of 150 lb and a radius of gyration $k_O = 2.25$ ft. If a cord is wrapped around its inner core and the end is pulled with a horizontal force of $P = 40$ lb, determine the angular velocity of the spool after the center O has moved 10 ft to the right. The spool starts from rest and does not slip at A as it rolls. Neglect the mass of the cord.

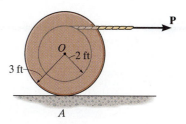

Prob. 18–9

18–10. A man having a weight of 180 lb sits in a chair of the Ferris wheel, which, excluding the man, has a weight of 15 000 lb and a radius of gyration $k_O = 37$ ft. If a torque $M = 80(10^3)$ lb·ft is applied about O, determine the angular velocity of the wheel after it has rotated 180°. Neglect the weight of the chairs and note that the man remains in an upright position as the wheel rotates. The wheel starts from rest in the position shown.

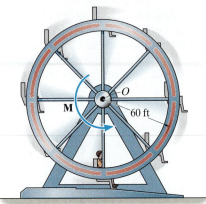

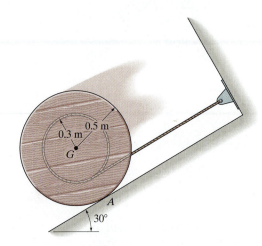

Prob. 18–10

18–11. A man having a weight of 150 lb crouches down on the end of a diving board as shown. In this position the radius of gyration about his center of gravity is $k_G = 1.2$ ft. While holding this position at $\theta = 0°$, he rotates about his toes at A until he loses contact with the board when $\theta = 90°$. If he remains rigid, determine approximately how many revolutions he makes before striking the water after falling 30 ft.

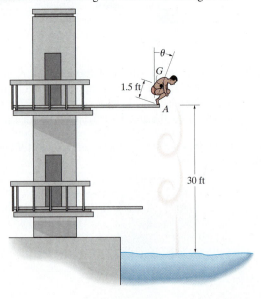

Prob. 18–11

*18–12. The spool has a mass of 60 kg and a radius of gyration $k_G = 0.3$ m. If it is released from rest, determine how far its center descends down the smooth plane before it attains an angular velocity of $\omega = 6$ rad/s. Neglect friction and the mass of the cord which is wound around the central core.

•18–13. Solve Prob. 18–12 if the coefficient of kinetic friction between the spool and plane at A is $\mu_k = 0.2$.

Probs. 18–12/13

18–14. The spool has a weight of 500 lb and a radius of gyration of $k_G = 1.75$ ft. A horizontal force of $P = 15$ lb is applied to the cable wrapped around its inner core. If the spool is originally at rest, determine its angular velocity after the mass center G has moved 6 ft to the left. The spool rolls without slipping. Neglect the mass of the cable.

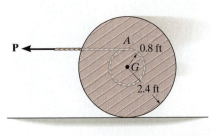

Prob. 18–14

18–15. If the system is released from rest, determine the speed of the 20-kg cylinders A and B after A has moved downward a distance of 2 m. The differential pulley has a mass of 15 kg with a radius of gyration about its center of mass of $k_O = 100$ mm.

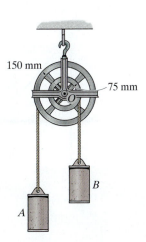

150 mm

75 mm

O

B

A

Prob. 18–15

•18–17. The 6-kg lid on the box is held in equilibrium by the torsional spring at $\theta = 60°$. If the lid is forced closed, $\theta = 0°$, and then released, determine its angular velocity at the instant it opens to $\theta = 45°$.

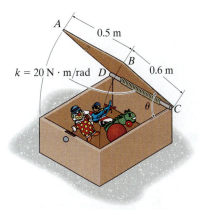

A

0.5 m

B

0.6 m

$k = 20$ N $\cdot$ m/rad D

θ

C

Prob. 18–17

18–18. The wheel and the attached reel have a combined weight of 50 lb and a radius of gyration about their center of $k_A = 6$ in. If pulley B attached to the motor is subjected to a torque of $M = 40(2 - e^{-0.1\theta})$ lb $\cdot$ ft, where θ is in radians, determine the velocity of the 200-lb crate after it has moved upwards a distance of 5 ft, starting from rest. Neglect the mass of pulley B.

18–19. The wheel and the attached reel have a combined weight of 50 lb and a radius of gyration about their center of $k_A = 6$ in. If pulley B that is attached to the motor is subjected to a torque of $M = 50$ lb $\cdot$ ft, determine the velocity of the 200-lb crate after the pulley has turned 5 revolutions. Neglect the mass of the pulley.

***18–16.** If the motor M exerts a constant force of $P = 300$ N on the cable wrapped around the reel's outer rim, determine the velocity of the 50-kg cylinder after it has traveled a distance of 2 m. Initially, the system is at rest. The reel has a mass of 25 kg, and the radius of gyration about its center of mass A is $k_A = 125$ mm.

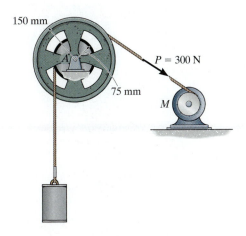

150 mm

A

$P = 300$ N

75 mm

M

Prob. 18–16

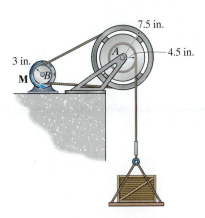

7.5 in.

4.5 in.

3 in.

M B

A

Probs. 18–18/19

18

*18–20.   The 30-lb ladder is placed against the wall at an angle of $\theta = 45°$ as shown. If it is released from rest, determine its angular velocity at the instant just before $\theta = 0°$. Neglect friction and assume the ladder is a uniform slender rod.

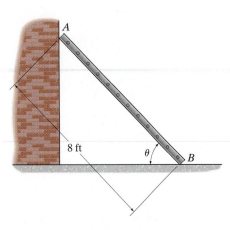

Prob. 18–20

•18–21.   Determine the angular velocity of the two 10-kg rods when $\theta = 180°$ if they are released from rest in the position $\theta = 60°$. Neglect friction.

18–22.   Determine the angular velocity of the two 10-kg rods when $\theta = 90°$ if they are released from rest in the position $\theta = 60°$. Neglect friction.

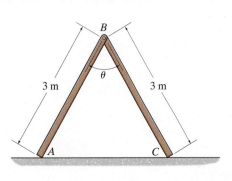

Probs. 18–21/22

18–23.   If the 50-lb bucket is released from rest, determine its velocity after it has fallen a distance of 10 ft. The windlass A can be considered as a 30-lb cylinder, while the spokes are slender rods, each having a weight of 2 lb. Neglect the pulley's weight.

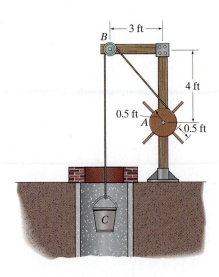

Prob. 18–23

*18–24.   If corner A of the 60-kg plate is subjected to a vertical force of $P = 500$ N, and the plate is released from rest when $\theta = 0°$, determine the angular velocity of the plate when $\theta = 45°$.

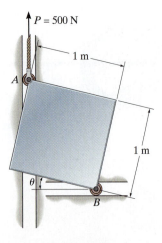

Prob. 18–24

•18–25.   The spool has a mass of 100 kg and a radius of gyration of 400 mm about its center of mass O. If it is released from rest, determine its angular velocity after its center O has moved down the plane a distance of 2 m. The contact surface between the spool and the inclined plane is smooth.

18–26.   The spool has a mass of 100 kg and a radius of gyration of 400 mm about its center of mass O. If it is released from rest, determine its angular velocity after its center O has moved down the plane a distance of 2 m. The coefficient of kinetic friction between the spool and the inclined plane is $\mu_k = 0.15$.

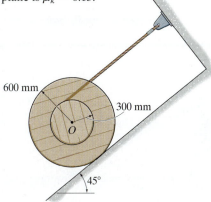

600 mm

300 mm

O

45°

Probs. 18–25/26

18–27.   The uniform door has a mass of 20 kg and can be treated as a thin plate having the dimensions shown. If it is connected to a torsional spring at A, which has a stiffness of $k = 80\ \text{N} \cdot \text{m/rad}$, determine the required initial twist of the spring in radians so that the door has an angular velocity of 12 rad/s when it closes at $\theta = 0°$ after being opened at $\theta = 90°$ and released from rest. *Hint:* For a torsional spring $M = k\theta$, when k is the stiffness and θ is the angle of twist.

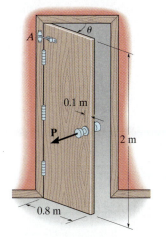

A

θ

0.1 m

P

2 m

0.8 m

Prob. 18–27

*18–28.   The 50-lb cylinder A is descending with a speed of 20 ft/s when the brake is applied. If wheel B must be brought to a stop after it has rotated 5 revolutions, determine the constant force **P** that must be applied to the brake arm. The coefficient of kinetic friction between the brake pad C and the wheel is $\mu_k = 0.5$. The wheel's weight is 25 lb, and the radius of gyration about its center of mass is $k = 0.6$ ft.

•18–29.   When a force of $P = 30$ lb is applied to the brake arm, the 50-lb cylinder A is descending with a speed of 20 ft/s. Determine the number of revolutions wheel B will rotate before it is brought to a stop. The coefficient of kinetic friction between the brake pad C and the wheel is $\mu_k = 0.5$. The wheel's weight is 25 lb, and the radius of gyration about its center of mass is $k = 0.6$ ft.

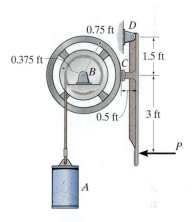

0.75 ft   D

0.375 ft

1.5 ft

B

C

0.5 ft

3 ft

P

A

Probs. 18–28/29

18

18–30.   The 100-lb block is transported a short distance by using two cylindrical rollers, each having a weight of 35 lb. If a horizontal force $P = 25$ lb is applied to the block, determine the block's speed after it has been displaced 2 ft to the left. Originally the block is at rest. No slipping occurs.

P = 25 lb

1.5 ft

1.5 ft

Prob. 18–30

18–31. The slender beam having a weight of 150 lb is supported by two cables. If the cable at end B is cut so that the beam is released from rest when $\theta = 30°$, determine the speed at which end A strikes the wall. Neglect friction at B.

18–33. The beam has a weight of 1500 lb and is being raised to a vertical position by pulling very slowly on its bottom end A. If the cord fails when $\theta = 60°$ and the beam is essentially at rest, determine the speed of A at the instant cord BC becomes vertical. Neglect friction and the mass of the cords, and treat the beam as a slender rod.

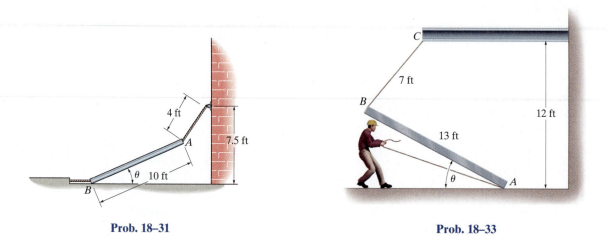

Prob. 18–31

Prob. 18–33

*18–32. The assembly consists of two 15-lb slender rods and a 20-lb disk. If the spring is unstretched when $\theta = 45°$ and the assembly is released from rest at this position, determine the angular velocity of rod AB at the instant $\theta = 0°$. The disk rolls without slipping.

18–34. The uniform slender bar that has a mass m and a length L is subjected to a uniform distributed load w_0, which is always directed perpendicular to the axis of the bar. If the bar is released from rest from the position shown, determine its angular velocity at the instant it has rotated 90°. Solve the problem for rotation in (a) the horizontal plane, and (b) the vertical plane.

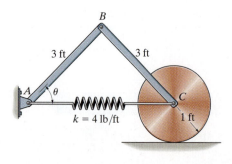

Prob. 18–32

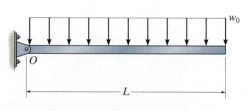

Prob. 18–34

18.5   Conservation of Energy

When a force system acting on a rigid body consists only of *conservative forces*, the conservation of energy theorem can be used to solve a problem that otherwise would be solved using the principle of work and energy. This theorem is often easier to apply since the work of a conservative force is *independent of the path* and depends only on the initial and final positions of the body. It was shown in Sec. 14.5 that the work of a conservative force can be expressed as the difference in the body's potential energy measured from an arbitrarily selected reference or datum.

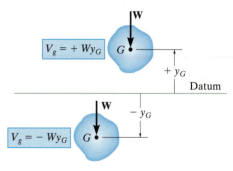

Gravitational potential energy

Fig. 18–16

Gravitational Potential Energy.   Since the total weight of a body can be considered concentrated at its center of gravity, the *gravitational potential energy* of the body is determined by knowing the height of the body's center of gravity above or below a horizontal datum.

$$V_g = W y_G \qquad\qquad (18\text{–}15)$$

Here the potential energy is *positive* when y_G is positive upward, since the weight has the ability to do *positive work* when the body moves back to the datum, Fig. 18–16. Likewise, if G is located *below* the datum $(-y_G)$, the gravitational potential energy is *negative*, since the weight does *negative work* when the body returns to the datum.

Elastic Potential Energy.   The force developed by an elastic spring is also a conservative force. The *elastic potential energy* which a spring imparts to an attached body when the spring is stretched or compressed from an initial undeformed position $(s = 0)$ to a final position s, Fig. 18–17, is

$$V_e = +\tfrac{1}{2}ks^2 \qquad\qquad (18\text{–}16)$$

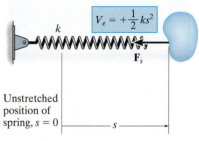

Elastic potential energy

Fig. 18–17

In the deformed position, the spring force acting *on the body* always has the ability for doing positive work when the spring returns back to its original undeformed position (see Sec. 14.5).

EXAMPLE 18.6

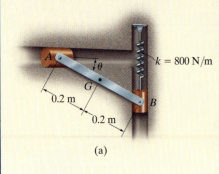

(a)

The 10-kg rod AB shown in Fig. 18–18a is confined so that its ends move in the horizontal and vertical slots. The spring has a stiffness of $k = 800$ N/m and is unstretched when $\theta = 0°$. Determine the angular velocity of AB when $\theta = 0°$, if the rod is released from rest when $\theta = 30°$. Neglect the mass of the slider blocks.

SOLUTION

Potential Energy. The two diagrams of the rod, when it is located at its initial and final positions, are shown in Fig. 18–18b. The datum, used to measure the gravitational potential energy, is placed in line with the rod when $\theta = 0°$.

When the rod is in position 1, the center of gravity G is located *below the datum* so its gravitational potential energy is *negative*. Furthermore, (positive) elastic potential energy is stored in the spring, since it is stretched a distance of $s_1 = (0.4 \sin 30°)$ m. Thus,

$$V_1 = -Wy_1 + \tfrac{1}{2}ks_1^2$$

$$= -(98.1\text{ N})(0.2\sin 30°\text{ m}) + \tfrac{1}{2}(800\text{ N/m})(0.4\sin 30°\text{ m})^2 = 6.19\text{ J}$$

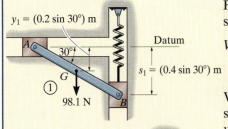

When the rod is in position 2, the potential energy of the rod is zero, since the center of gravity G is located at the datum, and the spring is unstretched, $s_2 = 0$. Thus,

$$V_2 = 0$$

Kinetic Energy. The rod is released from rest from position 1, thus $(\mathbf{v}_G)_1 = \boldsymbol{\omega}_1 = \mathbf{0}$, and so

$$T_1 = 0$$

In position 2, the angular velocity is $\boldsymbol{\omega}_2$ and the rod's mass center has a velocity of $(\mathbf{v}_G)_2$. Thus,

$$T_2 = \tfrac{1}{2}m(v_G)_2^2 + \tfrac{1}{2}I_G\omega_2^2$$

$$= \tfrac{1}{2}(10\text{ kg})(v_G)_2^2 + \tfrac{1}{2}\left[\tfrac{1}{12}(10\text{ kg})(0.4\text{ m})^2\right]\omega_2^2$$

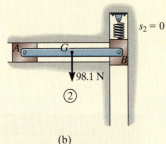

(c)

Using *kinematics*, $(\mathbf{v}_G)_2$ can be related to $\boldsymbol{\omega}_2$ as shown in Fig. 18–18c. At the instant considered, the instantaneous center of zero velocity (IC) for the rod is at point A; hence, $(v_G)_2 = (r_{G/IC})\omega_2 = (0.2\text{ m})\omega_2$. Substituting into the above expression and simplifying (or using $\tfrac{1}{2}I_{IC}\omega_2^2$), we get

$$T_2 = 0.2667\omega_2^2$$

Conservation of Energy.

$$\{T_1\} + \{V_1\} = \{T_2\} + \{V_2\}$$

$$\{0\} + \{6.19\text{ J}\} = \{0.2667\omega_2^2\} + \{0\}$$

$$\omega_2 = 4.82\text{ rad/s} \circlearrowleft \qquad\qquad Ans.$$

Fig. 18–18

18

EXAMPLE 18.7

The wheel shown in Fig. 18–19a has a weight of 30 lb and a radius of gyration of $k_G = 0.6$ ft. It is attached to a spring which has a stiffness $k = 2$ lb/ft and an unstretched length of 1 ft. If the disk is released from rest in the position shown and rolls without slipping, determine its angular velocity at the instant G moves 3 ft to the left.

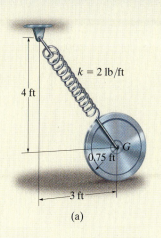

(a)

SOLUTION

Potential Energy. Two diagrams of the wheel, when it at the initial and final positions, are shown in Fig. 18–19b. A gravitational datum is not needed here since the weight is not displaced vertically. From the problem geometry the spring is stretched $s_1 = \left(\sqrt{3^2 + 4^2} - 1\right) = 4$ ft in the initial position, and $s_2 = (4 - 1) = 3$ ft in the final position. Hence,

$$V_1 = \tfrac{1}{2}ks_1^2 = \tfrac{1}{2}(2 \text{ lb/ft})(4 \text{ ft})^2 = 16 \text{ J}$$
$$V_2 = \tfrac{1}{2}ks_2^2 = \tfrac{1}{2}(2 \text{ lb/ft})(3 \text{ ft})^2 = 9 \text{ J}$$

Kinetic Energy. The disk is released from rest and so $(\mathbf{v}_G)_1 = \mathbf{0}$, $\boldsymbol{\omega}_1 = \mathbf{0}$. Therefore,

$$T_1 = 0$$

Since the instantaneous center of zero velocity is at the ground, Fig. 18–19c, we have

$$T_2 = \frac{1}{2}I_{IC}\omega_2^2$$

$$= \frac{1}{2}\left[\left(\frac{30 \text{ lb}}{32.2 \text{ ft/s}^2}\right)(0.6 \text{ ft})^2 + \left(\frac{30 \text{ lb}}{32.2 \text{ ft/s}^2}\right)(0.75 \text{ ft})^2\right]\omega_2^2$$

$$= 0.4297\omega_2^2$$

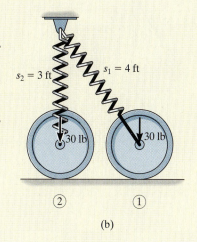

(b)

Conservation of Energy.

$$\{T_1\} + \{V_1\} = \{T_2\} + \{V_2\}$$
$$\{0\} + \{16 \text{ J}\} = \{0.4297\omega_2^2\} + \{9 \text{ J}\}$$
$$\omega_2 = 4.04 \text{ rad/s} \; \circlearrowright \qquad\qquad \textit{Ans.}$$

NOTE: If the principle of work and energy were used to solve this problem, then the work of the spring would have to be determined by considering both the change in magnitude and direction of the spring force.

(c)

Fig. 18–19

18

EXAMPLE 18.8

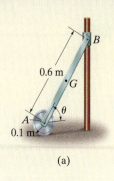

0.6 m

·G

θ

A

0.1 m

(a)

The 10-kg homogeneous disk shown in Fig. 18–20a is attached to a uniform 5-kg rod AB. If the assembly is released from rest when $\theta = 60°$, determine the angular velocity of the rod when $\theta = 0°$. Assume that the disk rolls without slipping. Neglect friction along the guide and the mass of the collar at B.

SOLUTION

Potential Energy. Two diagrams for the rod and disk, when they are located at their initial and final positions, are shown in Fig. 18–20b. For convenience the datum passes through point A.

When the system is in position 1, only the rod's weight has positive potential energy. Thus,

$$V_1 = W_r y_1 = (49.05 \text{ N})(0.3 \sin 60° \text{ m}) = 12.74 \text{ J}$$

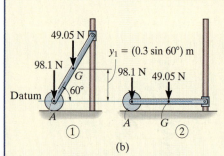

(b)

When the system is in position 2, both the weight of the rod and the weight of the disk have zero potential energy. Why? Thus,

$$V_2 = 0$$

Kinetic Energy. Since the entire system is at rest at the initial position,

$$T_1 = 0$$

In the final position the rod has an angular velocity $(\boldsymbol{\omega}_r)_2$ and its mass center has a velocity $(\mathbf{v}_G)_2$, Fig. 18–20c. Since the rod is *fully extended* in this position, the disk is momentarily at rest, so $(\boldsymbol{\omega}_d)_2 = \mathbf{0}$ and $(\mathbf{v}_A)_2 = \mathbf{0}$. For the rod $(\mathbf{v}_G)_2$ can be related to $(\boldsymbol{\omega}_r)_2$ from the instantaneous center of zero velocity, which is located at point A, Fig. 18–20c. Hence, $(v_G)_2 = r_{G/IC}(\omega_r)_2$ or $(v_G)_2 = 0.3(\omega_r)_2$. Thus,

$$T_2 = \frac{1}{2}m_r(v_G)_2^2 + \frac{1}{2}I_G(\omega_r)_2^2 + \frac{1}{2}m_d(v_A)_2^2 + \frac{1}{2}I_A(\omega_d)_2^2$$

$$= \frac{1}{2}(5 \text{ kg})[(0.3 \text{ m})(\omega_r)_2]^2 + \frac{1}{2}\left[\frac{1}{12}(5 \text{ kg})(0.6 \text{ m})^2\right](\omega_r)_2^2 + 0 + 0$$

$$= 0.3(\omega_r)_2^2$$

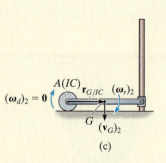

(c)

Fig. 18–20

Conservation of Energy.

$$\{T_1\} + \{V_1\} = \{T_2\} + \{V_2\}$$

$$\{0\} + \{12.74 \text{ J}\} = \{0.3(\omega_R)_2^2\} + \{0\}$$

$$(\omega_r)_2 = 6.52 \text{ rad/s} \circlearrowright \qquad \qquad Ans.$$

NOTE: We can also determine the final kinetic energy of the rod using $T_2 = \frac{1}{2}I_{IC}\omega_2^2$.

FUNDAMENTAL PROBLEMS

F18–7. If the 30-kg disk is released from rest when $\theta = 0°$, determine its angular velocity when $\theta = 90°$.

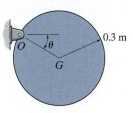

F18–7

F18–8. The 50-kg reel has a radius of gyration about its center O of $k_O = 300$ mm. If it is released from rest, determine its angular velocity when its center O has traveled 6 m down the smooth inclined plane.

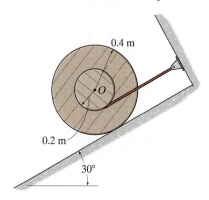

F18–8

F18–9. The 60-kg rod OA is released from rest when $\theta = 0°$. Determine its angular velocity when $\theta = 45°$. The spring remains vertical during the motion and is unstretched when $\theta = 0°$.

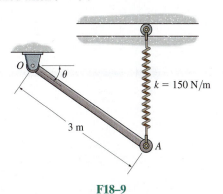

F18–9

F18–10. The 30-kg rod is released from rest when $\theta = 0°$. Determine the angular velocity of the rod when $\theta = 90°$. The spring is unstretched when $\theta = 0°$.

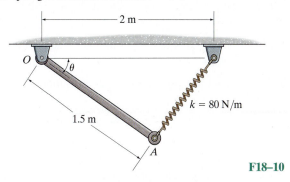

F18–10

F18–11. The 30-kg rod is released from rest when $\theta = 45°$. Determine the angular velocity of the rod when $\theta = 0°$. The spring is unstretched when $\theta = 45°$.

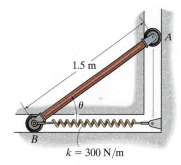

F18–11

F18–12. The 20-kg rod is released from rest when $\theta = 0°$. Determine its angular velocity when $\theta = 90°$. The spring has an unstretched length of 0.5 m.

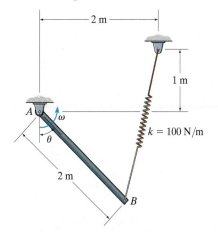

F18–12

PROBLEMS

18–35. Solve Prob. 18–5 using the conservation of energy equation.

***18–36.** Solve Prob. 18–12 using the conservation of energy equation.

•**18–37.** Solve Prob. 18–32 using the conservation of energy equation.

18–38. Solve Prob. 18–31 using the conservation of energy equation.

18–39. Solve Prob. 18–11 using the conservation of energy equation.

***18–40.** At the instant shown, the 50-lb bar rotates clockwise at 2 rad/s. The spring attached to its end always remains vertical due to the roller guide at C. If the spring has an unstretched length of 2 ft and a stiffness of $k = 6$ lb/ft, determine the angular velocity of the bar the instant it has rotated 30° clockwise.

•**18–41.** At the instant shown, the 50-lb bar rotates clockwise at 2 rad/s. The spring attached to its end always remains vertical due to the roller guide at C. If the spring has an unstretched length of 2 ft and a stiffness of $k = 12$ lb/ft, determine the angle θ, measured from the horizontal, to which the bar rotates before it momentarily stops.

18–42. A chain that has a negligible mass is draped over the sprocket which has a mass of 2 kg and a radius of gyration of $k_O = 50$ mm. If the 4-kg block A is released from rest from the position $s = 1$ m, determine the angular velocity of the sprocket at the instant $s = 2$ m.

18–43. Solve Prob. 18–42 if the chain has a mass per unit length of 0.8 kg/m. For the calculation neglect the portion of the chain that wraps over the sprocket.

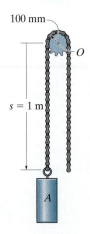

Probs. 18–42/43

***18–44.** The system consists of 60-lb and 20-lb blocks A and B, respectively, and 5-lb pulleys C and D that can be treated as thin disks. Determine the speed of block A after block B has risen 5 ft, starting from rest. Assume that the cord does not slip on the pulleys, and neglect the mass of the cord.

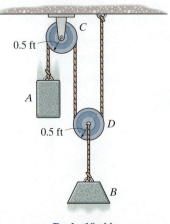

Prob. 18–44

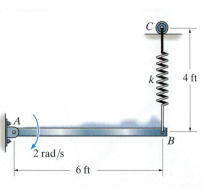

Probs. 18–40/41

•18–45. The system consists of a 20-lb disk A, 4-lb slender rod BC, and a 1-lb smooth collar C. If the disk rolls without slipping, determine the velocity of the collar at the instant the rod becomes horizontal, i.e., $\theta = 0°$. The system is released from rest when $\theta = 45°$.

18–46. The system consists of a 20-lb disk A, 4-lb slender rod BC, and a 1-lb smooth collar C. If the disk rolls without slipping, determine the velocity of the collar at the instant $\theta = 30°$. The system is released from rest when $\theta = 45°$.

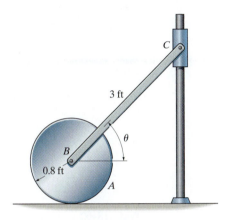

Probs. 18–45/46

18–47. The pendulum consists of a 2-lb rod BA and a 6-lb disk. The spring is stretched 0.3 ft when the rod is horizontal as shown. If the pendulum is released from rest and rotates about point D, determine its angular velocity at the instant the rod becomes vertical. The roller at C allows the spring to remain vertical as the rod falls.

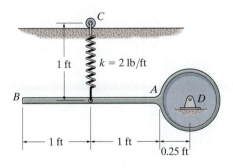

Prob. 18–47

*18–48. The uniform garage door has a mass of 150 kg and is guided along smooth tracks at its ends. Lifting is done using the two springs, each of which is attached to the anchor bracket at A and to the counterbalance shaft at B and C. As the door is raised, the springs begin to unwind from the shaft, thereby assisting the lift. If each spring provides a torsional moment of $M = (0.7\theta)$ N·m, where θ is in radians, determine the angle θ_0 at which both the left-wound and right-wound spring should be attached so that the door is completely balanced by the springs, i.e., when the door is in the vertical position and is given a slight force upwards, the springs will lift the door along the side tracks to the horizontal plane with no final angular velocity. *Note:* The elastic potential energy of a torsional spring is $V_e = \frac{1}{2}k\theta^2$, where $M = k\theta$ and in this case $k = 0.7$ N·m/rad.

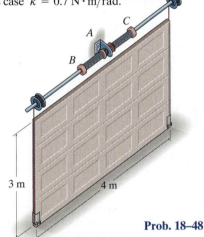

Prob. 18–48

•18–49. The garage door CD has a mass of 50 kg and can be treated as a thin plate. Determine the required unstretched length of each of the two side springs when the door is in the open position, so that when the door falls freely from the open position it comes to rest when it reaches the fully closed position, i.e., when AC rotates 180°. Each of the two side springs has a stiffness of $k = 350$ N/m. Neglect the mass of the side bars AC.

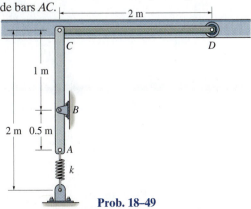

Prob. 18–49

18

18–50. The uniform rectangular door panel has a mass of 25 kg and is held in equilibrium above the horizontal at the position $\theta = 60°$ by rod BC. Determine the required stiffness of the torsional spring at A, so that the door's angular velocity becomes zero when the door reaches the closed position ($\theta = 0°$) once the supporting rod BC is removed. The spring is undeformed when $\theta = 60°$.

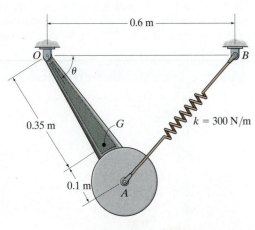

Prob. 18–50

18–51. The 30 kg pendulum has its mass center at G and a radius of gyration about point G of $k_G = 300$ mm. If it is released from rest when $\theta = 0°$, determine its angular velocity at the instant $\theta = 90°$. Spring AB has a stiffness of $k = 300$ N/m and is unstretched when $\theta = 0°$.

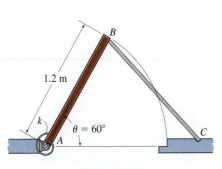

Prob. 18–51

*18–52. The 50-lb square plate is pinned at corner A and attached to a spring having a stiffness of $k = 20$ lb/ft. If the plate is released from rest when $\theta = 0°$, determine its angular velocity when $\theta = 90°$. The spring is unstretched when $\theta = 0°$.

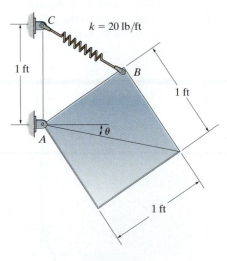

Prob. 18–52

•18–53. A spring having a stiffness of $k = 300$ N/m is attached to the end of the 15-kg rod, and it is unstretched when $\theta = 0°$. If the rod is released from rest when $\theta = 0°$, determine its angular velocity at the instant $\theta = 30°$. The motion is in the vertical plane.

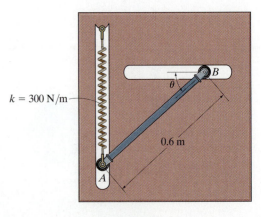

Prob. 18–53

18–54. If the 6-kg rod is released from rest at $\theta = 30°$, determine the angular velocity of the rod at the instant $\theta = 0°$. The attached spring has a stiffness of $k = 600$ N/m, with an unstretched length of 300 mm.

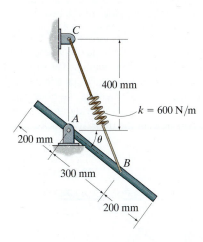

Prob. 18–54

18–55. The 50-kg rectangular door panel is held in the vertical position by rod CB. When the rod is removed, the panel closes due to its own weight. The motion of the panel is controlled by a spring attached to a cable that wraps around the half pulley. To reduce excessive slamming, the door panel's angular velocity is limited to 0.5 rad/s at the instant of closure. Determine the minimum stiffness k of the spring if the spring is unstretched when the panel is in the vertical position. Neglect the half pulley's mass.

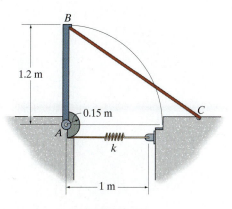

Prob. 18–55

*18–56. Rods AB and BC have weights of 15 lb and 30 lb, respectively. Collar C, which slides freely along the smooth vertical guide, has a weight of 5 lb. If the system is released from rest when $\theta = 0°$, determine the angular velocity of the rods when $\theta = 90°$. The attached spring is unstretched when $\theta = 0°$.

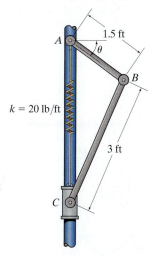

Prob. 18–56

•18–57. Determine the stiffness k of the torsional spring at A, so that if the bars are released from rest when $\theta = 0°$, bar AB has an angular velocity of 0.5 rad/s at the closed position, $\theta = 90°$. The spring is uncoiled when $\theta = 0°$. The bars have a mass per unit length of 10 kg/m.

18–58. The torsional spring at A has a stiffness of $k = 900$ N·m/rad and is uncoiled when $\theta = 0°$. Determine the angular velocity of the bars, AB and BC, when $\theta = 0°$, if they are released from rest at the closed position, $\theta = 90°$. The bars have a mass per unit length of 10 kg/m.

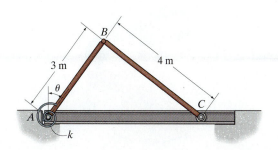

Probs. 18–57/58

18

18–59. The arm and seat of the amusement-park ride have a mass of 1.5 Mg, with the center of mass located at point G_1. The passenger seated at A has a mass of 125 kg, with the center of mass located at G_2 If the arm is raised to a position where $\theta = 150°$ and released from rest, determine the speed of the passenger at the instant $\theta = 0°$. The arm has a radius of gyration of $k_{G1} = 12$ m about its center of mass G_1. Neglect the size of the passenger.

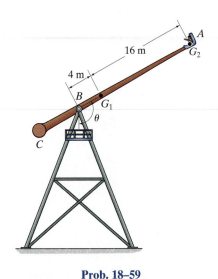

Prob. 18–59

18–60. The assembly consists of a 3-kg pulley A and 10-kg pulley B. If a 2-kg block is suspended from the cord, determine the block's speed after it descends 0.5 m starting from rest. Neglect the mass of the cord and treat the pulleys as thin disks. No slipping occurs.

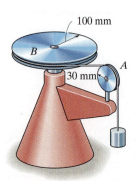

Prob. 18–60

•**18–61.** The motion of the uniform 80-lb garage door is guided at its ends by the track. Determine the required initial stretch in the spring when the door is open, $\theta = 0°$, so that when it falls freely it comes to rest when it just reaches the fully closed position, $\theta = 90°$. Assume the door can be treated as a thin plate, and there is a spring and pulley system on each of the two sides of the door.

18–62. The motion of the uniform 80-lb garage door is guided at its ends by the track. If it is released from rest at $\theta = 0°$, determine the door's angular velocity at the instant $\theta = 30°$. The spring is originally stretched 1 ft when the door is held open, $\theta = 0°$. Assume the door can be treated as a thin plate, and there is a spring and pulley system on each of the two sides of the door.

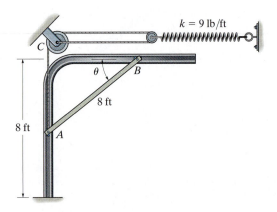

Probs. 18–61/62

18–63. The 500-g rod AB rests along the smooth inner surface of a hemispherical bowl. If the rod is released from rest from the position shown, determine its angular velocity at the instant it swings downward and becomes horizontal.

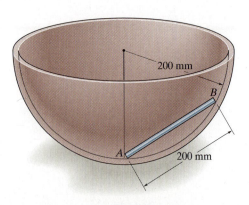

Prob. 18–63

*18–64. The 25-lb slender rod AB is attached to spring BC which has an unstretched length of 4 ft. If the rod is released from rest when $\theta = 30°$, determine its angular velocity at the instant $\theta = 90°$.

•18–65. The 25-lb slender rod AB is attached to spring BC which has an unstretched length of 4 ft. If the rod is released from rest when $\theta = 30°$, determine the angular velocity of the rod the instant the spring becomes unstretched.

*18–68. The uniform window shade AB has a total weight of 0.4 lb. When it is released, it winds up around the spring-loaded core O. Motion is caused by a spring within the core, which is coiled so that it exerts a torque $M = 0.3(10^{-3})\theta$ lb · ft, where θ is in radians, on the core. If the shade is released from rest, determine the angular velocity of the core at the instant the shade is completely rolled up, i.e., after 12 revolutions. When this occurs, the spring becomes uncoiled and the radius of gyration of the shade about the axle at O is $k_O = 0.9$ in. *Note:* The elastic potential energy of the torsional spring is $V_e = \frac{1}{2}k\theta^2$, where $M = k\theta$ and $k = 0.3(10^{-3})$ lb · ft/rad.

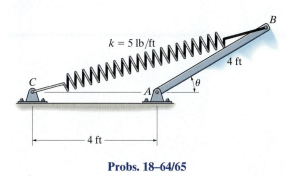

Probs. 18–64/65

Prob. 18–68

18–66. The assembly consists of two 8-lb bars which are pin connected to the two 10-lb disks. If the bars are released from rest when $\theta = 60°$, determine their angular velocities at the instant $\theta = 0°$. Assume the disks roll without slipping.

18–67. The assembly consists of two 8-lb bars which are pin connected to the two 10-lb disks. If the bars are released from rest when $\theta = 60°$, determine their angular velocities at the instant $\theta = 30°$. Assume the disks roll without slipping.

18–69. When the slender 10-kg bar AB is horizontal it is at rest and the spring is unstretched. Determine the stiffness k of the spring so that the motion of the bar is momentarily stopped when it has rotated clockwise 90°.

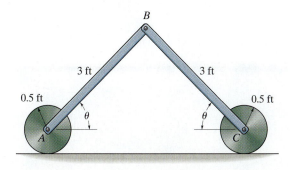

Probs. 18–66/67

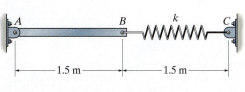

Prob. 18–69

18

CONCEPTUAL PROBLEMS

P18–1. The blade on the band saw wraps around the two large wheels *A* and *B*. When switched on, an electric motor turns the small pulley at *C* that then drives the larger pulley *D*, which is connected to *A* and turns with it. Explain why it is a good idea to use pulley *D*, and also use the larger wheels *A* and *B*. Use appropriate numerical values to explain your answer.

P18–1

P18–2. Two torsional springs, $M = k\theta$, are used to assist in opening and closing the hood of this truck. Assuming the springs are uncoiled ($\theta = 0°$) when the hood is opened, determine the stiffness *k* (N·m/rad) of each spring so that the hood can easily be lifted, i.e., practically no force applied to it, when it is closed. Use appropriate numerical values to explain your result.

P18–2

P18–3. The operation of this garage door is assisted using two springs *AB* and side members *BCD*, which are pinned at *C*. Assuming the springs are unstretched when the door is in the horizontal (open) position and *ABCD* is vertical, determine each spring stiffness *k* so that when the door falls to the vertical (closed) position, it will slowly come to a stop. Use appropriate numerical values to explain your result.

P18–3

P18–4. Determine the counterweight of *A* needed to balance the weight of the bridge deck when $\theta = 0°$. Show that this weight will maintain equilibrium of the deck by considering the potential energy of the system when the deck is in the arbitrary position θ. Both the deck and *AB* are horizontal when $\theta = 0°$. Neglect the weights of the other members. Use appropriate numerical values to explain this result.

P18–4

CHAPTER REVIEW

Kinetic Energy

The kinetic energy of a rigid body that undergoes planar motion can be referenced to its mass center. It includes a scalar sum of its translational and rotational kinetic energies.

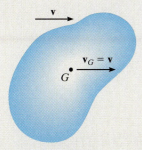

Translation

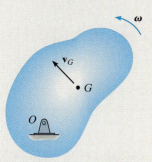

Rotation About a Fixed Axis

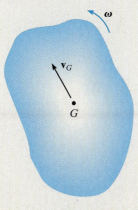

General Plane Motion

Translation

$$T = \tfrac{1}{2}mv_G^2$$

Rotation About a Fixed Axis

$$T = \tfrac{1}{2}mv_G^2 + \tfrac{1}{2}I_G\omega^2$$

or

$$T = \tfrac{1}{2}I_O\omega^2$$

General Plane Motion

$$T = \tfrac{1}{2}mv_G^2 + \tfrac{1}{2}I_G\omega^2$$

or

$$T = \tfrac{1}{2}I_{IC}\omega^2$$

Work of a Force and a Couple Moment

A force does work when it undergoes a displacement ds in the direction of the force. In particular, the frictional and normal forces that act on a cylinder or any circular body that rolls *without slipping* will do no work, since the normal force does not undergo a displacement and the frictional force acts on successive points on the surface of the body.

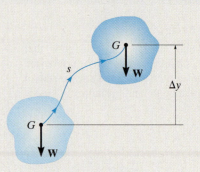

$$U_W = -W\,\Delta y$$

Weight

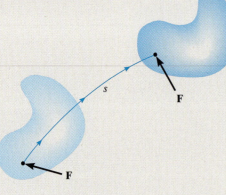

$$U_F = \int F \cos\theta\, ds$$

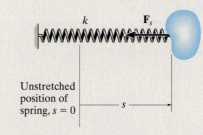

$$U = -\frac{1}{2}k\,s^2$$

Spring

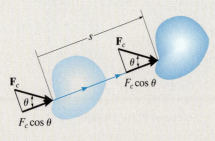

$$U_{F_c} = (F_c \cos\theta)s$$

Constant Force

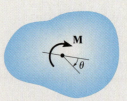

$$U_M = \int_{\theta_1}^{\theta_2} M\, d\theta$$

$$U_M = M(\theta_2 - \theta_1)$$

Constant magnitude

Principle of Work and Energy

Problems that involve velocity, force, and displacement can be solved using the principle of work and energy. The kinetic energy is the sum of both its rotational and translational parts. For application, a free-body diagram should be drawn in order to account for the work of all of the forces and couple moments that act on the body as it moves along the path.

$$T_1 = \Sigma U_{1-2} = T_2$$

Conservation of Energy

If a rigid body is subjected only to conservative forces, then the conservation-of-energy equation can be used to solve the problem. This equation requires that the sum of the potential and kinetic energies of the body remain the same at any two points along the path.

$$T_1 + V_1 = T_2 + V_2$$

$$\text{where } V = V_g + V_e$$

The potential energy is the sum of the body's gravitational and elastic potential energies. The gravitational potential energy will be positive if the body's center of gravity is located above a datum. If it is below the datum, then it will be negative. The elastic potential energy is always positive, regardless if the spring is stretched or compressed.

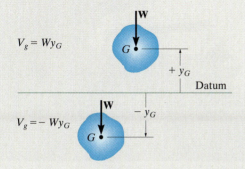

Gravitational potential energy

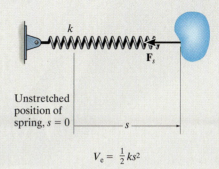

Elastic potential energy

The docking of the space shuttle to the international space station requires application of impulse and momentum principles to accurately predict their orbital motion and proper orientation.

Planar Kinetics of a Rigid Body: Impulse and Momentum

19

CHAPTER OBJECTIVES

- To develop formulations for the linear and angular momentum of a body.
- To apply the principles of linear and angular impulse and momentum to solve rigid-body planar kinetic problems that involve force, velocity, and time.
- To discuss application of the conservation of momentum.
- To analyze the mechanics of eccentric impact.

19.1   Linear and Angular Momentum

In this chapter we will use the principles of linear and angular impulse and momentum to solve problems involving force, velocity, and time as related to the planar motion of a rigid body. Before doing this, we will first formalize the methods for obtaining a body's linear and angular momentum, assuming the body is symmetric with respect to an inertial x–y reference plane.

Linear Momentum.   The linear momentum of a rigid body is determined by summing vectorially the linear momenta of all the particles of the body, i.e., $\mathbf{L} = \Sigma m_i \mathbf{v}_i$. Since $\Sigma m_i \mathbf{v}_i = m \mathbf{v}_G$ (see Sec. 15.2) we can also write

$$\mathbf{L} = m\mathbf{v}_G \tag{19–1}$$

This equation states that the body's linear momentum is a vector quantity having a *magnitude* mv_G, which is commonly measured in units of kg · m/s or slug · ft/s and a *direction* defined by $\mathbf{v}_G$ the velocity of the body's mass center.

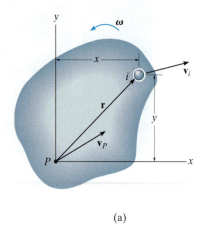

(a)

Angular Momentum. Consider the body in Fig. 19–1a, which is subjected to general plane motion. At the instant shown, the arbitrary point P has a known velocity $\mathbf{v}_P$, and the body has an angular velocity ω. Therefore the velocity of the ith particle of the body is

$$\mathbf{v}_i = \mathbf{v}_P + \mathbf{v}_{i/P} = \mathbf{v}_P + \boldsymbol{\omega} \times \mathbf{r}$$

The angular momentum of this particle about point P is equal to the "moment" of the particle's linear momentum about P, Fig. 19–1a. Thus,

$$(\mathbf{H}_P)_i = \mathbf{r} \times m_i \mathbf{v}_i$$

Expressing $\mathbf{v}_i$ in terms of $\mathbf{v}_P$ and using Cartesian vectors, we have

$$(H_P)_i \mathbf{k} = m_i(x\mathbf{i} + y\mathbf{j}) \times [(v_P)_x\mathbf{i} + (v_P)_y\mathbf{j} + \omega\mathbf{k} \times (x\mathbf{i} + y\mathbf{j})]$$
$$(H_P)_i = -m_i y(v_P)_x + m_i x(v_P)_y + m_i \omega r^2$$

Letting $m_i \rightarrow dm$ and integrating over the entire mass m of the body, we obtain

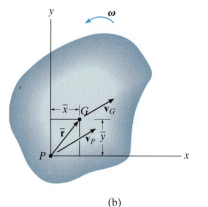

(b)

Fig. 19–1

$$H_P = -\left(\int_m y\, dm\right)(v_P)_x + \left(\int_m x\, dm\right)(v_P)_y + \left(\int_m r^2\, dm\right)\omega$$

Here H_P represents the angular momentum of the body about an axis (the z axis) perpendicular to the plane of motion that passes through point P. Since $\bar{y}m = \int y\, dm$ and $\bar{x}m = \int x\, dm$ the integrals for the first and second terms on the right are used to locate the body's center of mass G with respect to P, Fig. 19–1b. Also, the last integral represents the body's moment of inertia about point P. Thus,

$$H_P = -\bar{y}m(v_P)_x + \bar{x}m(v_P)_y + I_P\omega \qquad (19\text{–}2)$$

This equation reduces to a simpler form if P coincides with the mass center G for the body,* in which case $\bar{x} = \bar{y} = 0$. Hence,

---

*It also reduces to the same simple form, $H_P = I_P\omega$, if point P is a *fixed point* (see Eq. 19–9) or the velocity of P is directed along the line PG.

$$H_G = I_G\omega \qquad (19\text{–}3)$$

Here the angular momentum of the body about G is equal to the product of the moment of inertia of the body about an axis passing through G and the body's angular velocity. Realize that $\mathbf{H}_G$ is a vector quantity having a *magnitude* $I_G\omega$, which is commonly measured in units of $kg \cdot m^2/s$ or $slug \cdot ft^2/s$, and a *direction* defined by $\boldsymbol{\omega}$, which is always perpendicular to the plane of motion.

Equation 19–2 can also be rewritten in terms of the x and y components of the velocity of the body's mass center, $(\mathbf{v}_G)_x$ and $(\mathbf{v}_G)_y$, and the body's moment of inertia I_G. Since G is located at coordinates $(\bar{x},\bar{y})$, then by the parallel-axis theorem, $I_P = I_G + m(\bar{x}^2 + \bar{y}^2)$. Substituting into Eq. 19–2 and rearranging terms, we have

$$H_P = \bar{y}m[-(v_P)_x + \bar{y}\omega] + \bar{x}m[(v_P)_y + \bar{x}\omega] + I_G\omega \qquad (19\text{–}4)$$

From the kinematic diagram of Fig. 19–1b, $\mathbf{v}_G$ can be expressed in terms of $\mathbf{v}_P$ as

$$\mathbf{v}_G = \mathbf{v}_P + \boldsymbol{\omega} \times \bar{\mathbf{r}}$$

$$(v_G)_x\mathbf{i} + (v_G)_y\mathbf{j} = (v_P)_x\mathbf{i} + (v_P)_y\mathbf{j} + \omega\mathbf{k} \times (\bar{x}\mathbf{i} + \bar{y}\mathbf{j})$$

Carrying out the cross product and equating the respective $\mathbf{i}$ and $\mathbf{j}$ components yields the two scalar equations

$$(v_G)_x = (v_P)_x - \bar{y}\omega$$

$$(v_G)_y = (v_P)_y + \bar{x}\omega$$

Substituting these results into Eq. 19–4 yields

$$(\zeta+) \; H_P = -\bar{y}m(v_G)_x + \bar{x}m(v_G)_y + I_G\omega \qquad (19\text{–}5)$$

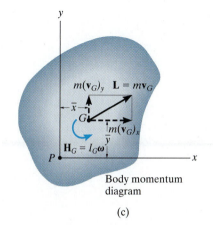

Body momentum diagram

(c)

Fig. 19–1 (cont.)

As shown in Fig. 19–1c, *this result indicates that when the angular momentum of the body is computed about point P, it is equivalent to the moment of the linear momentum* $m\mathbf{v}_G$, *or its components* $m(\mathbf{v}_G)_x$ *and* $m(\mathbf{v}_G)_y$, *about P plus the angular momentum* $I_G\boldsymbol{\omega}$. Using these results, we will now consider three types of motion.

19

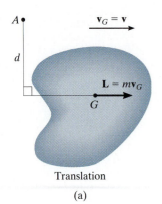

Translation

(a)

Translation. When a rigid body is subjected to either rectilinear or curvilinear *translation*, Fig. 19–2a, then $\boldsymbol{\omega} = \mathbf{0}$ and its mass center has a velocity of $\mathbf{v}_G = \mathbf{v}$. Hence, the linear momentum, and the angular momentum about G, become

$$\begin{aligned} L &= mv_G \\ H_G &= 0 \end{aligned} \qquad (19\text{–}6)$$

If the angular momentum is computed about some other point A, the "moment" of the linear momentum $\mathbf{L}$ must be found about the point. Since d is the "moment arm" as shown in Fig. 19–2a, then in accordance with Eq. 19–5, $H_A = (d)(mv_G)\,\text{↻}$.

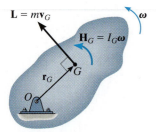

Rotation about a fixed axis

(b)

Fig. 19–2

Rotation About a Fixed Axis. When a rigid body is *rotating about a fixed axis*, Fig. 19–2b, the linear momentum, and the angular momentum about G, are

$$\begin{aligned} L &= mv_G \\ H_G &= I_G\omega \end{aligned} \qquad (19\text{–}7)$$

It is sometimes convenient to compute the angular momentum about point O. Noting that $\mathbf{L}$ (or $\mathbf{v}_G$) is always *perpendicular to* $\mathbf{r}_G$, we have

$$(\text{↺}+)\ H_O = I_G\omega + r_G(mv_G) \qquad (19\text{–}8)$$

Since $v_G = r_G\omega$, this equation can be written as $H_O = (I_G + mr_G^2)\omega$. Using the parallel-axis theorem,*

$$H_O = I_O\omega \qquad (19\text{–}9)$$

For the calculation, then, either Eq. 19–8 or 19–9 can be used.

---

*The similarity between this derivation and that of Eq. 17–16 ($\Sigma M_O = I_O\alpha$) and Eq. 18–5 $\left(T = \frac{1}{2}I_O\omega^2\right)$ should be noted. Also note that the same result can be obtained from Eq. 19–2 by selecting point P at O, realizing that $(v_O)_x = (v_O)_y = 0$.

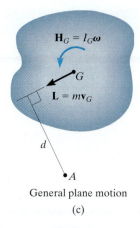

General plane motion

(c)

Fig 19–2

General Plane Motion

When a rigid body is subjected to general plane motion, Fig. 19–2c, the linear momentum, and the angular momentum about G, become

$$\boxed{\begin{aligned} L &= mv_G \\ H_G &= I_G\omega \end{aligned}} \qquad (19\text{–}10)$$

If the angular momentum is computed about point A, Fig. 19–2c, it is necessary to include the moment of $\mathbf{L}$ and $\mathbf{H}_G$ about this point. In this case,

$$(\zeta+) \qquad H_A = I_G\omega + (d)(mv_G)$$

Here d is the moment arm, as shown in the figure.

As a special case, if point A is the instantaneous center of zero velocity then, like Eq. 19–9, we can write the above equation as

$$\boxed{H_{IC} = I_{IC}\omega} \qquad (19\text{–}11)$$

where I_{IC} is the moment of inertia of the body about the IC. See Prob. 19–2.

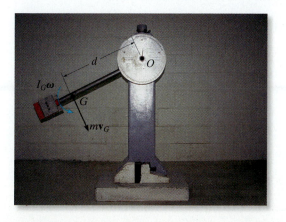

As the pendulum swings downward, its angular momentum about point O can be determined by computing the moment of $I_G\boldsymbol{\omega}$ and $m\mathbf{v}_G$ about O. This is $H_O = I_G\omega + (mv_G)d$. Since $v_G = \omega d$, then $H_O = I_G\omega + m(\omega d)d = (I_G + md^2)\omega = I_O\omega$.

EXAMPLE | 19.1

At a given instant the 5-kg slender bar has the motion shown in Fig. 19–3a. Determine its angular momentum about point G and about the IC at this instant.

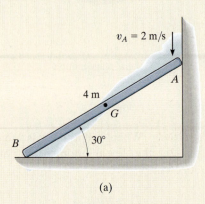

(a)

SOLUTION

Bar.   The bar undergoes *general plane motion*. The IC is established in Fig. 19–3b, so that

$$\omega = \frac{2 \text{ m/s}}{4 \text{ m cos } 30°} = 0.5774 \text{ rad/s}$$

$$v_G = (0.5774 \text{ rad/s})(2 \text{ m}) = 1.155 \text{ m/s}$$

Thus,

$$(\zeta+)\, H_G = I_G\omega = \left[\tfrac{1}{12}(5 \text{ kg})(4 \text{ m})^2\right](0.5774 \text{ rad/s}) = 3.85 \text{ kg} \cdot \text{m}^2/\text{s} \,\rangle \; \textit{Ans.}$$

Adding $I_G\omega$ and the moment of mv_G about the IC yields

$$(\zeta+)\, H_{IC} = I_G\omega + d(mv_G)$$

$$= \left[\tfrac{1}{12}(5 \text{ kg})(4 \text{ m})^2\right](0.5774 \text{ rad/s}) + (2 \text{ m})(5 \text{ kg})(1.155 \text{ m/s})$$

$$= 15.4 \text{ kg} \cdot \text{m}^2/\text{s} \,\rangle \qquad\qquad \textit{Ans.}$$

We can also use

$$(\zeta+)\, H_{IC} = I_{IC}\omega$$

$$= \left[\tfrac{1}{12}(5 \text{ kg})(4 \text{ m})^2 + (5 \text{ kg})(2 \text{ m})^2\right](0.5774 \text{ rad/s})$$

$$= 15.4 \text{ kg} \cdot \text{m}^2/\text{s} \,\rangle \qquad\qquad \textit{Ans.}$$

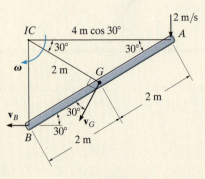

(b)

Fig. 19–3

19.2 Principle of Impulse and Momentum

Like the case for particle motion, the principle of impulse and momentum for a rigid body can be developed by *combining* the equation of motion with kinematics. The resulting equation will yield a *direct solution to problems involving force, velocity, and time.*

Principle of Linear Impulse and Momentum. The equation of translational motion for a rigid body can be written as $\Sigma\mathbf{F} = m\mathbf{a}_G = m(d\mathbf{v}_G/dt)$. Since the mass of the body is constant,

$$\Sigma\mathbf{F} = \frac{d}{dt}(m\mathbf{v}_G)$$

Multiplying both sides by dt and integrating from $t = t_1$, $\mathbf{v}_G = (\mathbf{v}_G)_1$ to $t = t_2$, $\mathbf{v}_G = (\mathbf{v}_G)_2$ yields

$$\Sigma\int_{t_1}^{t_2}\mathbf{F}\,dt = m(\mathbf{v}_G)_2 - m(\mathbf{v}_G)_1$$

This equation is referred to as the *principle of linear impulse and momentum*. It states that the sum of all the impulses created by the *external force system* which acts on the body during the time interval t_1 to t_2 is equal to the change in the linear momentum of the body during this time interval, Fig. 19–4.

Principle of Angular Impulse and Momentum. If the body has *general plane motion* then $\Sigma M_G = I_G\alpha = I_G(d\omega/dt)$. Since the moment of inertia is constant,

$$\Sigma M_G = \frac{d}{dt}(I_G\omega)$$

Multiplying both sides by dt and integrating from $t = t_1$, $\omega = \omega_1$ to $t = t_2$, $\omega = \omega_2$ gives

$$\Sigma\int_{t_1}^{t_2}M_G\,dt = I_G\omega_2 - I_G\omega_1 \tag{19–12}$$

In a similar manner, for *rotation about a fixed axis* passing through point O, Eq. 17–16 ($\Sigma M_O = I_O\alpha$) when integrated becomes

$$\Sigma\int_{t_1}^{t_2}M_O\,dt = I_O\omega_2 - I_O\omega_1 \tag{19–13}$$

Equations 19–12 and 19–13 are referred to as the *principle of angular impulse and momentum*. Both equations state that the sum of the angular impulses acting on the body during the time interval t_1 to t_2 is equal to the change in the body's angular momentum during this time interval.

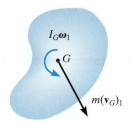

Initial momentum diagram

(a)

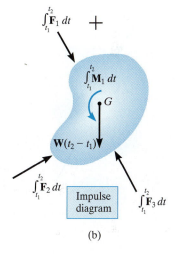

Impulse diagram

(b)

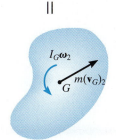

Final momentum diagram

(c)

Fig. 19–4

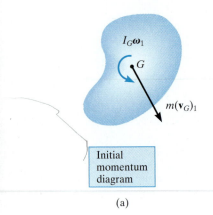

Initial
momentum
diagram

(a)

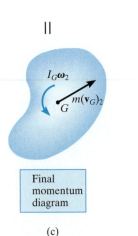

Impulse
diagram

(b)

||

Final
momentum
diagram

(c)

Fig. 19–4 (repeated)

To summarize these concepts, if motion occurs in the *x–y* plane, the following *three scalar equations* can be written to describe the *planar motion* of the body.

$$m(v_{Gx})_1 + \Sigma \int_{t_1}^{t_2} F_x \, dt = m(v_{Gx})_2$$

$$m(v_{Gy})_1 + \Sigma \int_{t_1}^{t_2} F_y \, dt = m(v_{Gy})_2 \qquad (19\text{–}14)$$

$$I_G \omega_1 + \Sigma \int_{t_1}^{t_2} M_G \, dt = I_G \omega_2$$

The terms in these equations can be shown graphically by drawing a set of impulse and momentum diagrams for the body, Fig. 19–4. Note that the linear momentum $m\mathbf{v}_G$ is applied at the body's mass center, Figs. 19–4a and 19–4c; whereas the angular momentum $I_G\boldsymbol{\omega}$ is a free vector, and therefore, like a couple moment, it can be applied at any point on the body. When the impulse diagram is constructed, Fig. 19–4b, the forces **F** and moment **M** vary with time, and are indicated by the integrals. However, if **F** and **M** are *constant* integration of the impulses yields $\mathbf{F}(t_2 - t_1)$ and $\mathbf{M}(t_2 - t_1)$, respectively. Such is the case for the body's weight **W**, Fig. 19–4b.

Equations 19–14 can also be applied to an entire system of connected bodies rather than to each body separately. This eliminates the need to include interaction impulses which occur at the connections since they are *internal* to the system. The resultant equations may be written in symbolic form as

$$\left(\Sigma \frac{\text{syst. linear}}{\text{momentum}} \right)_{x1} + \left(\Sigma \frac{\text{syst. linear}}{\text{impulse}} \right)_{x(1-2)} = \left(\Sigma \frac{\text{syst. linear}}{\text{momentum}} \right)_{x2}$$

$$\left(\Sigma \frac{\text{syst. linear}}{\text{momentum}} \right)_{y1} + \left(\Sigma \frac{\text{syst. linear}}{\text{impulse}} \right)_{y(1-2)} = \left(\Sigma \frac{\text{syst. linear}}{\text{momentum}} \right)_{y2}$$

$$\left(\Sigma \frac{\text{syst. angular}}{\text{momentum}} \right)_{O1} + \left(\Sigma \frac{\text{syst. angular}}{\text{impulse}} \right)_{O(1-2)} = \left(\Sigma \frac{\text{syst. angular}}{\text{momentum}} \right)_{O2}$$

(19–15)

As indicated by the third equation, the system's angular momentum and angular impulse must be computed with respect to the *same reference point O* for all the bodies of the system.

Procedure For Analysis

Impulse and momentum principles are used to solve kinetic problems that involve *velocity, force*, and *time* since these terms are involved in the formulation.

Free-Body Diagram.

- Establish the *x, y, z* inertial frame of reference and draw the free-body diagram in order to account for all the forces and couple moments that produce impulses on the body.
- The direction and sense of the initial and final velocity of the body's mass center, $\mathbf{v}_G$, and the body's angular velocity $\boldsymbol{\omega}$ should be established. If any of these motions is unknown, assume that the sense of its components is in the direction of the positive inertial coordinates.
- Compute the moment of inertia I_G or I_O.
- As an alternative procedure, draw the impulse and momentum diagrams for the body or system of bodies. Each of these diagrams represents an outlined shape of the body which graphically accounts for the data required for each of the three terms in Eqs. 19–14 or 19–15, Fig. 19–4. These diagrams are particularly helpful in order to visualize the "moment" terms used in the principle of angular impulse and momentum, if application is about the *IC* or another point other than the body's mass center *G* or a fixed point *O*.

Principle of Impulse and Momentum.

- Apply the three scalar equations of impulse and momentum.
- The angular momentum of a rigid body rotating about a fixed axis is the moment of $m\mathbf{v}_G$ plus $I_G\boldsymbol{\omega}$ about the axis. This is equal to $H_O = I_O\omega$, where I_O is the moment of inertia of the body about the axis.
- All the forces acting on the body's free-body diagram will create an impulse; however, some of these forces will do no work.
- Forces that are functions of time must be integrated to obtain the impulse.
- The principle of angular impulse and momentum is often used to eliminate unknown impulsive forces that are parallel or pass through a common axis, since the moment of these forces is zero about this axis.

Kinematics.

- If more than three equations are needed for a complete solution, it may be possible to relate the velocity of the body's mass center to the body's angular velocity using *kinematics*. If the motion appears to be complicated, kinematic (velocity) diagrams may be helpful in obtaining the necessary relation.

19

EXAMPLE 19.2

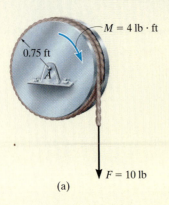

(a)

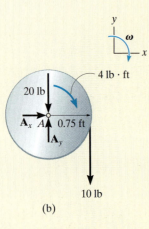

(b)

Fig. 19–5

The 20-lb disk shown in Fig. 19–5a is acted upon by a constant couple moment of 4 lb · ft and a force of 10 lb which is applied to a cord wrapped around its periphery. Determine the angular velocity of the disk two seconds after starting from rest. Also, what are the force components of reaction at the pin?

SOLUTION

Since angular velocity, force, and time are involved in the problems, we will apply the principles of impulse and momentum to the solution.

Free-Body Diagram. Fig. 19–5b. The disk's mass center does not move; however, the loading causes the disk to rotate clockwise. The moment of inertia of the disk about its fixed axis of rotation is

$$I_A = \frac{1}{2}mr^2 = \frac{1}{2}\left(\frac{20\text{ lb}}{32.2\text{ ft/s}^2}\right)(0.75\text{ ft})^2 = 0.1747\text{ slug} \cdot \text{ft}^2$$

Principle of Impulse and Momentum.

$$(\xrightarrow{+})\qquad\qquad m(v_{Ax})_1 + \Sigma \int_{t_1}^{t_2} F_x\,dt = m(v_{Ax})_2$$

$$0 + A_x(2\text{ s}) = 0$$

$$(+\uparrow)\qquad\qquad m(v_{Ay})_1 + \Sigma \int_{t_1}^{t_2} F_y\,dt = m(v_{Ay})_2$$

$$0 + A_y(2\text{ s}) - 20\text{ lb}(2\text{ s}) - 10\text{ lb}(2\text{ s}) = 0$$

$$(\zeta +)\qquad\qquad I_A\omega_1 + \Sigma \int_{t_1}^{t_2} M_A\,dt = I_A\omega_2$$

$$0 + 4\text{ lb} \cdot \text{ft}(2\text{ s}) + [10\text{ lb}(2\text{ s})](0.75\text{ ft}) = 0.1747\omega_2$$

Solving these equations yields

$$A_x = 0 \qquad\qquad\qquad\qquad\qquad Ans.$$

$$A_y = 30\text{ lb} \qquad\qquad\qquad\qquad Ans.$$

$$\omega_2 = 132\text{ rad/s}\,\zeta \qquad\qquad\qquad Ans.$$

EXAMPLE | 19.3

The 100-kg spool shown in Fig. 19–6a has a radius of gyration $k_G = 0.35$ m. A cable is wrapped around the central hub of the spool, and a horizontal force having a variable magnitude of $P = (t + 10)$ N is applied, where t is in seconds. If the spool is initially at rest, determine its angular velocity in 5 s. Assume that the spool rolls without slipping at A.

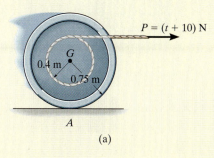

(a)

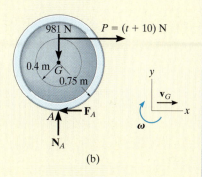

(b)

Fig. 19–6

SOLUTION

Free-Body Diagram.   From the free-body diagram, Fig. 19–6b, the *variable* force **P** will cause the friction force $\mathbf{F}_A$ to be variable, and thus the impulses created by both **P** and $\mathbf{F}_A$ must be determined by integration. Force **P** causes the mass center to have a velocity $\mathbf{v}_G$ to the right, and so the spool has a clockwise angular velocity $\boldsymbol{\omega}$.

Principle of Impulse and Momentum.   A direct solution for $\boldsymbol{\omega}$ can be obtained by applying the principle of angular impulse and momentum about point A, the IC, in order to eliminate the unknown friction impulse.

$(\zeta +)$ $I_A \omega_1 + \Sigma \int M_A \, dt = I_A \omega_2$

$$0 + \left[\int_0^{5\,s} (t + 10) \text{ N } dt \right] (0.75 \text{ m} + 0.4 \text{ m}) = [100 \text{ kg } (0.35 \text{ m})^2 + (100 \text{ kg})(0.75 \text{ m})^2] \omega_2$$

$$62.5(1.15) = 68.5 \omega_2$$

$$\omega_2 = 1.05 \text{ rad/s} \;\zeta \qquad\qquad Ans.$$

NOTE: Try solving this problem by applying the principle of impulse and momentum about G and using the principle of linear impulse and momentum in the x direction.

EXAMPLE 19.4

The cylinder shown in Fig. 19–7a has a mass of 6 kg. It is attached to a cord which is wrapped around the periphery of a 20-kg disk that has a moment of inertia $I_A = 0.40 \text{ kg} \cdot \text{m}^2$. If the cylinder is initially moving downward with a speed of 2 m/s, determine its speed in 3 s. Neglect the mass of the cord in the calculation.

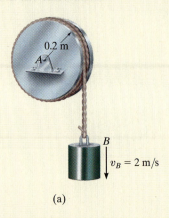

(a)

SOLUTION I

Free-Body Diagram. The free-body diagrams of the cylinder and disk are shown in Fig. 19–7b. All the forces are *constant* since the weight of the cylinder causes the motion. The downward motion of the cylinder, $\mathbf{v}_B$, causes $\boldsymbol{\omega}$ of the disk to be clockwise.

Principle of Impulse and Momentum. We can eliminate $\mathbf{A}_x$ and $\mathbf{A}_y$ from the analysis by applying the principle of angular impulse and momentum about point A. Hence

Disk

$$(\zeta+) \qquad I_A\omega_1 + \Sigma\int M_A\, dt = I_A\omega_2$$

$$0.40 \text{ kg} \cdot \text{m}^2(\omega_1) + T(3\text{ s})(0.2\text{ m}) = (0.40 \text{ kg} \cdot \text{m}^2)\omega_2$$

Cylinder

$$(+\uparrow) \qquad m_B(v_B)_1 + \Sigma\int F_y\, dt = m_B(v_B)_2$$

$$-6 \text{ kg}(2 \text{ m/s}) + T(3\text{ s}) - 58.86 \text{ N}(3\text{ s}) = -6 \text{ kg}(v_B)_2$$

Kinematics. Since $\omega = v_B/r$, then $\omega_1 = (2 \text{ m/s})/(0.2 \text{ m}) = 10$ rad/s and $\omega_2 = (v_B)_2/0.2 \text{ m} = 5(v_B)_2$. Substituting and solving the equations simultaneously for $(v_B)_2$ yields

$$(v_B)_2 = 13.0 \text{ m/s}\downarrow \qquad\qquad \textit{Ans.}$$

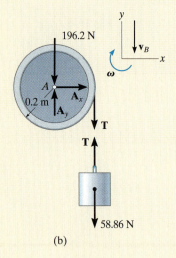

(b)

Fig. 19–7

SOLUTION II

Impulse and Momentum Diagrams. We can obtain $(v_B)_2$ *directly* by considering the *system* consisting of the cylinder, the cord, and the disk. The impulse and momentum diagrams have been drawn to clarify application of the principle of angular impulse and momentum about point A, Fig. 19–7c.

Principle of Angular Impulse and Momentum. Realizing that $\omega_1 = 10$ rad/s and $\omega_2 = 5(v_B)_2$, we have

$$(\circlearrowleft +)\left(\sum\begin{matrix}\text{syst. angular}\\\text{momentum}\end{matrix}\right)_{A1} + \left(\sum\begin{matrix}\text{syst. angular}\\\text{impulse}\end{matrix}\right)_{A(1-2)} = \left(\sum\begin{matrix}\text{syst. angular}\\\text{momentum}\end{matrix}\right)_{A2}$$

$$(6\text{ kg})(2\text{ m/s})(0.2\text{ m}) + (0.40\text{ kg}\cdot\text{m}^2)(10\text{ rad/s}) + (58.86\text{ N})(3\text{ s})(0.2\text{ m})$$

$$= (6\text{ kg})(v_B)_2(0.2\text{ m}) + (0.40\text{ kg}\cdot\text{m}^2)[5(v_B)_2(0.2\text{ m})]$$

$$(v_B)_2 = 13.0\text{ m/s} \downarrow \qquad\qquad Ans.$$

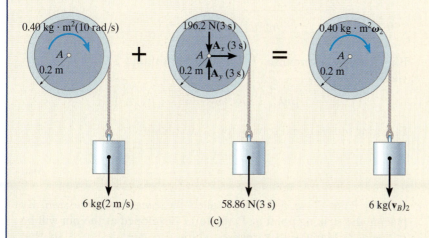

Fig. 19–7 (cont.)

EXAMPLE | 19.5

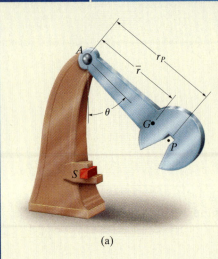

(a)

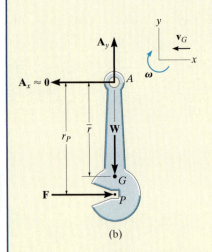

(b)

Fig. 19–8

The Charpy impact test is used in materials testing to determine the energy absorption characteristics of a material during impact. The test is performed using the pendulum shown in Fig. 19–8a, which has a mass m, mass center at G, and a radius of gyration k_G about G. Determine the distance r_P from the pin at A to the point P where the impact with the specimen S should occur so that the horizontal force at the pin A is essentially zero during the impact. For the calculation, assume the specimen absorbs all the pendulum's kinetic energy gained during the time it falls and thereby stops the pendulum from swinging when $\theta = 0°$.

SOLUTION

Free-Body Diagram. As shown on the free-body diagram, Fig. 19–8b, the conditions of the problem require the horizontal force at A to be zero. Just before impact, the pendulum has a clockwise angular velocity ω_1, and the mass center of the pendulum is moving to the left at $(v_G)_1 = \bar{r}\omega_1$.

Principle of Impulse and Momentum. We will apply the principle of angular impulse and momentum about point A. Thus,

$$I_A\omega_1 + \Sigma M_A\, dt = I_A\omega_2$$

$(\zeta +)$
$$I_A\omega_1 - \left(\int F\, dt\right)r_P = 0$$

$$m(v_G)_1 + \Sigma \int F\, dt = m(v_G)_2$$

$(\xrightarrow{+})$
$$-m(\bar{r}\omega_1) + \int F\, dt = 0$$

Eliminating the impulse $\int F\, dt$ and substituting $I_A = mk_G^2 + m\bar{r}^2$ yields

$$[mk_G^2 + m\bar{r}^2]\omega_1 - m(\bar{r}\omega_1)r_P = 0$$

Factoring out $m\omega_1$ and solving for r_P, we obtain

$$r_P = \bar{r} + \frac{k_G^2}{\bar{r}} \qquad\qquad Ans.$$

NOTE: Point P, so defined, is called the *center of percussion*. By placing the striking point at P, the force developed at the pin will be minimized. Many sports rackets, clubs, etc. are designed so that collision with the object being struck occurs at the center of percussion. As a consequence, no "sting" or little sensation occurs in the hand of the player. (Also see Probs. 17–66 and 19–1.)

FUNDAMENTAL PROBLEMS

F19–1. The 60-kg wheel has a radius of gyration about its center O of $k_O = 300$ mm. If it is subjected to a couple moment of $M = (3t^2)$ N·m, where t is in seconds, determine the angular velocity of the wheel when $t = 4$ s, starting from rest.

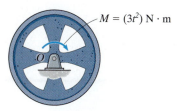

$M = (3t^2)$ N·m

F19–1

F19–2. The 300-kg wheel has a radius of gyration about its mass center O of $k_O = 400$ mm. If the wheel is subjected to a couple moment of $M = 300$ N·m, determine its angular velocity 6 s after it starts from rest and no slipping occurs. Also, determine the friction force that develops between the wheel and the ground.

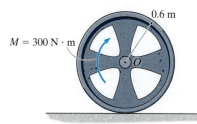

0.6 m

$M = 300$ N·m

F19–2

F19–3. If rod OA of negligible mass is subjected to the couple moment $M = 9$ N·m, determine the angular velocity of the 10-kg inner gear $t = 5$ s after it starts from rest. The gear has a radius of gyration about its mass center of $k_A = 100$ mm, and it rolls on the fixed outer gear. Motion occurs in the horizontal plane.

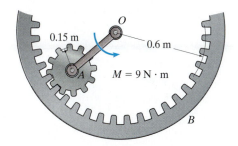

O

0.15 m

0.6 m

A

$M = 9$ N·m

B

F19–3

F19–4. Gears A and B of mass 10 kg and 50 kg have radii of gyration about their respective mass centers of $k_A = 80$ mm and $k_B = 150$ mm. If gear A is subjected to the couple moment $M = 10$ N·m, determine the angular velocity of gear B 5 s after it starts from rest.

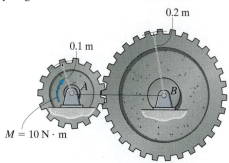

0.2 m

0.1 m

A

B

$M = 10$ N·m

F19–4

F19–5. The 50-kg spool is subjected to a horizontal force of $P = 150$ N. If the spool rolls without slipping, determine its angular velocity 3 s after it starts from rest. The radius of gyration of the spool about its center of mass is $k_G = 175$ mm.

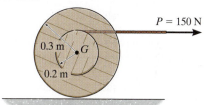

$P = 150$ N

0.3 m G

0.2 m

F19–5

F19–6. The reel has a weight of 150 lb and a radius of gyration about its center of gravity of $k_G = 1.25$ ft. If it is subjected to a torque of $M = 25$ lb·ft, and starts from rest when the torque is applied, determine its angular velocity in 3 seconds. The coefficient of kinetic friction between the reel and the horizontal plane is $\mu_k = 0.15$.

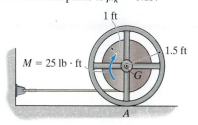

1 ft

1.5 ft

$M = 25$ lb·ft

G

A

F19–6

19

PROBLEMS

•19–1. The rigid body (slab) has a mass m and rotates with an angular velocity ω about an axis passing through the fixed point O. Show that the momenta of all the particles composing the body can be represented by a single vector having a magnitude mv_G and acting through point P, called the *center of percussion*, which lies at a distance $r_{P/G} = k_G^2/r_{G/O}$ from the mass center G. Here k_G is the radius of gyration of the body, computed about an axis perpendicular to the plane of motion and passing through G.

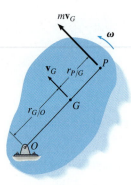

Prob. 19–1

19–2. At a given instant, the body has a linear momentum $\mathbf{L} = m\mathbf{v}_G$ and an angular momentum $\mathbf{H}_G = I_G\omega$ computed about its mass center. Show that the angular momentum of the body computed about the instantaneous center of zero velocity IC can be expressed as $\mathbf{H}_{IC} = I_{IC}\omega$, where I_{IC} represents the body's moment of inertia computed about the instantaneous axis of zero velocity. As shown, the IC is located at a distance $r_{G/IC}$ away from the mass center G.

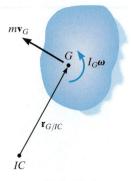

Prob. 19–2

19–3. Show that if a slab is rotating about a fixed axis perpendicular to the slab and passing through its mass center G, the angular momentum is the same when computed about any other point P.

Prob. 19–3

*19–4. The pilot of a crippled jet was able to control his plane by throttling the two engines. If the plane has a weight of 17 000 lb and a radius of gyration of $k_G = 4.7$ ft about the mass center G, determine the angular velocity of the plane and the velocity of its mass center G in $t = 5$ s if the thrust in each engine is altered to $T_1 = 5000$ lb and $T_2 = 800$ lb as shown. Originally the plane is flying straight at 1200 ft/s. Neglect the effects of drag and the loss of fuel.

Prob. 19–4

•19–5. The assembly weighs 10 lb and has a radius of gyration $k_G = 0.6$ ft about its center of mass G. The kinetic energy of the assembly is 31 ft · lb when it is in the position shown. If it rolls counterclockwise on the surface without slipping, determine its linear momentum at this instant.

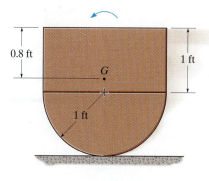

0.8 ft

1 ft

G

1 ft

Prob. 19–5

19–6. The impact wrench consists of a slender 1-kg rod AB which is 580 mm long, and cylindrical end weights at A and B that each have a diameter of 20 mm and a mass of 1 kg. This assembly is free to rotate about the handle and socket, which are attached to the lug nut on the wheel of a car. If the rod AB is given an angular velocity of 4 rad/s and it strikes the bracket C on the handle without rebounding, determine the angular impulse imparted to the lug nut.

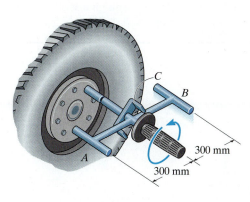

C

B

300 mm

A

300 mm

Prob. 19–6

19–7. The space shuttle is located in "deep space," where the effects of gravity can be neglected. It has a mass of 120 Mg, a center of mass at G, and a radius of gyration $(k_G)_x = 14$ m about the x axis. It is originally traveling forward at $v = 3$ km/s when the pilot turns on the engine at A, creating a thrust $T = 600(1 - e^{-0.3t})$ kN, where t is in seconds. Determine the shuttle's angular velocity 2 s later.

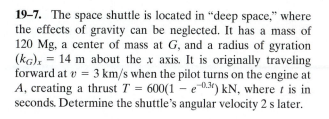

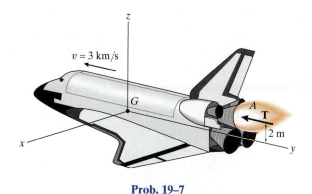

z

$v = 3$ km/s

G

A

T

2 m

x

y

Prob. 19–7

*19–8. The 50-kg cylinder has an angular velocity of 30 rad/s when it is brought into contact with the horizontal surface at C. If the coefficient of kinetic friction is $\mu_C = 0.2$, determine how long it will take for the cylinder to stop spinning. What force is developed in link AB during this time? The axle through the cylinder is connected to two symmetrical links. (Only AB is shown.) For the computation, neglect the weight of the links.

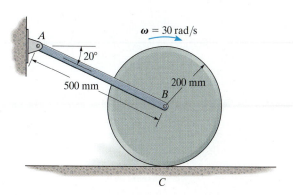

A

20°

500 mm

$\omega = 30$ rad/s

200 mm

B

C

Prob. 19–8

19

•**19–9.** If the cord is subjected to a horizontal force of $P = 150$ N, and the gear rack is fixed to the horizontal plane, determine the angular velocity of the gear in 4 s, starting from rest. The mass of the gear is 50 kg, and it has a radius of gyration about its center of mass O of $k_O = 125$ mm.

19–10. If the cord is subjected to a horizontal force of $P = 150$ N, and gear is supported by a fixed pin at O, determine the angular velocity of the gear and the velocity of the 20-kg gear rack in 4 s, starting from rest. The mass of the gear is 50 kg and it has a radius of gyration of $k_O = 125$ mm. Assume that the contact surface between the gear rack and the horizontal plane is smooth.

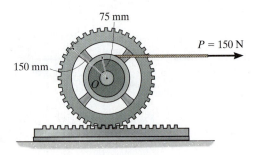

75 mm

$P = 150$ N

150 mm

O

Probs. 19–9/10

19–11. A motor transmits a torque of $M = 0.05$ N·m to the center of gear A. Determine the angular velocity of each of the three (equal) smaller gears in 2 s starting from rest. The smaller gears (B) are pinned at their centers, and the masses and centroidal radii of gyration of the gears are given in the figure.

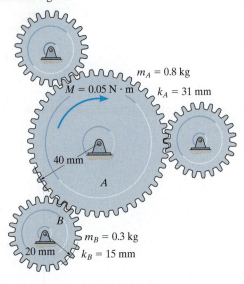

$m_A = 0.8$ kg
$M = 0.05$ N·m
$k_A = 31$ mm

40 mm

A

B
$m_B = 0.3$ kg
20 mm $k_B = 15$ mm

Prob. 19–11

*****19–12.** The 200-lb flywheel has a radius of gyration about its center of gravity O of $k_O = 0.75$ ft. If it rotates counterclockwise with an angular velocity of 1200 rev/min before the brake is applied, determine the time required for the wheel to come to rest when a force of $P = 200$ lb is applied to the handle. The coefficient of kinetic friction between the belt and the wheel rim is $\mu_k = 0.3$. (*Hint:* Recall from the statics text that the relation of the tension in the belt is given by $T_B = T_C e^{\mu\beta}$, where β is the angle of contact in radians.)

•**19–13.** The 200-lb flywheel has a radius of gyration about its center of gravity O of $k_O = 0.75$ ft. If it rotates counterclockwise with a constant angular velocity of 1200 rev/min before the brake is applied, determine the required force **P** that must be applied to the handle to stop the wheel in 2 s. The coefficient of kinetic friction between the belt and the wheel rim is $\mu_k = 0.3$. (*Hint:* Recall from the statics text that the relation of the tension in the belt is given by $T_B = T_C e^{\mu\beta}$, where β is the angle of contact in radians.)

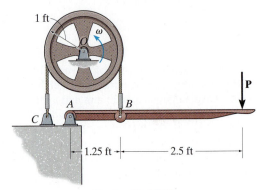

1 ft

ω

O

P

A B

C

1.25 ft   2.5 ft

Probs. 19–12/13

19–14. The 12-kg disk has an angular velocity of $\omega = 20$ rad/s. If the brake ABC is applied such that the magnitude of force **P** varies with time as shown, determine the time needed to stop the disk. The coefficient of kinetic friction at B is $\mu_k = 0.4$. Neglect the thickness of the brake.

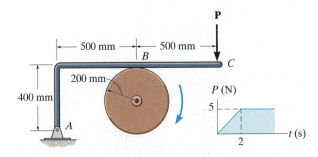

P

500 mm   500 mm

B C

200 mm

400 mm

P (N)

A

5

t (s)

2

Prob. 19–14

19–15. The 1.25-lb tennis racket has a center of gravity at G and a radius of gyration about G of $k_G = 0.625$ ft. Determine the position P where the ball must be hit so that 'no sting' is felt by the hand holding the racket, i.e., the horizontal force exerted by the racket on the hand is zero.

•19–17. The 5-kg ball is cast on the alley with a backspin of $\omega_0 = 10$ rad/s, and the velocity of its center of mass O is $v_0 = 5$ m/s. Determine the time for the ball to stop back spinning, and the velocity of its center of mass at this instant. The coefficient of kinetic friction between the ball and the alley is $\mu_k = 0.08$.

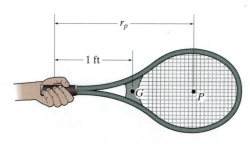

Prob. 19–15

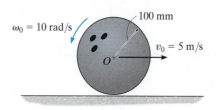

Prob. 19–17

*19–16. If the boxer hits the 75-kg punching bag with an impulse of $I = 20$ N·s, determine the angular velocity of the bag immediately after it has been hit. Also, find the location d of point B, about which the bag appears to rotate. Treat the bag as a uniform cylinder.

19–18. The smooth rod assembly shown is at rest when it is struck by a hammer at A with an impulse of 10 N·s. Determine the angular velocity of the assembly and the magnitude of velocity of its mass center immediately after it has been struck. The rods have a mass per unit length of 6 kg/m.

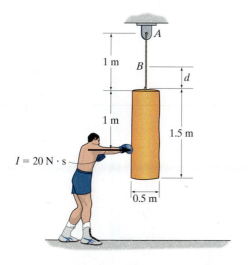

Prob. 19–16

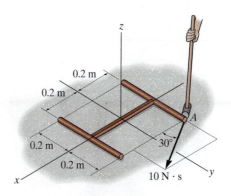

Prob. 19–18

19

19–19. The flywheel A has a mass of 30 kg and a radius of gyration of $k_C = 95$ mm. Disk B has a mass of 25 kg, is pinned at D, and is coupled to the flywheel using a belt which is subjected to a tension such that it does not slip at its contacting surfaces. If a motor supplies a counterclockwise torque or twist to the flywheel, having a magnitude of $M = (12t)$ N · m, where t is in seconds, determine the angular velocity of the disk 3 s after the motor is turned on. Initially, the flywheel is at rest.

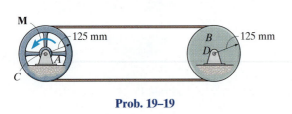

Prob. 19–19

•19–21. For safety reasons, the 20-kg supporting leg of a sign is designed to break away with negligible resistance at B when the leg is subjected to the impact of a car. Assuming that the leg is pinned at A and approximates a thin rod, determine the impulse the car bumper exerts on it, if after the impact the leg appears to rotate clockwise to a maximum angle of $\theta_{max} = 150°$.

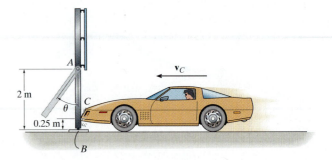

Prob. 19–21

*19–20. The 30-lb flywheel A has a radius of gyration about its center of 4 in. Disk B weighs 50 lb and is coupled to the flywheel by means of a belt which does not slip at its contacting surfaces. If a motor supplies a counterclockwise torque to the flywheel of $M = (50t)$ lb · ft, where t is in seconds, determine the time required for the disk to attain an angular velocity of 60 rad/s starting from rest.

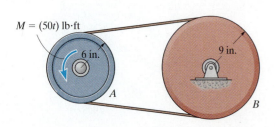

Prob. 19–20

19–22. The slender rod has a mass m and is suspended at its end A by a cord. If the rod receives a horizontal blow giving it an impulse $\mathbf{I}$ at its bottom B, determine the location y of the point P about which the rod appears to rotate during the impact.

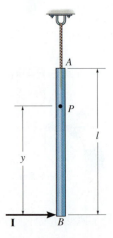

Prob. 19–22

19–23. The 25-kg circular disk is attached to the yoke by means of a smooth axle A. Screw C is used to lock the disk to the yoke. If the yoke is subjected to a torque of $M = (5t^2) \text{ N} \cdot \text{m}$, where t is in seconds, and the disk is unlocked, determine the angular velocity of the yoke when $t = 3$ s, starting from rest. Neglect the mass of the yoke.

*19–24. The 25-kg circular disk is attached to the yoke by means of a smooth axle A. Screw C is used to lock the disk to the yoke. If the yoke is subjected to a torque of $M = (5t^2) \text{ N} \cdot \text{m}$, where t is in seconds, and the disk is locked, determine the angular velocity of the yoke when $t = 3$ s, starting from rest. Neglect the mass of the yoke.

19–26. The body and bucket of a skid steer loader has a weight of 2000 lb, and its center of gravity is located at G. Each of the four wheels has a weight of 100 lb and a radius of gyration about its center of gravity of 1 ft. If the engine supplies a torque of $M = 100 \text{ lb} \cdot \text{ft}$ to each of the rear drive wheels, determine the speed of the loader in $t = 10$ s, starting from rest. The wheels roll without slipping.

19–27. The body and bucket of a skid steer loader has a weight of 2000 lb, and its center of gravity is located at G. Each of the four wheels has a weight of 100 lb and a radius of gyration about its center of gravity of 1 ft. If the loader attains a speed of 20 ft/s in 10 s, starting from rest, determine the torque **M** supplied to each of the rear drive wheels. The wheels roll without slipping.

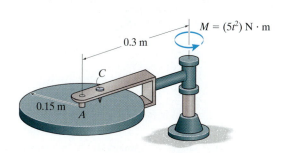

Probs. 19–23/24

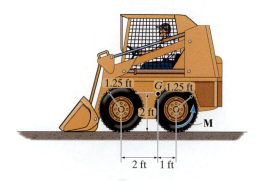

Probs. 19–26/27

•19–25. If the shaft is subjected to a torque of $M = (15t^2) \text{ N} \cdot \text{m}$, where t is in seconds, determine the angular velocity of the assembly when $t = 3$ s, starting from rest. Rods AB and BC each have a mass of 9 kg.

*19–28. The two rods each have a mass m and a length l, and lie on the smooth horizontal plane. If an impulse **I** is applied at an angle of 45° to one of the rods at midlength as shown, determine the angular velocity of each rod just after the impact. The rods are pin connected at B.

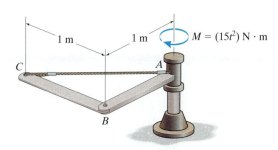

Prob. 19–25

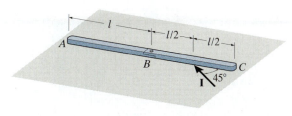

Prob. 19–28

•**19–29.**   The car strikes the side of a light pole, which is designed to break away from its base with negligible resistance. From a video taken of the collision it is observed that the pole was given an angular velocity of 60 rad/s when AC was vertical. The pole has a mass of 175 kg, a center of mass at G, and a radius of gyration about an axis perpendicular to the plane of the pole assembly and passing through G of $k_G = 2.25$ m. Determine the horizontal impulse which the car exerts on the pole at the instant AC is essentially vertical.

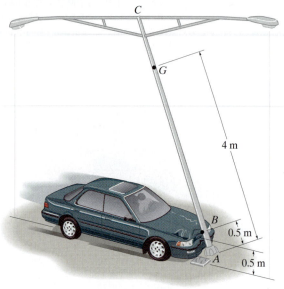
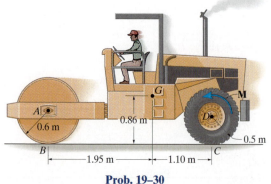

Prob. 19–29

19–30.   The frame of the roller has a mass of 5.5 Mg and a center of mass at G. The roller has a mass of 2 Mg and a radius of gyration about its mass center of $k_A = 0.45$ m. If a torque of $M = 600$ N·m is applied to the rear wheels, determine the speed of the compactor in $t = 4$ s, starting from rest. No slipping occurs. Neglect the mass of the driving wheels.

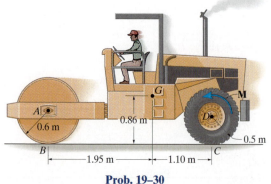

Prob. 19–30

19–31.   The 200-kg satellite has a radius of gyration about the centroidal z axis of $k_z = 1.25$ m. Initially it is rotating with a constant angular velocity of $\omega_0 = \{1500\,\mathbf{k}\}$ rev/min. If the two jets A and B are fired simultaneously and produce a thrust of $T = (5e^{-0.1t})$ kN, where t is in seconds, determine the angular velocity of the satellite, five seconds after firing.

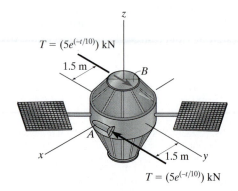

Prob. 19–31

*****19–32.**   If the shaft is subjected to a torque of $M = (30e^{-0.1t})$ N·m, where t is in seconds, determine the angular velocity of the assembly when $t = 5$ s, starting from rest. The rectangular plate has a mass of 25 kg. Rods AC and BC have the same mass of 5 kg.

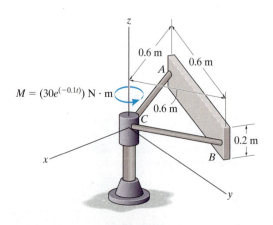

Prob. 19–32

19.3   Conservation of Momentum

Conservation of Linear Momentum   If the sum of all the *linear impulses* acting on a system of connected rigid bodies is *zero* in a specific direction, then the linear momentum of the system is constant, or conserved in this direction, that is,

$$\left(\sum \begin{array}{c} \text{syst. linear} \\ \text{momentum} \end{array} \right)_1 = \left(\sum \begin{array}{c} \text{syst. linear} \\ \text{momentum} \end{array} \right)_2 \qquad (19\text{–}16)$$

This equation is referred to as the *conservation of linear momentum.*

   Without inducing appreciable errors in the calculations, it may be possible to apply Eq. 19–16 in a specified direction for which the linear impulses are small or *nonimpulsive*. Specifically, nonimpulsive forces occur when small forces act over very short periods of time. Typical examples include the force of a slightly deformed spring, the initial contact force with soft ground, and in some cases the weight of the body.

Conservation of Angular Momentum   The angular momentum of a system of connected rigid bodies is conserved about the system's center of mass G, or a fixed point O, when the sum of all the angular impulses about these points is zero or appreciably small (nonimpulsive). The third of Eqs. 19–15 then becomes

$$\left(\sum \begin{array}{c} \text{syst. angular} \\ \text{momentum} \end{array} \right)_{O1} = \left(\sum \begin{array}{c} \text{syst. angular} \\ \text{momentum} \end{array} \right)_{O2} \qquad (19\text{–}17)$$

This equation is referred to as the *conservation of angular momentum.* In the case of a single rigid body, Eq. 19–17 applied to point G becomes $(I_G \omega)_1 = (I_G \omega)_2$. For example, consider a swimmer who executes a somersault after jumping off a diving board. By tucking his arms and legs in close to his chest, he *decreases* his body's moment of inertia and thus *increases* his angular velocity ($I_G \omega$ must be constant). If he straightens out just before entering the water, his body's moment of inertia is *increased*, and so his angular velocity *decreases*. Since the weight of his body creates a linear impulse during the time of motion, this example also illustrates how the angular momentum of a body can be conserved and yet the linear momentum is *not*. Such cases occur whenever the external forces creating the linear impulse pass through either the center of mass of the body or a fixed axis of rotation.

19

Procedure for Analysis

The conservation of linear or angular momentum should be applied using the following procedure.

Free-Body Diagram.

- Establish the x, y inertial frame of reference and draw the free-body diagram for the body or system of bodies during the time of impact. From this diagram classify each of the applied forces as being either "impulsive" or "nonimpulsive."

- By inspection of the free-body diagram, the *conservation of linear momentum* applies in a given direction when *no* external impulsive forces act on the body or system in that direction; whereas the *conservation of angular momentum* applies about a fixed point O or at the mass center G of a body or system of bodies when all the external impulsive forces acting on the body or system create zero moment (or zero angular impulse) about O or G.

- As an alternative procedure, draw the impulse and momentum diagrams for the body or system of bodies. These diagrams are particularly helpful in order to visualize the "moment" terms used in the conservation of angular momentum equation, when it has been decided that angular momenta are to be computed about a point other than the body's mass center G.

Conservation of Momentum.

- Apply the conservation of linear or angular momentum in the appropriate directions.

Kinematics.

- If the motion appears to be complicated, kinematic (velocity) diagrams may be helpful in obtaining the necessary kinematic relations.

19

EXAMPLE 19.6

The 10-kg wheel shown in Fig. 19–9a has a moment of inertia $I_G = 0.156$ kg·m². Assuming that the wheel does not slip or rebound, determine the minimum velocity $\mathbf{v}_G$ it must have to just roll over the obstruction at A.

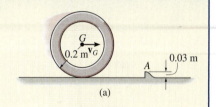

(a)

SOLUTION

Impulse and Momentum Diagrams. Since no slipping or rebounding occurs, the wheel essentially *pivots* about point A during contact. This condition is shown in Fig. 19–9b, which indicates, respectively, the momentum of the wheel *just before impact*, the impulses given to the wheel *during impact*, and the momentum of the wheel *just after impact*. Only two impulses (forces) act on the wheel. By comparison, the force at A is much greater than that of the weight, and since the time of impact is very short, the weight can be considered nonimpulsive. The impulsive force $\mathbf{F}$ at A has both an unknown magnitude and an unknown direction θ. To eliminate this force from the analysis, note that angular momentum about A is essentially *conserved* since $(98.1\Delta t)d \approx 0$.

Conservation of Angular Momentum. With reference to Fig. 19–9b,

$(\zeta +)$
$$(H_A)_1 = (H_A)_2$$

$$r'm(v_G)_1 + I_G\omega_1 = rm(v_G)_2 + I_G\omega_2$$

$$(0.2 \text{ m} - 0.03 \text{ m})(10 \text{ kg})(v_G)_1 + (0.156 \text{ kg} \cdot \text{m}^2)(\omega_1) =$$
$$(0.2 \text{ m})(10 \text{ kg})(v_G)_2 + (0.156 \text{ kg} \cdot \text{m}^2)(\omega_2)$$

Kinematics. Since no slipping occurs, in general $\omega = v_G/r = v_G/0.2 \text{ m} = 5v_G$. Substituting this into the above equation and simplifying yields

$$(v_G)_2 = 0.8921(v_G)_1 \tag{1}$$

Conservation of Energy. * In order to roll over the obstruction, the wheel must pass position 3 shown in Fig. 19–9c. Hence, if $(v_G)_2$ [or $(v_G)_1$] is to be a minimum, it is necessary that the kinetic energy of the wheel at position 2 be equal to the potential energy at position 3. Placing the datum through the center of gravity, as shown in the figure, and applying the conservation of energy equation, we have

$$\{T_2\} + \{V_2\} = \{T_3\} + \{V_3\}$$

$$\left\{\tfrac{1}{2}(10 \text{ kg})(v_G)_2^2 + \tfrac{1}{2}(0.156 \text{ kg} \cdot \text{m}^2)\omega_2^2\right\} + \{0\} =$$
$$\{0\} + \{(98.1 \text{ N})(0.03 \text{ m})\}$$

Substituting $\omega_2 = 5(v_G)_2$ and Eq. 1 into this equation, and solving,

$$(v_G)_1 = 0.729 \text{ m/s} \rightarrow \qquad Ans.$$

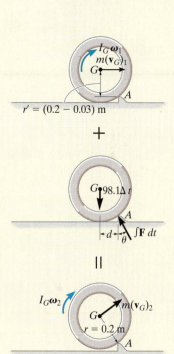

(b)

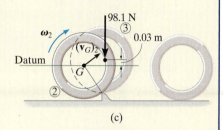

(c)

Fig. 19–9

*This principle *does not apply during impact*, since energy is *lost* during the collision. However, just after impact, as in Fig. 19–9c, it can be used.

EXAMPLE 19.7

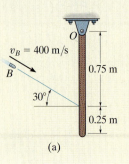

(a)

The 5-kg slender rod shown in Fig. 19–10a is pinned at O and is initially at rest. If a 4-g bullet is fired into the rod with a velocity of 400 m/s, as shown in the figure, determine the angular velocity of the rod just after the bullet becomes embedded in it.

SOLUTION

Impulse and Momentum Diagrams. The impulse which the bullet exerts on the rod can be eliminated from the analysis, and the angular velocity of the rod just after impact can be determined by considering the bullet and rod as a single system. To clarify the principles involved, the impulse and momentum diagrams are shown in Fig. 19–10b. The momentum diagrams are drawn *just before and just after impact*. During impact, the bullet and rod exert equal but *opposite internal impulses* at A. As shown on the impulse diagram, the impulses that are external to the system are due to the reactions at O and the weights of the bullet and rod. Since the time of impact, Δt, is very short, the rod moves only a slight amount, and so the "moments" of the weight impulses about point O are essentially zero. Therefore angular momentum is conserved about this point.

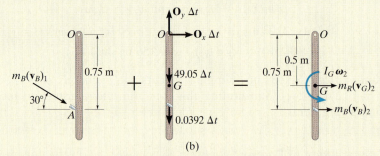

(b)

Conservation of Angular Momentum. From Fig. 19–10b, we have

$$(\zeta+) \qquad \Sigma(H_O)_1 = \Sigma(H_O)_2$$

$$m_B(v_B)_1 \cos 30°(0.75\text{ m}) = m_B(v_B)_2(0.75\text{ m}) + m_R(v_G)_2(0.5\text{ m}) + I_G\omega_2$$

$$(0.004\text{ kg})(400\cos 30°\text{ m/s})(0.75\text{ m}) =$$

$$(0.004\text{ kg})(v_B)_2(0.75\text{ m}) + (5\text{ kg})(v_G)_2(0.5\text{ m}) + \left[\tfrac{1}{12}(5\text{ kg})(1\text{ m})^2\right]\omega_2$$

or

$$1.039 = 0.003(v_B)_2 + 2.50(v_G)_2 + 0.4167\omega_2 \qquad (1)$$

Kinematics. Since the rod is pinned at O, from Fig. 19–10c we have

$$(v_G)_2 = (0.5\text{ m})\omega_2 \qquad (v_B)_2 = (0.75\text{ m})\omega_2$$

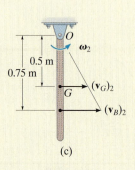

(c)

Fig. 19–10

Substituting into Eq. 1 and solving yields

$$\omega_2 = 0.623\text{ rad/s}\,\zeta \qquad\qquad Ans.$$

*19.4   Eccentric Impact

The concepts involving central and oblique impact of particles were presented in Sec. 15.4. We will now expand this treatment and discuss the eccentric impact of two bodies. *Eccentric impact* occurs when the line connecting the *mass centers* of the two bodies *does not* coincide with the line of impact.* This type of impact often occurs when one or both of the bodies are constrained to rotate about a fixed axis. Consider, for example, the collision at C between the two bodies A and B, shown in Fig. 19–11a. It is assumed that just before collision B is rotating counterclockwise with an angular velocity $(\boldsymbol{\omega}_B)_1$, and the velocity of the contact point C located on A is $(\mathbf{u}_A)_1$. Kinematic diagrams for both bodies just before collision are shown in Fig. 19–11b. Provided the bodies are smooth, the impulsive forces they exert on each other are directed along the line of impact. Hence, the component of velocity of point C on body B, which is directed along the line of impact, is $(v_B)_1 = (\omega_B)_1 r$, Fig. 19–11b. Likewise, on body A the component of velocity $(\mathbf{u}_A)_1$ along the line of impact is $(\mathbf{v}_A)_1$. In order for a collision to occur, $(v_A)_1 > (v_B)_1$.

During the impact an equal but opposite impulsive force $\mathbf{P}$ is exerted between the bodies which *deforms* their shapes at the point of contact. The resulting impulse is shown on the impulse diagrams for both bodies, Fig. 19–11c. Note that the impulsive force at point C on the rotating body creates impulsive pin reactions at O. On these diagrams it is assumed that the impact creates forces which are much larger than the nonimpulsive weights of the bodies, which are not shown. When the deformation at point C is a maximum, C on both the bodies moves with a common velocity $\mathbf{v}$ along the line of impact, Fig. 19–11d. A period of *restitution* then occurs in which the bodies tend to regain their original shapes. The restitution phase creates an equal but opposite impulsive force $\mathbf{R}$ acting between the bodies as shown on the impulse diagram, Fig. 19–11e. After restitution the bodies move apart such that point C on body B has a velocity $(\mathbf{v}_B)_2$ and point C on body A has a velocity $(\mathbf{u}_A)_2$, Fig. 19–11f, where $(v_B)_2 > (v_A)_2$.

In general, a problem involving the impact of two bodies requires determining the *two unknowns* $(v_A)_2$ and $(v_B)_2$, assuming $(v_A)_1$ and $(v_B)_1$ are known (or can be determined using kinematics, energy methods, the equations of motion, etc.). To solve such problems, two equations must be written. The *first equation* generally involves application of *the conservation of angular momentum to the two bodies*. In the case of both bodies A and B, we can state that angular momentum is conserved about point O since the impulses at C are internal to the system and the impulses at O create zero moment (or zero angular impulse) about O. The *second equation* can be obtained using the definition of the *coefficient of restitution, e*, which is a ratio of the restitution impulse to the deformation impulse.

* When these lines coincide, central impact occurs and the problem can be analyzed as discussed in Sec. 15.4.

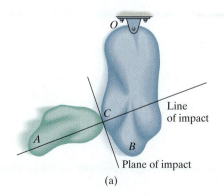

Line of impact

Plane of impact

(a)

Fig. 19–11

19

Here is an example of eccentric impact occurring between this bowling ball and pin.

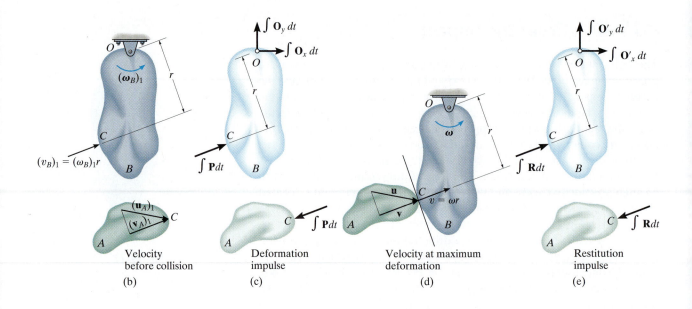

(b) Velocity before collision   (c) Deformation impulse   (d) Velocity at maximum deformation   (e) Restitution impulse

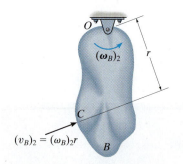

(f) Velocity after collision

Fig. 19–11 (cont.)

Is is important to realize, however, that *this analysis has only a very limited application in engineering, because values of e for this case have been found to be highly sensitive to the material, geometry, and the velocity of each of the colliding bodies.* To establish a useful form of the coefficient of restitution equation we must first apply the principle of angular impulse and momentum about point O to bodies B and A separately. Combining the results, we then obtain the necessary equation. Proceeding in this manner, the principle of impulse and momentum applied to body B from the time just before the collision to the instant of maximum deformation, Figs. 19–11b, 19–11c, and 19–11d, becomes

$$(\zeta+) \qquad\qquad I_O(\omega_B)_1 + r\int P\,dt = I_O\omega \qquad\qquad (19\text{–}18)$$

Here I_O is the moment of inertia of body B about point O. Similarly, applying the principle of angular impulse and momentum from the instant of maximum deformation to the time just after the impact, Figs. 19–11d, 19–11e, and 19–11f, yields

$$(\zeta+) \qquad\qquad I_O\omega + r\int R\,dt = I_O(\omega_B)_2 \qquad\qquad (19\text{–}19)$$

Solving Eqs. 19–18 and 19–19 for $\int P\,dt$ and $\int R\,dt$, respectively, and formulating e, we have

$$e = \frac{\displaystyle\int R\,dt}{\displaystyle\int P\,dt} = \frac{r(\omega_B)_2 - r\omega}{r\omega - r(\omega_B)_1} = \frac{(v_B)_2 - v}{v - (v_B)_1}$$

In the same manner, we can write an equation which relates the magnitudes of velocity $(v_A)_1$ and $(v_A)_2$ of body A. The result is

$$e = \frac{v - (v_A)_2}{(v_A)_1 - v}$$

Combining the above two equations by eliminating the common velocity v yields the desired result, i.e.,

$(+\nearrow)$

$$e = \frac{(v_B)_2 - (v_A)_2}{(v_A)_1 - (v_B)_1} \qquad (19\text{--}20)$$

This equation is identical to Eq. 15–11, which was derived for the central impact between two particles. It states that the coefficient of restitution is equal to the ratio of the relative velocity of *separation* of the points of contact (C) *just after impact* to the relative velocity at which the points *approach* one another *just* before impact. In deriving this equation, we assumed that the points of contact for both bodies move up and to the right *both* before and after impact. If motion of any one of the contacting points occurs down and to the left, the velocity of this point should be considered a negative quantity in Eq. 19–20.

During impact the columns of many highway signs are intended to break out of their supports and easily collapse at their joints. This is shown by the slotted connections at their base and the breaks at the column's midsection.

EXAMPLE 19.8

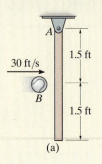

(a)

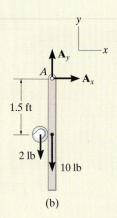

(b)

The 10-lb slender rod is suspended from the pin at A, Fig. 19–12a. If a 2-lb ball B is thrown at the rod and strikes its center with a velocity of 30 ft/s, determine the angular velocity of the rod just after impact. The coefficient of restitution is $e = 0.4$.

SOLUTION

Conservation of Angular Momentum. Consider the ball and rod as a system, Fig. 19–12b. Angular momentum is conserved about point A since the impulsive force between the rod and ball is *internal*. Also, the *weights* of the ball and rod are *nonimpulsive*. Noting the directions of the velocities of the ball and rod just after impact as shown on the kinematic diagram, Fig. 19–12c, we require

$$(\zeta+) \qquad\qquad (H_A)_1 = (H_A)_2$$

$$m_B(v_B)_1(1.5 \text{ ft}) = m_B(v_B)_2(1.5 \text{ ft}) + m_R(v_G)_2(1.5 \text{ ft}) + I_G\omega_2$$

$$\left(\frac{2 \text{ lb}}{32.2 \text{ ft/s}^2}\right)(30 \text{ ft/s})(1.5 \text{ ft}) = \left(\frac{2 \text{ lb}}{32.2 \text{ ft/s}^2}\right)(v_B)_2(1.5 \text{ ft}) +$$

$$\left(\frac{10 \text{ lb}}{32.2 \text{ ft/s}^2}\right)(v_G)_2(1.5 \text{ ft}) + \left[\frac{1}{12}\left(\frac{10 \text{ lb}}{32.2 \text{ ft/s}^2}\right)(3 \text{ ft})^2\right]\omega_2$$

Since $(v_G)_2 = 1.5\omega_2$ then

$$2.795 = 0.09317(v_B)_2 + 0.9317\omega_2 \qquad\qquad (1)$$

Coefficient of Restitution. With reference to Fig. 19–12c, we have

$$(\,\overset{+}{\rightarrow}\,) \qquad e = \frac{(v_G)_2 - (v_B)_2}{(v_B)_1 - (v_G)_1} \qquad 0.4 = \frac{(1.5 \text{ ft})\omega_2 - (v_B)_2}{30 \text{ ft/s} - 0}$$

$$12.0 = 1.5\omega_2 - (v_B)_2$$

Solving,

$$(v_B)_2 = -6.52 \text{ ft/s} = 6.52 \text{ ft/s} \leftarrow$$

$$\omega_2 = 3.65 \text{ rad/s} \;\zeta \qquad\qquad\qquad\qquad Ans.$$

(c)

Fig. 19–12

PROBLEMS

•19–33. The 75-kg gymnast lets go of the horizontal bar in a fully stretched position A, rotating with an angular velocity of $\omega_A = 3$ rad/s. Estimate his angular velocity when he assumes a tucked position B. Assume the gymnast at positions A and B as a uniform slender rod and a uniform circular disk, respectively.

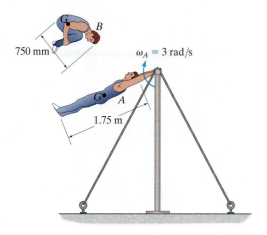

Prob. 19–33

19–34. A 75-kg man stands on the turntable A and rotates a 6-kg slender rod over his head. If the angular velocity of the rod is $\omega_r = 5$ rad/s measured relative to the man and the turntable is observed to be rotating in the opposite direction with an angular velocity of $\omega_t = 3$ rad/s, determine the radius of gyration of the man about the z axis. Consider the turntable as a thin circular disk of 300-mm radius and 5-kg mass.

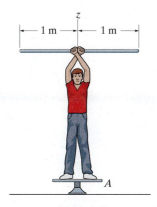

Prob. 19–34

19–35. A horizontal circular platform has a weight of 300 lb and a radius of gyration $k_z = 8$ ft about the z axis passing through its center O. The platform is free to rotate about the z axis and is initially at rest. A man having a weight of 150 lb begins to run along the edge in a circular path of radius 10 ft. If he maintains a speed of 4 ft/s relative to the platform, determine the angular velocity of the platform. Neglect friction.

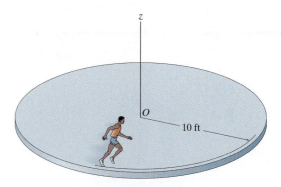

Prob. 19–35

*19–36. A horizontal circular platform has a weight of 300 lb and a radius of gyration $k_z = 8$ ft about the z axis passing through its center O. The platform is free to rotate about the z axis and is initially at rest. A man having a weight of 150 lb throws a 15-lb block off the edge of the platform with a horizontal velocity of 5 ft/s, *measured relative to the platform*. Determine the angular velocity of the platform if the block is thrown (a) tangent to the platform, along the $+t$ axis, and (b) outward along a radial line, or $+n$ axis. Neglect the size of the man.

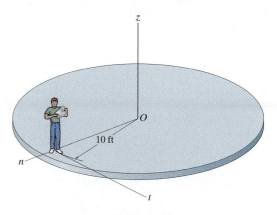

Prob. 19–36

•19–37. The man sits on the swivel chair holding two 5-lb weights with his arms outstretched. If he is rotating at 3 rad/s in this position, determine his angular velocity when the weights are drawn in and held 0.3 ft from the axis of rotation. Assume he weighs 160 lb and has a radius of gyration $k_z = 0.55$ ft about the z axis. Neglect the mass of his arms and the size of the weights for the calculation.

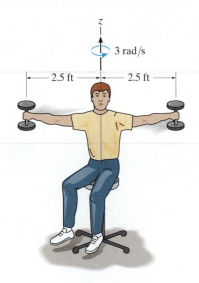

Prob. 19–37

19–38. The satellite's body C has a mass of 200 kg and a radius of gyration about the z axis of $k_z = 0.2$ m. If the satellite rotates about the z axis with an angular velocity of 5 rev/s, when the solar panels A and B are in a position of $\theta = 0°$, determine the angular velocity of the satellite when the solar panels are rotated to a position of $\theta = 90°$. Consider each solar panel to be a thin plate having a mass of 30 kg. Neglect the mass of the rods.

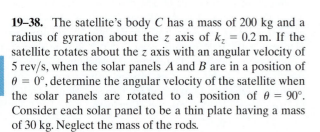

Prob. 19–38

19–39. A 150-lb man leaps off the circular platform with a velocity of $v_{m/p} = 5$ ft/s, relative to the platform. Determine the angular velocity of the platform afterwards. Initially the man and platform are at rest. The platform weighs 300 lb and can be treated as a uniform circular disk.

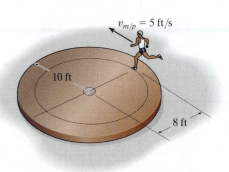

Prob. 19–39

*19–40. The 150-kg platform can be considered as a circular disk. Two men, A and B, of 60-kg and 75-kg mass, respectively, stand on the platform when it is at rest. If they start to walk around the circular paths with speeds of $v_{A/p} = 1.5$ m/s and $v_{B/p} = 2$ m/s, measured relative to the platform, determine the angular velocity of the platform.

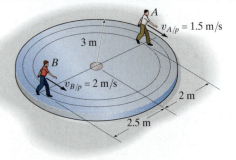

Prob. 19–40

•19–41. Two children A and B, each having a mass of 30 kg, sit at the edge of the merry-go-round which rotates at $\omega = 2$ rad/s. Excluding the children, the merry-go-round has a mass of 180 kg and a radius of gyration $k_z = 0.6$ m. Determine the angular velocity of the merry-go-round if A jumps off horizontally in the $-n$ direction with a speed of 2 m/s, measured relative to the merry-go-round. What is the merry-go-round's angular velocity if B then jumps off horizontally in the $-t$ direction with a speed of 2 m/s, measured relative to the merry-go-round? Neglect friction and the size of each child.

19–43. A ball having a mass of 8 kg and initial speed of $v_1 = 0.2$ m/s rolls over a 30-mm-long depression. Assuming that the ball rolls off the edges of contact first A, then B, without slipping, determine its final velocity v_2 when it reaches the other side.

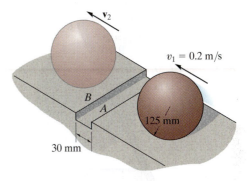

Prob. 19–43

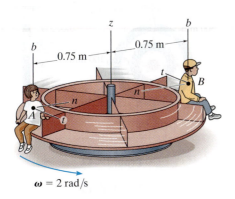

Prob. 19–41

19–42. A thin square plate of mass m rotates on the smooth surface with an angular velocity ω_1. Determine its new angular velocity just after the hook at its corner strikes the peg P and the plate starts to rotate about P without rebounding.

*19–44. The 15-kg thin ring strikes the 20-mm-high step. Determine the smallest angular velocity ω_1 the ring can have so that it will just roll over the step at A without slipping

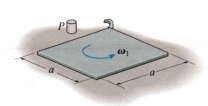

Prob. 19–42

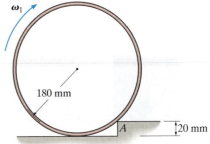

Prob. 19–44

R2–43. The disk rotates at a constant rate of 4 rad/s as it falls freely so that its center G has an acceleration of 32.2 ft/s². Determine the accelerations of points A and B on the rim of the disk at the instant shown.

R2–45. Shown is the internal gearing of a "spinner" used for drilling wells. With constant angular acceleration, the motor M rotates the shaft S to 100 rev/min in $t = 2$ s starting from rest. Determine the angular acceleration of the drill-pipe connection D and the number of revolutions it makes during the 2-s startup.

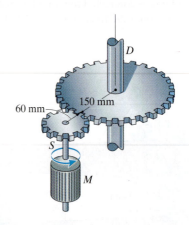

Prob. R2–45

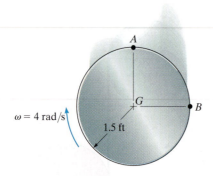

Prob. R2–43

*R2–44.** The operation of "reverse" for a three-speed automotive transmission is illustrated schematically in the figure. If the shaft G is turning with an angular velocity of $\omega_G = 60$ rad/s, determine the angular velocity of the drive shaft H. Each of the gears rotates about a fixed axis. Note that gears A and B, C and D, E and F are in mesh. The radius of each of these gears is reported in the figure.

R2–46. Gear A has a mass of 0.5 kg and a radius of gyration of $k_A = 40$ mm, and gear B has a mass of 0.8 kg and a radius of gyration of $k_B = 55$ mm. The link is pinned at C and has a mass of 0.35 kg. If the link can be treated as a slender rod, determine the angular velocity of the link after the assembly is released from rest when $\theta = 0°$ and falls to $\theta = 90°$.

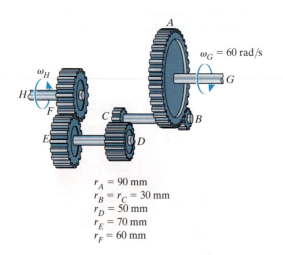

$r_A = 90$ mm
$r_B = r_C = 30$ mm
$r_D = 50$ mm
$r_E = 70$ mm
$r_F = 60$ mm

Prob. R2–44

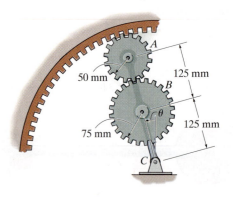

Prob. R2–46

R2–47. The 15-kg cylinder rotates with an angular velocity of $\omega = 40$ rad/s. If a force $F = 6$ N is applied to bar AB, as shown, determine the time needed to stop the rotation. The coefficient of kinetic friction between AB and the cylinder is $\mu_k = 0.4$. Neglect the thickness of the bar.

R2–49. If the thin hoop has a weight W and radius r and is thrown onto a *rough surface* with a velocity $\mathbf{v}_G$ parallel to the surface, determine the backspin, $\boldsymbol{\omega}$, it must be given so that it stops spinning at the same instant that its forward velocity is zero. It is not necessary to know the coefficient of kinetic friction at A for the calculation.

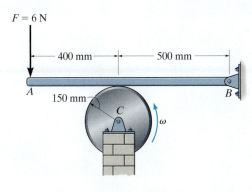

Prob. R2–47

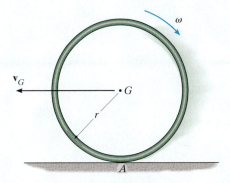

Prob. R2–49

*R2–48. If link AB rotates at $\omega_{AB} = 6$ rad/s, determine the angular velocities of links BC and CD at the instant shown.

R2–50. The wheel has a mass of 50 kg and a radius of gyration $k_G = 0.4$ m. If it rolls without slipping down the inclined plank, determine the horizontal and vertical components of reaction at A, and the normal reaction at the smooth support B at the instant the wheel is located at the midpoint of the plank. The plank has negligible thickness and has a mass of 20 kg.

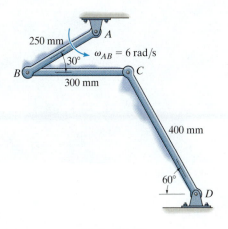

Prob. R2–48

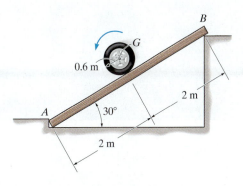

Prob. R2–50

The three-dimensional motion of this industrial robot must be accurately specified.

Three-Dimensional Kinematics of a Rigid Body

20

CHAPTER OBJECTIVES

- To analyze the kinematics of a body subjected to rotation about a fixed point and general plane motion.

- To provide a relative-motion analysis of a rigid body using translating and rotating axes.

20.1   Rotation About a Fixed Point

When a rigid body rotates about a fixed point, the distance r from the point to a particle located on the body is the *same* for *any position* of the body. Thus, the path of motion for the particle lies on the *surface of a sphere* having a radius r and centered at the fixed point. Since motion along this path occurs only from a series of rotations made during a finite time interval, we will first develop a familiarity with some of the properties of rotational displacements.

The boom can rotate up and down, and because it is hinged at a point on the vertical axis about which it turns, it is subjected to rotation about a fixed point.

Euler's Theorem.
Euler's theorem states that two "component" rotations about different axes passing through a point are equivalent to a single resultant rotation about an axis passing through the point. If more than two rotations are applied, they can be combined into pairs, and each pair can be further reduced and combined into one rotation.

Finite Rotations.
If component rotations used in Euler's theorem are *finite*, it is important that the *order* in which they are applied be maintained. To show this, consider the two finite rotations $\theta_1 + \theta_2$ applied to the block in Fig. 20–1a. Each rotation has a magnitude of 90° and a direction defined by the right-hand rule, as indicated by the arrow. The final position of the block is shown at the right. When these two rotations are applied in the order $\theta_2 + \theta_1$, as shown in Fig. 20–1b, the final position of the block is *not* the same as it is in Fig. 20–1a. Because *finite rotations* do not obey the commutative law of addition $(\theta_1 + \theta_2 \neq \theta_2 + \theta_1)$, *they cannot be classified as vectors.* If smaller, yet finite, rotations had been used to illustrate this point, e.g., 10° instead of 90°, the *final position* of the block after each combination of rotations would also be different; however, in this case, the difference is only a small amount.

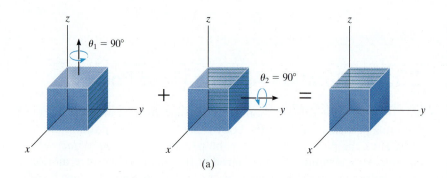

(a)

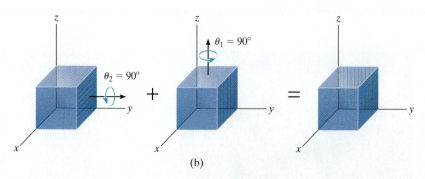

(b)

Fig. 20–1

Infinitesimal Rotations.

When defining the angular motions of a body subjected to three-dimensional motion, only rotations which are *infinitesimally small* will be considered. *Such rotations can be classified as vectors, since they can be added vectorially in any manner.* To show this, for purposes of simplicity let us consider the rigid body itself to be a sphere which is allowed to rotate about its central fixed point O, Fig. 20–2a. If we impose two infinitesimal rotations $d\boldsymbol{\theta}_1 + d\boldsymbol{\theta}_2$ on the body, it is seen that point P moves along the path $d\boldsymbol{\theta}_1 \times \mathbf{r} + d\boldsymbol{\theta}_2 \times \mathbf{r}$ and ends up at P'. Had the two successive rotations occurred in the order $d\boldsymbol{\theta}_2 + d\boldsymbol{\theta}_1$, then the resultant displacements of P would have been $d\boldsymbol{\theta}_2 \times \mathbf{r} + d\boldsymbol{\theta}_1 \times \mathbf{r}$. Since the vector cross product obeys the distributive law, by comparison $(d\boldsymbol{\theta}_1 + d\boldsymbol{\theta}_2) \times \mathbf{r} = (d\boldsymbol{\theta}_2 + d\boldsymbol{\theta}_1) \times \mathbf{r}$. Here infinitesimal rotations $d\boldsymbol{\theta}$ are vectors, since these quantities have both a magnitude and direction for which the order of (vector) addition is not important, i.e., $d\boldsymbol{\theta}_1 + d\boldsymbol{\theta}_2 = d\boldsymbol{\theta}_2 + d\boldsymbol{\theta}_1$. As a result, as shown in Fig. 20–2a, the two "component" rotations $d\boldsymbol{\theta}_1$ and $d\boldsymbol{\theta}_2$ are equivalent to a single resultant rotation $d\boldsymbol{\theta} = d\boldsymbol{\theta}_1 + d\boldsymbol{\theta}_2$, a consequence of Euler's theorem.

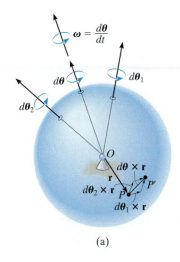

(a)

Angular Velocity.

If the body is subjected to an angular rotation $d\boldsymbol{\theta}$ about a fixed point, the angular velocity of the body is defined by the time derivative,

$$\boldsymbol{\omega} = \dot{\boldsymbol{\theta}} \qquad (20\text{–}1)$$

The line specifying the direction of $\boldsymbol{\omega}$, which is collinear with $d\boldsymbol{\theta}$, is referred to as the *instantaneous axis of rotation*, Fig. 20–2b. In general, this axis changes direction during each instant of time. Since $d\boldsymbol{\theta}$ is a vector quantity, so too is $\boldsymbol{\omega}$, and it follows from vector addition that if the body is subjected to two component angular motions, $\boldsymbol{\omega}_1 = \dot{\boldsymbol{\theta}}_1$ and $\boldsymbol{\omega}_2 = \dot{\boldsymbol{\theta}}_2$, the resultant angular velocity is $\boldsymbol{\omega} = \boldsymbol{\omega}_1 + \boldsymbol{\omega}_2$.

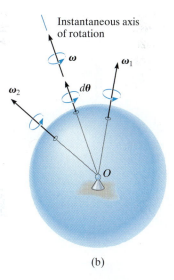

(b)

Fig. 20–2

Angular Acceleration.

The body's angular acceleration is determined from the time derivative of its angular velocity, i.e.,

$$\boldsymbol{\alpha} = \dot{\boldsymbol{\omega}} \qquad (20\text{–}2)$$

For motion about a fixed point, $\boldsymbol{\alpha}$ must account for a change in *both* the magnitude and direction of $\boldsymbol{\omega}$, so that, in general, $\boldsymbol{\alpha}$ is not directed along the instantaneous axis of rotation, Fig. 20–3.

As the direction of the instantaneous axis of rotation (or the line of action of $\boldsymbol{\omega}$) changes in space, the locus of the axis generates a fixed *space cone*, Fig. 20–4. If the change in the direction of this axis is viewed with respect to the rotating body, the locus of the axis generates a *body cone*.

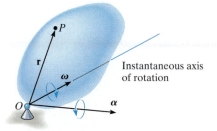

Fig. 20–3

20

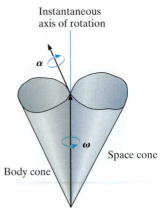

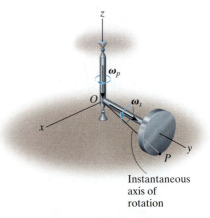

Instantaneous
axis of rotation

Space cone

Body cone

Fig. 20–4

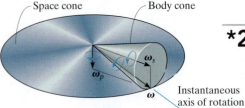

Instantaneous
axis of
rotation

(a)

Space cone   Body cone

Instantaneous
axis of rotation

(b)

Fig. 20–5

At any given instant, these cones meet along the instantaneous axis of rotation, and when the body is in motion, the body cone appears to roll either on the inside or the outside surface of the fixed space cone. Provided the paths defined by the open ends of the cones are described by the head of the $\boldsymbol{\omega}$ vector, then $\boldsymbol{\alpha}$ must act tangent to these paths at any given instant, since the time rate of change of $\boldsymbol{\omega}$ is equal to $\boldsymbol{\alpha}$. Fig. 20–4.

To illustrate this concept, consider the disk in Fig. 20–5a that spins about the rod at $\boldsymbol{\omega}_s$, while the rod and disk precess about the vertical axis at $\boldsymbol{\omega}_p$. The resultant angular velocity of the disk is therefore $\boldsymbol{\omega} = \boldsymbol{\omega}_s + \boldsymbol{\omega}_p$. Since both point O and the contact point P have zero velocity, then both $\boldsymbol{\omega}$ and the instantaneous axis of rotation are along OP. Therefore, as the disk rotates, this axis appears to move along the surface of the fixed space cone shown in Fig. 20–5b. If the axis is observed from the rotating disk, the axis then appears to move on the surface of the body cone. At any instant, though, these two cones meet each other along the axis OP. If $\boldsymbol{\omega}$ has a constant magnitude, then $\boldsymbol{\alpha}$ indicates only the change in the direction of $\boldsymbol{\omega}$, which is tangent to the cones at the tip of $\boldsymbol{\omega}$ as shown in Fig. 20–5b.

Velocity.   Once $\boldsymbol{\omega}$ is specified, the velocity of any point on a body rotating about a fixed point can be determined using the same methods as for a body rotating about a fixed axis. Hence, by the cross product,

$$\mathbf{v} = \boldsymbol{\omega} \times \mathbf{r} \tag{20–3}$$

Here $\mathbf{r}$ defines the position of the point measured from the fixed point O, Fig. 20–3.

Acceleration.   If $\boldsymbol{\omega}$ and $\boldsymbol{\alpha}$ are known at a given instant, the acceleration of a point can be obtained from the time derivative of Eq. 20–3, which yields

$$\mathbf{a} = \boldsymbol{\alpha} \times \mathbf{r} + \boldsymbol{\omega} \times (\boldsymbol{\omega} \times \mathbf{r}) \tag{20–4}$$

*20.2   The Time Derivative of a Vector Measured from Either a Fixed or Translating-Rotating System

In many types of problems involving the motion of a body about a fixed point, the angular velocity $\boldsymbol{\omega}$ is specified in terms of its components. Then, if the angular acceleration $\boldsymbol{\alpha}$ of such a body is to be determined, it is often easier to compute the time derivative of $\boldsymbol{\omega}$ using a coordinate system that has a *rotation* defined by one or more of the components of $\boldsymbol{\omega}$. For example, in the case of the disk in Fig. 20–5a, where $\boldsymbol{\omega} = \boldsymbol{\omega}_s + \boldsymbol{\omega}_p$, the x, y, z axes can be given an angular velocity of $\boldsymbol{\omega}_p$. For this reason, and for other uses later, an equation will now be derived, which relates the time derivative of any vector $\mathbf{A}$ defined from a translating-rotating reference to its time derivative defined from a fixed reference.

Consider the x, y, z axes of the moving frame of reference to be rotating with an angular velocity $\mathbf{\Omega}$, which is measured from the fixed X, Y, Z axes, Fig. 20–6a. In the following discussion, it will be convenient to express vector $\mathbf{A}$ in terms of its $\mathbf{i}$, $\mathbf{j}$, $\mathbf{k}$ components, which define the directions of the moving axes. Hence,

$$\mathbf{A} = A_x\mathbf{i} + A_y\mathbf{j} + A_z\mathbf{k}$$

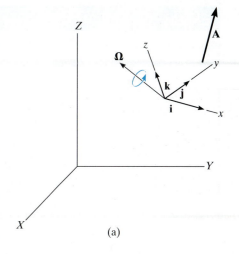

(a)

In general, the time derivative of $\mathbf{A}$ must account for the change in both its magnitude and direction. However, if this derivative is taken *with respect to the moving frame of reference*, only the change in the magnitudes of the components of $\mathbf{A}$ must be accounted for, since the directions of the components do not change with respect to the moving reference. Hence,

$$(\dot{\mathbf{A}})_{xyz} = \dot{A}_x\mathbf{i} + \dot{A}_y\mathbf{j} + \dot{A}_z\mathbf{k} \qquad (20\text{–}5)$$

When the time derivative of $\mathbf{A}$ is taken *with respect to the fixed frame of reference*, the *directions* of $\mathbf{i}$, $\mathbf{j}$, and $\mathbf{k}$ change only on account of the *rotation* $\mathbf{\Omega}$ of the axes and not their translation. Hence, in general,

$$\dot{\mathbf{A}} = \dot{A}_x\mathbf{i} + \dot{A}_y\mathbf{j} + \dot{A}_z\mathbf{k} + A_x\dot{\mathbf{i}} + A_y\dot{\mathbf{j}} + A_z\dot{\mathbf{k}}$$

The time derivatives of the unit vectors will now be considered. For example, $\dot{\mathbf{i}} = d\mathbf{i}/dt$ represents only the change in the *direction* of $\mathbf{i}$ with respect to time, since $\mathbf{i}$ always has a magnitude of 1 unit. As shown in Fig. 20–6b, the change, $d\mathbf{i}$, is *tangent to the path* described by the arrowhead of $\mathbf{i}$ as $\mathbf{i}$ swings due to the rotation $\mathbf{\Omega}$. Accounting for both the magnitude and direction of $d\mathbf{i}$, we can therefore define $\dot{\mathbf{i}}$ using the cross product, $\dot{\mathbf{i}} = \mathbf{\Omega} \times \mathbf{i}$. In general, then

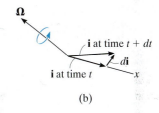

(b)

Fig. 20–6

$$\dot{\mathbf{i}} = \mathbf{\Omega} \times \mathbf{i} \qquad \dot{\mathbf{j}} = \mathbf{\Omega} \times \mathbf{j} \qquad \dot{\mathbf{k}} = \mathbf{\Omega} \times \mathbf{k}$$

These formulations were also developed in Sec. 16.8, regarding planar motion of the axes. Substituting these results into the above equation and using Eq. 20–5 yields

$$\boxed{\dot{\mathbf{A}} = (\dot{\mathbf{A}})_{xyz} + \mathbf{\Omega} \times \mathbf{A}} \qquad (20\text{–}6)$$

This result is important, and will be used throughout Sec. 20.4 and Chapter 21. It states that the time derivative of *any vector* $\mathbf{A}$ as observed from the fixed X, Y, Z frame of reference is equal to the time rate of change of $\mathbf{A}$ as observed from the x, y, z translating-rotating frame of reference, Eq. 20–5, plus $\mathbf{\Omega} \times \mathbf{A}$, the change of $\mathbf{A}$ caused by the rotation of the x, y, z frame. As a result, Eq. 20–6 should always be used whenever $\mathbf{\Omega}$ produces a change in the direction of $\mathbf{A}$ as seen from the X, Y, Z reference. If this change does not occur, i.e., $\mathbf{\Omega} = \mathbf{0}$, then $\dot{\mathbf{A}} = (\dot{\mathbf{A}})_{xyz}$, and so the time rate of change of $\mathbf{A}$ as observed from both coordinate systems will be the *same*.

20

EXAMPLE 20.1

The disk shown in Fig. 20–7 spins about its axle with a constant angular velocity $\omega_s = 3$ rad/s, while the horizontal platform on which the disk is mounted rotates about the vertical axis at a constant rate $\omega_p = 1$ rad/s. Determine the angular acceleration of the disk and the velocity and acceleration of point A on the disk when it is in the position shown.

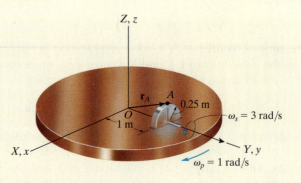

Fig. 20–7

SOLUTION

Point O represents a fixed point of rotation for the disk if one considers a hypothetical extension of the disk to this point. To determine the velocity and acceleration of point A, it is first necessary to determine the angular velocity $\boldsymbol{\omega}$ and angular acceleration $\boldsymbol{\alpha}$ of the disk, since these vectors are used in Eqs. 20–3 and 20–4.

Angular Velocity. The angular velocity, which is measured from X, Y, Z, is simply the vector addition of its two component motions. Thus,

$$\boldsymbol{\omega} = \boldsymbol{\omega}_s + \boldsymbol{\omega}_p = \{3\mathbf{j} - 1\mathbf{k}\} \text{ rad/s}$$

Angular Acceleration.   Since the magnitude of $\boldsymbol{\omega}$ is constant, only a change in its direction, as seen from the fixed reference, creates the angular acceleration $\boldsymbol{\alpha}$ of the disk. One way to obtain $\boldsymbol{\alpha}$ is to compute the time derivative of *each of the two components* of $\boldsymbol{\omega}$ using Eq. 20–6. At the instant shown in Fig. 20–7, imagine the fixed X, Y, Z and a rotating x, y, z frame to be coincident. If the rotating x, y, z frame is chosen to have an angular velocity of $\boldsymbol{\Omega} = \boldsymbol{\omega}_p = \{-1\mathbf{k}\}$ rad/s, then $\boldsymbol{\omega}_s$ will *always* be directed along the y (not Y) axis, and the time rate of change of $\boldsymbol{\omega}_s$ as seen from x, y, z is *zero*; i.e., $(\dot{\boldsymbol{\omega}}_s)_{xyz} = \mathbf{0}$ (the magnitude and direction of $\boldsymbol{\omega}_s$ is constant). Thus,

$$\dot{\boldsymbol{\omega}}_s = (\dot{\boldsymbol{\omega}}_s)_{xyz} + \boldsymbol{\omega}_p \times \boldsymbol{\omega}_s = \mathbf{0} + (-1\mathbf{k}) \times (3\mathbf{j}) = \{3\mathbf{i}\} \text{ rad/s}^2$$

By the same choice of axes rotation, $\boldsymbol{\Omega} = \boldsymbol{\omega}_p$, or even with $\boldsymbol{\Omega} = \mathbf{0}$, the time derivative $(\dot{\boldsymbol{\omega}}_p)_{xyz} = \mathbf{0}$, since $\boldsymbol{\omega}_p$ has a constant magnitude and direction with respect to x, y, z. Hence,

$$\dot{\boldsymbol{\omega}}_p = (\dot{\boldsymbol{\omega}}_p)_{xyz} + \boldsymbol{\omega}_p \times \boldsymbol{\omega}_p = \mathbf{0} + \mathbf{0} = \mathbf{0}$$

The angular acceleration of the disk is therefore

$$\boldsymbol{\alpha} = \dot{\boldsymbol{\omega}} = \dot{\boldsymbol{\omega}}_s + \dot{\boldsymbol{\omega}}_p = \{3\mathbf{i}\} \text{ rad/s}^2 \qquad \textit{Ans.}$$

Velocity and Acceleration.   Since $\boldsymbol{\omega}$ and $\boldsymbol{\alpha}$ have now been determined, the velocity and acceleration of point A can be found using Eqs. 20–3 and 20–4. Realizing that $\mathbf{r}_A = \{1\mathbf{j} + 0.25\mathbf{k}\}$ m, Fig. 20–7, we have

$$\mathbf{v}_A = \boldsymbol{\omega} \times \mathbf{r}_A = (3\mathbf{j} - 1\mathbf{k}) \times (1\mathbf{j} + 0.25\mathbf{k}) = \{1.75\mathbf{i}\} \text{ m/s} \qquad \textit{Ans.}$$

$$\mathbf{a}_A = \boldsymbol{\alpha} \times \mathbf{r}_A + \boldsymbol{\omega} \times (\boldsymbol{\omega} \times \mathbf{r}_A)$$

$$= (3\mathbf{i}) \times (1\mathbf{j} + 0.25\mathbf{k}) + (3\mathbf{j} - 1\mathbf{k}) \times [(3\mathbf{j} - 1\mathbf{k}) \times (1\mathbf{j} + 0.25\mathbf{k})]$$

$$= \{-2.50\mathbf{j} - 2.25\mathbf{k}\} \text{ m/s}^2 \qquad \textit{Ans.}$$

20

EXAMPLE 20.2

At the instant $\theta = 60°$, the gyrotop in Fig. 20–8 has three components of angular motion directed as shown and having magnitudes defined as:

$$Spin: \omega_s = 10 \text{ rad/s, increasing at the rate of } 6 \text{ rad/s}^2$$

$$Nutation: \omega_n = 3 \text{ rad/s, increasing at the rate of } 2 \text{ rad/s}^2$$

$$Precession: \omega_p = 5 \text{ rad/s, increasing at the rate of } 4 \text{ rad/s}^2$$

Determine the angular velocity and angular acceleration of the top.

SOLUTION

Angular Velocity. The top rotates about the fixed point O. If the fixed and rotating frames are coincident at the instant shown, then the angular velocity can be expressed in terms of $\mathbf{i}, \mathbf{j}, \mathbf{k}$ components, with reference to the x, y, z frame; i.e.,

$$\boldsymbol{\omega} = -\omega_n \mathbf{i} + \omega_s \sin \theta \mathbf{j} + (\omega_p + \omega_s \cos \theta)\mathbf{k}$$

$$= -3\mathbf{i} + 10 \sin 60°\mathbf{j} + (5 + 10 \cos 60°)\mathbf{k}$$

$$= \{-3\mathbf{i} + 8.66\mathbf{j} + 10\mathbf{k}\} \text{ rad/s} \qquad Ans.$$

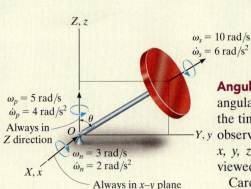

$\omega_s = 10 \text{ rad/s}$
$\dot{\omega}_s = 6 \text{ rad/s}^2$

$\omega_p = 5 \text{ rad/s}$
$\dot{\omega}_p = 4 \text{ rad/s}^2$
Always in
Z direction

$\omega_n = 3 \text{ rad/s}$
$\dot{\omega}_n = 2 \text{ rad/s}^2$

Always in x–y plane

Fig. 20–8

Angular Acceleration. As in the solution of Example 20.1, the angular acceleration $\boldsymbol{\alpha}$ will be determined by investigating separately the time rate of change of *each of the angular velocity components* as observed from the fixed X, Y, Z reference. We will choose an $\boldsymbol{\Omega}$ for the x, y, z reference so that the component of $\boldsymbol{\omega}$ being considered is viewed as having a *constant direction* when observed from x, y, z.

Careful examination of the motion of the top reveals that $\boldsymbol{\omega}_s$ has a *constant direction* relative to x, y, z if these axes rotate at $\boldsymbol{\Omega} = \boldsymbol{\omega}_n + \boldsymbol{\omega}_p$. Thus,

$$\dot{\boldsymbol{\omega}}_s = (\dot{\boldsymbol{\omega}}_s)_{xyz} + (\boldsymbol{\omega}_n + \boldsymbol{\omega}_p) \times \boldsymbol{\omega}_s$$

$$= (6 \sin 60°\mathbf{j} + 6 \cos 60°\mathbf{k}) + (-3\mathbf{i} + 5\mathbf{k}) \times (10 \sin 60°\mathbf{j} + 10 \cos 60°\mathbf{k})$$

$$= \{-43.30\mathbf{i} + 20.20\mathbf{j} - 22.98\mathbf{k}\} \text{ rad/s}^2$$

Since $\boldsymbol{\omega}_n$ *always* lies in the fixed X–Y plane, this vector has a *constant direction* if the motion is viewed from axes x, y, z having a rotation of $\boldsymbol{\Omega} = \boldsymbol{\omega}_p$ (not $\boldsymbol{\Omega} = \boldsymbol{\omega}_s + \boldsymbol{\omega}_p$). Thus,

$$\dot{\boldsymbol{\omega}}_n = (\dot{\boldsymbol{\omega}}_n)_{xyz} + \boldsymbol{\omega}_p \times \boldsymbol{\omega}_n = -2\mathbf{i} + (5\mathbf{k}) \times (-3\mathbf{i}) = \{-2\mathbf{i} - 15\mathbf{j}\} \text{ rad/s}^2$$

Finally, the component $\boldsymbol{\omega}_p$ is *always directed* along the Z axis so that here it is not necessary to think of x, y, z as rotating, i.e., $\boldsymbol{\Omega} = \mathbf{0}$. Expressing the data in terms of the $\mathbf{i}, \mathbf{j}, \mathbf{k}$ components, we therefore have

$$\dot{\boldsymbol{\omega}}_p = (\dot{\boldsymbol{\omega}}_p)_{xyz} + \mathbf{0} \times \boldsymbol{\omega}_p = \{4\mathbf{k}\} \text{ rad/s}^2$$

Thus, the angular acceleration of the top is

$$\boldsymbol{\alpha} = \dot{\boldsymbol{\omega}}_s + \dot{\boldsymbol{\omega}}_n + \dot{\boldsymbol{\omega}}_p = \{-45.3\mathbf{i} + 5.20\mathbf{j} - 19.0\mathbf{k}\} \text{ rad/s}^2 \qquad Ans.$$

20.3   General Motion

Shown in Fig. 20–9 is a rigid body subjected to general motion in three dimensions for which the angular velocity is $\boldsymbol{\omega}$ and the angular acceleration is $\boldsymbol{\alpha}$. If point A has a known motion of $\mathbf{v}_A$ and $\mathbf{a}_A$, the motion of any other point B can be determined by using a relative-motion analysis. In this section a *translating coordinate system* will be used to define the relative motion, and in the next section a reference that is both rotating and translating will be considered.

If the origin of the translating coordinate system x, y, z $(\boldsymbol{\Omega} = \mathbf{0})$ is located at the "base point" A, then, at the instant shown, the motion of the body can be regarded as the sum of an instantaneous translation of the body having a motion of $\mathbf{v}_A$, and $\mathbf{a}_A$, and a rotation of the body about an instantaneous axis passing through point A. Since the body is rigid, the motion of point B measured by an observer located at A is therefore the same as *the rotation of the body about a fixed point*. This relative motion occurs about the instantaneous axis of rotation and is defined by $\mathbf{v}_{B/A} = \boldsymbol{\omega} \times \mathbf{r}_{B/A}$, Eq. 20–3, and $\mathbf{a}_{B/A} = \boldsymbol{\alpha} \times \mathbf{r}_{B/A} + \boldsymbol{\omega} \times (\boldsymbol{\omega} \times \mathbf{r}_{B/A})$, Eq. 20–4. For translating axes, the relative motions are related to absolute motions by $\mathbf{v}_B = \mathbf{v}_A + \mathbf{v}_{B/A}$ and $\mathbf{a}_B = \mathbf{a}_A + \mathbf{a}_{B/A}$, Eqs. 16–15 and 16–17, so that the absolute velocity and acceleration of point B can be determined from the equations

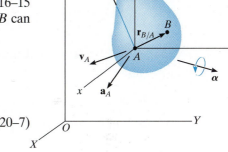

Fig. 20–9

$$\mathbf{v}_B = \mathbf{v}_A + \boldsymbol{\omega} \times \mathbf{r}_{B/A} \tag{20–7}$$

and

$$\mathbf{a}_B = \mathbf{a}_A + \boldsymbol{\alpha} \times \mathbf{r}_{B/A} + \boldsymbol{\omega} \times (\boldsymbol{\omega} \times \mathbf{r}_{B/A}) \tag{20–8}$$

These two equations are identical to those describing the general plane motion of a rigid body, Eqs. 16–16 and 16–18. However, difficulty in application arises for three-dimensional motion, because $\boldsymbol{\alpha}$ now measures the change in *both* the magnitude and direction of $\boldsymbol{\omega}$.

EXAMPLE | 20.3

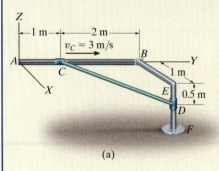

(a)

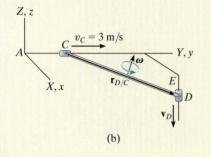

(b)

Fig. 20–10

If the collar at C in Fig. 20–10a moves towards B with a speed of 3 m/s, determine the velocity of the collar at D and the angular velocity of the bar at the instant shown. The bar is connected to the collars at its end points by ball-and-socket joints.

SOLUTION

Bar CD is subjected to general motion. Why? The velocity of point D on the bar can be related to the velocity of point C by the equation

$$\mathbf{v}_D = \mathbf{v}_C + \boldsymbol{\omega} \times \mathbf{r}_{D/C}$$

The fixed and translating frames of reference are assumed to coincide at the instant considered, Fig. 20–10b. We have

$$\mathbf{v}_D = -v_D \mathbf{k} \qquad \mathbf{v}_C = \{3\mathbf{j}\} \text{ m/s}$$

$$\mathbf{r}_{D/C} = \{1\mathbf{i} + 2\mathbf{j} - 0.5\mathbf{k}\} \text{ m} \qquad \boldsymbol{\omega} = \omega_x \mathbf{i} + \omega_y \mathbf{j} + \omega_z \mathbf{k}$$

Substituting into the above equation we get

$$-v_D \mathbf{k} = 3\mathbf{j} + \begin{vmatrix} \mathbf{i} & \mathbf{j} & \mathbf{k} \\ \omega_x & \omega_y & \omega_z \\ 1 & 2 & -0.5 \end{vmatrix}$$

Expanding and equating the respective $\mathbf{i}, \mathbf{j}, \mathbf{k}$ components yields

$$-0.5\omega_y - 2\omega_z = 0 \tag{1}$$
$$0.5\omega_x + 1\omega_z + 3 = 0 \tag{2}$$
$$2\omega_x - 1\omega_y + v_D = 0 \tag{3}$$

These equations contain four unknowns.* A fourth equation can be written if the direction of $\boldsymbol{\omega}$ is specified. In particular, any component of $\boldsymbol{\omega}$ acting along the bar's axis has no effect on moving the collars. This is because the bar is *free to rotate* about its axis. Therefore, if $\boldsymbol{\omega}$ is specified as acting *perpendicular* to the axis of the bar, then $\boldsymbol{\omega}$ must have a unique magnitude to satisfy the above equations. Perpendicularity is guaranteed provided the dot product of $\boldsymbol{\omega}$ and $\mathbf{r}_{D/C}$ is zero (see Eq. C–14 of Appendix C). Hence,

$$\boldsymbol{\omega} \cdot \mathbf{r}_{D/C} = (\omega_x \mathbf{i} + \omega_y \mathbf{j} + \omega_z \mathbf{k}) \cdot (1\mathbf{i} + 2\mathbf{j} - 0.5\mathbf{k}) = 0$$

$$1\omega_x + 2\omega_y - 0.5\omega_z = 0 \tag{4}$$

Solving Eqs. 1 through 4 simultaneously yields

$$\omega_x = -4.86 \text{ rad/s} \quad \omega_y = 2.29 \text{ rad/s} \quad \omega_z = -0.571 \text{ rad/s} \quad \textit{Ans.}$$

$$v_D = 12.0 \text{ m/s} \downarrow \qquad \textit{Ans.}$$

---

*Although this is the case, the magnitude of $\mathbf{v}_D$ can be obtained. For example, solve Eqs. 1 and 2 for ω_y and ω_x in terms of ω_z and substitute into Eq. 3. It will be noted that ω_z will cancel out, which will allow a solution for v_D.

PROBLEMS

•**20–1.** The anemometer located on the ship at A spins about its own axis at a rate ω_s, while the ship rolls about the x axis at the rate ω_x and about the y axis at the rate ω_y. Determine the angular velocity and angular acceleration of the anemometer at the instant the ship is level as shown. Assume that the magnitudes of all components of angular velocity are constant and that the rolling motion caused by the sea is independent in the x and y directions.

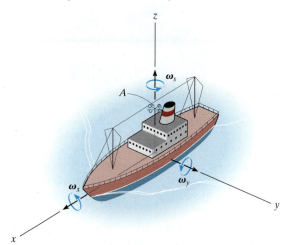

Prob. 20–1

20–2. The motion of the top is such that at the instant shown it rotates about the z axis at $\omega_1 = 0.6$ rad/s, while it spins at $\omega_2 = 8$ rad/s. Determine the angular velocity and angular acceleration of the top at this instant. Express the result as a Cartesian vector.

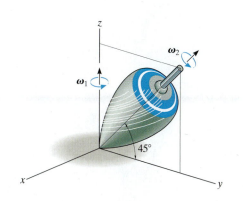

Prob. 20–2

20–3. At a given instant, the satellite dish has an angular motion $\omega_1 = 6$ rad/s and $\dot{\omega}_1 = 3$ rad/s² about the z axis. At this same instant $\theta = 25°$, the angular motion about the x axis is $\omega_2 = 2$ rad/s, and $\dot{\omega}_2 = 1.5$ rad/s². Determine the velocity and acceleration of the signal horn A at this instant.

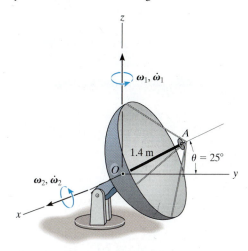

Prob. 20–3

*20–4. The fan is mounted on a swivel support such that at the instant shown it is rotating about the z axis at $\omega_1 = 0.8$ rad/s, which is increasing at 12 rad/s². The blade is spinning at $\omega_2 = 16$ rad/s, which is decreasing at 2 rad/s². Determine the angular velocity and angular acceleration of the blade at this instant.

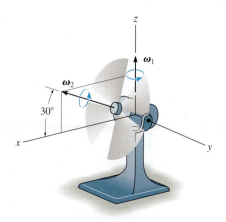

Prob. 20–4

20

•**20–5.** Gears A and B are fixed, while gears C and D are free to rotate about the shaft S. If the shaft turns about the z axis at a constant rate of $\omega_1 = 4$ rad/s, determine the angular velocity and angular acceleration of gear C.

20–7. If the top gear B rotates at a constant rate of ω, determine the angular velocity of gear A, which is free to rotate about the shaft and rolls on the bottom fixed gear C.

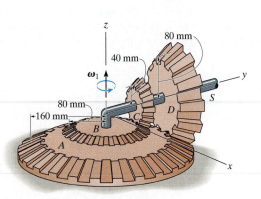

Prob. 20–5

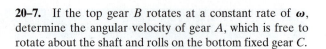

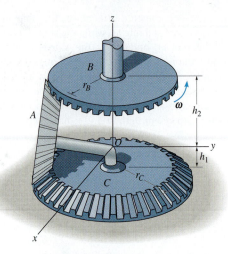

Prob. 20–7

20–6. The disk rotates about the z axis $\omega_z = 0.5$ rad/s without slipping on the horizontal plane. If at this same instant ω_z is increasing at $\dot{\omega}_z = 0.3$ rad/s², determine the velocity and acceleration of point A on the disk.

*20–8. The telescope is mounted on the frame F that allows it to be directed to any point in the sky. At the instant $\theta = 30°$, the frame has an angular acceleration of $\alpha_{y'} = 0.2$ rad/s² and an angular velocity of $\omega_{y'} = 0.3$ rad/s about the y' axis, and $\ddot{\theta} = 0.5$ rad/s² while $\dot{\theta} = 0.4$ rad/s. Determine the velocity and acceleration of the observing capsule at C at this instant.

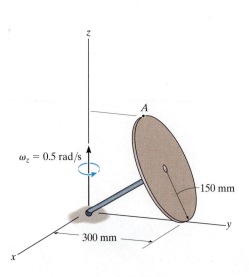

Prob. 20–6

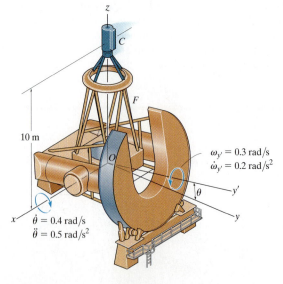

Prob. 20–8

•**20–9.** At the instant when $\theta = 90°$, the satellite's body is rotating with an angular velocity of $\omega_1 = 15$ rad/s and angular acceleration of $\dot{\omega}_1 = 3$ rad/s². Simultaneously, the solar panels rotate with an angular velocity of $\omega_2 = 6$ rad/s and angular acceleration of $\dot{\omega}_2 = 1.5$ rad/s². Determine the velocity and acceleration of point B on the solar panel at this instant.

20–10. At the instant when $\theta = 90°$, the satellite's body travels in the x direction with a velocity of $\mathbf{v}_O = \{500\mathbf{i}\}$ m/s and acceleration of $\mathbf{a}_O = \{50\mathbf{i}\}$ m/s². Simultaneously, the body also rotates with an angular velocity of $\omega_1 = 15$ rad/s and angular acceleration of $\dot{\omega}_1 = 3$ rad/s². At the same time, the solar panels rotate with an angular velocity of $\omega_2 = 6$ rad/s and angular acceleration of $\dot{\omega}_2 = 1.5$ rad/s² Determine the velocity and acceleration of point B on the solar panel.

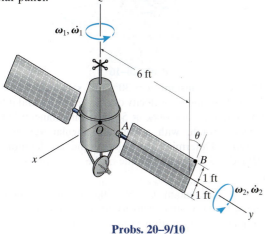

Probs. 20–9/10

20–11. The cone rolls in a circle and rotates about the z axis at a constant rate $\omega_z = 8$ rad/s. Determine the angular velocity and angular acceleration of the cone if it rolls without slipping. Also, what are the velocity and acceleration of point A?

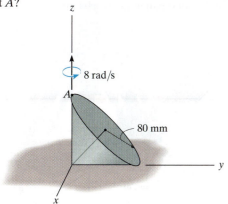

Prob. 20–11

***20–12.** At the instant shown, the motor rotates about the z axis with an angular velocity of $\omega_1 = 3$ rad/s and angular acceleration of $\dot{\omega}_1 = 1.5$ rad/s². Simultaneously, shaft OA rotates with an angular velocity of $\omega_2 = 6$ rad/s and angular acceleration of $\dot{\omega}_2 = 3$ rad/s², and collar C slides along rod AB with a velocity and acceleration of 6 m/s and 3 m/s². Determine the velocity and acceleration of collar C at this instant.

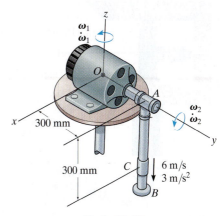

Prob. 20–12

•**20–13.** At the instant shown, the tower crane rotates about the z axis with an angular velocity $\omega_1 = 0.25$ rad/s, which is increasing at 0.6 rad/s². The boom OA rotates downward with an angular velocity $\omega_2 = 0.4$ rad/s, which is increasing at 0.8 rad/s². Determine the velocity and acceleration of point A located at the end of the boom at this instant.

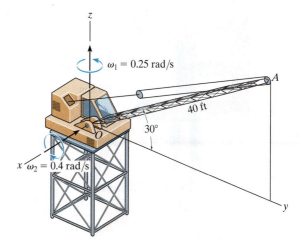

Prob. 20–13

EXAMPLE 20.4

A motor and attached rod AB have the angular motions shown in Fig. 20–12. A collar C on the rod is located 0.25 m from A and is moving downward along the rod with a velocity of 3 m/s and an acceleration of 2 m/s². Determine the velocity and acceleration of C at this instant.

SOLUTION

Coordinate Axes.
The origin of the fixed X, Y, Z reference is chosen at the center of the platform, and the origin of the moving x, y, z frame at point A, Fig. 20–12. Since the collar is subjected to two components of angular motion, ω_p and ω_M, it will be viewed as having an angular velocity of $\Omega_{xyz} = \omega_M$ in x, y, z. Therefore, the x, y, z axes will be attached to the platform so that $\Omega = \omega_p$.

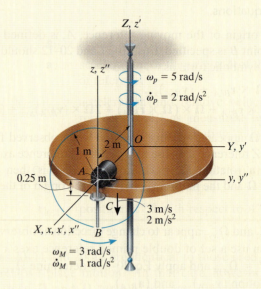

Fig. 20–12

Kinematic Equations. Equations 20–11 and 20–12, applied to points C and A, become

$$\mathbf{v}_C = \mathbf{v}_A + \mathbf{\Omega} \times \mathbf{r}_{C/A} + (\mathbf{v}_{C/A})_{xyz}$$

$$\mathbf{a}_C = \mathbf{a}_A + \dot{\mathbf{\Omega}} \times \mathbf{r}_{C/A} + \mathbf{\Omega} \times (\mathbf{\Omega} \times \mathbf{r}_{C/A}) + 2\mathbf{\Omega} \times (\mathbf{v}_{C/A})_{xyz} + (\mathbf{a}_{C/A})_{xyz}$$

Motion of A. Here $\mathbf{r}_A$ changes direction relative to X, Y, Z. To find the time derivatives of $\mathbf{r}_A$ we will use a set of x', y', z' axes coincident with the X, Y, Z axes that rotate at $\mathbf{\Omega}' = \boldsymbol{\omega}_p$. Thus,

$$\mathbf{\Omega} = \boldsymbol{\omega}_p = \{5\mathbf{k}\} \text{ rad/s } (\mathbf{\Omega} \text{ does not change direction relative to } X, Y, Z.)$$

$$\dot{\mathbf{\Omega}} = \dot{\boldsymbol{\omega}}_p = \{2\mathbf{k}\} \text{ rad/s}^2$$

$$\mathbf{r}_A = \{2\mathbf{i}\} \text{ m}$$

$$\mathbf{v}_A = \dot{\mathbf{r}}_A = (\dot{\mathbf{r}}_A)_{x'y'z'} + \boldsymbol{\omega}_p \times \mathbf{r}_A = \mathbf{0} + 5\mathbf{k} \times 2\mathbf{i} = \{10\mathbf{j}\} \text{ m/s}$$

$$\mathbf{a}_A = \ddot{\mathbf{r}}_A = [(\ddot{\mathbf{r}}_A)_{x'y'z'} + \boldsymbol{\omega}_p \times (\dot{\mathbf{r}}_A)_{x'y'z'}] + \dot{\boldsymbol{\omega}}_p \times \mathbf{r}_A + \boldsymbol{\omega}_p \times \dot{\mathbf{r}}_A$$

$$= [\mathbf{0} + \mathbf{0}] + 2\mathbf{k} \times 2\mathbf{i} + 5\mathbf{k} \times 10\mathbf{j} = \{-50\mathbf{i} + 4\mathbf{j}\} \text{ m/s}^2$$

Motion of C with Respect to A. Here $(\mathbf{r}_{C/A})_{xyz}$ changes direction relative to x, y, z. To find the time derivatives of $(\mathbf{r}_{C/A})_{xyz}$ use a set of x'', y'', z'' axes that rotate at $\mathbf{\Omega}'' = \mathbf{\Omega}_{xyz} = \boldsymbol{\omega}_M$. Thus,

$$\mathbf{\Omega}_{xyz} = \boldsymbol{\omega}_M = \{3\mathbf{i}\} \text{ rad/s } (\mathbf{\Omega}_{xyz} \text{ does not change direction relative to } x, y, z.)$$

$$\dot{\mathbf{\Omega}}_{xyz} = \dot{\boldsymbol{\omega}}_M = \{1\mathbf{i}\} \text{ rad/s}^2$$

$$(\mathbf{r}_{C/A})_{xyz} = \{-0.25\mathbf{k}\} \text{ m}$$

$$(\mathbf{v}_{C/A})_{xyz} = (\dot{\mathbf{r}}_{C/A})_{xyz} = (\dot{\mathbf{r}}_{C/A})_{x''y''z''} + \boldsymbol{\omega}_M \times (\mathbf{r}_{C/A})_{xyz}$$

$$= -3\mathbf{k} + [3\mathbf{i} \times (-0.25\mathbf{k})] = \{0.75\mathbf{j} - 3\mathbf{k}\} \text{ m/s}$$

$$(\mathbf{a}_{C/A})_{xyz} = (\ddot{\mathbf{r}}_{C/A})_{xyz} = [(\ddot{\mathbf{r}}_{C/A})_{x''y''z''} + \boldsymbol{\omega}_M \times (\dot{\mathbf{r}}_{C/A})_{x''y''z''}] + \dot{\boldsymbol{\omega}}_M \times (\mathbf{r}_{C/A})_{xyz} + \boldsymbol{\omega}_M \times (\dot{\mathbf{r}}_{C/A})_{xyz}$$

$$= [-2\mathbf{k} + 3\mathbf{i} \times (-3\mathbf{k})] + (1\mathbf{i}) \times (-0.25\mathbf{k}) + (3\mathbf{i}) \times (0.75\mathbf{j} - 3\mathbf{k})$$

$$= \{18.25\mathbf{j} + 0.25\mathbf{k}\} \text{ m/s}^2$$

Motion of C.

$$\mathbf{v}_C = \mathbf{v}_A + \mathbf{\Omega} \times \mathbf{r}_{C/A} + (\mathbf{v}_{C/A})_{xyz}$$

$$= 10\mathbf{j} + [5\mathbf{k} \times (-0.25\mathbf{k})] + (0.75\mathbf{j} - 3\mathbf{k})$$

$$= \{10.75\mathbf{j} - 3\mathbf{k}\} \text{ m/s} \qquad\qquad\qquad Ans.$$

$$\mathbf{a}_C = \mathbf{a}_A + \dot{\mathbf{\Omega}} \times \mathbf{r}_{C/A} + \mathbf{\Omega} \times (\mathbf{\Omega} \times \mathbf{r}_{C/A}) + 2\mathbf{\Omega} \times (\mathbf{v}_{C/A})_{xyz} + (\mathbf{a}_{C/A})_{xyz}$$

$$= (-50\mathbf{i} + 4\mathbf{j}) + [2\mathbf{k} \times (-0.25\mathbf{k})] + 5\mathbf{k} \times [5\mathbf{k} \times (-0.25\mathbf{k})]$$

$$+ 2[5\mathbf{k} \times (0.75\mathbf{j} - 3\mathbf{k})] + (18.25\mathbf{j} + 0.25\mathbf{k})$$

$$= \{-57.5\mathbf{i} + 22.25\mathbf{j} + 0.25\mathbf{k}\} \text{ m/s}^2 \qquad\qquad Ans.$$

20

EXAMPLE 20.5

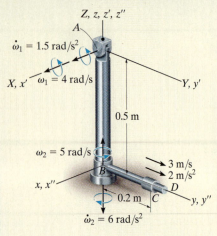

Fig 20–13

The pendulum shown in Fig. 20–13 consists of two rods; AB is pin supported at A and swings only in the Y–Z plane, whereas a bearing at B allows the attached rod BD to spin about rod AB. At a given instant, the rods have the angular motions shown. Also, a collar C, located 0.2 m from B, has a velocity of 3 m/s and an acceleration of 2 m/s^2 along the rod. Determine the velocity and acceleration of the collar at this instant.

SOLUTION I

Coordinate Axes.   The origin of the fixed X, Y, Z frame will be placed at A. Motion of the collar is conveniently observed from B, so the origin of the x, y, z frame is located at this point. We will choose $\boldsymbol{\Omega} = \boldsymbol{\omega}_1$ and $\boldsymbol{\Omega}_{xyz} = \boldsymbol{\omega}_2$.

Kinematic Equations.

$$\mathbf{v}_C = \mathbf{v}_B + \boldsymbol{\Omega} \times \mathbf{r}_{C/B} + (\mathbf{v}_{C/B})_{xyz}$$

$$\mathbf{a}_C = \mathbf{a}_B + \dot{\boldsymbol{\Omega}} \times \mathbf{r}_{C/B} + \boldsymbol{\Omega} \times (\boldsymbol{\Omega} \times \mathbf{r}_{C/B}) + 2\boldsymbol{\Omega} \times (\mathbf{v}_{C/B})_{xyz} + (\mathbf{a}_{C/B})_{xyz}$$

Motion of B.   To find the time derivatives of $\mathbf{r}_B$ let the x', y', z' axes rotate with $\boldsymbol{\Omega}' = \boldsymbol{\omega}_1$. Then

$$\boldsymbol{\Omega}' = \boldsymbol{\omega}_1 = \{4\mathbf{i}\} \text{ rad/s} \qquad \dot{\boldsymbol{\Omega}}' = \dot{\boldsymbol{\omega}}_1 = \{1.5\mathbf{i}\} \text{ rad/s}^2$$

$$\mathbf{r}_B = \{-0.5\mathbf{k}\} \text{ m}$$

$$\mathbf{v}_B = \dot{\mathbf{r}}_B = (\dot{\mathbf{r}}_B)_{x'y'z'} + \boldsymbol{\omega}_1 \times \mathbf{r}_B = 0 + 4\mathbf{i} \times (-0.5\mathbf{k}) = \{2\mathbf{j}\} \text{ m/s}$$

$$\mathbf{a}_B = \ddot{\mathbf{r}}_B = [(\ddot{\mathbf{r}}_B)_{x'y'z'} + \boldsymbol{\omega}_1 \times (\dot{\mathbf{r}}_B)_{x'y'z'}] + \dot{\boldsymbol{\omega}}_1 \times \mathbf{r}_B + \boldsymbol{\omega}_1 \times \dot{\mathbf{r}}_B$$

$$= [0 + 0] + 1.5\mathbf{i} \times (-0.5\mathbf{k}) + 4\mathbf{i} \times 2\mathbf{j} = \{0.75\mathbf{j} + 8\mathbf{k}\} \text{ m/s}^2$$

Motion of C with Respect to B.   To find the time derivatives of $(\mathbf{r}_{C/B})_{xyz}$, let the x'', y'', z'' axes rotate with $\boldsymbol{\Omega}_{xyz} = \boldsymbol{\omega}_2$. Then

$$\boldsymbol{\Omega}_{xyz} = \boldsymbol{\omega}_2 = \{5\mathbf{k}\} \text{ rad/s} \qquad \dot{\boldsymbol{\Omega}}_{xyz} = \dot{\boldsymbol{\omega}}_2 = \{-6\mathbf{k}\} \text{ rad/s}^2$$

$$(\mathbf{r}_{C/B})_{xyz} = \{0.2\mathbf{j}\} \text{ m}$$

$$(\mathbf{v}_{C/B})_{xyz} = (\dot{\mathbf{r}}_{C/B})_{xyz} = (\dot{\mathbf{r}}_{C/B})_{x''y''z''} + \boldsymbol{\omega}_2 \times (\mathbf{r}_{C/B})_{xyz} = 3\mathbf{j} + 5\mathbf{k} \times 0.2\mathbf{j} = \{-1\mathbf{i} + 3\mathbf{j}\} \text{ m/s}$$

$$(\mathbf{a}_{C/B})_{xyz} = (\ddot{\mathbf{r}}_{C/B})_{xyz} = [(\ddot{\mathbf{r}}_{C/B})_{x''y''z''} + \boldsymbol{\omega}_2 \times (\dot{\mathbf{r}}_{C/B})_{x''y''z''}] + \dot{\boldsymbol{\omega}}_2 \times (\mathbf{r}_{C/B})_{xyz} + \boldsymbol{\omega}_2 \times (\dot{\mathbf{r}}_{C/B})_{xyz}$$

$$= (2\mathbf{j} + 5\mathbf{k} \times 3\mathbf{j}) + (-6\mathbf{k} \times 0.2\mathbf{j}) + [5\mathbf{k} \times (-1\mathbf{i} + 3\mathbf{j})]$$

$$= \{-28.8\mathbf{i} - 3\mathbf{j}\} \text{ m/s}^2$$

Motion of C.

$$\mathbf{v}_C = \mathbf{v}_B + \boldsymbol{\Omega} \times \mathbf{r}_{C/B} + (\mathbf{v}_{C/B})_{xyz} = 2\mathbf{j} + 4\mathbf{i} \times 0.2\mathbf{j} + (-1\mathbf{i} + 3\mathbf{j})$$

$$= \{-1\mathbf{i} + 5\mathbf{j} + 0.8\mathbf{k}\} \text{ m/s} \qquad\qquad Ans.$$

$$\mathbf{a}_C = \mathbf{a}_B + \dot{\boldsymbol{\Omega}} \times \mathbf{r}_{C/B} + \boldsymbol{\Omega} \times (\boldsymbol{\Omega} \times \mathbf{r}_{C/B}) + 2\boldsymbol{\Omega} \times (\mathbf{v}_{C/B})_{xyz} + (\mathbf{a}_{C/B})_{xyz}$$

$$= (0.75\mathbf{j} + 8\mathbf{k}) + (1.5\mathbf{i} \times 0.2\mathbf{j}) + [4\mathbf{i} \times (4\mathbf{i} \times 0.2\mathbf{j})]$$

$$+ 2[4\mathbf{i} \times (-1\mathbf{i} + 3\mathbf{j})] + (-28.8\mathbf{i} - 3\mathbf{j})$$

$$= \{-28.8\mathbf{i} - 5.45\mathbf{j} + 32.3\mathbf{k}\} \text{ m/s}^2 \qquad\qquad Ans.$$

20

SOLUTION II

Coordinate Axes. Here we will let the x, y, z axes rotate at

$$\Omega = \omega_1 + \omega_2 = \{4\mathbf{i} + 5\mathbf{k}\} \text{ rad/s}$$

Then $\Omega_{xyz} = 0$.

Motion of B. From the constraints of the problem ω_1 does not change direction relative to X, Y, Z; however, the direction of ω_2 is changed by ω_1. Thus, to obtain $\dot{\Omega}$ consider x', y', z' axes coincident with the X, Y, Z axes at A, so that $\Omega' = \omega_1$. Then taking the derivative of the components of Ω,

$$\dot{\Omega} = \dot{\omega}_1 + \dot{\omega}_2 = [(\dot{\omega}_1)_{x'y'z'} + \omega_1 \times \omega_1] + [(\dot{\omega}_2)_{x'y'z'} + \omega_1 \times \omega_2]$$

$$= [1.5\mathbf{i} + 0] + [-6\mathbf{k} + 4\mathbf{i} \times 5\mathbf{k}] = \{1.5\mathbf{i} - 20\mathbf{j} - 6\mathbf{k}\} \text{ rad/s}^2$$

Also, ω_1 changes the direction of $\mathbf{r}_B$ so that the time derivatives of $\mathbf{r}_B$ can be found using the primed axes defined above. Hence,

$$\mathbf{v}_B = \dot{\mathbf{r}}_B = (\dot{\mathbf{r}}_B)_{x'y'z'} + \omega_1 \times \mathbf{r}_B$$

$$= 0 + 4\mathbf{i} \times (-0.5\mathbf{k}) = \{2\mathbf{j}\} \text{ m/s}$$

$$\mathbf{a}_B = \ddot{\mathbf{r}}_B = [(\ddot{\mathbf{r}}_B)_{x'y'z'} + \omega_1 \times (\dot{\mathbf{r}}_B)_{x'y'z'}] + \dot{\omega}_1 \times \mathbf{r}_B + \omega_1 \times \dot{\mathbf{r}}_B$$

$$= [0 + 0] + 1.5\mathbf{i} \times (-0.5\mathbf{k}) + 4\mathbf{i} \times 2\mathbf{j} = \{0.75\mathbf{j} + 8\mathbf{k}\} \text{ m/s}^2$$

Motion of C with Respect to B.

$$\Omega_{xyz} = 0$$

$$\dot{\Omega}_{xyz} = 0$$

$$(\mathbf{r}_{C/B})_{xyz} = \{0.2\mathbf{j}\} \text{ m}$$

$$(\mathbf{v}_{C/B})_{xyz} = \{3\mathbf{j}\} \text{ m/s}$$

$$(\mathbf{a}_{C/B})_{xyz} = \{2\mathbf{j}\} \text{ m/s}^2$$

Motion of C.

$$\mathbf{v}_C = \mathbf{v}_B + \Omega \times \mathbf{r}_{C/B} + (\mathbf{v}_{C/B})_{xyz}$$

$$= 2\mathbf{j} + [(4\mathbf{i} + 5\mathbf{k}) \times (0.2\mathbf{j})] + 3\mathbf{j}$$

$$= \{-1\mathbf{i} + 5\mathbf{j} + 0.8\mathbf{k}\} \text{ m/s} \qquad\qquad Ans.$$

$$\mathbf{a}_C = \mathbf{a}_B + \dot{\Omega} \times \mathbf{r}_{C/B} + \Omega \times (\Omega \times \mathbf{r}_{C/B}) + 2\Omega \times (\mathbf{v}_{C/B})_{xyz} + (\mathbf{a}_{C/B})_{xyz}$$

$$= (0.75\mathbf{j} + 8\mathbf{k}) + [(1.5\mathbf{i} - 20\mathbf{j} - 6\mathbf{k}) \times (0.2\mathbf{j})]$$

$$\qquad + (4\mathbf{i} + 5\mathbf{k}) \times [(4\mathbf{i} + 5\mathbf{k}) \times 0.2\mathbf{j}] + 2[(4\mathbf{i} + 5\mathbf{k}) \times 3\mathbf{j}] + 2\mathbf{j}$$

$$= \{-28.8\mathbf{i} - 5.45\mathbf{j} + 32.3\mathbf{k}\} \text{ m/s}^2 \qquad\qquad Ans.$$

20

PROBLEMS

20–39. Solve Example 20–5 such that the x, y, z axes move with curvilinear translation, $\Omega = 0$ in which case the collar appears to have both an angular velocity $\Omega_{xyz} = \omega_1 + \omega_2$ and radial motion.

*20–40. Solve Example 20–5 by fixing x, y, z axes to rod BD so that $\Omega = \omega_1 + \omega_2$. In this case the collar appears only to move radially outward along BD; hence $\Omega_{xyz} = 0$.

•20–41. At the instant shown, the shaft rotates with an angular velocity of $\omega_p = 6 \text{ rad/s}$ and has an angular acceleration of $\dot{\omega}_p = 3 \text{ rad/s}^2$. At the same instant, the disk spins about its axle with an angular velocity of $\omega_s = 12 \text{ rad/s}$, increasing at a constant rate of $\dot{\omega}_s = 6 \text{ rad/s}^2$. Determine the velocity of point C located on the rim of the disk at this instant.

20–42. At the instant shown, the shaft rotates with an angular velocity of $\omega_p = 6 \text{ rad/s}$ and has an angular acceleration of $\dot{\omega}_p = 3 \text{ rad/s}^2$. At the same instant, the disk spins about its axle with an angular velocity of $\omega_s = 12 \text{ rad/s}$, increasing at a constant rate of $\dot{\omega}_s = 6 \text{ rad/s}^2$. Determine the acceleration of point C located on the rim of the disk at this instant.

20–43. At the instant shown, the cab of the excavator rotates about the z axis with a constant angular velocity of $\omega_z = 0.3 \text{ rad/s}$. At the same instant $\theta = 60°$, and the boom OBC has an angular velocity of $\dot{\theta} = 0.6 \text{ rad/s}$, which is increasing at $\ddot{\theta} = 0.2 \text{ rad/s}^2$, both measured relative to the cab. Determine the velocity and acceleration of point C on the grapple at this instant.

*20–44. At the instant shown, the frame of the excavator travels forward in the y direction with a velocity of 2 m/s and an acceleration of 1 m/s^2, while the cab rotates about the z axis with an angular velocity of $\omega_z = 0.3 \text{ rad/s}$, which is increasing at $\alpha_z = 0.4 \text{ rad/s}^2$. At the same instant $\theta = 60°$, and the boom OBC has an angular velocity of $\dot{\theta} = 0.6 \text{ rad/s}$, which is increasing at $\ddot{\theta} = 0.2 \text{ rad/s}^2$, both measured relative to the cab. Determine the velocity and acceleration of point C on the grapple at this instant.

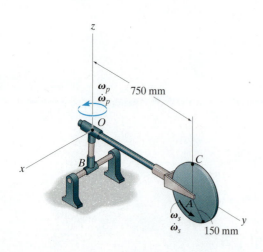

Probs. 20–41/42

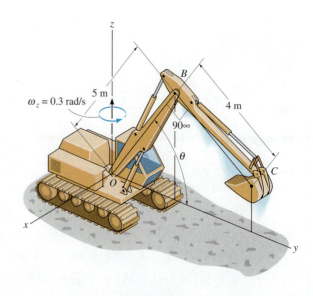

Probs. 20–43/44

•20–45.   The crane rotates about the z axis with a constant rate $\omega_1 = 0.6$ rad/s, while the boom rotates downward with a constant rate $\omega_2 = 0.2$ rad/s. Determine the velocity and acceleration of point A located at the end of the boom at the instant shown.

20–46.   The crane rotates about the z axis with a rate of $\omega_1 = 0.6$ rad/s, which is increasing at $\dot{\omega}_1 = 0.6$ rad/s^2. Also, the boom rotates downward at $\omega_2 = 0.2$ rad/s, which is increasing at $\dot{\omega}_2 = 0.3$ rad/s^2. Determine the velocity and acceleration of point A located at the end of the boom at the instant shown.

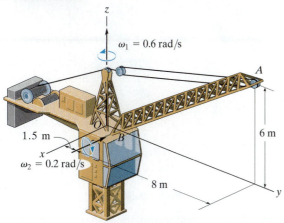

Probs. 20–45/46

20–47.   The motor rotates about the z axis with a constant angular velocity of $\omega_1 = 3$ rad/s. Simultaneously, shaft OA rotates with a constant angular velocity of $\omega_2 = 6$ rad/s. Also, collar C slides along rod AB with a velocity and acceleration of 6 m/s and 3 m/s^2. Determine the velocity and acceleration of collar C at the instant shown.

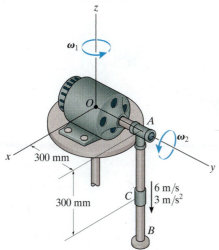

Prob. 20–47

*20–48.   At the instant shown, the helicopter is moving upwards with a velocity $v_H = 4$ ft/s and has an acceleration $a_H = 2$ ft/s^2. At the same instant the frame H, *not* the horizontal blade, rotates about a vertical axis with a constant angular velocity $\omega_H = 0.9$ rad/s. If the tail blade B rotates with a constant angular velocity $\omega_{B/H} = 180$ rad/s, measured relative to H, determine the velocity and acceleration of point P, located on the end of the blade, at the instant the blade is in the vertical position.

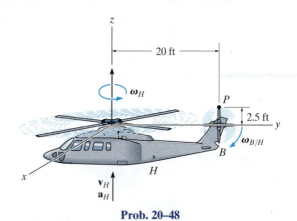

Prob. 20–48

•20–49.   At a given instant the boom AB of the tower crane rotates about the z axis with the motion shown. At this same instant, $\theta = 60°$ and the boom is rotating downward such that $\dot{\theta} = 0.4$ rad/s and $\ddot{\theta} = 0.6$ rad/s^2. Determine the velocity and acceleration of the end of the boom A at this instant. The boom has a length of $l_{AB} = 40$ m.

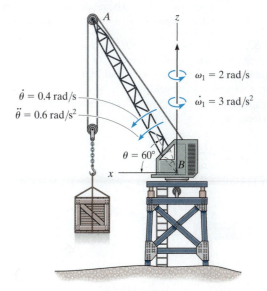

Prob. 20–49

20

20–50. At the instant shown, the tube rotates about the z axis with a constant angular velocity $\omega_1 = 2$ rad/s, while at the same instant the tube rotates upward at a constant rate $\omega_2 = 5$ rad/s. If the ball B is blown through the tube at a rate $\dot{r} = 7$ m/s, which is increasing at $\ddot{r} = 2$ m/s², determine the velocity and acceleration of the ball at the instant shown. Neglect the size of the ball.

20–51. At the instant shown, the tube rotates about the z axis with a constant angular velocity $\omega_1 = 2$ rad/s, while at the same instant the tube rotates upward at a constant rate $\omega_2 = 5$ rad/s. If the ball B is blown through the tube at a constant rate $\dot{r} = 7$ m/s, determine the velocity and acceleration of the ball at the instant shown. Neglect the size of the ball.

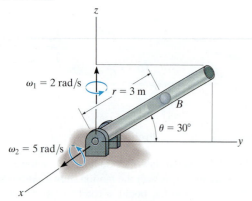

Probs. 20–50/51

*20–52.** At the instant $\theta = 30°$, the frame of the crane and the boom AB rotate with a constant angular velocity of $\omega_1 = 1.5$ rad/s and $\omega_2 = 0.5$ rad/s, respectively. Determine the velocity and acceleration of point B at this instant.

•20–53.** At the instant $\theta = 30°$, the frame of the crane is rotating with an angular velocity of $\omega_1 = 1.5$ rad/s and angular acceleration of $\dot{\omega}_1 = 0.5$ rad/s², while the boom AB rotates with an angular velocity of $\omega_2 = 0.5$ rad/s and angular acceleration of $\dot{\omega}_2 = 0.25$ rad/s². Determine the velocity and acceleration of point B at this instant.

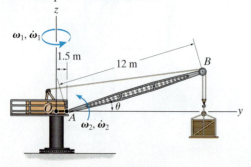

Probs. 20–52/53

20–54. At the instant shown, the base of the robotic arm rotates about the z axis with an angular velocity of $\omega_1 = 4$ rad/s, which is increasing at $\dot{\omega}_1 = 3$ rad/s². Also, the boom BC rotates at a constant rate of $\omega_{BC} = 8$ rad/s. Determine the velocity and acceleration of the part C held in its grip at this instant.

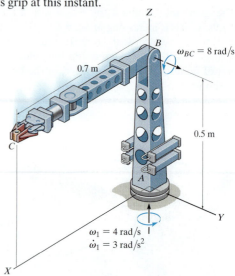

Prob. 20–54

20–55. At the instant shown, the base of the robotic arm rotates about the z axis with an angular velocity of $\omega_1 = 4$ rad/s, which is increasing at $\dot{\omega}_1 = 3$ rad/s². Also, the boom BC rotates at $\omega_{BC} = 8$ rad/s, which is increasing at $\dot{\omega}_{BC} = 2$ rad/s². Determine the velocity and acceleration of the part C held in its grip at this instant.

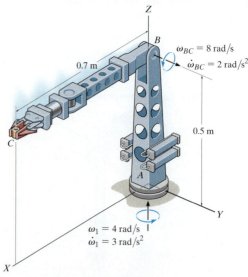

Prob. 20–55

CHAPTER REVIEW

Rotation About a Fixed Point

When a body rotates about a fixed point O, then points on the body follow a path that lies on the surface of a sphere centered at O.

Since the angular acceleration is a time rate of change in the angular velocity, then it is necessary to account for both the magnitude and directional changes of $\boldsymbol{\omega}$ when finding its time derivative. To do this, the angular velocity is often specified in terms of its component motions, such that the direction of some of these components will remain constant relative to rotating x, y, z axes. If this is the case, then the time derivative relative to the fixed axis can be determined using $\dot{\mathbf{A}} = (\dot{\mathbf{A}})_{xyz} + \boldsymbol{\Omega} \times \mathbf{A}$.

Once $\boldsymbol{\omega}$ and $\boldsymbol{\alpha}$ are known, the velocity and acceleration of any point P in the body can then be determined.

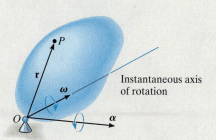

Instantaneous axis of rotation

$$\mathbf{v} = \boldsymbol{\omega} \times \mathbf{r}$$

$$\mathbf{a} = \boldsymbol{\alpha} \times \mathbf{r} + \boldsymbol{\omega} \times (\boldsymbol{\omega} \times \mathbf{r})$$

General Motion

If the body undergoes general motion, then the motion of a point B on the body can be related to the motion of another point A using a relative motion analysis, with translating axes attached to A.

$$\mathbf{v}_B = \mathbf{v}_A + \boldsymbol{\omega} \times \mathbf{r}_{B/A}$$

$$\mathbf{a}_B = \mathbf{a}_A + \boldsymbol{\alpha} \times \mathbf{r}_{B/A} + \boldsymbol{\omega} \times (\boldsymbol{\omega} \times \mathbf{r}_{B/A})$$

Relative Motion Analysis Using Translating and Rotating Axes

The motion of two points A and B on a body, a series of connected bodies, or each point located on two different paths, can be related using a relative motion analysis with rotating and translating axes at A.

When applying the equations, to find $\mathbf{v}_B$ and $\mathbf{a}_B$, it is important to account for both the magnitude and directional changes of $\mathbf{r}_A$, $(\mathbf{r}_{B/A})_{xyz}$, $\boldsymbol{\Omega}$, and $\boldsymbol{\Omega}_{xyz}$ when taking their time derivatives to find $\mathbf{v}_A$, $\mathbf{a}_A$, $(\mathbf{v}_{B/A})_{xyz}$, $(\mathbf{a}_{B/A})_{xyz}$, $\dot{\boldsymbol{\Omega}}$, and $\dot{\boldsymbol{\Omega}}_{xyz}$. To do this properly, one must use Eq. 20–6.

$$\mathbf{v}_B = \mathbf{v}_A + \boldsymbol{\Omega} \times \mathbf{r}_{B/A} + (\mathbf{v}_{B/A})_{xyz}$$

$$\mathbf{a}_B = \mathbf{a}_A + \dot{\boldsymbol{\Omega}} \times \mathbf{r}_{B/A} + \boldsymbol{\Omega} \times (\boldsymbol{\Omega} \times \mathbf{r}_{B/A}) + 2\boldsymbol{\Omega} \times (\mathbf{v}_{B/A})_{xyz} + (\mathbf{a}_{B/A})_{xyz}$$

20

The design of amusement-park rides requires a force analysis that depends upon their three-dimensional motion.

Three-Dimensional Kinetics of a Rigid Body

21

CHAPTER OBJECTIVES

- To introduce the methods for finding the moments of inertia and products of inertia of a body about various axes.

- To show how to apply the principles of work and energy and linear and angular momentum to a rigid body having three-dimensional motion.

- To develop and apply the equations of motion in three dimensions.

- To study gyroscopic and torque-free motion.

*21.1   Moments and Products of Inertia

When studying the planar kinetics of a body, it was necessary to introduce the moment of inertia I_G, which was computed about an axis perpendicular to the plane of motion and passing through the body's mass center G. For the kinetic analysis of three-dimensional motion it will sometimes be necessary to calculate six inertial quantities. These terms, called the moments and products of inertia, describe in a particular way the distribution of mass for a body relative to a given coordinate system that has a specified orientation and point of origin.

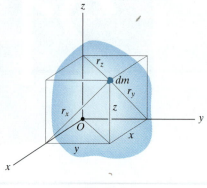

Fig. 21–1

Moment of Inertia.

Consider the rigid body shown in Fig. 21–1. The *moment of inertia* for a differential element dm of the body about any one of the three coordinate axes is defined as the product of the mass of the element and the square of the shortest distance from the axis to the element. For example, as noted in the figure, $r_x = \sqrt{y^2 + z^2}$, so that the mass moment of inertia of the element about the x axis is

$$dI_{xx} = r_x^2\, dm = (y^2 + z^2)\, dm$$

The moment of inertia I_{xx} for the body can be determined by integrating this expression over the entire mass of the body. Hence, for each of the axes, we can write

$$
\begin{aligned}
I_{xx} &= \int_m r_x^2\, dm = \int_m (y^2 + z^2)\, dm \\[6pt]
I_{yy} &= \int_m r_y^2\, dm = \int_m (x^2 + z^2)\, dm \\[6pt]
I_{zz} &= \int_m r_z^2\, dm = \int_m (x^2 + y^2)\, dm
\end{aligned}
\tag{21-1}
$$

Here it is seen that the moment of inertia is *always a positive quantity*, since it is the summation of the product of the mass dm, which is always positive, and the distances squared.

Product of Inertia.

The *product of inertia* for a differential element dm with respect to a set of *two orthogonal planes* is defined as the product of the mass of the element and the perpendicular (or shortest) distances from the planes to the element. For example, this distance is x to the y–z plane and it is y to the x–z plane, Fig. 21–1. The product of inertia dI_{xy} for the element is therefore

$$dI_{xy} = xy\, dm$$

Note also that $dI_{yx} = dI_{xy}$. By integrating over the entire mass, the products of inertia of the body with respect to each combination of planes can be expressed as

$$
\begin{aligned}
I_{xy} = I_{yx} &= \int_m xy\, dm \\[6pt]
I_{yz} = I_{zy} &= \int_m yz\, dm \\[6pt]
I_{xz} = I_{zx} &= \int_m xz\, dm
\end{aligned}
\tag{21-2}
$$

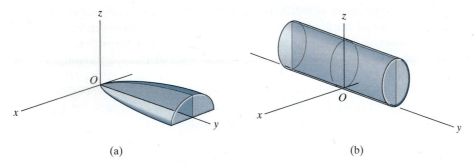

(a)   (b)

Fig. 21–2

Unlike the moment of inertia, which is always positive, the product of inertia may be positive, negative, or zero. The result depends on the algebraic signs of the two defining coordinates, which vary independently from one another. In particular, if either one or both of the orthogonal planes are *planes of symmetry* for the mass, the *product of inertia* with respect to these planes will be *zero*. In such cases, elements of mass will occur in *pairs* located on each side of the plane of symmetry. On one side of the plane the product of inertia for the element will be positive, while on the other side the product of inertia of the corresponding element will be negative, the sum therefore yielding zero. Examples of this are shown in Fig. 21–2. In the first case, Fig. 21–2a, the y–z plane is a plane of symmetry, and hence $I_{xy} = I_{xz} = 0$. Calculation of I_{yz} will yield a *positive* result, since all elements of mass are located using only positive y and z coordinates. For the cylinder, with the coordinate axes located as shown in Fig. 21–2b, the x–z and y–z planes are both planes of symmetry. Thus, $I_{xy} = I_{yz} = I_{zx} = 0$.

Parallel-Axis and Parallel-Plane Theorems.

The techniques of integration used to determine the moment of inertia of a body were described in Sec. 17.1. Also discussed were methods to determine the moment of inertia of a composite body, i.e., a body that is composed of simpler segments, as tabulated on the inside back cover. In both of these cases the *parallel-axis theorem* is often used for the calculations. This theorem, which was developed in Sec. 17.1, allows us to transfer the moment of inertia of a body from an axis passing through its mass center G to a parallel axis passing through some other point. If G has coordinates x_G, y_G, z_G defined with respect to the x, y, z axes, Fig. 21–3, then the parallel-axis equations used to calculate the moments of inertia about the x, y, z axes are

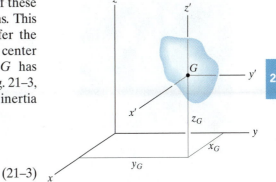

$$I_{xx} = (I_{x'x'})_G + m(y_G^2 + z_G^2)$$
$$I_{yy} = (I_{y'y'})_G + m(x_G^2 + z_G^2) \quad (21\text{–}3)$$
$$I_{zz} = (I_{z'z'})_G + m(x_G^2 + y_G^2)$$

Fig. 21–3

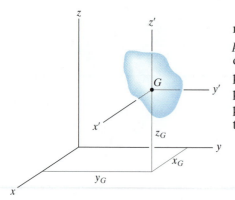

Fig. 21–3 (repeated)

The products of inertia of a composite body are computed in the same manner as the body's moments of inertia. Here, however, the *parallel-plane theorem* is important. This theorem is used to transfer the products of inertia of the body with respect to a set of three orthogonal planes passing through the body's mass center to a corresponding set of three parallel planes passing through some other point O. Defining the perpendicular distances between the planes as x_G, y_G and z_G, Fig. 21–3, the parallel-plane equations can be written as

$$
\begin{aligned}
I_{xy} &= (I_{x'y'})_G + mx_G y_G \\
I_{yz} &= (I_{y'z'})_G + my_G z_G \\
I_{zx} &= (I_{z'x'})_G + mz_G x_G
\end{aligned}
\tag{21–4}
$$

The derivation of these formulas is similar to that given for the parallel-axis equation, Sec. 17.1.

Inertia Tensor. The inertial properties of a body are therefore completely characterized by nine terms, six of which are independent of one another. This set of terms is defined using Eqs. 21–1 and 21–2 and can be written as

$$
\begin{pmatrix}
I_{xx} & -I_{xy} & -I_{xz} \\
-I_{yx} & I_{yy} & -I_{yz} \\
-I_{zx} & -I_{zy} & I_{zz}
\end{pmatrix}
$$

This array is called an *inertia tensor*.* It has a unique set of values for a body when it is determined for each location of the origin O and orientation of the coordinate axes.

In general, for point O we can specify a unique axes inclination for which the products of inertia for the body are zero when computed with respect to these axes. When this is done, the inertia tensor is said to be "diagonalized" and may be written in the simplified form

$$
\begin{pmatrix}
I_x & 0 & 0 \\
0 & I_y & 0 \\
0 & 0 & I_z
\end{pmatrix}
$$

Here $I_x = I_{xx}$, $I_y = I_{yy}$, and $I_z = I_{zz}$ are termed the *principal moments of inertia* for the body, which are computed with respect to the *principal axes of inertia*. Of these three principal moments of inertia, one will be a maximum and another a minimum of the body's moment of inertia.

The dynamics of the space shuttle while it orbits the earth can be predicted only if its moments and products of inertia are known relative to its mass center.

*The negative signs are here as a consequence of the development of angular momentum, Eqs. 21–10.

The mathematical determination of the directions of principal axes of inertia will not be discussed here (see Prob. 21–20). However, there are many cases in which the principal axes can be determined by inspection. From the previous discussion it was noted that if the coordinate axes are oriented such that *two* of the three orthogonal planes containing the axes are planes of *symmetry* for the body, then all the products of inertia for the body are zero with respect to these coordinate planes, and hence these coordinate axes are principal axes of inertia. For example, the x, y, z axes shown in Fig. 21–2b represent the principal axes of inertia for the cylinder at point O.

Moment of Inertia About an Arbitrary Axis.

Consider the body shown in Fig. 21–4, where the nine elements of the inertia tensor have been determined with respect to the x, y, z axes having an origin at O. Here we wish to determine the moment of inertia of the body about the Oa axis, which has a direction defined by the unit vector $\mathbf{u}_a$. By definition $I_{Oa} = \int b^2 \, dm$, where b is the *perpendicular distance* from dm to Oa. If the position of dm is located using $\mathbf{r}$, then $b = r \sin \theta$, which represents the *magnitude* of the cross product $\mathbf{u}_a \times \mathbf{r}$. Hence, the moment of inertia can be expressed as

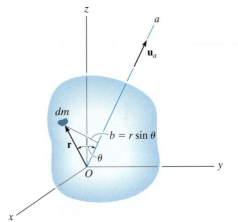

Fig. 21–4

$$I_{Oa} = \int_m |(\mathbf{u}_a \times \mathbf{r})|^2 dm = \int_m (\mathbf{u}_a \times \mathbf{r}) \cdot (\mathbf{u}_a \times \mathbf{r}) \, dm$$

Provided $\mathbf{u}_a = u_x \mathbf{i} + u_y \mathbf{j} + u_z \mathbf{k}$ and $\mathbf{r} = x\mathbf{i} + y\mathbf{j} + z\mathbf{k}$, then $\mathbf{u}_a \times \mathbf{r} = (u_y z - u_z y)\mathbf{i} + (u_z x - u_x z)\mathbf{j} + (u_x y - u_y x)\mathbf{k}$. After substituting and performing the dot-product operation, the moment of inertia is

$$I_{Oa} = \int_m [(u_y z - u_z y)^2 + (u_z x - u_x z)^2 + (u_x y - u_y x)^2] dm$$

$$= u_x^2 \int_m (y^2 + z^2) dm + u_y^2 \int_m (z^2 + x^2) dm + u_z^2 \int_m (x^2 + y^2) \, dm$$

$$- 2u_x u_y \int_m xy \, dm - 2u_y u_z \int_m yz \, dm - 2u_z u_x \int_m zx \, dm$$

Recognizing the integrals to be the moments and products of inertia of the body, Eqs. 21–1 and 21–2, we have

$$I_{Oa} = I_{xx}u_x^2 + I_{yy}u_y^2 + I_{zz}u_z^2 - 2I_{xy}u_x u_y - 2I_{yz}u_y u_z - 2I_{zx}u_z u_x \quad (21\text{–}5)$$

Thus, if the inertia tensor is specified for the x, y, z axes, the moment of inertia of the body about the inclined Oa axis can be found. For the calculation, the direction cosines u_x, u_y, u_z of the axes must be determined. These terms specify the cosines of the coordinate direction angles α, β, γ made between the positive Oa axis and the positive x, y, z axes, respectively (see Appendix C).

21

EXAMPLE 21.1

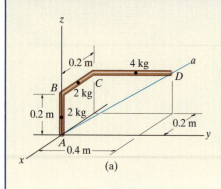

(a)

Determine the moment of inertia of the bent rod shown in Fig. 21–5a about the Aa axis. The mass of each of the three segments is given in the figure.

SOLUTION

Before applying Eq. 21–5, it is first necessary to determine the moments and products of inertia of the rod with respect to the x, y, z axes. This is done using the formula for the moment of inertia of a slender rod, $I = \frac{1}{12}ml^2$, and the parallel-axis and parallel-plane theorems, Eqs. 21–3 and 21–4. Dividing the rod into three parts and locating the mass center of each segment, Fig. 21–5b, we have

$$I_{xx} = \left[\tfrac{1}{12}(2)(0.2)^2 + 2(0.1)^2\right] + \left[0 + 2(0.2)^2\right]$$
$$+ \left[\tfrac{1}{12}(4)(0.4)^2 + 4((0.2)^2 + (0.2)^2)\right] = 0.480 \text{ kg} \cdot \text{m}^2$$

$$I_{yy} = \left[\tfrac{1}{12}(2)(0.2)^2 + 2(0.1)^2\right] + \left[\tfrac{1}{12}(2)(0.2)^2 + 2((-0.1)^2 + (0.2)^2)\right]$$
$$+ \left[0 + 4((-0.2)^2 + (0.2)^2)\right] = 0.453 \text{ kg} \cdot \text{m}^2$$

$$I_{zz} = [0 + 0] + \left[\tfrac{1}{12}(2)(0.2)^2 + 2(-0.1)^2\right] + \left[\tfrac{1}{12}(4)(0.4)^2 + \right.$$
$$\left. 4((-0.2)^2 + (0.2)^2)\right] = 0.400 \text{ kg} \cdot \text{m}^2$$

$$I_{xy} = [0 + 0] + [0 + 0] + [0 + 4(-0.2)(0.2)] = -0.160 \text{ kg} \cdot \text{m}^2$$

$$I_{yz} = [0 + 0] + [0 + 0] + [0 + 4(0.2)(0.2)] = 0.160 \text{ kg} \cdot \text{m}^2$$

$$I_{zx} = [0 + 0] + [0 + 2(0.2)(-0.1)] +$$
$$[0 + 4(0.2)(-0.2)] = -0.200 \text{ kg} \cdot \text{m}^2$$

The Aa axis is defined by the unit vector

$$\mathbf{u}_{Aa} = \frac{\mathbf{r}_D}{r_D} = \frac{-0.2\mathbf{i} + 0.4\mathbf{j} + 0.2\mathbf{k}}{\sqrt{(-0.2)^2 + (0.4)^2 + (0.2)^2}} = -0.408\mathbf{i} + 0.816\mathbf{j} + 0.408\mathbf{k}$$

Thus,

$$u_x = -0.408 \qquad u_y = 0.816 \qquad u_z = 0.408$$

Substituting these results into Eq. 21–5 yields

$$I_{Aa} = I_{xx}u_x^2 + I_{yy}u_y^2 + I_{zz}u_z^2 - 2I_{xy}u_xu_y - 2I_{yz}u_yu_z - 2I_{zx}u_zu_x$$
$$= 0.480(-0.408)^2 + (0.453)(0.816)^2 + 0.400(0.408)^2$$
$$- 2(-0.160)(-0.408)(0.816) - 2(0.160)(0.816)(0.408)$$
$$- 2(-0.200)(0.408)(-0.408)$$
$$= 0.169 \text{ kg} \cdot \text{m}^2 \qquad\qquad\qquad\qquad\qquad\qquad \textit{Ans.}$$

(b)

Fig. 21–5

PROBLEMS

•**21–1.** Show that the sum of the moments of inertia of a body, $I_{xx} + I_{yy} + I_{zz}$, is independent of the orientation of the x, y, z axes and thus depends only on the location of its origin.

21–2. Determine the moment of inertia of the cone with respect to a vertical $\bar{y}$ axis that passes through the cone's center of mass. What is the moment of inertia about a parallel axis y' that passes through the diameter of the base of the cone? The cone has a mass m.

***21–4.** Determine by direct integration the product of inertia I_{yz} for the homogeneous prism. The density of the material is ρ. Express the result in terms of the total mass m of the prism.

•**21–5.** Determine by direct integration the product of inertia I_{xy} for the homogeneous prism. The density of the material is ρ. Express the result in terms of the total mass m of the prism.

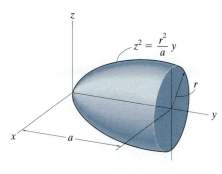

Prob. 21–2

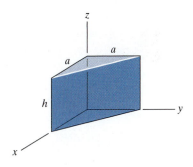

Probs. 21–4/5

21–3. Determine the moments of inertia I_x and I_y of the paraboloid of revolution. The mass of the paraboloid is m.

21–6. Determine the product of inertia I_{xy} for the homogeneous tetrahedron. The density of the material is ρ. Express the result in terms of the total mass m of the solid. *Suggestion:* Use a triangular element of thickness dz and then express dI_{xy} in terms of the size and mass of the element using the result of Prob. 21–5.

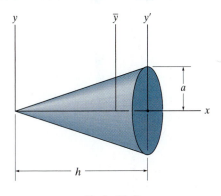

Prob. 21–3

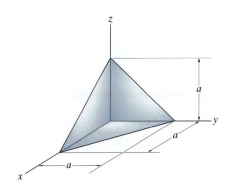

Prob. 21–6

21–7. Determine the moments of inertia for the homogeneous cylinder of mass m about the x', y', z' axes.

•21–9. The slender rod has a mass per unit length of 6 kg/m. Determine its moments and products of inertia with respect to the x, y, z axes.

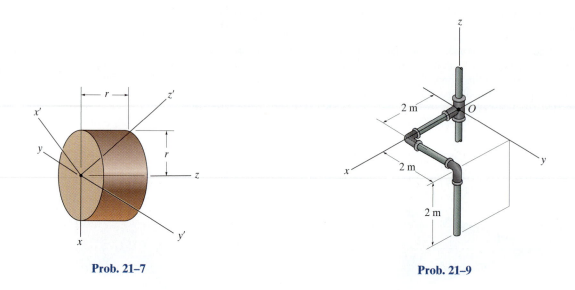

Prob. 21–7

Prob. 21–9

*21–8. Determine the product of inertia I_{xy} of the homogeneous triangular block. The material has a density of ρ. Express the result in terms of the total mass m of the block.

21–10. Determine the products of inertia I_{xy}, I_{yz}, and I_{xz} of the homogeneous solid. The material has a density of 7.85 Mg/m³.

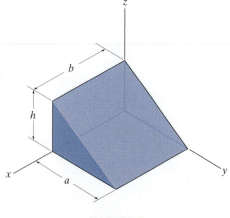

Prob. 21–8

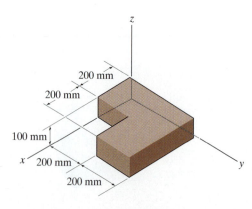

Prob. 21–10

21–11. The assembly consists of two thin plates A and B which have a mass of 3 kg each and a thin plate C which has a mass of 4.5 kg. Determine the moments of inertia I_x, I_y and I_z.

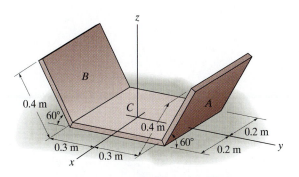

Prob. 21–11

•21–13. The bent rod has a weight of 1.5 lb/ft. Locate the center of gravity $G(\bar{x}, \bar{y})$ and determine the principal moments of inertia $I_{x'}$, $I_{y'}$, and $I_{z'}$ of the rod with respect to the x', y', z' axes.

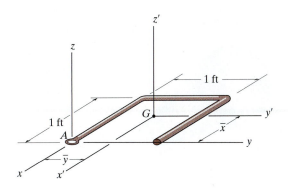

Prob. 21–13

*21–12. Determine the products of inertia I_{xy}, I_{yz}, and I_{xz}, of the thin plate. The material has a density per unit area of 50 kg/m^2.

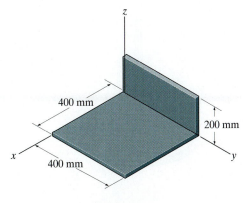

Prob. 21–12

21–14. The assembly consists of a 10-lb slender rod and a 30-lb thin circular disk. Determine its moment of inertia about the y' axis.

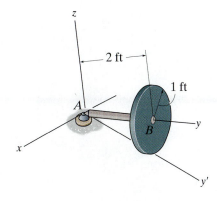

Prob. 21–14

21

21–15. The top consists of a cone having a mass of 0.7 kg and a hemisphere of mass 0.2 kg. Determine the moment of inertia I_z when the top is in the position shown.

•21–17. Determine the product of inertia I_{xy} for the bent rod. The rod has a mass per unit length of 2 kg/m.

21–18. Determine the moments of inertia I_{xx}, I_{yy}, I_{zz} for the bent rod. The rod has a mass per unit length of 2 kg/m.

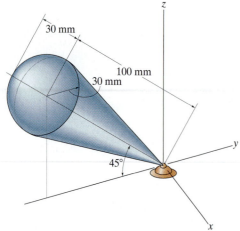

Prob. 21–15

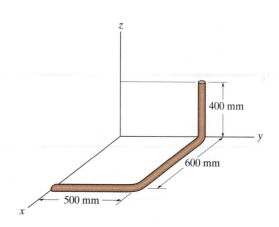

Probs. 21–17/18

*****21–16.** Determine the products of inertia I_{xy}, I_{yz}, and I_{xz} of the thin plate. The material has a mass per unit area of 50 kg/m².

21–19. Determine the moment of inertia of the rod-and-thin-ring assembly about the z axis. The rods and ring have a mass per unit length of 2 kg/m.

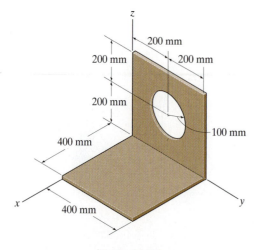

Prob. 21–16

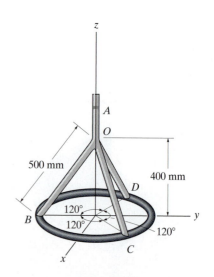

Prob. 21–19

21

21.2 Angular Momentum

In this section we will develop the necessary equations used to determine the angular momentum of a rigid body about an arbitrary point. These equations will provide a means for developing both the principle of impulse and momentum and the equations of rotational motion for a rigid body.

Consider the rigid body in Fig. 21–6, which has a mass m and center of mass at G. The X, Y, Z coordinate system represents an inertial frame of reference, and hence, its axes are fixed or translate with a constant velocity. The angular momentum as measured from this reference will be determined relative to the arbitrary point A. The position vectors $\mathbf{r}_A$ and $\boldsymbol{\rho}_A$ are drawn from the origin of coordinates to point A and from A to the ith particle of the body. If the particle's mass is m_i, the angular momentum about point A is

$$(\mathbf{H}_A)_i = \boldsymbol{\rho}_A \times m_i \mathbf{v}_i$$

where $\mathbf{v}_i$ represents the particle's velocity measured from the X, Y, Z coordinate system. If the body has an angular velocity $\boldsymbol{\omega}$ at the instant considered, $\mathbf{v}_i$ may be related to the velocity of A by applying Eq. 20–7, i.e.,

$$\mathbf{v}_i = \mathbf{v}_A + \boldsymbol{\omega} \times \boldsymbol{\rho}_A$$

Thus,

$$(\mathbf{H}_A)_i = \boldsymbol{\rho}_A \times m_i(\mathbf{v}_A + \boldsymbol{\omega} \times \boldsymbol{\rho}_A)$$

$$= (\boldsymbol{\rho}_A m_i) \times \mathbf{v}_A + \boldsymbol{\rho}_A \times (\boldsymbol{\omega} \times \boldsymbol{\rho}_A)m_i$$

Summing the moments of all the particles of the body requires an integration. Since $m_i \rightarrow dm$, we have

$$\mathbf{H}_A = \left(\int_m \boldsymbol{\rho}_A \, dm\right) \times \mathbf{v}_A + \int_m \boldsymbol{\rho}_A \times (\boldsymbol{\omega} \times \boldsymbol{\rho}_A)dm \qquad (21\text{–}6)$$

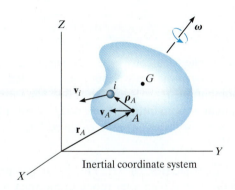

Inertial coordinate system

Fig. 21–6

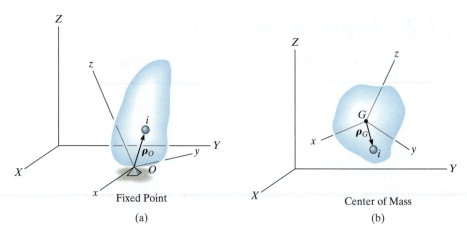

Fixed Point

(a)

Center of Mass

(b)

Fig. 21–7

Fixed Point O.

If A becomes a *fixed point O* in the body, Fig. 21–7a, then $\mathbf{v}_A = \mathbf{0}$ and Eq. 21–6 reduces to

$$\mathbf{H}_O = \int_m \boldsymbol{\rho}_O \times (\boldsymbol{\omega} \times \boldsymbol{\rho}_O)\,dm \tag{21–7}$$

Center of Mass G.

If A is located at the *center of mass G* of the body, Fig. 21–7b, then $\int_m \boldsymbol{\rho}_A\,dm = \mathbf{0}$ and

$$\mathbf{H}_G = \int_m \boldsymbol{\rho}_G \times (\boldsymbol{\omega} \times \boldsymbol{\rho}_G)\,dm \tag{21–8}$$

Arbitrary Point A.

In general, A can be a point other than O or G, Fig. 21–7c, in which case Eq. 21–6 may nevertheless be simplified to the following form (see Prob. 21–21).

$$\mathbf{H}_A = \boldsymbol{\rho}_{G/A} \times m\mathbf{v}_G + \mathbf{H}_G \tag{21–9}$$

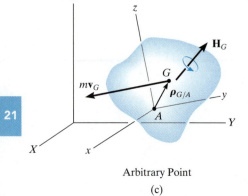

Arbitrary Point

(c)

Here the angular momentum consists of two parts—the moment of the linear momentum $m\mathbf{v}_G$ of the body about point A added (vectorially) to the angular momentum $\mathbf{H}_G$. Equation 21–9 can also be used to determine the angular momentum of the body about a fixed point O. The results, of course, will be the same as those found using the more convenient Eq. 21–7.

Rectangular Components of H.

To make practical use of Eqs. 21–7 through 21–9, the angular momentum must be expressed in terms of its scalar components. For this purpose, it is convenient to choose a second set of x, y, z axes having an arbitrary orientation relative to the X, Y, Z axes, Fig. 21–7, and for a general formulation, note that Eqs. 21–7 and 21–8 are both of the form

$$\mathbf{H} = \int_m \boldsymbol{\rho} \times (\boldsymbol{\omega} \times \boldsymbol{\rho})\, dm$$

Expressing $\mathbf{H}$, $\boldsymbol{\rho}$, and $\boldsymbol{\omega}$ in terms of x, y, z components, we have

$$H_x\mathbf{i} + H_y\mathbf{j} + H_z\mathbf{k} = \int_m (x\mathbf{i} + y\mathbf{j} + z\mathbf{k}) \times [(\omega_x\mathbf{i} + \omega_y\mathbf{j} + \omega_z\mathbf{k})$$
$$\times (x\mathbf{i} + y\mathbf{j} + z\mathbf{k})]\, dm$$

Expanding the cross products and combining terms yields

$$H_x\mathbf{i} + H_y\mathbf{j} + H_z\mathbf{k} = \left[\omega_x \int_m (y^2 + z^2)\, dm - \omega_y \int_m xy\, dm - \omega_z \int_m xz\, dm\right]\mathbf{i}$$
$$+ \left[-\omega_x \int_m xy\, dm + \omega_y \int_m (x^2 + z^2)\, dm - \omega_z \int_m yz\, dm\right]\mathbf{j}$$
$$+ \left[-\omega_x \int_m zx\, dm - \omega_y \int_m yz\, dm + \omega_z \int_m (x^2 + y^2)\, dm\right]\mathbf{k}$$

Equating the respective $\mathbf{i}$, $\mathbf{j}$, $\mathbf{k}$ components and recognizing that the integrals represent the moments and products of inertia, we obtain

$$\begin{aligned} H_x &= I_{xx}\omega_x - I_{xy}\omega_y - I_{xz}\omega_z \\ H_y &= -I_{yx}\omega_x + I_{yy}\omega_y - I_{yz}\omega_z \\ H_z &= -I_{zx}\omega_x - I_{zy}\omega_y + I_{zz}\omega_z \end{aligned} \qquad (21\text{–}10)$$

These equations can be simplified further if the x, y, z coordinate axes are oriented such that they become *principal axes of inertia* for the body at the point. When these axes are used, the products of inertia $I_{xy} = I_{yz} = I_{zx} = 0$, and if the principal moments of inertia about the x, y, z axes are represented as $I_x = I_{xx}$, $I_y = I_{yy}$, and $I_z = I_{zz}$, the three components of angular momentum become

$$H_x = I_x\omega_x \qquad H_y = I_y\omega_y \qquad H_z = I_z\omega_z \qquad (21\text{–}11)$$

21

The motion of the astronaut is controlled by use of small directional jets attached to his or her space suit. The impulses these jets provide must be carefully specified in order to prevent tumbling and loss of orientation.

Principle of Impulse and Momentum. Now that the formulation of the angular momentum for a body has been developed, the *principle of impulse and momentum*, as discussed in Sec. 19.2, can be used to solve kinetic problems which involve *force, velocity, and time*. For this case, the following two vector equations are available:

$$m(\mathbf{v}_G)_1 + \Sigma \int_{t_1}^{t_2} \mathbf{F}\, dt = m(\mathbf{v}_G)_2 \qquad (21\text{–}12)$$

$$(\mathbf{H}_O)_1 + \Sigma \int_{t_1}^{t_2} \mathbf{M}_O\, dt = (\mathbf{H}_O)_2 \qquad (21\text{–}13)$$

In three dimensions each vector term can be represented by three scalar components, and therefore a total of *six scalar equations* can be written. Three equations relate the linear impulse and momentum in the *x, y, z* directions, and the other three equations relate the body's angular impulse and momentum about the *x, y, z* axes. Before applying Eqs. 21–12 and 21–13 to the solution of problems, the material in Secs. 19.2 and 19.3 should be reviewed.

21.3   Kinetic Energy

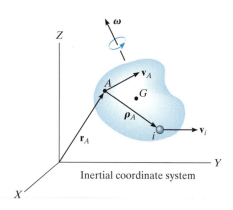

Inertial coordinate system

Fig. 21–8

In order to apply the principle of work and energy to solve problems involving general rigid body motion, it is first necessary to formulate expressions for the kinetic energy of the body. To do this, consider the rigid body shown in Fig. 21–8, which has a mass m and center of mass at G. The kinetic energy of the ith particle of the body having a mass m_i and velocity $\mathbf{v}_i$, measured relative to the inertial X, Y, Z frame of reference, is

$$T_i = \tfrac{1}{2}m_i v_i^2 = \tfrac{1}{2}m_i(\mathbf{v}_i \cdot \mathbf{v}_i)$$

Provided the velocity of an arbitrary point A in the body is known, $\mathbf{v}_i$ can be related to $\mathbf{v}_A$ by the equation $\mathbf{v}_i = \mathbf{v}_A + \boldsymbol{\omega} \times \boldsymbol{\rho}_A$, where $\boldsymbol{\omega}$ is the angular velocity of the body, measured from the X, Y, Z coordinate system, and $\boldsymbol{\rho}_A$ is a position vector extending from A to i. Using this expression, the kinetic energy for the particle can be written as

$$T_i = \tfrac{1}{2}m_i(\mathbf{v}_A + \boldsymbol{\omega} \times \boldsymbol{\rho}_A) \cdot (\mathbf{v}_A + \boldsymbol{\omega} \times \boldsymbol{\rho}_A)$$
$$= \tfrac{1}{2}(\mathbf{v}_A \cdot \mathbf{v}_A)m_i + \mathbf{v}_A \cdot (\boldsymbol{\omega} \times \boldsymbol{\rho}_A)m_i + \tfrac{1}{2}(\boldsymbol{\omega} \times \boldsymbol{\rho}_A) \cdot (\boldsymbol{\omega} \times \boldsymbol{\rho}_A)m_i$$

The kinetic energy for the entire body is obtained by summing the kinetic energies of all the particles of the body. This requires an integration. Since $m_i \rightarrow dm$, we get

$$T = \tfrac{1}{2}m(\mathbf{v}_A \cdot \mathbf{v}_A) + \mathbf{v}_A \cdot \left(\boldsymbol{\omega} \times \int_m \boldsymbol{\rho}_A\, dm\right) + \tfrac{1}{2}\int_m (\boldsymbol{\omega} \times \boldsymbol{\rho}_A) \cdot (\boldsymbol{\omega} \times \boldsymbol{\rho}_A)\, dm$$

21

The last term on the right can be rewritten using the vector identity $\mathbf{a} \times \mathbf{b} \cdot \mathbf{c} = \mathbf{a} \cdot \mathbf{b} \times \mathbf{c}$, where $\mathbf{a} = \boldsymbol{\omega}$, $\mathbf{b} = \boldsymbol{\rho}_A$, and $\mathbf{c} = \boldsymbol{\omega} \times \boldsymbol{\rho}_A$. The final result is

$$T = \tfrac{1}{2}m(\mathbf{v}_A \cdot \mathbf{v}_A) + \mathbf{v}_A \cdot \left(\boldsymbol{\omega} \times \int_m \boldsymbol{\rho}_A \, dm \right)$$
$$+ \tfrac{1}{2}\boldsymbol{\omega} \cdot \int_m \boldsymbol{\rho}_A \times (\boldsymbol{\omega} \times \boldsymbol{\rho}_A) dm \qquad (21\text{–}14)$$

This equation is rarely used because of the computations involving the integrals. Simplification occurs, however, if the reference point A is either a fixed point or the center of mass.

Fixed Point O. If A is a *fixed point* O in the body, Fig. 21–7a, then $\mathbf{v}_A = \mathbf{0}$, and using Eq. 21–7, we can express Eq. 21–14 as

$$T = \tfrac{1}{2}\boldsymbol{\omega} \cdot \mathbf{H}_O$$

If the x, y, z axes represent the principal axes of inertia for the body, then $\boldsymbol{\omega} = \omega_x \mathbf{i} + \omega_y \mathbf{j} + \omega_z \mathbf{k}$ and $\mathbf{H}_O = I_x \omega_x \mathbf{i} + I_y \omega_y \mathbf{j} + I_z \omega_z \mathbf{k}$. Substituting into the above equation and performing the dot-product operations yields

$$\boxed{T = \tfrac{1}{2}I_x \omega_x^2 + \tfrac{1}{2}I_y \omega_y^2 + \tfrac{1}{2}I_z \omega_z^2} \qquad (21\text{–}15)$$

Center of Mass G. If A is located at the *center of mass* G of the body, Fig. 21–7b, then $\int \boldsymbol{\rho}_A \, dm = \mathbf{0}$ and, using Eq. 21–8, we can write Eq. 21–14 as

$$T = \tfrac{1}{2}mv_G^2 + \tfrac{1}{2}\boldsymbol{\omega} \cdot \mathbf{H}_G$$

In a manner similar to that for a fixed point, the last term on the right side may be represented in scalar form, in which case

$$\boxed{T = \tfrac{1}{2}mv_G^2 + \tfrac{1}{2}I_x \omega_x^2 + \tfrac{1}{2}I_y \omega_y^2 + \tfrac{1}{2}I_z \omega_z^2} \qquad (21\text{–}16)$$

Here it is seen that the kinetic energy consists of two parts; namely, the translational kinetic energy of the mass center, $\tfrac{1}{2}mv_G^2$, and the body's rotational kinetic energy.

Principle of Work and Energy. Having formulated the kinetic energy for a body, the *principle of work and energy* can be applied to solve kinetics problems which involve *force, velocity, and displacement*. For this case only one scalar equation can be written for each body, namely,

$$\boxed{T_1 + \Sigma U_{1\text{-}2} = T_2} \qquad (21\text{–}17)$$

Before applying this equation, the material in Chapter 18 should be reviewed.

21

EXAMPLE 21.2

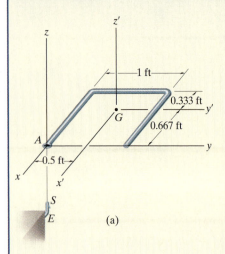

(a)

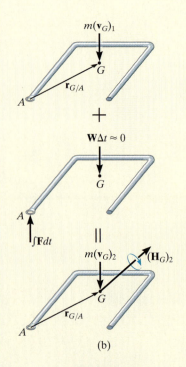

(b)

Fig. 21–9

The rod in Fig. 21–9a has a weight per unit length of 1.5 lb/ft. Determine its angular velocity just after the end A falls onto the hook at E. The hook provides a permanent connection for the rod due to the spring-lock mechanism S. Just before striking the hook the rod is falling downward with a speed $(v_G)_1 = 10$ ft/s.

SOLUTION

The principle of impulse and momentum will be used since impact occurs.

Impulse and Momentum Diagrams. Fig. 21–9b. During the short time Δt, the impulsive force $\mathbf{F}$ acting at A changes the momentum of the rod. (The impulse created by the rod's weight $\mathbf{W}$ during this time is small compared to $\int \mathbf{F}\, dt$, so that it can be neglected, i.e., the weight is a nonimpulsive force.) Hence, the angular momentum of the rod is *conserved* about point A since the moment of $\int \mathbf{F}\, dt$ about A is zero.

Conservation of Angular Momentum. Equation 21–9 must be used to find the angular momentum of the rod, since A does not become a *fixed* point until *after* the impulsive interaction with the hook. Thus, with reference to Fig. 21–9b, $(\mathbf{H}_A)_1 = (\mathbf{H}_A)_2$, or

$$\mathbf{r}_{G/A} \times m(\mathbf{v}_G)_1 = \mathbf{r}_{G/A} \times m(\mathbf{v}_G)_2 + (\mathbf{H}_G)_2 \qquad (1)$$

From Fig. 21–9a, $\mathbf{r}_{G/A} = \{-0.667\mathbf{i} + 0.5\mathbf{j}\}$ ft. Furthermore, the primed axes are principal axes of inertia for the rod because $I_{x'y'} = I_{x'z'} = I_{z'y'} = 0$. Hence, from Eqs. 21–11, $(\mathbf{H}_G)_2 = I_{x'}\omega_x\mathbf{i} + I_{y'}\omega_y\mathbf{j} + I_{z'}\omega_z\mathbf{k}$. The principal moments of inertia are $I_{x'} = 0.0272$ slug·ft², $I_{y'} = 0.0155$ slug·ft², $I_{z'} = 0.0427$ slug·ft² (see Prob. 21–13). Substituting into Eq. 1, we have

$$(-0.667\mathbf{i} + 0.5\mathbf{j}) \times \left[\left(\frac{4.5}{32.2}\right)(-10\mathbf{k})\right] = (-0.667\mathbf{i} + 0.5\mathbf{j}) \times \left[\left(\frac{4.5}{32.2}\right)(-v_G)_2\mathbf{k}\right]$$
$$+ 0.0272\omega_x\mathbf{i} + 0.0155\omega_y\mathbf{j} + 0.0427\omega_z\mathbf{k}$$

Expanding and equating the respective $\mathbf{i}, \mathbf{j}, \mathbf{k}$ components yields

$$-0.699 = -0.0699(v_G)_2 + 0.0272\omega_x \qquad (2)$$
$$-0.932 = -0.0932(v_G)_2 + 0.0155\omega_y \qquad (3)$$
$$0 = 0.0427\omega_z \qquad (4)$$

Kinematics. There are four unknowns in the above equations; however, another equation may be obtained by relating $\boldsymbol{\omega}$ to $(\mathbf{v}_G)_2$ using *kinematics*. Since $\omega_z = 0$ (Eq. 4) and after impact the rod rotates about the fixed point A, Eq. 20–3 can be applied, in which case $(\mathbf{v}_G)_2 = \boldsymbol{\omega} \times \mathbf{r}_{G/A}$, or

$$-(v_G)_2\mathbf{k} = (\omega_x\mathbf{i} + \omega_y\mathbf{j}) \times (-0.667\mathbf{i} + 0.5\mathbf{j})$$
$$-(v_G)_2 = 0.5\omega_x + 0.667\omega_y \qquad (5)$$

Solving Eqs. 2, 3 and 5 simultaneously yields

$$(\mathbf{v}_G)_2 = \{-8.41\mathbf{k}\} \text{ ft/s} \qquad \boldsymbol{\omega} = \{-4.09\mathbf{i} - 9.55\mathbf{j}\} \text{ rad/s } \textit{Ans.}$$

21

EXAMPLE 21.3

A 5-N·m torque is applied to the vertical shaft CD shown in Fig. 21–10a, which allows the 10-kg gear A to turn freely about CE. Assuming that gear A starts from rest, determine the angular velocity of CD after it has turned two revolutions. Neglect the mass of shaft CD and axle CE and assume that gear A can be approximated by a thin disk. Gear B is fixed.

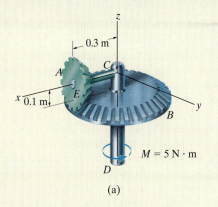

(a)

SOLUTION
The principle of work and energy may be used for the solution. Why?

Work. If shaft CD, the axle CE, and gear A are considered as a system of connected bodies, only the applied torque $\mathbf{M}$ does work. For two revolutions of CD, this work is $\Sigma U_{1-2} = (5\text{ N}\cdot\text{m})(4\pi\text{ rad}) = 62.83\text{ J}$.

Kinetic Energy. Since the gear is initially at rest, its initial kinetic energy is zero. A kinematic diagram for the gear is shown in Fig. 21–10b. If the angular velocity of CD is taken as $\boldsymbol{\omega}_{CD}$, then the angular velocity of gear A is $\boldsymbol{\omega}_A = \boldsymbol{\omega}_{CD} + \boldsymbol{\omega}_{CE}$. The gear may be imagined as a portion of a massless extended body which is rotating about the *fixed point C*. The instantaneous axis of rotation for this body is along line CH, because both points C and H on the body (gear) have zero velocity and must therefore lie on this axis. This requires that the components $\boldsymbol{\omega}_{CD}$ and $\boldsymbol{\omega}_{CE}$ be related by the equation $\omega_{CD}/0.1\text{ m} = \omega_{CE}/0.3\text{ m}$ or $\omega_{CE} = 3\omega_{CD}$. Thus,

$$\boldsymbol{\omega}_A = -\omega_{CE}\mathbf{i} + \omega_{CD}\mathbf{k} = -3\omega_{CD}\mathbf{i} + \omega_{CD}\mathbf{k} \qquad (1)$$

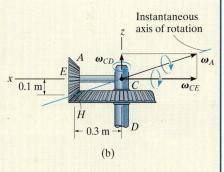

(b)

Fig. 21–10

The x, y, z axes in Fig. 21–10a represent *principal axes of inertia at C* for the gear. Since point C is a fixed point of rotation, Eq. 21–15 may be applied to determine the kinetic energy, i.e.,

$$T = \tfrac{1}{2}I_x\omega_x^2 + \tfrac{1}{2}I_y\omega_y^2 + \tfrac{1}{2}I_z\omega_z^2 \qquad (2)$$

Using the parallel-axis theorem, the moments of inertia of the gear about point C are as follows:

$$I_x = \tfrac{1}{2}(10\text{ kg})(0.1\text{ m})^2 = 0.05\text{ kg}\cdot\text{m}^2$$

$$I_y = I_z = \tfrac{1}{4}(10\text{ kg})(0.1\text{ m})^2 + 10\text{ kg}(0.3\text{ m})^2 = 0.925\text{ kg}\cdot\text{m}^2$$

Since $\omega_x = -3\omega_{CD}$, $\omega_y = 0$, $\omega_z = \omega_{CD}$, Eq. 2 becomes

$$T_A = \tfrac{1}{2}(0.05)(-3\omega_{CD})^2 + 0 + \tfrac{1}{2}(0.925)(\omega_{CD})^2 = 0.6875\omega_{CD}^2$$

Principle of Work and Energy. Applying the principle of work and energy, we obtain

$$T_1 + \Sigma U_{1-2} = T_2$$
$$0 + 62.83 = 0.6875\omega_{CD}^2$$
$$\omega_{CD} = 9.56\text{ rad/s} \qquad \qquad \textit{Ans.}$$

21

PROBLEMS

*21–20. If a body contains *no planes of symmetry,* the principal moments of inertia can be determined mathematically. To show how this is done, consider the rigid body which is spinning with an angular velocity $\boldsymbol{\omega}$, directed along one of its principal axes of inertia. If the principal moment of inertia about this axis is I, the angular momentum can be expressed as $\mathbf{H} = I\boldsymbol{\omega} = I\omega_x\mathbf{i} + I\omega_y\mathbf{j} + I\omega_z\mathbf{k}$. The components of $\mathbf{H}$ may also be expressed by Eqs. 21–10, where the inertia tensor is assumed to be known. Equate the $\mathbf{i}, \mathbf{j},$ and $\mathbf{k}$ components of both expressions for $\mathbf{H}$ and consider $\omega_x, \omega_y,$ and ω_z to be unknown. The solution of these three equations is obtained provided the determinant of the coefficients is zero. Show that this determinant, when expanded, yields the cubic equation

$$I^3 - (I_{xx} + I_{yy} + I_{zz})I^2$$
$$+ (I_{xx}I_{yy} + I_{yy}I_{zz} + I_{zz}I_{xx} - I_{xy}^2 - I_{yz}^2 - I_{zx}^2)I$$
$$- (I_{xx}I_{yy}I_{zz} - 2I_{xy}I_{yz}I_{zx} - I_{xx}I_{yz}^2$$
$$- I_{yy}I_{zx}^2 - I_{zz}I_{xy}^2) = 0$$

The three positive roots of I, obtained from the solution of this equation, represent the principal moments of inertia $I_x, I_y,$ and I_z.

•21–21. Show that if the angular momentum of a body is determined with respect to an arbitrary point A, then $\mathbf{H}_A$ can be expressed by Eq. 21–9. This requires substituting $\boldsymbol{\rho}_A = \boldsymbol{\rho}_G + \boldsymbol{\rho}_{G/A}$ into Eq. 21–6 and expanding, noting that $\int \boldsymbol{\rho}_G \, dm = \mathbf{0}$ by definition of the mass center and $\mathbf{v}_G = \mathbf{v}_A + \boldsymbol{\omega} \times \boldsymbol{\rho}_{G/A}$.

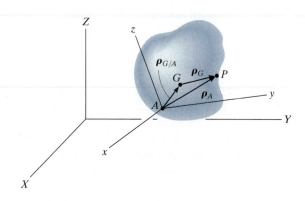

Prob. 21–21

21–22. The 4-lb rod AB is attached to the disk and collar using ball-and-socket joints. If the disk has a constant angular velocity of 2 rad/s, determine the kinetic energy of the rod when it is in the position shown. Assume the angular velocity of the rod is directed perpendicular to the axis of the rod.

21–23. Determine the angular momentum of rod AB in Prob. 21–22 about its mass center at the instant shown. Assume the angular velocity of the rod is directed perpendicular to the axis of the rod.

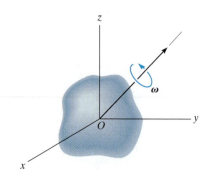

Prob. 21–20

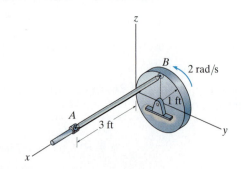

Probs. 21–22/23

*21–24. The uniform thin plate has a mass of 15 kg. Just before its corner A strikes the hook, it is falling with a velocity of $\mathbf{v}_G = \{-5\mathbf{k}\}$ m/s with no rotational motion. Determine its angular velocity immediately after corner A strikes the hook without rebounding.

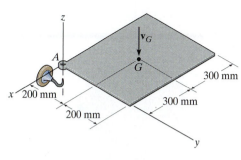

Prob. 21–24

21–27. The space capsule has a mass of 5 Mg and the radii of gyration are $k_x = k_z = 1.30$ m and $k_y = 0.45$ m. If it travels with a velocity $\mathbf{v}_G = \{400\mathbf{j} + 200\mathbf{k}\}$ m/s, compute its angular velocity just after it is struck by a meteoroid having a mass of 0.80 kg and a velocity $\mathbf{v}_m = \{-300\mathbf{i} + 200\mathbf{j} - 150\mathbf{k}\}$ m/s. Assume that the meteoroid embeds itself into the capsule at point A and that the capsule initially has no angular velocity.

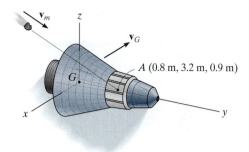

Prob. 21–27

•21–25. The 5-kg disk is connected to the 3-kg slender rod. If the assembly is attached to a ball-and-socket joint at A and the 5-N·m couple moment is applied, determine the angular velocity of the rod about the z axis after the assembly has made two revolutions about the z axis starting from rest. The disk rolls without slipping.

21–26. The 5-kg disk is connected to the 3-kg slender rod. If the assembly is attached to a ball-and-socket joint at A and the 5-N·m couple moment gives it an angular velocity about the z axis of $\omega_z = 2$ rad/s, determine the magnitude of the angular momentum of the assembly about A.

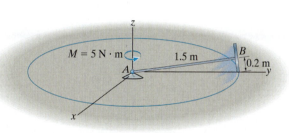

Probs. 21–25/26

*21–28. Each of the two disks has a weight of 10 lb. The axle AB weighs 3 lb. If the assembly rotates about the z axis at $\omega_z = 6$ rad/s, determine its angular momentum about the z axis and its kinetic energy. The disks roll without slipping.

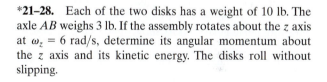

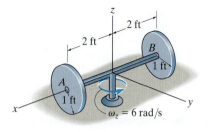

Prob. 21–28

•21–29.   The 10-kg circular disk spins about its axle with a constant angular velocity of $\omega_1 = 15$ rad/s. Simultaneously, arm OB and shaft OA rotate about their axes with constant angular velocities of $\omega_2 = 0$ and $\omega_3 = 6$ rad/s, respectively. Determine the angular momentum of the disk about point O, and its kinetic energy.

21–30.   The 10-kg circular disk spins about its axle with a constant angular velocity of $\omega_1 = 15$ rad/s. Simultaneously, arm OB and shaft OA rotate about their axes with constant angular velocities of $\omega_2 = 10$ rad/s and $\omega_3 = 6$ rad/s, respectively. Determine the angular momentum of the disk about point O, and its kinetic energy.

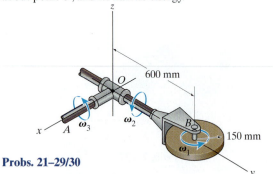

Probs. 21–29/30

21–31.   The 200-kg satellite has its center of mass at point G. Its radii of gyration about the z', x', y' axes are $k_{z'} = 300$ mm, $k_{x'} = k_{y'} = 500$ mm, respectively. At the instant shown, the satellite rotates about the x', y', and z' axes with the angular velocity shown, and its center of mass G has a velocity of $\mathbf{v}_G = \{-250\mathbf{i} + 200\mathbf{j} + 120\mathbf{k}\}$ m/s. Determine the angular momentum of the satellite about point A at this instant.

*21–32.   The 200-kg satellite has its center of mass at point G. Its radii of gyration about the z', x', y' axes are $k_{z'} = 300$ mm, $k_{x'} = k_{y'} = 500$ mm, respectively. At the instant shown, the satellite rotates about the x', y', and z' axes with the angular velocity shown, and its center of mass G has a velocity of $\mathbf{v}_G = \{-250\mathbf{i} + 200\mathbf{j} + 120\mathbf{k}\}$ m/s. Determine the kinetic energy of the satellite at this instant.

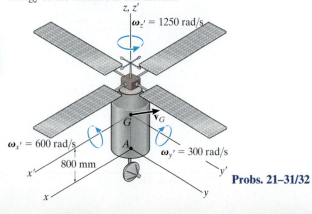

Probs. 21–31/32

•21–33.   The 25-lb thin plate is suspended from a ball-and-socket joint at O. A 0.2-lb projectile is fired with a velocity of $\mathbf{v} = \{-300\mathbf{i} - 250\mathbf{j} + 300\mathbf{k}\}$ ft/s into the plate and becomes embedded in the plate at point A. Determine the angular velocity of the plate just after impact and the axis about which it begins to rotate. Neglect the mass of the projectile after it embeds into the plate.

21–34.   Solve Prob. 21–33 if the projectile emerges from the plate with a velocity of 275 ft/s in the same direction.

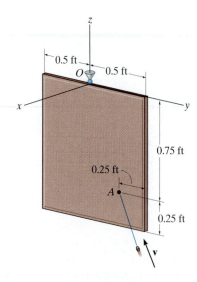

Probs. 21–33/34

21–35.   A thin plate, having a mass of 4 kg, is suspended from one of its corners by a ball-and-socket joint O. If a stone strikes the plate perpendicular to its surface at an adjacent corner A with an impulse of $\mathbf{I}_s = \{-60\mathbf{i}\}$ N·s, determine the instantaneous axis of rotation for the plate and the impulse created at O.

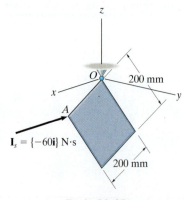

Prob. 21–35

*21–36. The 15-lb plate is subjected to a force $F = 8$ lb which is always directed perpendicular to the face of the plate. If the plate is originally at rest, determine its angular velocity after it has rotated one revolution (360°). The plate is supported by ball-and-socket joints at A and B.

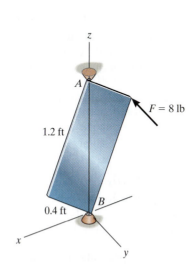

Prob. 21–36

•21–37. The plate has a mass of 10 kg and is suspended from parallel cords. If the plate has an angular velocity of 1.5 rad/s about the z axis at the instant shown, determine how high the center of the plate rises at the instant the plate momentarily stops swinging.

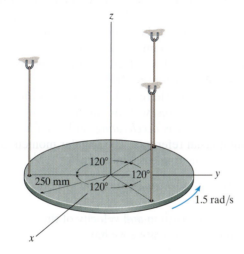

Prob. 21–37

21–38. The satellite has a mass of 200 kg and radii of gyration of $k_x = k_y = 400$ mm and $k_z = 250$ mm. When it is not rotating, the two small jets A and B are ignited simultaneously, and each jet provides an impulse of $I = 1000$ N·s on the satellite. Determine the satellite's angular velocity immediately after the ignition.

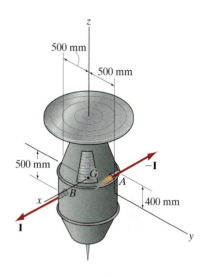

Prob. 21–38

21–39. The bent rod has a mass per unit length of 6 kg/m, and its moments and products of inertia have been calculated in Prob. 21–9. If shaft AB rotates with a constant angular velocity of $\omega_z = 6$ rad/s, determine the angular momentum of the rod about point O, and the kinetic energy of the rod.

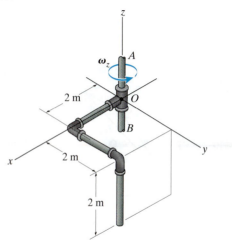

Prob. 21–39

By defin
since the

Similar e
When the

Here $\dot{\mathbf{H}}_G$
body con
  The rel
$\mathbf{a}_{i/G} = \mathbf{a}_i$
of the *itl*
frame of
property

By defini
to zero, s
term in t
product *r*
the *i*th p
written a

  The rot
from eith
the angu
principal
these con
an angula
ω, then tl
must acc
inertial X
Eqs. 21–2

Here $(\dot{\mathbf{H}})$
reference
  There a
axes. Obv
yield the
particular

21

EXAMPLE 21.5

The airplane shown in Fig. 21–13a is in the process of making a steady *horizontal* turn at the rate of ω_p. During this motion, the propeller is spinning at the rate of ω_s. If the propeller has two blades, determine the moments which the propeller shaft exerts on the propeller at the instant the blades are in the vertical position. For simplicity, assume the blades to be a uniform slender bar having a moment of inertia I about an axis perpendicular to the blades passing through the center of the bar, and having zero moment of inertia about a longitudinal axis.

(a)

SOLUTION

Free-Body Diagram. Fig. 21–13b. The reactions of the connecting shaft on the propeller are indicated by the resultants $\mathbf{F}_R$ and $\mathbf{M}_R$. (The propeller's weight is assumed to be negligible.) The x, y, z axes will be taken fixed to the propeller, since these axes always represent the principal axes of inertia for the propeller. Thus, $\mathbf{\Omega} = \boldsymbol{\omega}$. The moments of inertia I_x and I_y are equal ($I_x = I_y = I$) and $I_z = 0$.

Kinematics. The angular velocity of the propeller observed from the X, Y, Z axes, coincident with the x, y, z axes, Fig. 21–13c, is $\boldsymbol{\omega} = \boldsymbol{\omega}_s + \boldsymbol{\omega}_p = \omega_s \mathbf{i} + \omega_p \mathbf{k}$, so that the x, y, z components of $\boldsymbol{\omega}$ are

$$\omega_x = \omega_s \qquad \omega_y = 0 \qquad \omega_z = \omega_p$$

Since $\mathbf{\Omega} = \boldsymbol{\omega}$, then $\dot{\boldsymbol{\omega}} = (\dot{\boldsymbol{\omega}})_{xyz}$. To find $\dot{\boldsymbol{\omega}}$, which is the time derivative with respect to the fixed X, Y, Z axes, we can use Eq. 20–6 since $\boldsymbol{\omega}$ changes direction relative to X, Y, Z. The time rate of change of each of these components $\dot{\boldsymbol{\omega}} = \dot{\boldsymbol{\omega}}_s + \dot{\boldsymbol{\omega}}_p$ relative to the X, Y, Z axes can be obtained by introducing a third coordinate system x', y', z', which has an angular velocity $\mathbf{\Omega}' = \boldsymbol{\omega}_p$ and is coincident with the X, Y, Z axes at the instant shown. Thus

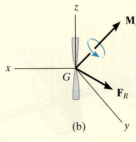

(b)

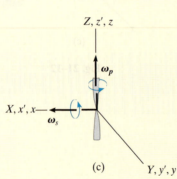

(c)

Fig. 21–13

$$\dot{\boldsymbol{\omega}} = (\dot{\boldsymbol{\omega}})_{x'y'z'} + \boldsymbol{\omega}_p \times \boldsymbol{\omega}$$

$$= (\dot{\boldsymbol{\omega}}_s)_{x'y'z'} + (\dot{\boldsymbol{\omega}}_p)_{x'y'z'} + \boldsymbol{\omega}_p \times (\boldsymbol{\omega}_s + \boldsymbol{\omega}_p)$$

$$= \mathbf{0} + \mathbf{0} + \boldsymbol{\omega}_p \times \boldsymbol{\omega}_s + \boldsymbol{\omega}_p \times \boldsymbol{\omega}_p$$

$$= \mathbf{0} + \mathbf{0} + \omega_p \mathbf{k} \times \omega_s \mathbf{i} + \mathbf{0} = \omega_p \omega_s \mathbf{j}$$

Since the X, Y, Z axes are coincident with the x, y, z axes at the instant shown, the components of $\dot{\boldsymbol{\omega}}$ along x, y, z are therefore

$$\dot{\omega}_x = 0 \qquad \dot{\omega}_y = \omega_p \omega_s \qquad \dot{\omega}_z = 0$$

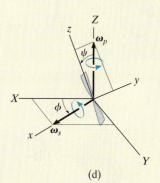

(d)

These same results can also be determined by direct calculation of $(\dot{\boldsymbol{\omega}})_{xyz}$; however, this will involve a bit more work. To do this, it will be necessary to view the propeller (or the x, y, z axes) in some *general position* such as shown in Fig. 21–13d. Here the plane has turned through an angle ϕ (phi) and the propeller has turned through an angle ψ (psi) relative to the plane. Notice that $\boldsymbol{\omega}_p$ is always directed along the fixed Z axis and $\boldsymbol{\omega}_s$ follows the x axis. Thus the general components of $\boldsymbol{\omega}$ are

$$\omega_x = \omega_s \qquad \omega_y = \omega_p \sin \psi \qquad \omega_z = \omega_p \cos \psi$$

Since ω_s and ω_p are constant, the time derivatives of these components become

$$\dot{\omega}_x = 0 \qquad \dot{\omega}_y = \omega_p \cos \psi \, \dot{\psi} \qquad \dot{\omega}_z = -\omega_p \sin \psi \, \dot{\psi}$$

But $\phi = \psi = 0°$ and $\dot{\psi} = \omega_s$ at the instant considered. Thus,

$$\omega_x = \omega_s \qquad \omega_y = 0 \qquad \omega_z = \omega_p$$

$$\dot{\omega}_x = 0 \qquad \dot{\omega}_y = \omega_p \omega_s \qquad \dot{\omega}_z = 0$$

which are the same results as those obtained previously.

Equations of Motion.   Using Eqs. 21–25, we have

$$\Sigma M_x = I_x \dot{\omega}_x - (I_y - I_z)\omega_y\omega_z = I(0) - (I - 0)(0)\omega_p$$

$$M_x = 0 \qquad\qquad Ans.$$

$$\Sigma M_y = I_y \dot{\omega}_y - (I_z - I_x)\omega_z\omega_x = I(\omega_p\omega_s) - (0 - I)\omega_p\omega_s$$

$$M_y = 2I\omega_p\omega_s \qquad\qquad Ans.$$

$$\Sigma M_z = I_z \dot{\omega}_z - (I_x - I_y)\omega_x\omega_y = 0(0) - (I - I)\omega_s(0)$$

$$M_z = 0 \qquad\qquad Ans.$$

21

EXAMPLE | 21.6

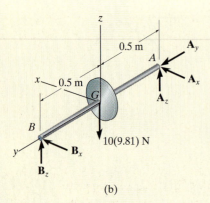

(a)

The 10-kg flywheel (or thin disk) shown in Fig. 21–14a rotates (spins) about the shaft at a constant angular velocity of $\omega_s = 6$ rad/s. At the same time, the shaft rotates (precessing) about the bearing at A with an angular velocity of $\omega_p = 3$ rad/s. If A is a thrust bearing and B is a journal bearing, determine the components of force reaction at each of these supports due to the motion.

SOLUTION I

Free-Body Diagram. Fig. 21–14b. The origin of the x, y, z coordinate system is located at the center of mass G of the flywheel. Here we will let these coordinates have an angular velocity of $\Omega = \omega_p = \{3\mathbf{k}\}$ rad/s. Although the wheel spins relative to these axes, the moments of inertia *remain constant,** i.e.,

$$I_x = I_z = \tfrac{1}{4}(10 \text{ kg})(0.2 \text{ m})^2 = 0.1 \text{ kg} \cdot \text{m}^2$$
$$I_y = \tfrac{1}{2}(10 \text{ kg})(0.2 \text{ m})^2 = 0.2 \text{ kg} \cdot \text{m}^2$$

Kinematics. From the coincident inertial X, Y, Z frame of reference, Fig. 21–14c, the flywheel has an angular velocity of $\omega = \{6\mathbf{j} + 3\mathbf{k}\}$ rad/s, so that

$$\omega_x = 0 \quad \omega_y = 6 \text{ rad/s} \quad \omega_z = 3 \text{ rad/s}$$

The time derivative of ω must be determined relative to the x, y, z axes. In this case both ω_p and ω_s do not change their magnitude or direction, and so

$$\dot{\omega}_x = 0 \quad \dot{\omega}_y = 0 \quad \dot{\omega}_z = 0$$

Equations of Motion. Applying Eqs. 21–26 ($\Omega \neq \omega$) yields

$$\Sigma M_x = I_x \dot{\omega}_x - I_y \Omega_z \omega_y + I_z \Omega_y \omega_z$$
$$-A_z(0.5) + B_z(0.5) = 0 - (0.2)(3)(6) + 0 = -3.6$$
$$\Sigma M_y = I_y \dot{\omega}_y - I_z \Omega_x \omega_z + I_x \Omega_z \omega_x$$
$$0 = 0 - 0 + 0$$
$$\Sigma M_z = I_z \dot{\omega}_z - I_x \Omega_y \omega_x + I_y \Omega_x \omega_y$$
$$A_x(0.5) - B_x(0.5) = 0 - 0 + 0$$

(b)

Fig. 21–14

---

* This would not be true for the propeller in Example 21.5.

Applying Eqs. 21–19, we have

$$\Sigma F_X = m(a_G)_X; \qquad A_x + B_x = 0$$

$$\Sigma F_Y = m(a_G)_Y; \qquad A_y = -10(0.5)(3)^2$$

$$\Sigma F_Z = m(a_G)_Z; \qquad A_z + B_z - 10(9.81) = 0$$

Solving these equations, we obtain

$$A_x = 0 \quad A_y = -45.0 \text{ N} \quad A_z = 52.6 \text{ N} \qquad \textit{Ans.}$$
$$B_x = 0 \qquad\qquad\qquad B_z = 45.4 \text{ N} \qquad \textit{Ans.}$$

NOTE: If the precession ω_p had not occurred, the z component of force at A and B would be equal to 49.05 N. In this case, however, the difference in these components is caused by the "gyroscopic moment" created whenever a spinning body precesses about another axis. We will study this effect in detail in the next section.

SOLUTION II

This example can also be solved using Euler's equations of motion, Eqs. 21–25. In this case $\Omega = \omega = \{6\mathbf{j} + 3\mathbf{k}\}$ rad/s, and the time derivative $(\dot{\boldsymbol{\omega}})_{xyz}$ can be conveniently obtained with reference to the fixed X, Y, Z axes since $\dot{\boldsymbol{\omega}} = (\dot{\boldsymbol{\omega}})_{xyz}$. This calculation can be performed by choosing x', y', z' axes to have an angular velocity of $\Omega' = \omega_p$, Fig. 21–14c, so that

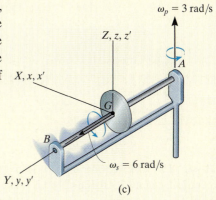

$\omega_p = 3$ rad/s

Z, z, z'

X, x, x'

G

A

B

$\omega_s = 6$ rad/s

Y, y, y'

(c)

Fig. 21–14

$$\dot{\boldsymbol{\omega}} = (\dot{\boldsymbol{\omega}})_{x'y'z'} + \boldsymbol{\omega}_p \times \boldsymbol{\omega} = 0 + 3\mathbf{k} \times (6\mathbf{j} + 3\mathbf{k}) = \{-18\mathbf{i}\} \text{ rad/s}^2$$

$$\dot{\omega}_x = -18 \text{ rad/s} \quad \dot{\omega}_y = 0 \quad \dot{\omega}_z = 0$$

The moment equations then become

$$\Sigma M_x = I_x \dot{\omega}_x - (I_y - I_z)\omega_y\omega_z$$
$$-A_z(0.5) + B_z(0.5) = 0.1(-18) - (0.2 - 0.1)(6)(3) = -3.6$$
$$\Sigma M_y = I_y\dot{\omega}_y - (I_z - I_x)\omega_z\omega_x$$
$$0 = 0 - 0$$
$$\Sigma M_z = I_z\dot{\omega}_z - (I_x - I_y)\omega_x\omega_y$$
$$A_x(0.5) - B_x(0.5) = 0 - 0$$

The solution then proceeds as before.

PROBLEMS

*21–40. Derive the scalar form of the rotational equation of motion about the x axis if $\Omega \neq \omega$ and the moments and products of inertia of the body are *not constant* with respect to time.

•21–41. Derive the scalar form of the rotational equation of motion about the x axis if $\Omega \neq \omega$ and the moments and products of inertia of the body are *constant* with respect to time.

21–42. Derive the Euler equations of motion for $\Omega \neq \omega$, i.e., Eqs. 21–26.

21–43. The uniform rectangular plate has a mass of $m = 2$ kg and is given a rotation of $\omega = 4$ rad/s about its bearings at A and B. If $a = 0.2$ m and $c = 0.3$ m, determine the vertical reactions at A and B at the instant the plate is vertical as shown. Use the x, y, z axes shown and note that
$$I_{zx} = -\left(\frac{mac}{12}\right)\left(\frac{c^2 - a^2}{c^2 + a^2}\right).$$

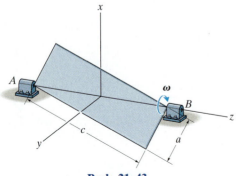

Prob. 21–43

*21–44. The disk, having a mass of 3 kg, is mounted eccentrically on shaft AB. If the shaft is rotating at a constant rate of 9 rad/s, determine the reactions at the journal bearing supports when the disk is in the position shown.

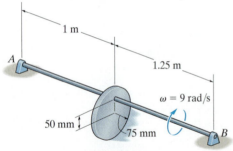

Prob. 21–44

•21–45. The slender rod AB has a mass m and it is connected to the bracket by a smooth pin at A. The bracket is rigidly attached to the shaft. Determine the required constant angular velocity of ω of the shaft, in order for the rod to make an angle of θ with the vertical.

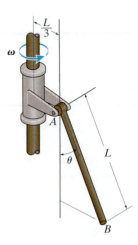

Prob. 21–45

21–46. The 5-kg rod AB is supported by a rotating arm. The support at A is a journal bearing, which develops reactions normal to the rod. The support at B is a thrust bearing, which develops reactions both normal to the rod and along the axis of the rod. Neglecting friction, determine the x, y, z components of reaction at these supports when the frame rotates with a constant angular velocity of $\omega = 10$ rad/s.

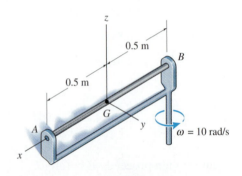

Prob. 21–46

21

21–47. The car travels around the curved road of radius ρ such that its mass center has a constant speed v_G. Write the equations of rotational motion with respect to the x, y, z axes. Assume that the car's six moments and products of inertia with respect to these axes are known.

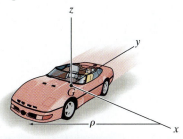

Prob. 21–47

***21–48.** The shaft is constructed from a rod which has a mass per unit length of 2 kg/m. Determine the x, y, z components of reaction at the bearings A and B if at the instant shown the shaft spins freely and has an angular velocity of $\omega = 30$ rad/s. What is the angular acceleration of the shaft at this instant? Bearing A can support a component of force in the y direction, whereas bearing B cannot.

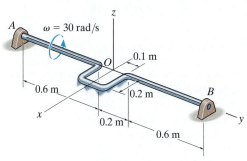

Prob. 21–48

•**21–49.** Four spheres are connected to shaft AB. If $m_C = 1$ kg and $m_E = 2$ kg, determine the mass of spheres D and F and the angles of the rods, θ_D and θ_F, so that the shaft is dynamically balanced, that is, so that the bearings at A and B exert only vertical reactions on the shaft as it rotates. Neglect the mass of the rods.

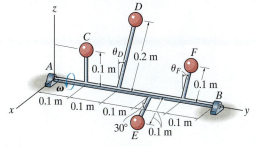

Prob. 21–49

21–50. A man stands on a turntable that rotates about a vertical axis with a constant angular velocity of $\omega_p = 10$ rad/s. If the wheel that he holds spins with a constant angular speed of $\omega_s = 30$ rad/s, determine the magnitude of moment that he must exert on the wheel to hold it in the position shown. Consider the wheel as a thin circular hoop (ring) having a mass of 3 kg and a mean radius of 300 mm.

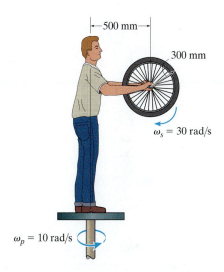

Prob. 21–50

21–51. The 50-lb disk spins with a constant angular rate of $\omega_1 = 50$ rad/s about its axle. Simultaneously, the shaft rotates with a constant angular rate of $\omega_2 = 10$ rad/s. Determine the x, y, z components of the moment developed in the arm at A at the instant shown. Neglect the weight of arm AB.

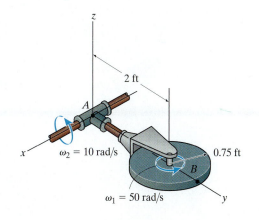

Prob. 21–51

*21–52. The man stands on a turntable that rotates about a vertical axis with a constant angular velocity of $\omega_1 = 6$ rad/s. If he tilts his head forward at a constant angular velocity of $\omega_2 = 1.5$ rad/s about point O, determine the magnitude of the moment that must be resisted by his neck at O at the instant $\theta = 30°$. Assume that his head can be considered as a uniform 10-lb sphere, having a radius of 4.5 in. and center of gravity located at G, and point O is on the surface of the sphere.

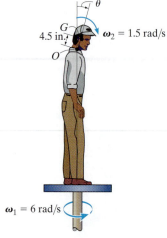

Prob. 21–52

•21–53. The blades of a wind turbine spin about the shaft S with a constant angular speed of ω_s, while the frame precesses about the vertical axis with a constant angular speed of ω_p. Determine the x, y, and z components of moment that the shaft exerts on the blades as a function of θ. Consider each blade as a slender rod of mass m and length l.

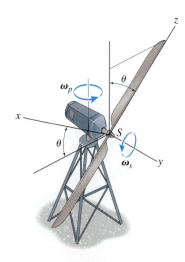

Prob. 21–53

21–54. Rod CD of mass m and length L is rotating with a constant angular rate of ω_1 about axle AB, while shaft EF rotates with a constant angular rate of ω_2. Determine the X, Y, and Z components of reaction at thrust bearing E and journal bearing F at the instant shown. Neglect the mass of the other members.

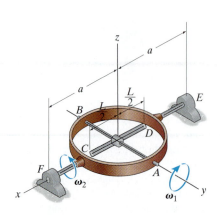

Prob. 21–54

21–55. If shaft AB is driven by the motor with an angular velocity of $\omega_1 = 50$ rad/s and angular acceleration of $\dot{\omega}_1 = 20$ rad/s² at the instant shown, and the 10-kg wheel rolls without slipping, determine the frictional force and the normal reaction on the wheel, and the moment **M** that must be supplied by the motor at this instant. Assume that the wheel is a uniform circular disk.

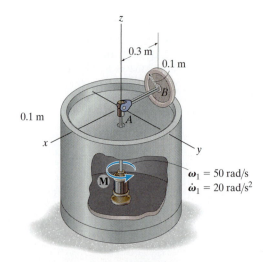

Prob. 21–55

*21–56.   A stone crusher consists of a large thin disk which is pin connected to a horizontal axle. If the axle rotates at a constant rate of 8 rad/s, determine the normal force which the disk exerts on the stones. Assume that the disk rolls without slipping and has a mass of 25 kg. Neglect the mass of the axle.

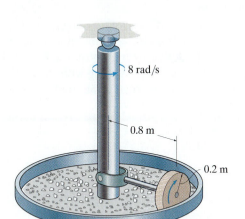

Prob. 21–56

21–59.   If shaft AB rotates with a constant angular velocity of $\omega = 50$ rad/s, determine the X, Y, Z components of reaction at journal bearing A and thrust bearing B at the instant shown. The thin plate has a mass of 10 kg. Neglect the mass of shaft AB.

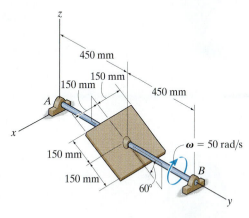

Prob. 21–59

•21–57.   The 25-lb disk is *fixed* to rod BCD, which has negligible mass. Determine the torque **T** which must be applied to the vertical shaft so that the shaft has an angular acceleration of $\alpha = 6$ rad/s². The shaft is free to turn in its bearings.

21–58.   Solve Prob. 21–57, assuming rod BCD has a weight per unit length of 2 lb/ft.

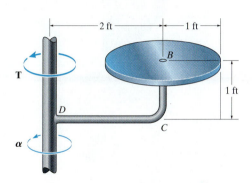

Probs. 21–57/58

*21–60.   A thin uniform plate having a mass of 0.4 kg spins with a constant angular velocity ω about its diagonal AB. If the person holding the corner of the plate at B releases his finger, the plate will fall downward on its side AC. Determine the necessary couple moment **M** which if applied to the plate would prevent this from happening.

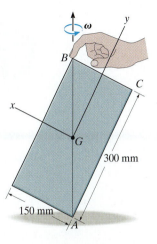

Prob. 21–60

21

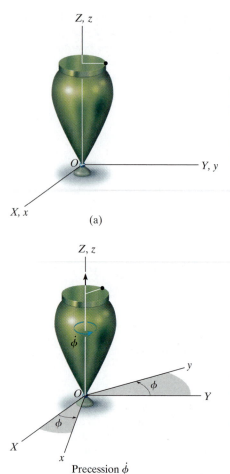

(a)

Precession $\dot{\phi}$

(b)

*21.5 Gyroscopic Motion

In this section we will develop the equations defining the motion of a body (top) which is symmetrical with respect to an axis and rotating about a fixed point. These equations also apply to the motion of a particularly interesting device, the gyroscope.

The body's motion will be analyzed using *Euler angles* ϕ, θ, ψ (phi, theta, psi). To illustrate how they define the position of a body, consider the top shown in Fig. 21–15a. To define its final position, Fig. 21–15d, a second set of x, y, z axes is fixed in the top. Starting with the X, Y, Z and x, y, z axes in coincidence, Fig. 21–15a, the final position of the top can be determined using the following three steps:

1. Rotate the top about the Z (or z) axis through an angle ϕ ($0 \le \phi < 2\pi$), Fig. 21–15b.

2. Rotate the top about the x axis through an angle θ ($0 \le \theta \le \pi$), Fig. 21–15c.

3. Rotate the top about the z axis through an angle ψ ($0 \le \psi < 2\pi$) to obtain the final position, Fig. 20–15d.

The sequence of these three angles, ϕ, θ, then ψ, must be maintained, since finite rotations are *not vectors* (see Fig. 20–1). Although this is the case, the differential rotations $d\phi$, $d\theta$, and $d\psi$ are vectors, and thus the angular velocity $\boldsymbol{\omega}$ of the top can be expressed in terms of the time derivatives of the Euler angles. The angular-velocity components $\dot{\phi}$, $\dot{\theta}$, and $\dot{\psi}$ are known as the *precession*, *nutation*, and *spin*, respectively.

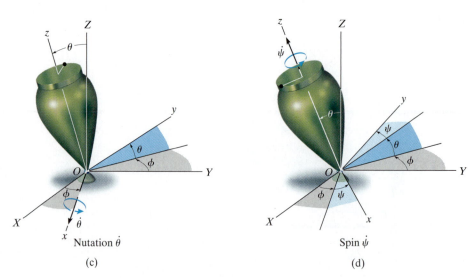

Nutation $\dot{\theta}$

(c)

Spin $\dot{\psi}$

(d)

Fig. 21–15

21

Their positive directions are shown in Fig. 21–16. It is seen that these vectors are not all perpendicular to one another; however, $\boldsymbol{\omega}$ of the top can still be expressed in terms of these three components.

Since the body (top) is symmetric with respect to the z or spin axis, there is no need to attach the x, y, z axes to the top since the inertial properties of the top will remain constant with respect to this frame during the motion. Therefore $\boldsymbol{\Omega} = \boldsymbol{\omega}_p + \boldsymbol{\omega}_n$, Fig. 21–16. Hence, the angular velocity of the body is

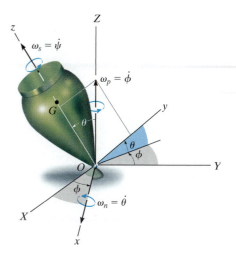

$$\boldsymbol{\omega} = \omega_x \mathbf{i} + \omega_y \mathbf{j} + \omega_z \mathbf{k}$$

$$= \dot{\theta}\mathbf{i} + (\dot{\phi}\sin\theta)\mathbf{j} + (\dot{\phi}\cos\theta + \dot{\psi})\mathbf{k} \qquad (21\text{--}27)$$

And the angular velocity of the axes is

$$\boldsymbol{\Omega} = \Omega_x \mathbf{i} + \Omega_y \mathbf{j} + \Omega_z \mathbf{k}$$

$$= \dot{\theta}\mathbf{i} + (\dot{\phi}\sin\theta)\mathbf{j} + (\dot{\phi}\cos\theta)\mathbf{k} \qquad (21\text{--}28)$$

Fig. 21–16

Have the x, y, z axes represent principal axes of inertia for the top, and so the moments of inertia will be represented as $I_{xx} = I_{yy} = I$ and $I_{zz} = I_z$. Since $\boldsymbol{\Omega} \neq \boldsymbol{\omega}$, Eqs. 21–26 are used to establish the rotational equations of motion. Substituting into these equations the respective angular-velocity components defined by Eqs. 21–27 and 21–28, their corresponding time derivatives, and the moment of inertia components, yields

$$\Sigma M_x = I(\ddot{\theta} - \dot{\phi}^2 \sin\theta \cos\theta) + I_z\dot{\phi}\sin\theta(\dot{\phi}\cos\theta + \dot{\psi})$$

$$\Sigma M_y = I(\ddot{\phi}\sin\theta + 2\dot{\phi}\dot{\theta}\cos\theta) - I_z\dot{\theta}(\dot{\phi}\cos\theta + \dot{\psi}) \qquad (21\text{--}29)$$

$$\Sigma M_z = I_z(\ddot{\psi} + \ddot{\phi}\cos\theta - \dot{\phi}\dot{\theta}\sin\theta)$$

Each moment summation applies only at the fixed point O or the center of mass G of the body. Since the equations represent a coupled set of nonlinear second-order differential equations, in general a closed-form solution may not be obtained. Instead, the Euler angles ϕ, θ, and ψ may be obtained graphically as functions of time using numerical analysis and computer techniques.

A special case, however, does exist for which simplification of Eqs. 21–29 is possible. Commonly referred to as *steady precession*, it occurs when the nutation angle θ, precession $\dot{\phi}$, and spin $\dot{\psi}$ all remain *constant*. Equations 21–29 then reduce to the form

$$\boxed{\Sigma M_x = -I\dot{\phi}^2 \sin\theta \cos\theta + I_z\dot{\phi}\sin\theta(\dot{\phi}\cos\theta + \dot{\psi})} \qquad (21\text{--}30)$$

$$\Sigma M_y = 0$$

$$\Sigma M_z = 0$$

21

Equation 21–30 can be further simplified by noting that, from Eq. 21–27, $\omega_z = \dot{\phi} \cos \theta + \dot{\psi}$, so that

$$\Sigma M_x = -I\dot{\phi}^2 \sin \theta \cos \theta + I_z \dot{\phi}(\sin \theta)\omega_z$$

or

$$\boxed{\Sigma M_x = \dot{\phi} \sin \theta(I_z \omega_z - I\dot{\phi} \cos \theta)} \qquad (21\text{–}31)$$

It is interesting to note what effects the spin $\dot{\psi}$ has on the moment about the x axis. To show this, consider the spinning rotor in Fig. 21–17. Here $\theta = 90°$, in which case Eq. 21–30 reduces to the form

$$\Sigma M_x = I_z \dot{\phi}\dot{\psi}$$

or

$$\boxed{\Sigma M_x = I_z \Omega_y \omega_z} \qquad (21\text{–}32)$$

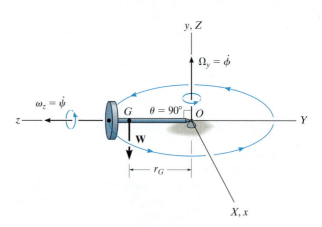

Fig. 21–17

From the figure it can be seen that $\boldsymbol{\Omega}_y$ and $\boldsymbol{\omega}_z$ act along their respective *positive axes* and therefore are mutually perpendicular. Instinctively, one would expect the rotor to fall down under the influence of gravity! However, this is not the case at all, provided the product $I_z \Omega_y \omega_z$ is correctly chosen to counterbalance the moment $\Sigma M_x = Wr_G$ of the rotor's weight about O. This unusual phenomenon of rigid-body motion is often referred to as the *gyroscopic effect*.

Perhaps a more intriguing demonstration of the gyroscopic effect comes from studying the action of a *gyroscope*, frequently referred to as a *gyro*. A gyro is a rotor which spins at a very high rate about its axis of symmetry. This rate of spin is considerably greater than its precessional rate of rotation about the vertical axis. Hence, for all practical purposes, the angular momentum of the gyro can be assumed directed along its axis of spin. Thus, for the gyro rotor shown in Fig. 21–18, $\omega_z \gg \Omega_y$, and the magnitude of the angular momentum about point O, as determined from Eqs. 21–11, reduces to the form $H_O = I_z\omega_z$. Since both the magnitude and direction of $\mathbf{H}_O$ are constant as observed from x, y, z, direct application of Eq. 21–22 yields

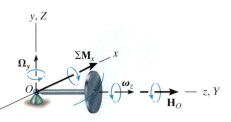

Fig. 21–18

$$\boxed{\Sigma\mathbf{M}_x = \mathbf{\Omega}_y \times \mathbf{H}_O} \qquad (21\text{–}33)$$

Using the right-hand rule applied to the cross product, it can be seen that $\mathbf{\Omega}_y$ always swings $\mathbf{H}_O$ (or $\boldsymbol{\omega}_z$) toward the sense of $\Sigma\mathbf{M}_x$. In effect, the *change in direction* of the gyro's angular momentum, $d\mathbf{H}_O$, is equivalent to the angular impulse caused by the gyro's weight about O, i.e., $d\mathbf{H}_O = \Sigma\mathbf{M}_x \, dt$, Eq. 21–20. Also, since $H_O = I_z\omega_z$ and $\Sigma\mathbf{M}_x$, $\mathbf{\Omega}_y$, and $\mathbf{H}_O$ are mutually perpendicular, Eq. 21–33 reduces to Eq. 21–32.

When a gyro is mounted in gimbal rings, Fig. 21–19, it becomes *free* of external moments applied to its base. Thus, in theory, its angular momentum $\mathbf{H}$ will never precess but, instead, maintain its same fixed orientation along the axis of spin when the base is rotated. This type of gyroscope is called a *free gyro* and is useful as a gyrocompass when the spin axis of the gyro is directed north. In reality, the gimbal mechanism is never completely free of friction, so such a device is useful only for the local navigation of ships and aircraft. The gyroscopic effect is also useful as a means of stabilizing both the rolling motion of ships at sea and the trajectories of missiles and projectiles. Furthermore, this effect is of significant importance in the design of shafts and bearings for rotors which are subjected to forced precessions.

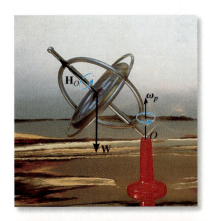

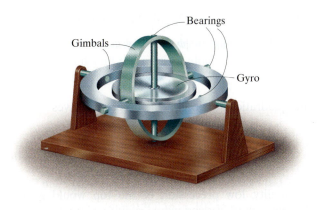

Fig. 21–19

The spinning of the gyro within the frame of this toy gyroscope produces angular momentum $\mathbf{H}_O$, which is changing direction as the frame precesses $\boldsymbol{\omega}_p$ about the vertical axis. The gyroscope will not fall down since the moment of its weight $\mathbf{W}$ about the support is balanced by the change in the direction of $\mathbf{H}_O$.

21.6   Torque-Free Motion

When the only external force acting on a body is caused by gravity, the general motion of the body is referred to as *torque-free motion*. This type of motion is characteristic of planets, artificial satellites, and projectiles—provided air friction is neglected.

In order to describe the characteristics of this motion, the distribution of the body's mass will be assumed *axisymmetric*. The satellite shown in Fig. 21–22 is an example of such a body, where the z axis represents an axis of symmetry. The origin of the x, y, z coordinates is located at the mass center G, such that $I_{zz} = I_z$ and $I_{xx} = I_{yy} = I$. Since gravity is the only external force present, the summation of moments about the mass center is zero. From Eq. 21–21, this requires the angular momentum of the body to be constant, i.e.,

$$\mathbf{H}_G = \text{constant}$$

At the instant considered, it will be assumed that the inertial frame of reference is oriented so that the positive Z axis is directed along $\mathbf{H}_G$ and the y axis lies in the plane formed by the z and Z axes, Fig. 21–22. The Euler angle formed between Z and z is θ, and therefore, with this choice of axes the angular momentum can be expressed as

$$\mathbf{H}_G = H_G \sin\theta\, \mathbf{j} + H_G \cos\theta\, \mathbf{k}$$

Furthermore, using Eqs. 21–11, we have

$$\mathbf{H}_G = I\omega_x \mathbf{i} + I\omega_y \mathbf{j} + I_z\omega_z \mathbf{k}$$

Equating the respective $\mathbf{i}$, $\mathbf{j}$, and $\mathbf{k}$ components of the above two equations yields

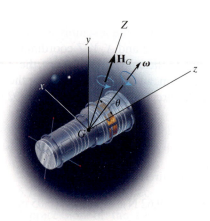

Fig. 21–22

$$\omega_x = 0 \qquad \omega_y = \frac{H_G \sin \theta}{I} \qquad \omega_z = \frac{H_G \cos \theta}{I_z} \qquad (21\text{--}34)$$

or

$$\boldsymbol{\omega} = \frac{H_G \sin \theta}{I}\mathbf{j} + \frac{H_G \cos \theta}{I_z}\mathbf{k} \qquad (21\text{--}35)$$

In a similar manner, equating the respective **i**, **j**, **k** components of Eq. 21–27 to those of Eq. 21–34, we obtain

$$\dot{\theta} = 0$$

$$\dot{\phi} \sin \theta = \frac{H_G \sin \theta}{I}$$

$$\dot{\phi} \cos \theta + \dot{\psi} = \frac{H_G \cos \theta}{I_z}$$

Solving, we get

$$\theta = \text{constant}$$
$$\dot{\phi} = \frac{H_G}{I}$$
$$\dot{\psi} = \frac{I - I_z}{I I_z} H_G \cos \theta \qquad (21\text{--}36)$$

Thus, for torque-free motion of an axisymmetrical body, the angle θ formed between the angular-momentum vector and the spin of the body remains constant. Furthermore, the angular momentum $\mathbf{H}_G$, precession $\dot{\phi}$, and spin $\dot{\psi}$ for the body remain constant at all times during the motion.

Eliminating H_G from the second and third of Eqs. 21–36 yields the following relation between the spin and precession:

$$\dot{\psi} = \frac{I - I_z}{I_z} \dot{\phi} \cos \theta \qquad (21\text{--}37)$$

21

•21–73.   At the moment of take off, the landing gear of an airplane is retracted with a constant angular velocity of $\omega_p = 2$ rad/s, while the wheel continues to spin. If the plane takes off with a speed of $v = 320$ km/h, determine the torque at A due to the gyroscopic effect. The wheel has a mass of 50 kg, and the radius of gyration about its spinning axis is $k = 300$ mm.

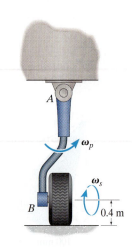

Prob. 21–73

21–74.   The projectile shown is subjected to torque-free motion. The transverse and axial moments of inertia are I and I_z, respectively. If θ represents the angle between the precessional axis Z and the axis of symmetry z, and β is the angle between the angular velocity ω and the z axis, show that β and θ are related by the equation $\tan \theta = (I/I_z) \tan \beta$.

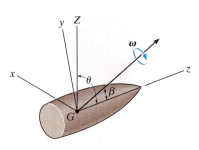

Prob. 21–74

21–75.   The space capsule has a mass of 3.2 Mg, and about axes passing through the mass center G the axial and transverse radii of gyration are $k_z = 0.90$ m and $k_t = 1.85$ m, respectively. If it spins at $\omega_s = 0.8$ rev/s, determine its angular momentum. Precession occurs about the Z axis.

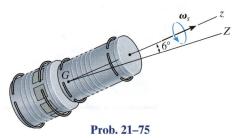

Prob. 21–75

*21–76.   The radius of gyration about an axis passing through the axis of symmetry of the 2.5-Mg satellite is $k_z = 2.3$ m, and about any transverse axis passing through the center of mass G, $k_t = 3.4$ m. If the satellite has a steady-state precession of two revolutions per hour about the Z axis, determine the rate of spin about the z axis.

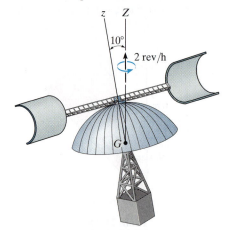

Prob. 21–76

•21–77.   The 4-kg disk is thrown with a spin $\omega_z = 6$ rad/s. If the angle θ is measured as 160°, determine the precession about the Z axis.

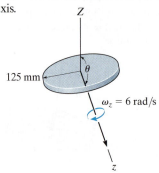

Prob. 21–77

21–78. The projectile precesses about the Z axis at a constant rate of $\dot{\phi} = 15$ rad/s when it leaves the barrel of a gun. Determine its spin $\dot{\psi}$ and the magnitude of its angular momentum $\mathbf{H}_G$. The projectile has a mass of 1.5 kg and radii of gyration about its axis of symmetry (z axis) and about its transverse axes (x and y axes) of $k_z = 65$ mm and $k_x = k_y = 125$ mm, respectively.

*21–80. The football has a mass of 450 g and radii of gyration about its axis of symmetry (z axis) and its transverse axes (x or y axis) of $k_z = 30$ mm and $k_x = k_y = 50$ mm, respectively. If the football has an angular momentum of $H_G = 0.02$ kg·m^2/s, determine its precession $\dot{\phi}$ and spin $\dot{\psi}$. Also, find the angle β that the angular velocity vector makes with the z axis.

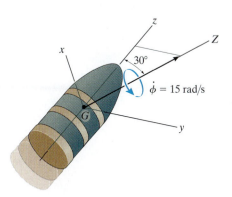

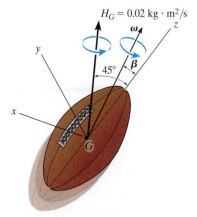

$H_G = 0.02$ kg·m^2/s

Prob. 21–78

Prob. 21–80

21–79. The satellite has a mass of 100 kg and radii of gyration about its axis of symmetry (z axis) and its transverse axes (x or y axis) of $k_z = 300$ mm and $k_x = k_y = 900$ mm, respectively. If the satellite spins about the z axis at a constant rate of $\dot{\psi} = 200$ rad/s, and precesses about the Z axis, determine the precession $\dot{\phi}$ and the magnitude of its angular momentum $\mathbf{H}_G$.

•21–81. The space capsule has a mass of 2 Mg, center of mass at G, and radii of gyration about its axis of symmetry (z axis) and its transverse axes (x or y axis) of $k_z = 2.75$ m and $k_x = k_y = 5.5$ m, respectively. If the capsule has the angular velocity shown, determine its precession $\dot{\phi}$ and spin $\dot{\psi}$. Indicate whether the precession is regular or retrograde. Also, draw the space cone and body cone for the motion.

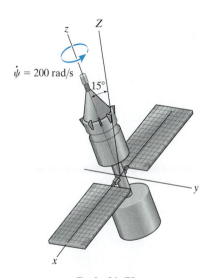

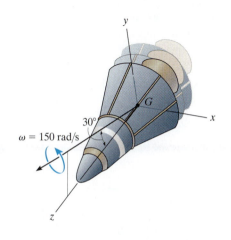

Prob. 21–79

Prob. 21–81

CHAPTER REVIEW

Moments and Products of Inertia

A body has six components of inertia for any specified x, y, z axes. Three of these are moments of inertia about each of the axes, I_{xx}, I_{yy}, I_{zz}, and three are products of inertia, each defined from two orthogonal planes, I_{xy}, I_{yz}, I_{xz}. If either one or both of these planes are planes of symmetry, then the product of inertia with respect to these planes will be zero.

$$I_{xx} = \int_m r_x^2 \, dm = \int_m (y^2 + z^2) \, dm \qquad I_{xy} = I_{yx} = \int_m xy \, dm$$

$$I_{yy} = \int_m r_y^2 \, dm = \int_m (x^2 + z^2) \, dm \qquad I_{yz} = I_{zy} = \int_m yz \, dm$$

$$I_{zz} = \int_m r_z^2 \, dm = \int_m (x^2 + y^2) \, dm \qquad I_{xz} = I_{zx} = \int_m xz \, dm$$

The moments and products of inertia can be determined by direct integration or by using tabulated values. If these quantities are to be determined with respect to axes or planes that do not pass through the mass center, then parallel-axis and parallel-plane theorems must be used.

Provided the six components of inertia are known, then the moment of inertia about any axis can be determined using the inertia transformation equation.

$$I_{Oa} = I_{xx}u_x^2 + I_{yy}u_y^2 + I_{zz}u_z^2 - 2I_{xy}u_xu_y - 2I_{yz}u_yu_z - 2I_{zx}u_zu_x$$

Principal Moments of Inertia

At any point on or off the body, the x, y, z axes can be oriented so that the products of inertia will be zero. The resulting moments of inertia are called the principal moments of inertia, one of which will be a maximum and the other a minimum.

$$\begin{pmatrix} I_x & 0 & 0 \\ 0 & I_y & 0 \\ 0 & 0 & I_z \end{pmatrix}$$

Principle of Impulse and Momentum

The angular momentum for a body can be determined about any arbitrary point A.

Once the linear and angular momentum for the body have been formulated, then the principle of impulse and momentum can be used to solve problems that involve force, velocity, and time.

$$m(\mathbf{v}_G)_1 + \Sigma \int_{t_1}^{t_2} \mathbf{F} \, dt = m(\mathbf{v}_G)_2 \qquad (\mathbf{H}_O)_1 + \Sigma \int_{t_1}^{t_2} \mathbf{M}_O \, dt = (\mathbf{H}_O)_2$$

$$\mathbf{H}_O = \int_m \boldsymbol{\rho}_O \times (\boldsymbol{\omega} \times \boldsymbol{\rho}_O) \, dm$$

Fixed Point O

$$\mathbf{H}_G = \int_m \boldsymbol{\rho}_G \times (\boldsymbol{\omega} \times \boldsymbol{\rho}_G) \, dm$$

Center of Mass

$$\mathbf{H}_A = \boldsymbol{\rho}_{G/A} \times m\mathbf{v}_G + \mathbf{H}_G$$

Arbitrary Point

where

$$H_x = I_{xx}\omega_x - I_{xy}\omega_y - I_{xz}\omega_z$$
$$H_y = -I_{yx}\omega_x + I_{yy}\omega_y - I_{yz}\omega_z$$
$$H_z = -I_{zx}\omega_x - I_{zy}\omega_y + I_{zz}\omega_z$$

Principle of Work and Energy

The kinetic energy for a body is usually determined relative to a fixed point or the body's mass center.

$$T = \tfrac{1}{2}I_x\omega_x^2 + \tfrac{1}{2}I_y\omega_y^2 + \tfrac{1}{2}I_z\omega_z^2 \qquad T = \tfrac{1}{2}mv_G^2 + \tfrac{1}{2}I_x\omega_x^2 + \tfrac{1}{2}I_y\omega_y^2 + \tfrac{1}{2}I_z\omega_z^2$$

Fixed Point   Center of Mass

These formulations can be used with the principle of work and energy to solve problems that involve force, velocity, and displacement.

$$T_1 + \Sigma U_{1-2} = T_2$$

Equations of Motion

There are three scalar equations of translational motion for a rigid body that moves in three dimensions.

$$\Sigma F_x = m(a_G)_x$$
$$\Sigma F_y = m(a_G)_y$$
$$\Sigma F_z = m(a_G)_z$$

The three scalar equations of rotational motion depend upon the motion of the x, y, z reference. Most often, these axes are oriented so that they are principal axes of inertia. If the axes are fixed in and move with the body so that $\Omega = \omega$, then the equations are referred to as the Euler equations of motion.

$$\Sigma M_x = I_x \dot{\omega}_x - (I_y - I_z)\omega_y\omega_z$$
$$\Sigma M_y = I_y \dot{\omega}_y - (I_z - I_x)\omega_z\omega_x$$
$$\Sigma M_z = I_z \dot{\omega}_z - (I_x - I_y)\omega_x\omega_y$$

$$\Omega = \omega$$

$$\Sigma M_x = I_x \dot{\omega}_x - I_y\Omega_z\omega_y + I_z\Omega_y\omega_z$$
$$\Sigma M_y = I_y \dot{\omega}_y - I_z\Omega_x\omega_z + I_x\Omega_z\omega_x$$
$$\Sigma M_z = I_z \dot{\omega}_z - I_x\Omega_y\omega_x + I_y\Omega_x\omega_y$$

$$\Omega \neq \omega$$

A free-body diagram should always accompany the application of the equations of motion.

Gyroscopic Motion

The angular motion of a gyroscope is best described using the three Euler angles ϕ, θ, and ψ. The angular velocity components are called the precession $\dot{\phi}$, the nutation $\dot{\theta}$, and the spin $\dot{\psi}$.

If $\dot{\theta} = 0$ and $\dot{\phi}$ and $\dot{\psi}$ are constant, then the motion is referred to as steady precession.

It is the spin of a gyro rotor that is responsible for holding a rotor from falling downward, and instead causing it to process about a vertical axis. This phenomenon is called the gyroscopic effect.

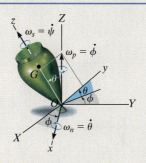

$$\Sigma M_x = -I\dot{\phi}^2 \sin\theta\cos\theta + I_z\dot{\phi}\sin\theta(\dot{\phi}\cos\theta + \dot{\psi})$$

$$\Sigma M_y = 0, \ \Sigma M_z = 0$$

Torque-Free Motion

A body that is only subjected to a gravitational force will have no moments on it about its mass center, and so the motion is described as torque-free motion. The angular momentum for the body about its mass center will remain constant. This causes the body to have both a spin and a precession. The motion depends upon the magnitude of the moment of inertia of a symmetric body about the spin axis, I_z, versus that about a perpendicular axis, I.

$$\theta = \text{constant}$$

$$\dot{\phi} = \frac{H_G}{I}$$

$$\dot{\psi} = \frac{I - I_z}{II_z}H_G\cos\theta$$

21

Spring suspensions can induce vibrations in moving vehicles, such as this railroad car. In order to predict the behavior we must use a vibrational analysis.

Vibrations

22

CHAPTER OBJECTIVES

- To discuss undamped one-degree-of-freedom vibration of a rigid body using the equation of motion and energy methods.

- To study the analysis of undamped forced vibration and viscous damped forced vibration.

*22.1 Undamped Free Vibration

A *vibration* is the periodic motion of a body or system of connected bodies displaced from a position of equilibrium. In general, there are two types of vibration, free and forced. *Free vibration* occurs when the motion is maintained by gravitational or elastic restoring forces, such as the swinging motion of a pendulum or the vibration of an elastic rod. *Forced vibration* is caused by an external periodic or intermittent force applied to the system. Both of these types of vibration can either be damped or undamped. *Undamped* vibrations can continue indefinitely because frictional effects are neglected in the analysis. Since in reality both internal and external frictional forces are present, the motion of all vibrating bodies is actually *damped*.

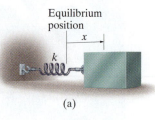

Equilibrium position

x

k

(a)

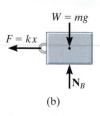

$W = mg$

$F = kx$

N_B

(b)

Fig. 22–1

The simplest type of vibrating motion is undamped free vibration, represented by the block and spring model shown in Fig. 22–1a. Vibrating motion occurs when the block is released from a displaced position x so that the spring pulls on the block. The block will attain a velocity such that it will proceed to move out of equilibrium when $x = 0$, and provided the supporting surface is smooth, the block will oscillate back and forth.

The time-dependent path of motion of the block can be determined by applying the equation of motion to the block when it is in the displaced position x. The free-body diagram is shown in Fig. 22–1b. The elastic restoring force $F = kx$ is always directed toward the equilibrium position, whereas the acceleration **a** is assumed to act in the direction of *positive displacement*. Since $a = d^2x/dt^2 = \ddot{x}$, we have

$$\overset{+}{\rightarrow} \Sigma F_x = ma_x; \qquad\qquad -kx = m\ddot{x}$$

Note that the acceleration is proportional to the block's displacement. Motion described in this manner is called *simple harmonic motion*. Rearranging the terms into a "standard form" gives

$$\ddot{x} + \omega_n^2 x = 0 \qquad\qquad (22\text{–}1)$$

The constant ω_n is called the *natural frequency*, and in this case

$$\boxed{\omega_n = \sqrt{\frac{k}{m}}} \qquad\qquad (22\text{–}2)$$

k

Equilibrium position

y

(a)

Equation 22–1 can also be obtained by considering the block to be suspended so that the displacement y is measured from the block's *equilibrium position*, Fig. 22–2a. When the block is in equilibrium, the spring exerts an upward force of $F = W = mg$ on the block. Hence, when the block is displaced a distance y downward from this position, the magnitude of the spring force is $F = W + ky$, Fig. 22–2b. Applying the equation of motion gives

$$+\downarrow \Sigma F_y = ma_y; \qquad\qquad -W - ky + W = m\ddot{y}$$

or

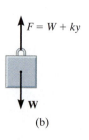

$F = W + ky$

W

(b)

Fig. 22–2

$$\ddot{y} + \omega_n^2 y = 0$$

which is the same form as Eq. 22–1 and ω_n is defined by Eq. 22–2.

Equation 22–1 is a homogeneous, second-order, linear, differential equation with constant coefficients. It can be shown, using the methods of differential equations, that the general solution is

$$x = A \sin \omega_n t + B \cos \omega_n t \qquad (22\text{–}3)$$

Here A and B represent two constants of integration. The block's velocity and acceleration are determined by taking successive time derivatives, which yields

$$v = \dot{x} = A\omega_n \cos \omega_n t - B\omega_n \sin \omega_n t \qquad (22\text{–}4)$$

$$a = \ddot{x} = -A\omega_n^2 \sin \omega_n t - B\omega_n^2 \cos \omega_n t \qquad (22\text{–}5)$$

When Eqs. 22–3 and 22–5 are substituted into Eq. 22–1, the differential equation will be satisfied, showing that Eq. 22–3 is indeed the solution to Eq. 22–1.

The integration constants in Eq. 22–3 are generally determined from the initial conditions of the problem. For example, suppose that the block in Fig. 22–1a has been displaced a distance x_1 to the right from its equilibrium position and given an initial (positive) velocity $\mathbf{v}_1$ directed to the right. Substituting $x = x_1$ when $t = 0$ into Eq. 22–3 yields $B = x_1$. And since $v = v_1$ when $t = 0$, using Eq. 22–4 we obtain $A = v_1/\omega_n$. If these values are substituted into Eq. 22–3, the equation describing the motion becomes

$$x = \frac{v_1}{\omega_n} \sin \omega_n t + x_1 \cos \omega_n t \qquad (22\text{–}6)$$

Equation 22–3 may also be expressed in terms of simple sinusoidal motion. To show this, let

$$A = C \cos \phi \qquad (22\text{–}7)$$

and

$$B = C \sin \phi \qquad (22\text{–}8)$$

where C and ϕ are new constants to be determined in place of A and B. Substituting into Eq. 22–3 yields

$$x = C \cos \phi \sin \omega_n t + C \sin \phi \cos \omega_n t$$

And since $\sin(\theta + \phi) = \sin \theta \cos \phi + \cos \theta \sin \phi$, then

$$x = C \sin(\omega_n t + \phi) \qquad (22\text{–}9)$$

If this equation is plotted on an x versus $\omega_n t$ axis, the graph shown in Fig. 22–3 is obtained. The maximum displacement of the block from its

EXAMPLE 22.1

(a)

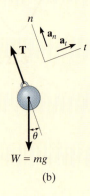

(b)

Fig. 22–4

Determine the period of oscillation for the simple pendulum shown in Fig. 22–4a. The bob has a mass m and is attached to a cord of length l. Neglect the size of the bob.

SOLUTION

Free-Body Diagram. Motion of the system will be related to the position coordinate $(q =) \theta$, Fig. 22–4b. When the bob is displaced by a small angle θ, the *restoring force* acting on the bob is created by the tangential component of its weight, $mg \sin \theta$. Furthermore, $\mathbf{a}_t$ acts in the direction of *increasing s* (or θ).

Equation of Motion. Applying the equation of motion in the *tangential direction*, since it involves the restoring force, yields

$$+\nearrow \Sigma F_t = ma_t; \qquad -mg \sin \theta = ma_t \qquad (1)$$

Kinematics. $a_t = d^2s/dt^2 = \ddot{s}$. Furthermore, s can be related to θ by the equation $s = l\theta$, so that $a_t = l\ddot{\theta}$. Hence, Eq. 1 reduces to

$$\ddot{\theta} + \frac{g}{l}\sin \theta = 0 \qquad (2)$$

The solution of this equation involves the use of an elliptic integral. For *small displacements*, however, $\sin \theta \approx \theta$, in which case

$$\ddot{\theta} + \frac{g}{l}\theta = 0 \qquad (3)$$

Comparing this equation with Eq. 22–16 ($\ddot{x} + \omega_n^2 x = 0$), it is seen that $\omega_n = \sqrt{g/l}$. From Eq. 22–12, the period of time required for the bob to make one complete swing is therefore

$$\tau = \frac{2\pi}{\omega_n} = 2\pi \sqrt{\frac{l}{g}} \qquad Ans.$$

This interesting result, originally discovered by Galileo Galilei through experiment, indicates that the period depends only on the length of the cord and not on the mass of the pendulum bob or the angle θ.

NOTE: The solution of Eq. 3 is given by Eq. 22–3, where $\omega_n = \sqrt{g/l}$ and θ is substituted for x. Like the block and spring, the constants A and B in this problem can be determined if, for example, one knows the displacement and velocity of the bob at a given instant.

EXAMPLE 22.2

The 10-kg rectangular plate shown in Fig. 22–5a is suspended at its center from a rod having a torsional stiffness $k = 1.5 \, \text{N} \cdot \text{m/rad}$. Determine the natural period of vibration of the plate when it is given a small angular displacement θ in the plane of the plate.

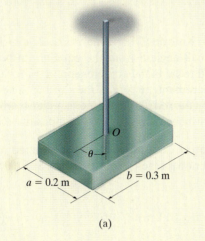

$a = 0.2$ m $b = 0.3$ m

(a)

SOLUTION

Free-Body Diagram.   Fig. 22–5b. Since the plate is displaced in its own plane, the torsional *restoring* moment created by the rod is $M = k\theta$. This moment acts in the direction opposite to the angular displacement θ. The angular acceleration $\ddot{\theta}$ acts in the direction of *positive* θ.

Equation of Motion.

$$\Sigma M_O = I_O \alpha; \qquad -k\theta = I_O \ddot{\theta}$$

or

$$\ddot{\theta} + \frac{k}{I_O}\theta = 0$$

Since this equation is in the "standard form," the natural frequency is $\omega_n = \sqrt{k/I_O}$.

From the table on the inside back cover, the moment of inertia of the plate about an axis coincident with the rod is $I_O = \frac{1}{12}m(a^2 + b^2)$. Hence,

$$I_O = \frac{1}{12}(10 \text{ kg})[(0.2 \text{ m})^2 + (0.3 \text{ m})^2] = 0.1083 \text{ kg} \cdot \text{m}^2$$

The natural period of vibration is therefore,

$$\tau = \frac{2\pi}{\omega_n} = 2\pi\sqrt{\frac{I_O}{k}} = 2\pi\sqrt{\frac{0.1083}{1.5}} = 1.69 \text{ s} \qquad \textit{Ans.}$$

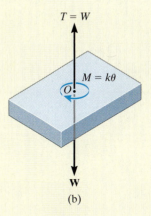

$T = W$

$M = k\theta$

O

W

(b)

Fig. 22–5

22

EXAMPLE 22.3

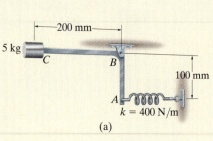

(a)

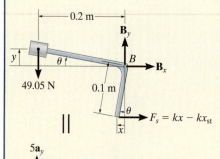

(b)

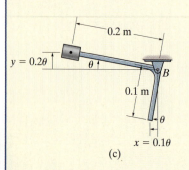

(c)

Fig. 22–6

The bent rod shown in Fig. 22–6a has a negligible mass and supports a 5-kg collar at its end. If the rod is in the equilibrium position shown, determine the natural period of vibration for the system.

SOLUTION

Free-Body and Kinetic Diagrams. Fig. 22–6b. Here the rod is displaced by a small angle θ from the equilibrium position. Since the spring is subjected to an initial compression of x_{st} for equilibrium, then when the displacement $x > x_{st}$ the spring exerts a force of $F_s = kx - kx_{st}$ on the rod. To obtain the "standard form," Eq. 22–16, $5\mathbf{a}_y$ must act *upward*, which is in accordance with positive θ displacement.

Equation of Motion. Moments will be summed about point B to eliminate the unknown reaction at this point. Since θ is small,

$$\zeta + \Sigma M_B = \Sigma (\mathcal{M}_k)_B;$$

$$kx(0.1 \text{ m}) - kx_{st}(0.1 \text{ m}) + 49.05 \text{ N}(0.2 \text{ m}) = -(5 \text{ kg})a_y(0.2 \text{ m})$$

The second term on the left side, $-kx_{st}(0.1 \text{ m})$, represents the moment created by the spring force which is necessary to hold the collar in *equilibrium*, i.e., at $x = 0$. Since this moment is equal and opposite to the moment 49.05 N(0.2 m) created by the weight of the collar, these two terms cancel in the above equation, so that

$$kx(0.1) = -5a_y(0.2) \tag{1}$$

Kinematics. The deformation of the spring and the position of the collar can be related to the angle θ, Fig. 22–6c. Since θ is small, $x = (0.1 \text{ m})\theta$ and $y = (0.2 \text{ m})\theta$. Therefore, $a_y = \ddot{y} = 0.2\ddot{\theta}$. Substituting into Eq. 1 yields

$$400(0.1\theta) \, 0.1 = -5(0.2\ddot{\theta})0.2$$

Rewriting this equation in the "standard form" gives

$$\ddot{\theta} + 200\theta = 0$$

Compared with $\ddot{x} + \omega_n^2 x = 0$ (Eq. 22–16), we have

$$\omega_n^2 = 20 \qquad \omega_n = 4.47 \text{ rad/s}$$

The natural period of vibration is therefore

$$\tau = \frac{2\pi}{\omega_n} = \frac{2\pi}{4.47} = 1.40 \text{ s} \qquad \textit{Ans.}$$

EXAMPLE 22.4

A 10-lb block is suspended from a cord that passes over a 15-lb disk, as shown in Fig. 22–7a. The spring has a stiffness $k = 200$ lb/ft. Determine the natural period of vibration for the system.

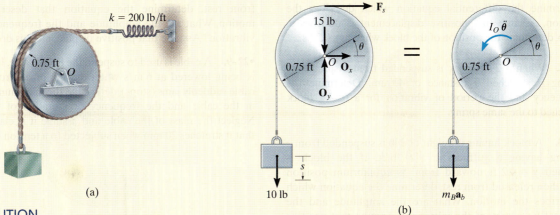

(a)   (b)

10 lb

SOLUTION

Free-Body and Kinetic Diagrams.   Fig. 22–7b. The *system* consists of the disk, which undergoes a rotation defined by the angle θ, and the block, which translates by an amount s. The vector $I_O\ddot{\theta}$ acts in the direction of *positive* θ, and consequently $m_B a_b$ acts downward in the direction of *positive* s.

Equation of Motion.   Summing moments about point O to eliminate the reactions $\mathbf{O}_x$ and $\mathbf{O}_y$, realizing that $I_O = \frac{1}{2}mr^2$, yields

$$\zeta + \Sigma M_O = \Sigma(\mathcal{M}_k)_O;$$

$$10 \text{ lb}(0.75 \text{ ft}) - F_s(0.75 \text{ ft})$$

$$= \frac{1}{2}\left(\frac{15 \text{ lb}}{32.2 \text{ ft/s}^2}\right)(0.75 \text{ ft})^2 \ddot{\theta} + \left(\frac{10 \text{ lb}}{32.2 \text{ ft/s}^2}\right)a_b(0.75 \text{ ft}) \quad (1)$$

Kinematics.   As shown on the kinematic diagram in Fig. 22–7c, a small positive displacement θ of the disk causes the block to lower by an amount $s = 0.75\theta$; hence, $a_b = \ddot{s} = 0.75\ddot{\theta}$. When $\theta = 0°$, the spring force required for *equilibrium* of the disk is 10 lb, acting to the right. For position θ, the spring force is $F_s = (200 \text{ lb/ft})(0.75\theta \text{ ft}) + 10 \text{ lb}$. Substituting these results into Eq. 1 and simplifying yields

$$\ddot{\theta} + 368\theta = 0$$

Hence,

$$\omega_n^2 = 368 \qquad \omega_n = 19.18 \text{ rad/s}$$

Therefore, the natural period of vibration is

$$\tau = \frac{2\pi}{\omega_n} = \frac{2\pi}{19.18} = 0.328 \text{ s} \qquad \textit{Ans.}$$

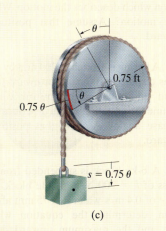

(c)

Fig. 22–7

*22–24. If the spool undergoes a small angular displacement of θ and is then released, determine the frequency of oscillation. The spool has a mass of 50 kg and a radius of gyration about its center of mass O of $k_O = 250$ mm. The spool rolls without slipping.

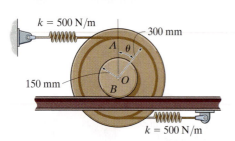

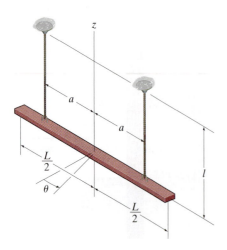

Prob. 22–24

22–26. A wheel of mass m is suspended from two equal-length cords as shown. When it is given a small angular displacement of θ about the z axis and released, it is observed that the period of oscillation is τ. Determine the radius of gyration of the wheel about the z axis.

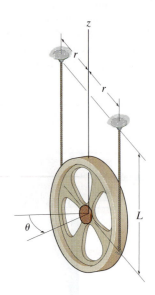

Prob. 22–26

•22–25. The slender bar of mass m is supported by two equal-length cords. If it is given a small angular displacement of θ about the vertical axis and released, determine the natural period of oscillation.

22–27. A wheel of mass m is suspended from three equal-length cords. When it is given a small angular displacement of θ about the z axis and released, it is observed that the period of oscillation is τ. Determine the radius of gyration of the wheel about the z axis.

Prob. 22–25

Prob. 22–27

*22.2 Energy Methods

The simple harmonic motion of a body, discussed in the previous section, is due only to gravitational and elastic restoring forces acting on the body. Since these forces are *conservative*, it is also possible to use the conservation of energy equation to obtain the body's natural frequency or period of vibration. To show how to do this, consider again the block and spring model in Fig. 22–8. When the block is displaced x from the equilibrium position, the kinetic energy is $T = \frac{1}{2}mv^2 = \frac{1}{2}m\dot{x}^2$ and the potential energy is $V = \frac{1}{2}kx^2$. Since energy is conserved, it is necessary that

$$T + V = \text{constant}$$

$$\tfrac{1}{2}m\dot{x}^2 + \tfrac{1}{2}kx^2 = \text{constant} \tag{22-17}$$

The differential equation describing the *accelerated motion* of the block can be obtained by *differentiating* this equation with respect to time; i.e.,

$$m\dot{x}\ddot{x} + kx\dot{x} = 0$$

$$\dot{x}(m\ddot{x} + kx) = 0$$

Since the velocity $\dot{x}$ is not *always* zero in a vibrating system,

$$\ddot{x} + \omega_n^2 x = 0 \qquad \omega_n = \sqrt{k/m}$$

which is the same as Eq. 22–1.

If the conservation of energy equation is written for a *system of connected bodies*, the natural frequency or the equation of motion can also be determined by time differentation. It is *not necessary* to dismember the system to account for the internal forces because they do no work.

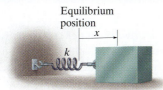

Equilibrium
position

x

k

Fig. 22–8

The suspension of a railroad car consists of a set of springs which are mounted between the frame of the car and the wheel truck. This will give the car a natural frequency of vibration which can be determined.

Procedure for Analysis

The natural frequency ω_n of a body or system of connected bodies can be determined by applying the conservation of energy equation using the following procedure.

Energy Equation.

- Draw the body when it is displaced by a *small amount* from its equilibrium position and define the location of the body from its equilibrium position by an appropriate position coordinate q.

- Formulate the conservation of energy for the body, $T + V =$ constant, in terms of the position coordinate.

- In general, the kinetic energy must account for both the body's translational and rotational motion, $T = \frac{1}{2}mv_G^2 + \frac{1}{2}I_G\omega^2$, Eq. 18–2.

- The potential energy is the sum of the gravitational and elastic potential energies of the body, $V = V_g + V_e$, Eq. 18–17. In particular, V_g should be measured from a datum for which $q = 0$ (equilibrium position).

Time Derivative.

- Take the time derivative of the energy equation using the chain rule of calculus and factor out the common terms. The resulting differential equation represents the equation of motion for the system. The natural frequency of ω_n is obtained after rearranging the terms in the "standard form," $\ddot{q} + \omega_n^2 q = 0$.

22

EXAMPLE 22.5

The thin hoop shown in Fig. 22–9a is supported by the peg at O. Determine the natural period of oscillation for small amplitudes of swing. The hoop has a mass m.

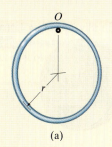

(a)

SOLUTION

Energy Equation. A diagram of the hoop when it is displaced a small amount $(q =) \theta$ from the equilibrium position is shown in Fig. 22–9b. Using the table on the inside back cover and the parallel-axis theorem to determine I_O, the kinetic energy is

$$T = \tfrac{1}{2}I_O\omega_n^2 = \tfrac{1}{2}[mr^2 + mr^2]\dot{\theta}^2 = mr^2\dot{\theta}^2$$

If a horizontal datum is placed through point O, then in the displaced position, the potential energy is

$$V = -mg(r \cos \theta)$$

The total energy in the system is

$$T + V = mr^2\dot{\theta}^2 - mgr \cos \theta$$

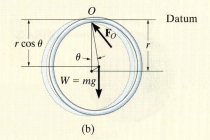

(b)

Fig. 22–9

Time Derivative.

$$mr^2(2\dot{\theta})\ddot{\theta} + mgr \sin \theta \dot{\theta} = 0$$
$$mr\dot{\theta}(2r\ddot{\theta} + g \sin \theta) = 0$$

Since $\dot{\theta}$ is not always equal to zero, from the terms in parentheses,

$$\ddot{\theta} + \frac{g}{2r}\sin \theta = 0$$

For small angle θ, $\sin \theta \approx \theta$.

$$\ddot{\theta} + \frac{g}{2r}\theta = 0$$

$$\omega_n = \sqrt{\frac{g}{2r}}$$

so that

$$\tau = \frac{2\pi}{\omega_n} = 2\pi\sqrt{\frac{2r}{g}} \qquad \textit{Ans.}$$

22

•**22–37.** A torsional spring of stiffness k is attached to a wheel that has a mass of M. If the wheel is given a small angular displacement of θ about the z axis determine the natural period of oscillation. The wheel has a radius of gyration about the z axis of k_z.

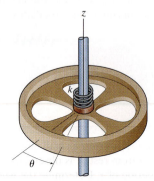

Prob. 22–37

22–38. Determine the frequency of oscillation of the cylinder of mass m when it is pulled down slightly and released. Neglect the mass of the small pulley.

Prob. 22–38

22–39. Determine the frequency of oscillation of the cylinder of mass m when it is pulled down slightly and released. Neglect the mass of the small pulleys.

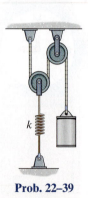

Prob. 22–39

*__22–40.__ The gear of mass m has a radius of gyration about its center of mass O of k_O. The springs have stiffnesses of k_1 and k_2, respectively, and both springs are unstretched when the gear is in an equilibrium position. If the gear is given a small angular displacement of θ and released, determine its natural period of oscillation.

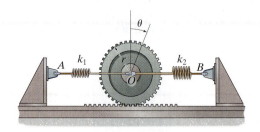

Prob. 22–40

22–41. The bar has a mass of 8 kg and is suspended from two springs such that when it is in equilibrium, the springs make an angle of 45° with the horizontal as shown. Determine the natural period of vibration if the bar is pulled down a short distance and released. Each spring has a stiffness of $k = 40$ N/m.

Prob. 22–41

22

*22.3   Undamped Forced Vibration

Undamped forced vibration is considered to be one of the most important types of vibrating motion in engineering. Its principles can be used to describe the motion of many types of machines and structures.

Periodic Force. The block and spring shown in Fig. 22–11a provide a convenient model which represents the vibrational characteristics of a system subjected to a periodic force $F = F_0 \sin \omega_0 t$. This force has an amplitude of F_0 and a *forcing frequency* ω_0. The free-body diagram for the block when it is displaced a distance x is shown in Fig. 22–11b. Applying the equation of motion, we have

$$\xrightarrow{+} \Sigma F_x = ma_x; \qquad F_0 \sin \omega_0 t - kx = m\ddot{x}$$

or

$$\ddot{x} + \frac{k}{m}x = \frac{F_0}{m}\sin \omega_0 t \tag{22–18}$$

This equation is a nonhomogeneous second-order differential equation. The general solution consists of a complementary solution, x_c, *plus* a particular solution, x_p.

The *complementary solution* is determined by setting the term on the right side of Eq. 22–18 equal to zero and solving the resulting homogeneous equation. The solution is defined by Eq. 22–9, i.e.,

$$x_c = C \sin(\omega_n t + \phi) \tag{22–19}$$

where ω_n is the natural frequency, $\omega_n = \sqrt{k/m}$, Eq. 22–2.

Since the motion is periodic, the *particular solution* of Eq. 22–18 can be determined by assuming a solution of the form

$$x_p = X \sin \omega_0 t \tag{22–20}$$

where X is a constant. Taking the second time derivative and substituting into Eq. 22–18 yields

$$-X\omega_0^2 \sin \omega_0 t + \frac{k}{m}(X \sin \omega_0 t) = \frac{F_0}{m}\sin \omega_0 t$$

Factoring out $\sin \omega_0 t$ and solving for X gives

$$X = \frac{F_0/m}{(k/m) - \omega_0^2} = \frac{F_0/k}{1 - (\omega_0/\omega_n)^2} \tag{22–21}$$

Substituting into Eq. 22–20, we obtain the particular solution

$$\boxed{x_p = \frac{F_0/k}{1 - (\omega_0/\omega_n)^2}\sin \omega_0 t} \tag{22–22}$$

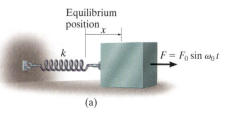

Equilibrium position

(a)

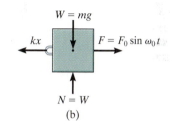

(b)

Fig. 22–11

Shaker tables provide forced vibration and are used to separate out granular materials.

22

EXAMPLE | 22.7

The instrument shown in Fig. 22–14 is rigidly attached to a platform P, which in turn is supported by *four* springs, each having a stiffness $k = 800 \text{ N/m}$. If the floor is subjected to a vertical displacement $\delta = 10 \sin(8t)$ mm, where t is in seconds, determine the amplitude of steady-state vibration. What is the frequency of the floor vibration required to cause resonance? The instrument and platform have a total mass of 20 kg.

Fig. 22–14

SOLUTION

The natural frequency is

$$\omega_n = \sqrt{\frac{k}{m}} = \sqrt{\frac{4(800 \text{ N/m})}{20 \text{ kg}}} = 12.65 \text{ rad/s}$$

The amplitude of steady state vibration is found using Eq. 22–21, with $k\delta_0$ replacing F_0.

$$X = \frac{\delta_0}{1 - (\omega_0/\omega_n)^2} = \frac{10}{1 - [(8 \text{ rad/s})/(12.65 \text{ rad/s})]^2} = 16.7 \text{ mm}$$ *Ans.*

Resonance will occur when the amplitude of vibration X caused by the floor displacement approaches infinity. This requires

$$\omega_0 = \omega_n = 12.6 \text{ rad/s}$$ *Ans.*

*22.4   Viscous Damped Free Vibration

The vibration analysis considered thus far has not included the effects of friction or damping in the system, and as a result, the solutions obtained are only in close agreement with the actual motion. Since all vibrations die out in time, the presence of damping forces should be included in the analysis.

In many cases damping is attributed to the resistance created by the substance, such as water, oil, or air, in which the system vibrates. Provided the body moves slowly through this substance, the resistance to motion is directly proportional to the body's speed. The type of force developed under these conditions is called a *viscous damping force*. The magnitude of this force is expressed by an equation of the form

$$F = c\dot{x} \qquad (22\text{–}26)$$

where the constant c is called the *coefficient of viscous damping* and has units of $N \cdot s/m$ or $lb \cdot s/ft$.

The vibrating motion of a body or system having viscous damping can be characterized by the block and spring shown in Fig. 22–15a. The effect of damping is provided by the *dashpot* connected to the block on the right side. Damping occurs when the piston P moves to the right or left within the enclosed cylinder. The cylinder contains a fluid, and the motion of the piston is retarded since the fluid must flow around or through a small hole in the piston. The dashpot is assumed to have a coefficient of viscous damping c.

If the block is displaced a distance x from its equilibrium position, the resulting free-body diagram is shown in Fig. 22–15b. Both the spring and damping force oppose the forward motion of the block, so that applying the equation of motion yields

$$\xrightarrow{+} \Sigma F_x = ma_x; \qquad -kx - c\dot{x} = m\ddot{x}$$

or

$$m\ddot{x} + c\dot{x} + kx = 0 \qquad (22\text{–}27)$$

This linear, second-order, homogeneous, differential equation has a solution of the form

$$x = e^{\lambda t}$$

where e is the base of the natural logarithm and λ (lambda) is a constant. The value of λ can be obtained by substituting this solution and its time derivatives into Eq. 22–27, which yields

$$m\lambda^2 e^{\lambda t} + c\lambda e^{\lambda t} + k e^{\lambda t} = 0$$

or

$$e^{\lambda t}(m\lambda^2 + c\lambda + k) = 0$$

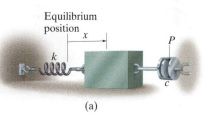

Equilibrium position

(a)

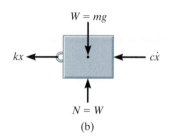

$W = mg$

kx

$c\dot{x}$

$N = W$

(b)

Fig. 22–15

Since $e^{\lambda t}$ can never be zero, a solution is possible provided

$$m\lambda^2 + c\lambda + k = 0$$

Hence, by the quadratic formula, the two values of λ are

$$\lambda_1 = -\frac{c}{2m} + \sqrt{\left(\frac{c}{2m}\right)^2 - \frac{k}{m}}$$

$$\lambda_2 = -\frac{c}{2m} - \sqrt{\left(\frac{c}{2m}\right)^2 - \frac{k}{m}} \tag{22–28}$$

The general solution of Eq. 22–27 is therefore a combination of exponentials which involves both of these roots. There are three possible combinations of λ_1 and λ_2 which must be considered. Before discussing these combinations, however, we will first define the critical damping coefficient c_c as the value of c which makes the radical in Eqs. 22–28 equal to zero; i.e.,

$$\left(\frac{c_c}{2m}\right)^2 - \frac{k}{m} = 0$$

or

$$c_c = 2m\sqrt{\frac{k}{m}} = 2m\omega_n \tag{22–29}$$

Overdamped System.   When $c > c_c$, the roots λ_1 and λ_2 are both real. The general solution of Eq. 22–27 can then be written as

$$x = Ae^{\lambda_1 t} + Be^{\lambda_2 t} \tag{22–30}$$

Motion corresponding to this solution is *nonvibrating*. The effect of damping is so strong that when the block is displaced and released, it simply creeps back to its original position without oscillating. The system is said to be *overdamped*.

Critically Damped System.   If $c = c_c$, then $\lambda_1 = \lambda_2 = -c_c/2m = -\omega_n$. This situation is known as *critical damping*, since it represents a condition where c has the smallest value necessary to cause the system to be nonvibrating. Using the methods of differential equations, it can be shown that the solution to Eq. 22–27 for critical damping is

$$x = (A + Bt)e^{-\omega_n t} \tag{22–31}$$

Underdamped System.

Most often $c < c_c$, in which case the system is referred to as *underdamped*. In this case the roots λ_1 and λ_2 are complex numbers, and it can be shown that the general solution of Eq. 22–27 can be written as

$$x = D[e^{-(c/2m)t} \sin(\omega_d t + \phi)] \qquad (22\text{–}32)$$

where D and ϕ are constants generally determined from the initial conditions of the problem. The constant ω_d is called the *damped natural frequency* of the system. It has a value of

$$\omega_d = \sqrt{\frac{k}{m} - \left(\frac{c}{2m}\right)^2} = \omega_n \sqrt{1 - \left(\frac{c}{c_c}\right)^2} \qquad (22\text{–}33)$$

where the ratio c/c_c is called the *damping factor*.

The graph of Eq. 22–32 is shown in Fig. 22–16. The initial limit of motion, D, diminishes with each cycle of vibration, since motion is confined within the bounds of the exponential curve. Using the damped natural frequency ω_d, the period of damped vibration can be written as

$$\tau_d = \frac{2\pi}{\omega_d} \qquad (22\text{–}34)$$

Since $\omega_d < \omega_n$, Eq. 22–33, the period of damped vibration, τ_d, will be greater than that of free vibration, $\tau = 2\pi/\omega_n$.

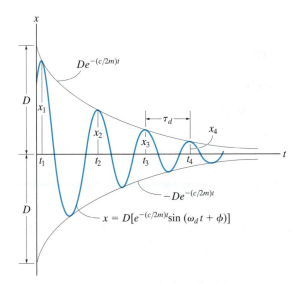

$$x = D[e^{-(c/2m)t} \sin(\omega_d t + \phi)]$$

Fig. 22–16

*22.5   Viscous Damped Forced Vibration

The most general case of single-degree-of-freedom vibrating motion occurs when the system includes the effects of forced motion and induced damping. The analysis of this particular type of vibration is of practical value when applied to systems having significant damping characteristics.

If a dashpot is attached to the block and spring shown in Fig. 22–11a, the differential equation which describes the motion becomes

$$m\ddot{x} + c\dot{x} + kx = F_0 \sin \omega_0 t \qquad (22\text{–}35)$$

A similar equation can be written for a block and spring having a periodic support displacement, Fig. 22–13a, which includes the effects of damping. In that case, however, F_0 is replaced by $k\delta_0$. Since Eq. 22–35 is nonhomogeneous, the general solution is the sum of a complementary solution, x_c, and a particular solution, x_p. The complementary solution is determined by setting the right side of Eq. 22–35 equal to zero and solving the homogeneous equation, which is equivalent to Eq. 22–27. The solution is therefore given by Eq. 22–30, 22–31, or 22–32, depending on the values of λ_1 and λ_2. Because all systems are subjected to friction, then this solution will dampen out with time. Only the particular solution, which describes the *steady-state vibration* of the system, will remain. Since the applied forcing function is harmonic, the steady-state motion will also be harmonic. Consequently, the particular solution will be of the form

$$X_P = X' \sin(\omega_0 t - \phi') \qquad (22\text{–}36)$$

The constants X' and ϕ' are determined by taking the first and second time derivatives and substituting them into Eq. 22–35, which after simplification yields

$$-X'm\omega_0^2 \sin(\omega_0 t - \phi') +$$
$$X'c\omega_0 \cos(\omega_0 t - \phi') + X'k \sin(\omega_0 t - \phi') = F_0 \sin \omega_0 t$$

Since this equation holds for all time, the constant coefficients can be obtained by setting $\omega_0 t - \phi' = 0$ and $\omega_0 t - \phi' = \pi/2$, which causes the above equation to become

$$X'c\omega_0 = F_0 \sin \phi'$$
$$-X'm\omega_0^2 + X'k = F_0 \cos \phi'$$

22

The amplitude is obtained by squaring these equations, adding the resuts, and using the identity $\sin^2\phi' + \cos^2\phi' = 1$, which gives

$$X' = \frac{F_0}{\sqrt{(k - m\omega_0^2)^2 + c^2\omega_0^2}} \qquad (22\text{--}37)$$

Dividing the first equation by the second gives

$$\phi' = \tan^{-1}\left[\frac{c\omega_0}{k - m\omega_0^2}\right] \qquad (22\text{--}38)$$

Since $\omega_n = \sqrt{k/m}$ and $c_c = 2m\omega_n$, then the above equations can also be written as

$$X' = \frac{F_0/k}{\sqrt{[1 - (\omega_0/\omega_n)^2]^2 + [2(c/c_c)(\omega_0/\omega_n)]^2}}$$

$$(22\text{--}39)$$

$$\phi' = \tan^{-1}\left[\frac{2(c/c_c)(\omega_0/\omega_n)}{1 - (\omega_0/\omega_n)^2}\right]$$

The angle ϕ' represents the phase difference between the applied force and the resulting steady-state vibration of the damped system.

The *magnification factor* MF has been defined in Sec. 22.3 as the ratio of the amplitude of deflection caused by the forced vibration to the deflection caused by a static force F_0. Thus,

$$\text{MF} = \frac{X'}{F_0/k} = \frac{1}{\sqrt{[1 - (\omega_0/\omega_n)^2]^2 + [2(c/c_c)(\omega_0/\omega_n)]^2}} \qquad (22\text{--}40)$$

The MF is plotted in Fig. 22–17 versus the frequency ratio ω_0/ω_n for various values of the damping factor c/c_c. It can be seen from this graph that the magnification of the amplitude increases as the damping factor decreases. Resonance obviously occurs only when the damping factor is zero and the frequency ratio equals 1.

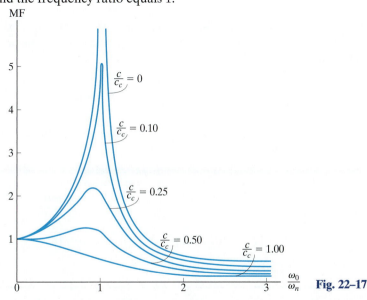

Fig. 22–17

EXAMPLE | 22.8

The 30-kg electric motor shown in Fig. 22–18 is supported by *four* springs, each spring having a stiffness of 200 N/m. If the rotor is unbalanced such that its effect is equivalent to a 4-kg mass located 60 mm from the axis of rotation, determine the amplitude of vibration when the rotor is turning at $\omega_0 = 10$ rad/s. The damping factor is $c/c_c = 0.15$.

Fig. 22–18

SOLUTION

The periodic force which causes the motor to vibrate is the centrifugal force due to the unbalanced rotor. This force has a constant magnitude of

$$F_0 = ma_n = mr\omega_0^2 = 4 \text{ kg}(0.06 \text{ m})(10 \text{ rad/s})^2 = 24 \text{ N}$$

Since $F = F_0 \sin \omega_0 t$, where $\omega_0 = 10$ rad/s, then

$$F = 24 \sin 10t$$

The stiffness of the entire system of four springs is $k = 4(200 \text{ N/m}) = 800$ N/m. Therefore, the natural frequency of vibration is

$$\omega_n = \sqrt{\frac{k}{m}} = \sqrt{\frac{800 \text{ N/m}}{30 \text{ kg}}} = 5.164 \text{ rad/s}$$

Since the damping factor is known, the steady-state amplitude can be determined from the first of Eqs. 22–39, i.e.,

$$X' = \frac{F_0/k}{\sqrt{[1 - (\omega_0/\omega_n)^2]^2 + [2(c/c_c)(\omega_0/\omega_n)]^2}}$$

$$= \frac{24/800}{\sqrt{[1 - (10/5.164)^2]^2 + [2(0.15)(10/5.164)]^2}}$$

$$= 0.0107 \text{ m} = 10.7 \text{ mm} \qquad \qquad \textit{Ans.}$$

22

*22.6 Electrical Circuit Analogs

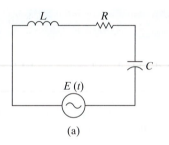

(a)

The characteristics of a vibrating mechanical system can be represented by an electric circuit. Consider the circuit shown in Fig. 22–19a, which consists of an inductor L, a resistor R, and a capacitor C. When a voltage $E(t)$ is applied, it causes a current of magnitude i to flow through the circuit. As the current flows past the inductor the voltage drop is $L(di/dt)$, when it flows across the resistor the drop is Ri, and when it arrives at the capacitor the drop is $(1/C)\int i\, dt$. Since current cannot flow past a capacitor, it is only possible to measure the charge q acting on the capacitor. The charge can, however, be related to the current by the equation $i = dq/dt$. Thus, the voltage drops which occur across the inductor, resistor, and capacitor becomes $L\, d^2q/dt^2$, $R\, dq/dt$, and q/C, respectively. According to Kirchhoff's voltage law, the applied voltage balances the sum of the voltage drops around the circuit. Therefore,

$$L\frac{d^2q}{dt^2} + R\frac{dq}{dt} + \frac{1}{C}q = E(t) \qquad (22\text{--}41)$$

Consider now the model of a single-degree-of-freedom mechanical system, Fig. 22–19b, which is subjected to both a general forcing function $F(t)$ and damping. The equation of motion for this system was established in the previous section and can be written as

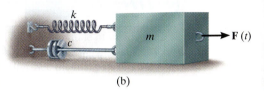

(b)

Fig. 22–19

$$m\frac{d^2x}{dt^2} + c\frac{dx}{dt} + kx = F(t) \qquad (22\text{--}42)$$

By comparison, it is seen that Eqs. 22–41 and 22–42 have the same form, and hence mathematically the procedure of analyzing an electric circuit is the same as that of analyzing a vibrating mechanical system. The analogs between the two equations are given in Table 22–1.

This analogy has important application to experimental work, for it is much easier to simulate the vibration of a complex mechanical system using an electric circuit, which can be constructed on an analog computer, than to make an equivalent mechanical spring-and-dashpot model.

Table 22–1
Electrical–Mechanical Analogs

Electrical		Mechanical	
Electric charge	q	Displacement	x
Electric current	i	Velocity	dx/dt
Voltage	$E(t)$	Applied force	$F(t)$
Inductance	L	Mass	m
Resistance	R	Viscous damping coefficient	c
Reciprocal of capacitance	$1/C$	Spring stiffness	k

22

EXAMPLE C–1

If $y = x^3$ and $x = t^4$, find $\ddot{y}$, the second derivative of y with respect to time.

SOLUTION.
Using the chain rule, Eq. C–1,

$$\dot{y} = 3x^2\dot{x}$$

To obtain the second time derivative we must use the product rule since x and $\dot{x}$ are both functions of time, and also, for $3x^2$ the chain rule must be applied. Thus, with $u = 3x^2$ and $v = \dot{x}$, we have

$$\ddot{y} = [6x\dot{x}]\dot{x} + 3x^2[\ddot{x}]$$
$$= 3x[2\dot{x}^2 + x\ddot{x}]$$

Since $x = t^4$, then $\dot{x} = 4t^3$ and $\ddot{x} = 12t^2$ so that

$$\ddot{y} = 3(t^4)[2(4t^3)^2 + t^4(12t^2)]$$
$$= 132t^{10}$$

Note that this result can also be obtained by combining the functions, then taking the time derivatives, that is,

$$y = x^3 = (t^4)^3 = t^{12}$$
$$\dot{y} = 12t^{11}$$
$$\ddot{y} = 132t^{10}$$

EXAMPLE C–2

If $y = xe^x$, find $\ddot{y}$.

SOLUTION
Since x and e^x are both functions of time the product and chain rules must be applied. Have $u = x$ and $v = e^x$.

$$\dot{y} = [\dot{x}]e^x + x[e^x\dot{x}]$$

The second time derivative also requires application of the product and chain rules. Note that the product rule applies to three time variables in the last term, i.e., x, e^x, and $\dot{x}$.

$$\ddot{y} = \{[\ddot{x}]e^x + \dot{x}[e^x\dot{x}]\} + \{[\dot{x}]e^x\dot{x} + x[e^x\dot{x}]\dot{x} + xe^x[\ddot{x}]\}$$
$$= e^x[\ddot{x}(1 + x) + \dot{x}^2(2 + x)]$$

If $x = t^2$ then $\dot{x} = 2t$, $\ddot{x} = 2$ so that in terms in t, we have

$$\ddot{y} = e^{t^2}[2(1 + t^2) + 4t^2(2 + t^2)]$$

C

EXAMPLE | C–3

If the path in radical coordinates is given as $r = 5\theta^2$, where θ is a known function of time, find $\ddot{r}$.

SOLUTION
First, using the chain rule then the chain and product rules where $u = 10\theta$ and $v = \dot{\theta}$, we have

$$r = 5\theta^2$$

$$\dot{r} = 10\theta\dot{\theta}$$

$$\ddot{r} = 10[(\dot{\theta})\dot{\theta} + \theta(\ddot{\theta})]$$

$$= 10\dot{\theta}^2 + 10\theta\ddot{\theta}$$

EXAMPLE | C–4

If $r^2 = 6\theta^3$, find $\ddot{r}$.

SOLUTION
Here the chain and product rules are applied as follows.

$$r^2 = 6\theta^3$$

$$2r\dot{r} = 18\theta^2\dot{\theta}$$

$$2[(\dot{r})\dot{r} + r(\ddot{r})] = 18[(2\theta\dot{\theta})\dot{\theta} + \theta^2(\ddot{\theta})]$$

$$\dot{r}^2 + r\ddot{r} = 9(2\theta\dot{\theta}^2 + \theta^2\ddot{\theta})$$

To find $\ddot{r}$ at a specified value of θ which is a known function of time, we can first find $\dot{\theta}$ and $\ddot{\theta}$. Then using these values, evaluate r from the first equation, $\dot{r}$ from the second equation and $\ddot{r}$ using the last equation.

C

Fundamental Problems
Partial Solutions And Answers

Chapter 12

F12–1. $v = v_0 + a_c t$

$10 = 35 + a_c(15)$

$a_c = -1.67 \text{ m/s}^2 = 1.67 \text{ m/s}^2 \leftarrow$ *Ans.*

F12–2. $s = s_0 + v_0 t + \frac{1}{2} a_c t^2$

$0 = 0 + 15t + \frac{1}{2}(-9.81)t^2$

$t = 3.06 \text{ s}$ *Ans.*

F12–3. $ds = v \, dt$

$\int_0^s ds = \int_0^t \left(4t - 3t^2\right) dt$

$s = \left(2t^2 - t^3\right) \text{m}$

$s = 2\left(4^2\right) - 4^3$

$= -32 \text{ m} = 32 \text{ m} \leftarrow$ *Ans.*

F12–4. $a = \frac{dv}{dt} = \frac{d}{dt}\left(0.5t^3 - 8t\right)$

$a = \left(1.5t^2 - 8\right) \text{m/s}^2$

When $t = 2$ s,

$a = 1.5\left(2^2\right) - 8 = -2 \text{ m/s}^2 = 2 \text{ m/s}^2 \leftarrow$ *Ans.*

F12–5. $v = \frac{ds}{dt} = \frac{d}{dt}\left(2t^2 - 8t + 6\right) = (4t - 8) \text{ m/s}$

$v = 0 = (4t - 8)$

$t = 2 \text{ s}$ *Ans.*

$s|_{t=0} = 2\left(0^2\right) - 8(0) + 6 = 6 \text{ m}$

$s|_{t=2} = 2\left(2^2\right) - 8(2) + 6 = -2 \text{ m}$

$s|_{t=3} = 2\left(3^2\right) - 8(3) + 6 = 0 \text{ m}$

$(\Delta s)_{\text{Tot}} = 8 \text{ m} + 2 \text{ m} = 10 \text{ m}$ *Ans.*

F12–6. $\int v \, dv = \int a \, ds$

$\int_{5 \text{ m/s}}^v v \, dv = \int_0^s (10 - 0.2s) ds$

$v = \left(\sqrt{20s - 0.2s^2 + 25}\right) \text{m/s}$

At $s = 10$ m,

$v = \sqrt{20(10) - 0.2(10^2) + 25} = 14.3 \text{ m/s}$ *Ans.*

F12–7. $v = \int \left(4t^2 - 2\right) dt$

$v = \frac{4}{3}t^3 - 2t + C_1$

680

$s = \int \left(\frac{4}{3}t^3 - 2t + C_1\right) dt$

$s = \frac{1}{3}t^4 - t^2 + C_1 t + C_2$

$t = 0, s = -2, C_2 = -2$

$t = 2, s = -20, C_1 = -9.67$

$t = 4, s = 28.7 \text{ m}$ *Ans.*

F12–8. $a = v\frac{dv}{ds}$

$= \left(20 - 0.05s^2\right)(-0.1s)$

At $s = 15$ m,

$a = -13.1 \text{ m/s}^2 = 13.1 \text{ m/s}^2 \leftarrow$ *Ans.*

F12–9. $v = \frac{ds}{dt} = \frac{d}{dt}\left(0.5t^3\right) = 1.5t^2$

$v|_{t=6 \text{ s}} = 1.5(6^2) = 54 \text{ m/s}$

$v = \frac{ds}{dt} = \frac{d}{dt}(108) = 0$ *Ans.*

F12–10. $ds = v \, dt$

$\int_0^s ds = \int_0^t (-4t + 80) \, dt$

$s = -2t^2 + 80t$

$s = -2(20)^2 + 80(20) = 800 \text{ ft}$

$a = \frac{dv}{dt} = \frac{d}{dt}(-4t + 80) = -4 \text{ ft/s}^2 = 4 \text{ ft/s}^2 \leftarrow$

Also,

$a = \frac{\Delta v}{\Delta t} = \frac{0 - 80 \text{ ft/s}}{20 \text{ s} - 0} = -4 \text{ ft/s}^2$

F12–11. $a \, ds = v \, dv$

$a = v\frac{dv}{ds} = 0.25s\frac{d}{ds}(0.25s) = 0.0625s$

$a|_{s=40 \text{ m}} = 0.0625(40 \text{ m}) = 2.5 \text{ m/s}^2 \rightarrow$

F12–12. $0 \le t < 5$ s,

$v = \frac{ds}{dt} = \frac{d}{dt}\left(3t^2\right) = (6t) \text{ m/s}$

$v|_{t=5 \text{ s}} = 6(5) = 30 \text{ m/s}$

$5 \text{ s} < t \le 10$ s,

$v = \frac{ds}{dt} = \frac{d}{dt}(30t - 75) = 30 \text{ m/s}$

$v = \frac{\Delta s}{\Delta t} = \frac{225 \text{ m} - 75 \text{ m}}{10 \text{ m} - 5 \text{ m}} = 30 \text{ m/s}$

$0 \le t < 5$ s,

$a = \frac{dv}{dt} = \frac{d}{dt}(6t) = 6 \text{ m/s}^2$

$5 \text{ s} < t \le 10$ s,

$a = \frac{dv}{dt} = \frac{d}{dt}(30) = 0$

$0 \le t < 5 \text{ s}, a = \Delta v/\Delta t = 6 \text{ m/s}^2$

$5 \text{ s} < t \le 10 \text{ s}, a = \Delta v/\Delta t = 0$

F12–13. $0 \le t < 5$ s,

$$dv = a \, dt \qquad \int_0^v dv = \int_0^t 20 \, dt$$

$v = (20t)$ m/s

$v = 20(5) = 100$ m/s

5 s $< t \le t'$,

$$(\overset{+}{\rightarrow}) \qquad dv = a \, dt \qquad \int_{100 \, m/s}^v dv = \int_{5s}^t -10 \, dt$$

$v = (150 - 10t)$ m/s,

$0 = 150 - 10t'$

$t' = 15$ s

Also,

$\Delta v = 0 =$ Area under the a–t graph

$0 = (20 \text{ m/s}^2)(5 \text{ s}) + [-(10 \text{ m/s})(t' - 5) \text{ s}]$

$t' = 15$ s

F12–14. $0 \le t \le 5$ s,

$$ds = v \, dt \qquad \int_0^s ds = \int_0^t 30t \, dt$$

$s|_0^s = 15t^2|_0^t$

$s = (15t^2)$ m

$s = 15(5^2) = 375$ m

5 s $< t \le 15$ s,

$$(\overset{+}{\rightarrow}) \qquad ds = v \, dt; \qquad \int_{375 \, m}^s ds = \int_{5s}^t (-15t + 225) dt$$

$s = (-7.5t^2 + 225t - 562.5)$ m

$s = (-7.5)(15)^2 + 225(15) - 562.5$ m

$\quad = 1125$ m   *Ans.*

Also,

$\Delta s =$ Area under the v–t graph

$\quad = \frac{1}{2}(150 \text{ m/s})(15 \text{ s})$

$\quad = 1125$ m   *Ans.*

F12–15. $\int_0^x dx = \int_0^t 32t \, dt$

$x = (16t^2)$ m   (1)

$\int_0^y dy = \int_0^t 8 \, dt$

$t = \frac{y}{8}$ (2)

Substituting Eq. (2) into Eq. (1),

$y^2 = 4x$ *Ans.*

F12–16. $y = 0.75(8t) = 6t$

$v_x = \dot{x} = \frac{dx}{dt} = \frac{d}{dt}(8t) = 8$ m/s $\rightarrow$

$v_y = \dot{y} = \frac{dy}{dt} = \frac{d}{dt}(6t) = 6$ m/s $\uparrow$

The magnitude of the particle's velocity is

$v = \sqrt{v_x^2 + v_y^2} = \sqrt{(8 \text{ m/s})^2 + (6 \text{ m/s})^2}$

$\quad = 10$ m/s   *Ans.*

F12–17. $y = (4t^2)$ m

$v_x = \dot{x} = \frac{d}{dt}(4t^4) = (16t^3)$ m/s $\rightarrow$

$v_y = \dot{y} = \frac{d}{dt}(4t^2) = (8t)$ m/s $\uparrow$

When $t = 0.5$ s,

$v = \sqrt{v_x^2 + v_y^2} = \sqrt{(2 \text{ m/s})^2 + (4 \text{ m/s})^2}$

$\quad = 4.47$ m/s   *Ans.*

$a_x = \dot{v}_x = \frac{d}{dt}(16t^3) = (48t^2)$ m/s^2

$a_y = \dot{v}_y = \frac{d}{dt}(8t) = 8$ m/s^2

When $t = 0.5$ s,

$a = \sqrt{a_x^2 + a_y^2} = \sqrt{(12 \text{ m/s}^2)^2 + (8 \text{ m/s}^2)^2}$

$\quad = 14.4$ m/s^2 *Ans.*

F12–18. $y = 0.5x$

$\dot{y} = 0.5\dot{x}$

$v_y = t^2$

When $t = 4$ s,

$v_x = 32$ m/s $v_y = 16$ m/s

$v = \sqrt{v_x^2 + v_y^2} = 35.8$ m/s   *Ans.*

$a_x = \dot{v}_x = 4t$

$a_y = \dot{v}_y = 2t$

When $t = 4$ s,

$a_x = 16$ m/s^2 $a_y = 8$ m/s^2

$a = \sqrt{a_x^2 + a_y^2} = \sqrt{16^2 + 8^2} = 17.9$ m/s^2 *Ans.*

F12–19. $y = (t^4)$ m

$v_x = \dot{x} = (4t)$ m/s $v_y = \dot{y} = (4t^3)$ m/s

When $t = 2$ s,

$v_x = 8$ m/s $v_y = 32$ m/s

$v = \sqrt{v_x^2 + v_y^2} = 33.0$ m/s   *Ans.*

$a_x = \dot{v}_x = 4$ m/s^2

$a_y = \dot{v}_y = (12t^2)$ m/s^2

When $t = 2$ s,

$a_x = 4$ m/s^2 $a_y = 48$ m/s^2

$a = \sqrt{a_x^2 + a_y^2} = \sqrt{4^2 + 48^2} = 48.2$ m/s^2 *Ans.*

F12–20. $\mathbf{v} = \frac{d\mathbf{r}}{dt} = [4\cos 2t\,\mathbf{i} - 2\sin t\,\mathbf{j} - 4t\mathbf{k}]$ ft/s

When $t = 2$ s,

$= \{-2.61\mathbf{i} - 1.82\mathbf{j} - 8\mathbf{k}\}$ ft/s   *Ans.*

$\mathbf{a} = \frac{d\mathbf{v}}{dt} = \{-8\sin 2t\,\mathbf{i} - 2\cos t\,\mathbf{j} - 4\,\mathbf{k}\}$ ft/s^2

When $t = 2$ s,

$\mathbf{a} = [6.05\mathbf{i} + 0.832\mathbf{j} - 4\mathbf{k}]$ ft/s^2 *Ans.*

F12–21. $(v_B)_y^2 = (v_A)_y^2 + 2a_y(y_B - y_A)$

$0^2 = (5 \text{ m/s})^2 + 2(-9.81 \text{ m/s}^2)(h - 0)$

$h = 1.27$ m   *Ans.*

F12–22. $y_C = y_A + (v_A)_y t_{AC} + \frac{1}{2}a_y t_{AC}^2$

$0 = 0 + (5 \text{ m/s})t_{AC} + \frac{1}{2}(-9.81 \text{ m/s}^2)t_{AC}^2$

$t_{AC} = 1.0194$ s

$(v_C)_y = (v_A)_y + a_y t_{AC}$

$(v_C)_y = 5 \text{ m/s} + (-9.81 \text{ m/s}^2)(1.0194 \text{ s})$

$= -5 \text{ m/s} = 5 \text{ m/s} \downarrow$

$v_C = \sqrt{(v_C)_x^2 + (v_C)_y^2}$

$= \sqrt{(8.660 \text{ m/s})^2 + (5 \text{ m/s})^2} = 10$ m/s   *Ans.*

$R = x_A + (v_A)_x t_{AC} = 0 + (8.660 \text{ m/s})(1.0194 \text{ s})$

$= 8.83$ m   *Ans.*

F12–23. $s = s_0 + v_0 t$

$10 = 0 + v_A \cos 30° t$

$s = s_0 + v_0 t + \frac{1}{2}a_c t^2$

$3 = 1.5 + v_A \sin 30° t + \frac{1}{2}(-9.81)t^2$

$t = 0.9334, \quad v_A = 12.4$ m/s   *Ans.*

F12–24. $s = s_0 + v_0 t$

$R\left(\frac{4}{5}\right) = 0 + 20\left(\frac{3}{5}\right)t$

$s = s_0 + v_0 t + \frac{1}{2}a_c t^2$

$-R\left(\frac{3}{5}\right) = 0 + 20\left(\frac{4}{5}\right)t + \frac{1}{2}(-9.81)t^2$

$t = 5.10$ s

$R = 76.5$ m   *Ans.*

F12–25. $x_B = x_A + (v_A)_x t_{AB}$

$12 \text{ ft} = 0 + (0.8660\,v_A)t_{AB}$

$v_A t_{AB} = 13.856$ (1)

$y_B = y_A + (v_A)_y t_{AB} + \frac{1}{2}a_y t_{AB}^2$

$(8 - 3) \text{ ft} = 0 + 0.5 v_A t_{AB} + \frac{1}{2}(-32.2 \text{ ft/s}^2)t_{AB}^2$

Using Eq. (1),

$5 = 0.5(13.856) - 16.1\,t_{AB}^2$

$t_{AB} = 0.3461$ s

$v_A = 40.0$ ft/s   *Ans.*

F12–26. $y_B = y_A + (v_A)_y t_{AB} + \frac{1}{2}a_y t_{AB}^2$

$-150 \text{ m} = 0 + (90 \text{ m/s})t_{AB} + \frac{1}{2}(-9.81 \text{ m/s}^2)t_{AB}^2$

$t_{AB} = 19.89$ s

$x_B = x_A + (v_A)_x t_{AB}$

$R = 0 + 120 \text{ m/s}(19.89 \text{ s}) = 2386.37$ m

$= 2.39$ km   *Ans.*

F12–27. $a_t = \dot{v} = \frac{dv}{dt} = \frac{d}{dt}(0.0625t^2) = (0.125t) \text{ m/s}^2\big|_{t=10 \text{ s}}$

$= 1.25$ m/s^2

$a_n = \frac{v^2}{\rho} = \frac{(0.0625t^2)^2}{40 \text{ m}} = \left[97.656(10^{-6})t^4\right] \text{m/s}^2\big|_{t=10 \text{ s}}$

$= 0.9766$ m/s^2

$a = \sqrt{a_t^2 + a_n^2} = \sqrt{(1.25 \text{ m/s}^2)^2 + (0.9766 \text{ m/s}^2)^2}$

$= 1.59$ m/s^2 *Ans.*

F12–28. $dt = \frac{ds}{v}$ $\int_0^{3 \text{ s}} dt = \int_0^s \frac{ds}{(300/s)}$

$t\big|_0^{3 \text{ s}} = \frac{1}{600}s^2\big|_0^s$

$3 = \frac{1}{600}s^2$

$s = 42.43$ m

$v = \left(\frac{300}{42.43}\right) \text{ m/s} = 7.071$ m/s

$a_n = \frac{v^2}{\rho} = \frac{(7.071 \text{ m/s})^2}{100 \text{ m}} = 0.5$ m/s^2

$a_t = v\frac{dv}{ds} = \left(\frac{300}{s}\right)\left(-\frac{300}{s^2}\right)$ m/s^2

At $s = 43.43$ m,

$a_t = -1.179$ m/s^2

$a = \sqrt{a_t^2 + a_n^2} = \sqrt{(0.5 \text{ m/s}^2)^2 + (-1.179 \text{ m/s}^2)^2}$

$= 1.28$ m/s^2 *Ans.*

F12–29. $v_C^2 = v_A^2 + 2a_t(s_C - s_A)$

$(15 \text{ m/s})^2 = (25 \text{ m/s})^2 + 2a_t(300 \text{ m} - 0)$

$a_t = -0.6667$ m/s^2

$v_B^2 = v_A^2 + 2a_t(s_B - s_A)$

$v_B^2 = (25 \text{ m/s})^2 + 2(-0.6667 \text{ m/s}^2)(250 \text{ m} - 0)$

$v_B = 17.08$ m/s

$(a_B)_n = \frac{v_B^2}{\rho} = \frac{(17.08 \text{ m/s})^2}{300 \text{ m}} = 0.9722$ m/s^2

$a_B = \sqrt{(a_B)_t^2 + (a_B)_n^2}$

$= \sqrt{(-0.6667 \text{ m/s}^2)^2 + (0.9722 \text{ m/s}^2)^2}$

$= 1.18$ m/s^2 *Ans.*

F12–30. $\tan\theta = \dfrac{dy}{dx} = \dfrac{d}{dx}\left(\dfrac{1}{24}x^2\right) = \dfrac{1}{12}x$

$\theta = \tan^{-1}\left(\dfrac{1}{12}x\right)\Big|_{x=10\,\text{ft}}$

$\qquad = \tan^{-1}\left(\dfrac{10}{12}\right) = 39.81° = 39.8°$ *Ans.*

$\rho = \dfrac{\left[1 + (dy/dx)^2\right]^{3/2}}{|d^2y/dx^2|} = \dfrac{\left[1 + \left(\frac{1}{12}x\right)^2\right]^{3/2}}{\left|\frac{1}{12}\right|}\Bigg|_{x=10\,\text{ft}}$

$\qquad = 26.468\,\text{ft}$

$a_n = \dfrac{v^2}{\rho} = \dfrac{(20\,\text{ft/s})^2}{26.468\,\text{ft}} = 15.11\,\text{ft/s}^2$

$a = \sqrt{(a_t)^2 + (a_n)^2} = \sqrt{(6\,\text{ft/s}^2)^2 + (15.11\,\text{ft/s}^2)^2}$

$\qquad = 16.3\,\text{ft/s}^2$ *Ans.*

F12–31. $(a_B)_t = -0.001s = (-0.001)(300\,\text{m})\left(\dfrac{\pi}{2}\,\text{rad}\right)\text{m/s}^2$

$\qquad = -0.4712\,\text{m/s}^2$

$v\,dv = a_t\,ds$

$\displaystyle\int_{25\,\text{m/s}}^{v_B} v\,dv = \int_0^{150\pi\ \text{m}} -0.001s\,ds$

$v_B = 20.07\,\text{m/s}$

$(a_B)_n = \dfrac{v_B{}^2}{\rho} = \dfrac{(20.07\,\text{m/s})^2}{300\,\text{m}} = 1.343\,\text{m/s}^2$

$a_B = \sqrt{(a_B)_t{}^2 + (a_B)_n{}^2}$

$\qquad = \sqrt{(-0.4712\,\text{m/s}^2)^2 + (1.343\,\text{m/s}^2)^2}$

$\qquad = 1.42\,\text{m/s}^2$ *Ans.*

F12–32. $a_t\,ds = v\,dv$

$a_t = v\dfrac{dv}{ds} = (0.2s)(0.2) = (0.04s)\,\text{m/s}^2$

$a_t = 0.04(50\,\text{m}) = 2\,\text{m/s}^2$

$v = 0.2\,(50\,\text{m}) = 10\,\text{m/s}$

$a_n = \dfrac{v^2}{\rho} = \dfrac{(10\,\text{m/s})^2}{500\,\text{m}} = 0.2\,\text{m/s}^2$

$a = \sqrt{a_t^2 + a_n^2} = \sqrt{(2\,\text{m/s}^2)^2 + (0.2\,\text{m/s}^2)^2}$

$\qquad = 2.01\,\text{m/s}^2$ *Ans.*

F12–33. $v_r = \dot{r} = 0$

$v_\theta = r\dot\theta = (400\dot\theta)\,\text{ft/s}$

$v = \sqrt{v_r^2 + v_\theta^2}$

$55\,\text{ft/s} = \sqrt{0^2 + [(400\dot\theta)\,\text{ft/s}]^2}$

$\dot\theta = 0.1375\,\text{rad/s}$ *Ans.*

F12–34. $r = 0.1t^3\big|_{t=1.5\,\text{s}} = 0.3375\,\text{m}$

$\dot{r} = 0.3t^2\big|_{t=1.5\,\text{s}} = 0.675\,\text{m/s}$

$\ddot{r} = 0.6t\big|_{t=1.5\,\text{s}} = 0.900\,\text{m/s}^2$

$\theta = 4t^{3/2}\big|_{t=1.5\,\text{s}} = 7.348\,\text{rad}$

$\dot\theta = 6t^{1/2}\big|_{t=1.5\,\text{s}} = 7.348\,\text{rad/s}$

$\ddot\theta = 3t^{-1/2}\big|_{t=1.5\,\text{s}} = 2.449\,\text{rad/s}^2$

$v_r = \dot{r} = 0.675\,\text{m/s}$

$v_\theta = r\dot\theta = (0.3375\,\text{m})(7.348\,\text{rad/s}) = 2.480\,\text{m/s}$

$a_r = \ddot{r} - r\dot\theta^2$

$\qquad = (0.900\,\text{m/s}^2) - (0.3375\,\text{m})(7.348\,\text{rad/s})^2$

$\qquad = -17.325\,\text{m/s}^2$

$a_\theta = r\ddot\theta + 2\dot{r}\dot\theta = (0.3375\,\text{m})(2.449\,\text{rad/s}^2)$

$\qquad\quad + 2(0.675\,\text{m/s})(7.348\,\text{rad/s}) = 10.747\,\text{m/s}^2$

$v = \sqrt{v_r{}^2 + v_\theta{}^2}$

$\qquad = \sqrt{(0.675\,\text{m/s})^2 + (2.480\,\text{m/s})^2}$

$\qquad = 2.57\,\text{m/s}$ *Ans.*

$a = \sqrt{a_r^2 + a_\theta^2}$

$\qquad = \sqrt{(-17.325\,\text{m/s}^2)^2 + (10.747\,\text{m/s}^2)^2}$

$\qquad = 20.4\,\text{m/s}^2$ *Ans.*

F12–35. $r = 2\theta$

$\dot{r} = 2\dot\theta$

$\ddot{r} = 2\ddot\theta$

At $\theta = \pi/4$ rad,

$r = 2\left(\dfrac{\pi}{4}\right) = \dfrac{\pi}{2}$

$\dot{r} = 2(3\,\text{rad/s}) = 6\,\text{ft/s}$

$\ddot{r} = 2(1\,\text{rad/s}) = 2\,\text{ft/s}^2$

$a_r = \ddot{r} - r\dot\theta^2 = 2\,\text{ft/s}^2 - \left(\dfrac{\pi}{2}\,\text{ft}\right)(3\,\text{rad/s})^2$

$\qquad = -12.14\,\text{ft/s}^2$

$a_\theta = r\ddot\theta + 2\dot{r}\dot\theta$

$\qquad = \left(\dfrac{\pi}{2}\,\text{ft}\right)(1\,\text{rad/s}^2) + 2(6\,\text{ft/s})(3\,\text{rad/s})$

$\qquad = 37.57\,\text{ft/s}^2$

$a = \sqrt{a_r{}^2 + a_\theta{}^2}$

$\qquad = \sqrt{(-12.14\,\text{ft/s}^2)^2 + (37.57\,\text{ft/s}^2)^2}$

$\qquad = 39.5\,\text{ft/s}^2$ *Ans.*

F12–36. $r = e^{\theta}$

$\dot{r} = e^{\theta}\dot{\theta}$

$\ddot{r} = e^{\theta}\ddot{\theta} + e^{\theta}\dot{\theta}^2$

$a_r = \ddot{r} - r\dot{\theta}^2 = (e^{\theta}\ddot{\theta} + e^{\theta}\dot{\theta}^2) - (e^{\theta}\dot{\theta}^2 = e^{\pi/4}(4)$

$\quad = 8.77 \text{ m/s}^2$ *Ans.*

$a_{\theta} = r\ddot{\theta} + 2\dot{r}\dot{\theta} = (e^{\theta}\ddot{\theta}) + (2(e^{\theta}\dot{\theta})\dot{\theta}) = e^{\theta}(\ddot{\theta} + 2\dot{\theta}^2)$

$\quad = e^{\pi/4}(4 + 2(2)^2)$

$\quad = 26.3 \text{ m/s}^2$ *Ans.*

F12–37. $r = [0.2(1 + \cos\theta)] \text{ m}\big|_{\theta=30°} = 0.3732 \text{ m}$

$\dot{r} = \left[-0.2(\sin\theta)\dot{\theta}\right] \text{ m/s}\big|_{\theta=30°}$

$\quad = -0.2\sin 30°(3 \text{ rad/s})$

$\quad = -0.3 \text{ m/s}$

$v_r = \dot{r} = -0.3 \text{ m/s}$

$v_{\theta} = r\dot{\theta} = (0.3732 \text{ m})(3 \text{ rad/s}) = 1.120 \text{ m/s}$

$v = \sqrt{v_r^2 + v_{\theta}^2} = \sqrt{(-0.3 \text{ m/s})^2 + (1.120 \text{ m/s})^2}$

$\quad = 1.16\text{m}$ *Ans.*

F12–38. $30 \text{ m} = r\sin\theta$

$r = \left(\frac{30 \text{ m}}{\sin\theta}\right) = (30\csc\theta) \text{ m}$

$r = (30\csc\theta)\big|_{\theta=45°} = 42.426 \text{ m}$

$\dot{r} = -30\csc\theta\ctn\theta\,\dot{\theta}\big|_{\theta=45°} = -(42.426\dot{\theta}) \text{ m/s}$

$v_r = \dot{r} = -(42.426\dot{\theta}) \text{ m/s}$

$v_{\theta} = r\dot{\theta} = (42.426\dot{\theta}) \text{ m/s}$

$v = \sqrt{v_r^2 + v_{\theta}^2}$

$2 = \sqrt{(-42.426\dot{\theta})^2 + (42.426\dot{\theta})^2}$

$\dot{\theta} = 0.0333 \text{ rad/s}$ *Ans.*

F12–39. $l_T = 3s_D + s_A$

$0 = 3v_D + v_A$

$0 = 3v_D + 3 \text{ m/s}$

$v_D = -1 \text{ m/s} = 1 \text{ m/s} \uparrow$ *Ans.*

F12–40. $s_B + 2s_A + 2h = l$

$v_B + 2v_A = 0$

$6 + 2v_A = 0 \qquad v_A = -3 \text{ ft/s} = 3 \text{ ft/s} \uparrow$ *Ans.*

F12–41. $3s_A + s_B = l$

$3v_A + v_A = 0$

$3v_A + 1.5 = 0 \quad v_A = -0.5 \text{ m/s} = 0.5 \text{ m/s} \uparrow$ *Ans.*

F12–42. $l_T = 4s_A + s_F$

$0 = 4v_A + v_F$

$0 = 4v_A + 3 \text{ m/s}$

$v_A = -0.75 \text{ m/s} = 0.75 \text{ m/s} \uparrow$ *Ans.*

F12–43. $s_A + 2(s_A - a) + (s_A - s_P) = l$

$4s_A - s_P = l + 2a$

$4v_A - v_P = 0$

$4v_A - 4 = 0 \qquad\qquad v_A = 1 \text{ m/s}$ *Ans.*

F12–44. $s_C + s_B = l_{CED}$ (1)

$(s_A - s_C) + (s_B - s_C) + s_B = l_{ACDF}$

$s_A + 2s_B - 2s_C = l_{ACDF}$ (2)

Thus

$v_C + v_B = 0$

$v_A + 2v_B - 2v_C = 0$

Eliminating v_C,

$v_A + 4v_B = 0$

Thus,

$4 \text{ ft/s} + 4v_B = 0$

$v_B = -1 \text{ ft/s} = 1 \text{ ft/s} \uparrow$ *Ans.*

F12–45. $\mathbf{v}_B = \mathbf{v}_A + \mathbf{v}_{B/A}$

$100\mathbf{i} = 80\mathbf{j} + \mathbf{v}_{B/A}$

$\mathbf{v}_{B/A} = 100\mathbf{i} - 80\mathbf{j}$

$\mathbf{v}_{B/A} = \sqrt{(v_{B/A})_x^2 + (v_{B/A})_y^2}$

$\quad = \sqrt{(100 \text{ km/h})^2 + (-80 \text{ km/h})^2}$

$\quad = 128 \text{ km/h}$ *Ans.*

$\theta = \tan^{-1}\left[\frac{(v_{B/A})_y}{(v_{B/A})_x}\right] = \tan^{-1}\left(\frac{80 \text{ km/h}}{100 \text{ km/h}}\right) = 38.7°\ \searailing$ *Ans.*

F12–46. $\mathbf{v}_B = \mathbf{v}_A + \mathbf{v}_{B/A}$

$(-400\mathbf{i} - 692.82\mathbf{j}) = (650\mathbf{i}) + \mathbf{v}_{B/A}$

$\mathbf{v}_{B/A} = [-1050\mathbf{i} - 692.82\mathbf{j}] \text{ km/h}$

$\mathbf{v}_{B/A} = \sqrt{(v_{B/A})_x^2 + (v_{B/A})_y^2}$

$\quad = \sqrt{(1050 \text{ km/h})^2 + (692.82 \text{ km/h})^2}$

$\quad = 1258 \text{ km/h}$ *Ans.*

$\theta = \tan^{-1}\left[\frac{(v_{B/A})_y}{(v_{B/A})_x}\right] = \tan^{-1}\left(\frac{692.82 \text{ km/h}}{1050 \text{ km/h}}\right) = 33.4°\ \swarrow$

Ans.

F12–47. $\mathbf{v}_B = \mathbf{v}_A + \mathbf{v}_{B/A}$

$(5\mathbf{i} + 8.660\mathbf{j}) = (12.99\mathbf{i} + 7.5\mathbf{j}) + \mathbf{v}_{B/A}$

$\mathbf{v}_{B/A} = [-7.990\mathbf{i} + 1.160\mathbf{j}] \text{ m/s}$

$v_{B/A} = \sqrt{(-7.990 \text{ m/s})^2 + (1.160 \text{ m/s})^2}$

$\quad = 8.074 \text{ m/s}$ *Ans.*

$d_{AB} = v_{B/A}t = (8.074 \text{ m/s})(4 \text{ s}) = 32.3 \text{ m}$ *Ans.*

F12–48. $\mathbf{v}_A = \mathbf{v}_B + \mathbf{v}_{A/B}$

$-20 \cos 45°\mathbf{i} + 20 \sin 45°\mathbf{j} = 65\mathbf{i} + \mathbf{v}_{A/B}$

$\mathbf{v}_{A/B} = -79.14\mathbf{i} + 14.14\mathbf{j}$

$\mathbf{v}_{A/B} = \sqrt{(-79.14)^2 + (14.14)^2}$

$\quad = 80.4 \text{ km/h}$ *Ans.*

$\mathbf{a}_A = \mathbf{a}_B + \mathbf{a}_{A/B}$

$\dfrac{(20)^2}{0.1} \cos 45°\mathbf{i} + \dfrac{(20)^2}{0.1} \sin 45°\mathbf{j} = 1200\mathbf{i} + \mathbf{a}_{A/B}$

$\mathbf{a}_{A/B} = 1628\mathbf{i} + 2828\mathbf{j}$

$a_{A/B} = \sqrt{(1628)^2 + (2828)^2}$

$\quad = 3.26(10^3) \text{ km/h}^2$ *Ans.*

Chapter 13

F13–1. $\quad s = s_0 + v_0t + \frac{1}{2} a_c t^2$

$6 \text{ m} = 0 + 0 + \frac{1}{2} a(3 \text{ s})^2$

$a = 1.333 \text{ m/s}^2$

$\Sigma F_y = ma_y; \qquad N_A - 20(9.81) \text{ N} \cos 30° = 0$

$\qquad\qquad N_A = 169.91 \text{ N}$

$\Sigma F_x = ma_x; \qquad T - 20(9.81) \text{ N} \sin 30°$

$- 0.3(169.91 \text{ N}) = (20 \text{ kg})(1.333 \text{ m/s}^2)$

$T = 176 \text{ N}$ *Ans.*

F13–2. $(F_f)_{max} = \mu_s N_A = 0.3(245.25 \text{ N}) = 73.575 \text{ N}.$

Since $F = 100 \text{ N} > (F_f)_{max}$ when $t = 0$, the crate will start to move immediately after $\mathbf{F}$ is applied.

$+\uparrow \Sigma F_y = ma_y; \qquad N_A - 25(9.81) \text{ N} = 0$

$\qquad\qquad N_A = 245.25 \text{ N}$

$\xrightarrow{+} \Sigma F_x = ma_x;$

$10t^2 + 100 - 0.25(245.25 \text{ N}) = (25 \text{ kg})a$

$a = (0.4t^2 + 1.5475) \text{ m/s}^2$

$dv = a \, dt$

$\displaystyle\int_0^v dv = \int_0^{4 \text{ s}} \left(0.4 \, t^2 + 1.5475\right)dt$

$v = 14.7 \text{ m/s} \rightarrow$ *Ans.*

F13–3. $\xrightarrow{+} \Sigma F_x = ma_x;$

$\left(\frac{4}{5}\right)500 \text{ N} - (500s)\text{N} = (10 \text{ kg})a$

$a = (40 - 50s) \text{ m/s}^2$

$v \, dv = a \, ds$

$\displaystyle\int_0^v v \, dv = \int_0^{0.5 \text{ m}} (40 - 50s) \, ds$

$\dfrac{v^2}{2}\Big|_0^v = \left(40s - 25s^2\right)\Big|_0^{0.5 \text{ m}}$

$v = 5.24 \text{ m/s}$ *Ans.*

F13–4. $\xrightarrow{+} \Sigma F_x = ma_x \qquad (100s) \text{ N} = (2000 \text{ kg})a$

$\qquad\qquad a = (0.05s) \text{ m/s}^2$

$v \, dv = a \, ds$

$\displaystyle\int_0^v v \, dv = \int_0^{10 \text{ m}} 0.05 \, s \, ds$

$v = 2.24 \text{ m/s}$

F13–5. $F_{sp} = k(l - l_0) = (200 \text{ N/m})(0.5 \text{ m} - 0.3 \text{ m})$

$\quad = 40 \text{ N}$

$\theta = \tan^{-1}\left(\frac{0.3 \text{ m}}{0.4 \text{ m}}\right) = 36.86°$

$\xrightarrow{+} \Sigma F_x = ma_x;$

$100 \text{ N} - (40 \text{ N})\cos 36.86° = (25 \text{ kg})a$

$a = 2.72 \text{ m/s}^2$

F13–6. Blocks A and B:

$\xrightarrow{+} \Sigma F_x = ma_x; \quad 6 = \frac{70}{32.2} a; \; a = 2.76 \text{ ft/s}^2$

Check if slipping occurs between A and B.

$\xrightarrow{+} \Sigma F_x = ma_x; \quad 6 - F = \frac{20}{32.2} (2.76);$

$F = 4.29 \text{ lb} < 0.4(20) = 8 \text{ lb}$

$a_A = a_B = 2.76 \text{ ft/s}^2$ *Ans.*

F13–7. $\Sigma F_n = m\frac{v^2}{\rho}; \; (0.3)m(9.81) = m\frac{v^2}{2}$

$v = 2.43 \text{ m/s}$ *Ans.*

F13–8. $+\downarrow \Sigma F_n = ma_n; \; m(32.2) = m\left(\frac{v^2}{250}\right)$

$v = 89.7 \text{ ft/s}$ *Ans.*

F13–9. $+\downarrow \Sigma F_n = ma_n; \; 150 + N_p = \dfrac{150}{32.2}\left(\dfrac{(120)^2}{400}\right)$

$N_p = 17.7 \text{ lb}$ *Ans.*

F13–10. $\xleftarrow{+} \Sigma F_n = ma_n;$

$N_c \sin 30° + 0.2 N_c \cos 30° = m\dfrac{v^2}{500}$

$+\uparrow \Sigma F_b = 0;$

$N_c \cos 30° - 0.2 N_c \sin 30° - m(32.2) = 0$

$v = 119 \text{ ft/s}$ *Ans.*

F13–11. $\Sigma F_t = ma_t; \qquad 10(9.81) \text{ N} \cos 45° = (10 \text{ kg})a_t$

$\qquad\qquad a_t = 6.94 \text{ m/s}^2$ *Ans.*

$\Sigma F_n = ma_n;$

$T - 10(9.81) \text{ N} \sin 45° = (10 \text{ kg})\dfrac{(3 \text{ m/s})^2}{2 \text{ m}}$

$T = 114 \text{ N}$ *Ans.*

F13–12. $\Sigma F_n = ma_n$;

$$F_n = (500 \text{ kg}) \frac{(15 \text{ m/s})^2}{200 \text{ m}} = 562.5 \text{ N}$$

$\Sigma F_t = ma_t$;

$$F_t = (500 \text{ kg})(1.5 \text{ m/s}^2) = 750 \text{ N}$$

$$F = \sqrt{F_n^2 + F_t^2} = \sqrt{(562.5 \text{ N})^2 + (750 \text{ N})^2}$$

$$= 938 \text{ N} \qquad \qquad \qquad Ans.$$

F13–13. $a_r = \ddot{r} - r\dot{\theta}^2 = 0 - (1.5 \text{ m} + (8 \text{ m})\sin 45°)\dot{\theta}^2$

$$= (-7.157 \,\dot{\theta}^2) \text{ m/s}^2$$

$\Sigma F_z = ma_z$;

$$T\cos 45° - m(9.81) = m(0) \quad T = 13.87 \, m$$

$\Sigma F_r = ma_r$;

$$-(13.87m)\sin 45° = m(-7.157 \,\dot{\theta}^2)$$

$$\dot{\theta} = 1.17 \text{ rad/s} \qquad \qquad Ans.$$

F13–14. $\theta = \pi t^2 \big|_{t=0.5 \text{ s}} = (\pi/4) \text{ rad}$

$$\dot{\theta} = 2\pi t \big|_{t=0.5 \text{ s}} = \pi \text{ rad/s}$$

$$\ddot{\theta} = 2\pi \text{ rad/s}^2$$

$$r = 0.6 \sin\theta \big|_{\theta=\pi/4 \text{ rad}} = 0.4243 \text{ m}$$

$$\dot{r} = 0.6 (\cos\theta)\dot{\theta} \big|_{\theta=\pi/4 \text{ rad}} = 1.3329 \text{ m/s}$$

$$\ddot{r} = 0.6 (\cos\theta)\ddot{\theta} - (\sin\theta)\dot{\theta}^2 \big|_{\theta=\pi/4 \text{ rad}} = -1.5216 \text{ m/s}^2$$

$a_r = \ddot{r} - r\dot{\theta}^2 = -1.5216 \text{ m/s}^2 - (0.4243 \text{ m})(\pi \text{ rad/s})^2$

$$= -5.7089 \text{ m/s}^2$$

$a_\theta = r\ddot{\theta} + 2\dot{r}\dot{\theta} = 0.4243 \text{ m}(2\pi \text{ rad/s}^2)$

$$+ 2(1.3329 \text{ m/s})(\pi \text{ rad/s})$$

$$= 11.0404 \text{m/s}^2$$

$\Sigma F_r = ma_r$;

$F\cos 45° - N\cos 45° - 0.2(9.81)\cos 45°$

$$= 0.2(-5.7089)$$

$\Sigma F_\theta = ma_\theta$;

$F\sin 45° + N\sin 45° - 0.2(9.81)\sin 45°$

$$= 0.2(11.0404)$$

$$N = 2.37 \text{ N} \qquad F = 2.72 \text{ N} \qquad Ans.$$

F13–15. $r = 50e^{2\theta} \big|_{\theta=\pi/6 \text{ rad}} = \left[50e^{2(\pi/6)}\right] \text{ m} = 142.48 \text{ m}$

$$\dot{r} = 50\left(2e^{2\theta}\,\dot{\theta}\right) = 100e^{2\theta}\,\dot{\theta}\big|_{\theta=\pi/6 \text{ rad}}$$

$$= \left[100e^{2(\pi/6)}(0.05)\right] = 14.248 \text{ m/s}$$

$$\ddot{r} = 100\left((2e^{2\theta}\dot{\theta})\dot{\theta} + e^{2\theta}(\ddot{\theta})\right)\big|_{\theta=\pi/6 \text{ rad}}$$

$$= 100\left[2e^{2(\pi/6)}(0.05^2) + e^{2(\pi/6)}(0.01)\right]$$

$$= 4.274 \text{ m/s}^2$$

$a_r = \ddot{r} - r\dot{\theta}^2 = 4.274 \text{ m/s}^2 - 142.48 \text{ m}(0.05 \text{ rad/s})^2$

$$= 3.918 \text{ m/s}^2$$

$a_\theta = r\ddot{\theta} + 2\dot{r}\dot{\theta} = 142.48 \text{ m}(0.01 \text{ rad/s}^2)$

$$+ 2(14.248 \text{ m/s})(0.05 \text{ rad/s})$$

$$= 2.850 \text{ m/s}^2$$

$\Sigma F_r = ma_r$;

$$F_r = (2000 \text{ kg})(3.918 \text{ m/s}^2) = 7836.55 \text{ N}$$

$\Sigma F_\theta = ma_\theta$;

$$F_\theta = (2000 \text{ kg})(2.850 \text{ m/s}^2) = 5699.31 \text{ N}$$

$$F = \sqrt{F_r^2 + F_\theta^2}$$

$$= \sqrt{(7836.55 \text{ N})^2 + (5699.31 \text{ N})^2}$$

$$= 9689.87 \text{ N} = 9.69 \text{ kN}$$

F13–16. $r = (0.6 \cos 2\theta) \text{ m}\big|_{\theta=0°} = [0.6 \cos 2(0°)] \text{ m} = 0.6 \text{ m}$

$$\dot{r} = (-1.2 \sin 2\theta\dot{\theta}) \text{ m/s}\big|_{\theta=0°}$$

$$= \left[-1.2 \sin 2(0°)(-3)\right] \text{ m/s} = 0$$

$$\ddot{r} = -1.2\left(\sin 2\theta\ddot{\theta} + 2\cos 2\theta\dot{\theta}^2\right) \text{ m/s}^2\big|_{\theta=0°}$$

$$= -21.6 \text{ m/s}^2$$

Thus,

$a_r = \ddot{r} - r\dot{\theta}^2 = -21.6 \text{ m/s}^2 - 0.6 \text{ m}(-3 \text{ rad/s})^2$

$$= -27 \text{ m/s}^2$$

$a_\theta = r\ddot{\theta} + 2\dot{r}\dot{\theta} = 0.6 \text{ m}(0) + 2(0)(-3 \text{ rad/s}) = 0$

$\Sigma F_\theta = ma_\theta$; $\quad F - 0.2(9.81) \text{ N} = 0.2 \text{ kg}(0)$

$$F = 1.96 \text{ N} \qquad \qquad Ans.$$

Chapter 14

F14–1. $T_1 + \Sigma U_{1-2} = T_2$

$$0 + \left(\tfrac{4}{5}\right)(500 \text{ N})(0.5 \text{ m}) - \tfrac{1}{2}(500 \text{ N/m})(0.5 \text{ m})^2$$

$$= \tfrac{1}{2}(10 \text{ kg})v^2$$

$$v = 5.24 \text{ m/s} \qquad \qquad Ans.$$

F14–2. $\Sigma F_y = ma_y$; $\quad N_A - 20(9.81) \text{ N}\cos 30° = 0$

$$N_A = 169.91 \text{ N}$$

$$T_1 + \Sigma U_{1-2} = T_2$$

$$0 + 300 \text{ N}(10 \text{ m}) - 0.3(169.91 \text{ N})(10 \text{ m})$$

$$- 20(9.81)\text{N}(10 \text{ m})\sin 30°$$

$$= \tfrac{1}{2}(20 \text{ kg})v^2$$

$$v = 12.3 \text{ m/s} \qquad \qquad Ans.$$

F14–3. $T_1 + \Sigma U_{1-2} = T_2$

$$0 + 2\left[\int_0^{15\,\text{m}} (600 + 2s^2)\,\text{N}\,ds\right] - 100(9.81)\,\text{N}(15\,\text{m})$$

$$= \tfrac{1}{2}(100\,\text{kg})v^2$$

$v = 12.5\,\text{m/s}$ *Ans.*

F14–4. $T_1 + \Sigma U_{1-2} = T_2$

$$\tfrac{1}{2}(1800\,\text{kg})(125\,\text{m/s})^2 - \left[\frac{(50\,000\,\text{N} + 20\,000\,\text{N})}{2}(400\,\text{m})\right]$$

$$= \tfrac{1}{2}(1800\,\text{kg})v^2$$

$v = 8.33\,\text{m/s}$ *Ans.*

F14–5. $T_1 + \Sigma U_{1-2} = T_2$

$$\tfrac{1}{2}(10\,\text{kg})(5\,\text{m/s})^2 + 100\,\text{N}s' + [10(9.81)\,\text{N}]\,s'\sin 30°$$

$$-\tfrac{1}{2}(200\,\text{N/m})(s')^2 = 0$$

$s' = 2.09\,\text{m}$

$s = 0.6\,\text{m} + 2.09\,\text{m} = 2.69\,\text{m}$ *Ans.*

F14–6. $T_A + \Sigma U_{A-B} = T_B$

Consider difference in cord length $AC - BC$, which is distance F moves.

$$0 + 10\,\text{lb}(\sqrt{(3\,\text{ft})^2 + (4\,\text{ft})^2} - 3\,\text{ft})$$

$$= \tfrac{1}{2}\left(\tfrac{5}{32.2}\,\text{slug}\right)v_B^2$$

$v_B = 16.0\,\text{ft/s}$ *Ans.*

F14–7. $\overset{+}{\rightarrow}\Sigma F_x = ma_x;$

$30\left(\tfrac{4}{5}\right) = 20a$ $a = 1.2\,\text{m/s}^2 \rightarrow$

$v = v_0 + a_c t$

$v = 0 + 1.2(4) = 4.8\,\text{m/s}$

$P = \mathbf{F} \cdot \mathbf{v} = F(\cos\theta)v$

$$= 30\left(\tfrac{4}{5}\right)(4.8)$$

$$= 115\,\text{W}$$ *Ans.*

F14–8. $\overset{+}{\rightarrow}\Sigma F_x = ma_x;$

$10s = 20a$ $a = 0.5s\,\text{m/s}^2 \rightarrow$

$v\,dv = a\,ds$

$$\int_0^v v\,dv = \int_0^{5\,\text{m}} 0.5\,s\,ds$$

$v = 3.536\,\text{m/s}$

$P = \mathbf{F} \cdot \mathbf{v} = 10(5)(3.536) = 177\,\text{W}$ *Ans.*

F14–9. $(+\uparrow)\Sigma F_y = 0;$

$T_1 - 100\,\text{lb} = 0$ $T_1 = 100\,\text{lb}$

$(+\uparrow)\Sigma F_y = 0;$

$100\,\text{lb} + 100\,\text{lb} - T_2 = 0$ $T_2 = 200\,\text{lb}$

$P_{\text{out}} = \mathbf{T}_B \cdot \mathbf{v}_B = (200\,\text{lb})(3\,\text{ft/s}) = 1.091\,\text{hp}$

$$P_{\text{in}} = \frac{P_{\text{out}}}{\varepsilon} = \frac{1.091\,\text{hp}}{0.8} = 1.36\,\text{hp}$$ *Ans.*

F14–10. $\Sigma F_{y'} = ma_{y'};$ $N - 20(9.81)\cos 30° = 20(0)$

$$N = 169.91\,\text{N}$$

$\Sigma F_{x'} = ma_{x'};$

$F - 20(9.81)\sin 30° - 0.2(169.91) = 0$

$F = 132.08\,\text{N}$

$P = \mathbf{F} \cdot \mathbf{v} = 132.08(5) = 660\,\text{W}$ *Ans.*

F14–11. $+\uparrow\Sigma F_y = ma_y;$

$T - 50(9.81) = 50(0)$ $T = 490.5\,\text{N}$

$P_{\text{out}} = \mathbf{T} \cdot \mathbf{v} = 490.5(1.5) = 735.75\,\text{W}$

$$P_{\text{in}} = \frac{P_{\text{out}}}{\varepsilon} = \frac{735.75}{0.8} = 920\,\text{W}$$ *Ans.*

F14–12. $2s_A + s_P = l$

$2a_A + a_P = 0$

$2a_A + 6 = 0$

$a_A = -3\,\text{m/s}^2 = 3\,\text{m/s}^2 \uparrow$

$\Sigma F_y = ma_y;$ $T_A - 490.5\,\text{N} = (50\,\text{kg})(3\,\text{m/s}^2)$

$$T_A = 640.5\,\text{N}$$

$P_{\text{out}} = \mathbf{T} \cdot \mathbf{v} = (640.5\,\text{N}/2)(12) = 3843\,\text{W}$

$$P_{\text{in}} = \frac{P_{\text{out}}}{\varepsilon} = \frac{3843}{0.8} = 4803.75\,\text{W} = 4.80\,\text{kW}$$ *Ans.*

F14–13. $T_A + V_A = T_B + V_B$

$0 + 2(9.81)(1.5) = \tfrac{1}{2}(2)(v_B)^2 + 0$

$v_B = 5.42\,\text{m/s}$ *Ans.*

$$+\uparrow\Sigma F_n = ma_n; T - 2(9.81) = 2\left(\frac{(5.42)^2}{1.5}\right)$$

$T = 58.9\,\text{N}$ *Ans.*

F14–14. $T_A + V_A = T_B + V_B$

$\tfrac{1}{2}m_A v_A^2 + mgh_A = \tfrac{1}{2}m_B v_B^2 + mgh_B$

$\left[\tfrac{1}{2}(2\,\text{kg})(1\,\text{m/s})^2\right] + [2\,(9.81)\,\text{N}(4\,\text{m})]$

$= \left[\tfrac{1}{2}(2\,\text{kg})v_B^2\right] + [0]$

$v_B = 8.915\,\text{m/s} = 8.92\,\text{m/s}$ *Ans.*

$+\uparrow\Sigma F_n = ma_n;$ $N_B - 2(9.81)\text{N}$

$$= (2\,\text{kg})\left(\frac{(8.915\,\text{m/s})^2}{2\,\text{m}}\right)$$

$N_B = 99.1\,\text{N}$ *Ans.*

F14–15. $T_1 + V_1 = T_2 + V_2$

$\frac{1}{2}(2)(4)^2 + \frac{1}{2}(30)(2-1)^2$

$\quad = \frac{1}{2}(2)(v)^2 - 2(9.81)(1) + \frac{1}{2}(30)(\sqrt{5}-1)^2$

$v = 5.26$ m/s   *Ans.*

F14–16. $T_A + V_A = T_B + V_B$

$0 + \frac{1}{2}(4)(2.5 - 0.5)^2 + 5(2.5)$

$\quad = \frac{1}{2}\left(\frac{5}{32.2}\right)v_B^2 + \frac{1}{2}(4)(1-0.5)^2$

$v_B = 16.0$ ft/s   *Ans.*

F14–17. $T_1 + V_1 = T_2 + V_2$

$\frac{1}{2}mv_1^2 + mgy_1 + \frac{1}{2}ks_1^2$

$\quad = \frac{1}{2}mv_2^2 + mgy_2 + \frac{1}{2}ks_2^2$

$[0] + [0] + [0] = [0] +$

$[-75\text{ lb}(5\text{ ft} + s)] + [2(\frac{1}{2}(1000\text{ lb/ft})s^2)$

$\quad + \frac{1}{2}(1500\text{ lb/ft})(s - 0.25\text{ ft})^2]$

$s = s_A = s_C = 0.580$ ft   *Ans.*

Also,

$s_B = 0.5803$ ft $-$ 0.25 ft $=$ 0.330 ft   *Ans.*

F14–18. $T_A + V_A = T_B + V_B$

$\frac{1}{2}mv_A^2 + \left(\frac{1}{2}ks_A^2 + mgy_A\right)$

$\quad = \frac{1}{2}mv_B^2 + \left(\frac{1}{2}ks_B^2 + mgy_B\right)$

$\frac{1}{2}(4\text{ kg})(2\text{ m/s})^2 + \frac{1}{2}(400\text{ N/m})(0.1\text{ m} - 0.2\text{ m})^2 + 0$

$= \frac{1}{2}(4\text{kg})v_B^2 + \frac{1}{2}(400\text{ N/m})(\sqrt{(0.4\text{ m})^2 + (0.3\text{ m})^2} - 0.2\text{ m})^2 +$

$\quad [4(9.81)\text{ N}](-(0.1\text{ m} + 0.3\text{ m}))$

$v_B = 1.962$ m/s $=$ 1.96 m/s   *Ans.*

Chapter 15

F15–1. $(\overset{+}{\rightarrow})$ $m(v_1)_x + \Sigma \int_{t1}^{t2} F_x\, dt = m(v_2)_x$

$(0.5\text{ kg})(25\text{ m/s})\cos 45° - \int F_x\, dt$

$\quad = (0.5\text{ kg})(10\text{ m/s})\cos 30°$

$I_x = \int F_x\, dt = 4.509$ N·s

$(+\uparrow)$ $m(v_1)_y + \Sigma \int_{t1}^{t2} F_y\, dt = m(v_2)_y$

$- (0.5\text{ kg})(25\text{ m/s})\sin 45° + \int F_y\, dt$

$\quad = (0.5\text{ kg})(10\text{ m/s})\sin 30°$

$I_y = \int F_y\, dt = 11.339$ N·s

$I = \int F\, dt = \sqrt{(4.509\text{ N·s})^2 + (11.339\text{ N·s})^2}$

$\quad = 12.2$ N·s   *Ans.*

F15–2. $(+\uparrow)$ $m(v_1)_y + \Sigma \int_{t1}^{t2} F_y\, dt = m(v_2)_y$

$0 + N(4\text{ s}) + (100\text{ lb})(4\text{ s})\sin 30°$

$\quad - (150\text{ lb})(4\text{ s}) = 0$

$N = 100$ lb

$(\overset{+}{\rightarrow})$ $m(v_1)_x + \Sigma \int_{t1}^{t2} F_x\, dt = m(v_2)_x$

$0 + (100\text{ lb})(4\text{ s})\cos 30° - 0.2(100\text{ lb})(4\text{ s})$

$\quad = \left(\frac{150}{32.2}\text{ slug}\right)v$

$v = 57.2$ ft/s   *Ans.*

F15–3. Time to start motion,

$+\uparrow\Sigma F_y = 0;\quad N - 25(9.81)\text{ N} = 0\quad N = 245.25$ N

$\overset{+}{\rightarrow}\Sigma F_x = 0;\quad 20t^2 - 0.3(245.25\text{ N}) = 0\quad t = 1.918$ s

$(\overset{+}{\rightarrow})$ $m(v_1)_x + \Sigma \int_{t1}^{t2} F_x\, dt = m(v_2)_x$

$0 + \int_{1.918\text{ s}}^{4\text{ s}} 20t^2\, dt - (0.25(245.25\text{ N}))(4\text{ s} - 1.918\text{ s})$

$\quad = (25\text{ kg})v$

$v = 10.1$ m/s   *Ans.*

F15–4. $(\overset{+}{\rightarrow})$ $m(v_1)_x + \Sigma \int_{t1}^{t2} F_x\, dt = m(v_2)_x$

$(1500\text{ kg})(0) + \lfloor\frac{1}{2}(6000\text{ N})(2\text{ s}) + (6000\text{ N})(6\text{ s} - 2\text{ s})\rfloor$

$\quad = (1500\text{ kg})\, v$

$v = 20$ m/s   *Ans.*

F15–5. SUV and trailer,

$m(v_1)_{x'} + \Sigma \int_{t1}^{t2} F_{x'}\, dt = m(v_2)_{x'}$

$0 + (9000\text{ N})(20\text{ s}) = (1500\text{ kg} + 2500\text{ kg})v$

$v = 45.0$ m/s   *Ans.*

Trailer,

$m(v_1)_{x'} + \Sigma \int_{t1}^{t2} F_{x'}\, dt = m(v_2)_{x'}$

$0 + T(20\text{ s}) = (1500\text{ kg})(45.0\text{ m/s})$

$T = 3375$ N $=$ 3.375 kN   *Ans.*

F15–6. Block B:

$(+\downarrow) \; mv_1 + \int F \, dt = mv_2$

$0 + 8(5) - T(5) = \frac{8}{32.2}(1)$

$T = 7.95 \, \text{lb}$ *Ans.*

Block A:

$(\xrightarrow{+}) \; mv_1 + \int F \, dt = mv_2$

$0 + 7.95(5) - \mu_k(10)(5) = \frac{10}{32.2}(1)$

$\mu_k = 0.789$ *Ans.*

F15–7. $(\xrightarrow{+}) \; m_A(v_A)_1 + m_B(v_B)_1 = m_A(v_A)_2 + m_B(v_B)_2$

$(20(10^3) \, \text{kg})(3 \, \text{m/s}) + (15(10^3) \, \text{kg})(-1.5 \, \text{m/s})$

$= (20(10^3) \, \text{kg})(v_A)_2 + (15(10^3) \, \text{kg})(2 \, \text{m/s})$

$(v_A)_2 = 0.375 \, \text{m/s} \rightarrow$ *Ans.*

$(\xrightarrow{+}) \; m(v_B)_1 + \Sigma \int_{t1}^{t2} F \, dt = m(v_B)_2$

$(15(10^3) \, \text{kg})(-1.5 \, \text{m/s}) + F_{avg}(0.5 \, \text{s})$

$= (15(10^3) \, \text{kg}(2 \, \text{m/s})$

$F_{avg} = 105(10^3) \, \text{N} = 105 \, \text{kN}$ *Ans.*

F15–8. $(\xrightarrow{+}) \; m_p[(v_p)_1]_x + m_c[(v_c)_1]_x = (m_p + m_c)v_2$

$5\left[10\left(\frac{4}{5}\right)\right] + 0 = (5 + 20)v_2$

$v_2 = 1.6 \, \text{m/s}$ *Ans.*

F15–9. $T_1 + V_1 = T_2 + V_2$

$\frac{1}{2}m_A(v_A)_1^2 + (v_g)_1 = \frac{1}{2}m_A(v_A)_2^2 + (v_g)_2$

$\frac{1}{2}(5)(5)^2 + 5(9.81)(1.5) = \frac{1}{2}(5)(v_A)_2^2$

$(v_A)_2 = 7.378 \, \text{m/s}$

$(\xrightarrow{+}) \; m_A(v_A)_2 + m_B(v_B)_2 = (m_A + m_B)v$

$5(7.378) + 0 = (5 + 8)v$

$v = 2.84 \, \text{m/s}$ *Ans.*

F15–10. $(\xrightarrow{+}) \; m_A(v_A)_1 + m_B(v_B)_1 = m_A(v_A)_2 + m_A(v_B)_2$

$0 + 0 = 10(v_A)_2 + 15(v_B)_2$ (1)

$T_1 + V_1 = T_2 + V_2$

$\frac{1}{2}m_A(v_A)_1^2 + \frac{1}{2}m_B(v_B)_1^2 + (V_e)_1$

$= \frac{1}{2}m_A(v_A)_2^2 + \frac{1}{2}m_B(v_B)_2^2 + (V_e)_2$

$0 + 0 + \frac{1}{2}\left[5\left(10^3\right)\right]\left(0.2^2\right)$

$= \frac{1}{2}(10)(v_A)_2^2 + \frac{1}{2}(15)(v_B)_2^2 + 0$

$5(v_A)_2^2 + 7.5\,(v_B)_2^2 = 100$ (2)

Solving Eqs. (1) and (2),

$(v_B)_2 = 2.31 \, \text{m/s} \rightarrow$ *Ans.*

$(v_A)_2 = -3.464 \, \text{m/s} = 3.46 \, \text{m/s} \leftarrow$ *Ans.*

F15–11. $(\xleftarrow{+}) \; m_A(v_A)_1 + m_B(v_B)_1 = (m_A + m_B)v_2$

$0 + 10(15) = (15 + 10)v_2$

$v_2 = 6 \, \text{m/s}$

$T_1 + V_1 = T_2 + V_2$

$\frac{1}{2}(m_A + m_B)v_2^2 + (V_e)_2 = \frac{1}{2}(m_A + m_B)v_3^2 + (V_e)_3$

$\frac{1}{2}(15 + 10)\left(6^2\right) + 0 = 0 + \frac{1}{2}\left[10\left(10^3\right)\right]s_{max}^2$

$s_{max} = 0.3 \, \text{m} = 300 \, \text{mm}$ *Ans.*

F15–12. $(\xrightarrow{+}) \; 0 + 0 = m_p(v_p)_x - m_c v_c$

$0 = (20 \, \text{kg})(v_p)_x - (250 \, \text{kg})v_c$

$(v_p)_x = 12.5 \, v_c$ (1)

$\mathbf{v}_p = \mathbf{v}_c + \mathbf{v}_{p/c}$

$(v_p)_x \mathbf{i} + (v_p)_y \mathbf{j} = -v_c \mathbf{i} + [(400 \, \text{m/s}) \cos 30° \mathbf{i}$

$+ (400 \, \text{m/s}) \sin 30° \mathbf{j}]$

$(v_p)_x \mathbf{i} + (v_p)_y \mathbf{j} = (346.41 - v_c)\mathbf{i} + 200\mathbf{j}$

$(v_p)_x = 346.41 - v_c$

$(v_p)_y = 200 \, \text{m/s}$

$(v_p)_x = 320.75 \, \text{m/s} \qquad v_c = 25.66 \, \text{m/s}$

$v_p = \sqrt{(v_p)_x^2 + (v_p)_y^2}$

$= \sqrt{(320.75 \, \text{m/s})^2 + (200 \, \text{m/s})^2}$

$= 378 \, \text{m/s}$ *Ans.*

F15–13. $(\xrightarrow{+}) \; e = \dfrac{(v_B)_2 - (v_A)_2}{(v_A)_1 - (v_B)_1}$

$= \dfrac{(9 \, \text{m/s}) - (1 \, \text{m/s})}{(8 \, \text{m/s}) - (-2 \, \text{m/s})} = 0.8$

F15–14. $(\xrightarrow{+}) \; m_A(v_A)_1 + m_B(v_B)_1 = m_A(v_A)_2 + m_B(v_B)_2$

$[15(10^3) \, \text{kg}](5 \, \text{m/s}) + [25(10^3)](-7 \, \text{m/s})$

$= [15(10^3) \, \text{kg}](v_A)_2 + [25(10^3)](v_B)_2$

$15(v_A)_2 + 25(v_B)_2 = -100$ (1)

Using the coefficient of restitution equation,

$(\xrightarrow{+}) \qquad e = \dfrac{(v_B)_2 - (v_A)_2}{(v_A)_1 - (v_B)_1}$

$0.6 = \dfrac{(v_B)_2 - (v_A)_2}{5 \, \text{m/s} - (-7 \, \text{m/s})}$

$(v_B)_2 - (v_A)_2 = 7.2$ (2)

Solving,

$(v_B)_2 = 0.2 \, \text{m/s} \rightarrow$ *Ans.*

$(v_A)_2 = -7 \, \text{m/s} = 7 \, \text{m/s} \leftarrow$ *Ans.*

F15–15. $T_1 + V_1 = T_2 + V_2$

$\frac{1}{2}m(v_A)_1^2 + mg(h_A)_1 = \frac{1}{2}m(v_A)_2^2 + mg(h_A)_2$

$\frac{1}{2}\left(\frac{30}{32.2}\text{ slug}\right)(5\text{ ft/s})^2 + (30\text{ lb})(10\text{ft})$

$\qquad = \frac{1}{2}\left(\frac{30}{32.2}\text{ slug}\right)(v_A)_2^2 + 0$

$(v_A)_2 = 25.87\text{ ft/s} \leftarrow$

$(\overset{+}{\leftarrow})\quad m_A(v_A)_2 + m_B(v_B)_2 = m_A(v_A)_3 + m_B(v_B)_3$

$\left(\frac{30}{32.2}\text{ slug}\right)(25.87\text{ ft/s}) + 0$

$\qquad = \left(\frac{30}{32.2}\text{ slug}\right)(v_A)_3 + \left(\frac{80}{32.2}\text{ slug}\right)(v_B)_3$

$30(v_A)_3 + 80(v_B)_3 = 775.95 \qquad (1)$

$(\overset{+}{\leftarrow})\qquad e = \dfrac{(v_B)_3 - (v_A)_3}{(v_A)_2 - (v_B)_2}$

$0.6 = \dfrac{(v_B)_3 - (v_A)_3}{25.87\text{ ft/s} - 0}$

$(v_B)_3 - (v_A)_3 = 15.52 \qquad (2)$

Solving Eqs. (1) and (2), yields

$(v_B)_3 = 11.3\text{ ft/s} \leftarrow$

$(v_A)_3 = -4.23\text{ ft/s} = 4.23\text{ ft/s} \rightarrow \qquad$ *Ans.*

F15–16. After collision: $T_1 + \Sigma U_{1-2} = T_2$

$\frac{1}{2}\left(\frac{5}{32.2}\right)(v_A)_2^2 - 0.2(5)\left(\frac{2}{12}\right) = 0$

$(v_A)_2 = 1.465\text{ ft/s}$

$\frac{1}{2}\left(\frac{10}{32.2}\right)(v_B)_2^2 - 0.2(10)\left(\frac{3}{12}\right) = 0$

$(v_B)_2 = 1.794\text{ ft/s}$

$\Sigma mv_1 = \Sigma mv_2$

$\frac{5}{32.2}(v_A)_1 + 0 = \frac{5}{32.2}(1.465) + \frac{10}{32.2}(1.794)$

$(v_A)_1 = 5.054$

$e = \dfrac{(v_B)_2 - (v_A)_2}{(v_A)_1 - (v_B)_1} = \dfrac{1.794 - 1.465}{5.054 - 0}$

$\quad = 0.0652 \qquad$ *Ans.*

F15–17. $(+\uparrow)\quad m[(v_b)_1]_y = m[(v_b)_2]_y$

$[(v_b)_2]_y = [(v_b)_1]_y = (20\text{ m/s})\sin 30° = 10\text{ m/s} \uparrow$

$(\overset{+}{\rightarrow})\qquad e = \dfrac{(v_w)_2 - [(v_b)_2]_x}{[(v_b)_1]_x - (v_w)_1}$

$0.75 = \dfrac{0 - [(v_b)_2]_x}{(20\text{ m/s})\cos 30° - 0}$

$[(v_b)_2]_x = -12.99\text{ m/s} = 12.99\text{ m/s} \leftarrow$

$(v_b)_2 = \sqrt{[(v_b)_2]_x^2 + [(v_b)_2]_y^2}$

$\qquad = \sqrt{(12.99\text{ m/s})^2 + (10\text{ m/s})^2}$

$\qquad = 16.4\text{ m/s} \qquad$ *Ans.*

$\theta = \tan^{-1}\left(\dfrac{[(v_b)_2]_y}{[(v_b)_2]_x}\right) = \tan^{-1}\left(\dfrac{10\text{ m/s}}{12.99\text{ m/s}}\right)$

$\quad = 37.6° \qquad$ *Ans.*

F15–18. $\Sigma m(v_x)_1 = \Sigma m(v_x)_2$

$0 + 0 = \frac{2}{32.2}(1) + \frac{11}{32.2}(v_{Bx})_2$

$(v_{Bx})_2 = -0.1818\text{ ft/s}$

$\Sigma m(v_y)_1 = \Sigma m(v_y)_2$

$\frac{2}{32.2}(3) + 0 = 0 + \frac{11}{32.2}(v_{By})_2$

$(v_{By})_2 = 0.545\text{ ft/s}$

$(v_B)_2 = \sqrt{(-0.1818)^2 + (0.545)^2}$

$\qquad = 0.575\text{ ft/s} \qquad$ *Ans.*

F15–19. $H_O = \Sigma mvd;$

$H_O = \left[2(10)\left(\frac{4}{5}\right)\right](4) - \left[2(10)\left(\frac{3}{5}\right)\right](3)$

$\quad = 28\text{ kg}\cdot\text{m}^2/\text{s} \,\circlearrowright$

F15–20. $H_P = \Sigma mvd;$

$H_P = [2(15)\sin 30°](2) - [2(15)\cos 30°](5)$

$\quad = -99.9\text{ kg}\cdot\text{m}^2/\text{s} = 99.9\text{ kg}\cdot\text{m}^2/\text{s}\,\circlearrowright$

F15–21. $(H_z)_1 + \Sigma\displaystyle\int M_z\,dt = (H_z)_2$

$5(2)(1.5) + 5(1.5)(3) = 5v(1.5)$

$v = 5\text{ m/s} \qquad$ *Ans.*

F15–22. $(H_z)_1 + \Sigma\displaystyle\int M_z\,dt = (H_z)_2$

$0 + \displaystyle\int_0^{4\text{ s}}(10t)\left(\frac{4}{5}\right)(1.5)\,dt = 5v(1.5)$

$v = 12.8\text{ m/s} \qquad$ *Ans.*

F15–23. $(H_z)_1 + \Sigma\displaystyle\int M_z\,dt = (H_z)_2$

$0 + \displaystyle\int_0^{5\text{ s}} 0.9t^2\,dt = 2v(0.6)$

$v = 31.2\text{ m/s} \qquad$ *Ans.*

F15–24. $(H_z)_1 + \Sigma\displaystyle\int M_z\,dt = (H_z)_2$

$0 + \displaystyle\int_0^{4\text{ s}} 8t\,dt + 2(10)(0.5)(4) = 2[10v(0.5)]$

$v = 10.4\text{ m/s} \qquad$ *Ans.*

Chapter 16

F16–1. $\theta = (20 \text{ rev})\left(\frac{2\pi \text{ rad}}{1 \text{ rev}}\right) = 40\pi \text{ rad}$

$\omega^2 = \omega_0^2 + 2\alpha_c(\theta - \theta_0)$

$(30 \text{ rad/s})^2 = 0^2 + 2\alpha_c[(40\pi \text{ rad}) - 0]$

$\alpha_c = 3.581 \text{ rad/s}^2 = 3.58 \text{ rad/s}^2$ *Ans.*

$\omega = \omega_0 + \alpha_c t$

$30 \text{ rad/s} = 0 + (3.581 \text{ rad/s}^2)t$

$t = 8.38 \text{ s}$ *Ans.*

F16–2. $\frac{d\omega}{d\theta} = 2(0.005\theta) = (0.01\theta)$

$\alpha = \omega \frac{d\omega}{d\theta} = \left(0.005 \theta^2\right)(0.01\theta) = 50\left(10^{-6}\right)\theta^3 \text{ rad/s}^2$

When $\theta = 20 \text{ rev}(2\pi \text{ rad/1 rev}) = 40\pi \text{ rad}$,

$\alpha = \left[50\left(10^{-6}\right)(40\pi)^3\right] \text{ rad/s}^2$

$= 99.22 \text{ rad/s}^2 = 99.2 \text{ rad/s}^2$ *Ans.*

F16–3. $\omega = 4\theta^{1/2}$

$150 \text{ rad/s} = 4 \theta^{1/2}$

$\theta = 1406.25 \text{ rad}$

$dt = \frac{d\theta}{\omega}$

$\int_0^t dt = \int_0^\theta \frac{d\theta}{4\theta^{1/2}}$

$t\big|_0^t = \frac{1}{2}\theta^{1/2}\big|_0^\theta$

$t = \frac{1}{2}\theta^{1/2}$

$t = \frac{1}{2}(1406.25)^{1/2} = 18.75 \text{ s}$ *Ans.*

F16–4. $\omega = \frac{d\theta}{dt} = (1.5t^2 + 15) \text{ rad/s}$

$\alpha = \frac{d\omega}{dt} = (3t) \text{ rad/s}$

$\omega = [1.5(3^2) + 15] \text{ rad/s} = 28.5 \text{ rad/s}$

$\alpha = 3(3) \text{ rad/s}^2 = 9 \text{ rad/s}^2.$

$v = \omega r = (28.5 \text{ rad/s})(0.75 \text{ ft}) = 21.4 \text{ ft/s}$ *Ans.*

$a = \alpha r = (9 \text{ rad/s}^2)(0.75 \text{ ft}) = 6.75 \text{ ft/s}^2$ *Ans.*

F16–5. $\omega \, d\omega = \alpha \, d\theta$

$\int_0^\omega \omega \, d\omega = \int_0^\theta 0.5\theta \, d\theta$

$\frac{\omega^2}{2}\big|_0^\omega = 0.25\theta^2\big|_0^\theta$

$\omega = (0.7071\theta) \text{ rad/s}$

When $\theta = 2 \text{ rev} = 4\pi \text{ rad}$,

$\omega = [0.7071(4\pi)] \text{ rad/s} = 8.886 \text{ rad/s}$

$v_P = \omega r = (8.886 \text{ rad/s})(0.2 \text{ m}) = 1.78 \text{ m/s}$ *Ans.*

$(a_P)_t = \alpha r = (0.5\theta \text{ rad/s}^2)(0.2 \text{ m})\big|_{\theta = 4\pi \text{ rad}}$

$= 1.257 \text{ m/s}^2$

$(a_P)_n = \omega^2 r = (8.886 \text{ rad/s})^2(0.2 \text{ m}) = 15.79 \text{ m/s}^2$

$a_P = \sqrt{(a_P)_t^2 + (a_P)_n^2}$

$= \sqrt{(1.257 \text{ m/s}^2)^2 + (15.79 \text{ m/s}^2)^2}$

$= 15.8 \text{ m/s}^2$ *Ans.*

F16–6. $\alpha_B = \alpha_A\left(\frac{r_A}{r_B}\right)$

$= (4.5 \text{ rad/s}^2)\left(\frac{0.075 \text{ m}}{0.225 \text{ m}}\right) = 1.5 \text{ rad/s}^2$

$\omega_B = (\omega_B)_0 + \alpha_B t$

$\omega_B = 0 + (1.5 \text{ rad/s}^2)(3 \text{ s}) = 4.5 \text{ rad/s}$

$\theta_B = (\theta_B)_0 + (\omega_B)_0 t + \frac{1}{2}\alpha_B t^2$

$\theta_B = 0 + 0 + \frac{1}{2}(1.5 \text{ rad/s}^2)(3 \text{ s})^2$

$\theta_B = 6.75 \text{ rad}$

$v_C = \omega_B r_D = (4.5 \text{ rad/s})(0.125 \text{ m})$

$= 0.5625 \text{ m/s}$ *Ans.*

$s_C = \theta_B r_D = (6.75 \text{ rad})(0.125 \text{ m}) = 0.84375 \text{ m}$

$= 844 \text{ mm}$ *Ans.*

F16–7. $\mathbf{v}_B = \mathbf{v}_A + \boldsymbol{\omega} \times \mathbf{r}_{B/A}$

$-v_B\mathbf{j} = (3\mathbf{i})\text{m/s}$

$\quad + (\omega\mathbf{k}) \times (-1.5 \cos 30°\mathbf{i} + 1.5 \sin 30°\mathbf{j})$

$-v_B\mathbf{j} = [3 - \omega_{AB}(1.5 \sin 30°)]\mathbf{i} - \omega(1.5 \cos 30°)\mathbf{j}$

$0 = 3 - \omega(1.5 \sin 30°)$ (1)

$-v_B = 0 - \omega(1.5 \cos 30°)$ (2)

$\omega = 4 \text{ rad/s}$ $v_B = 5.20 \text{ m/s}$ *Ans.*

F16–8. $\mathbf{v}_B = \mathbf{v}_A + \boldsymbol{\omega} \times \mathbf{r}_{B/A}$

$(v_B)_x\mathbf{i} + (v_B)_y\mathbf{j} = 0 + (-10\mathbf{k}) \times (-0.6\mathbf{i} + 0.6\mathbf{j})$

$(v_B)_x\mathbf{i} + (v_B)_y\mathbf{j} = 6\mathbf{i} + 6\mathbf{j}$

$(v_B)_x = 6 \text{ m/s}$ and $(v_B)_y = 6 \text{ m/s}$

$v_B = \sqrt{(v_B)_x^2 + (v_B)_y^2}$

$= \sqrt{(6 \text{ m/s})^2 + (6 \text{ m/s})^2}$

$= 8.49 \text{ m/s}$ *Ans.*

F16–9. $\mathbf{v}_B = \mathbf{v}_A + \boldsymbol{\omega} \times \mathbf{r}_{B/A}$

$(4 \text{ ft/s})\mathbf{i} = (-2 \text{ ft/s})\mathbf{i} + (-\omega\mathbf{k}) \times (3 \text{ ft})\mathbf{j}$

$4\mathbf{i} = (-2 + 3\omega)\mathbf{i}$

$\omega = 2 \text{ rad/s}$ *Ans.*

F16–10. $\mathbf{v}_A = \boldsymbol{\omega}_{OA} \times \mathbf{r}_A$

$\qquad = (12 \text{ rad/s})\mathbf{k} \times (0.3 \text{ m})\mathbf{j}$

$\qquad = [-3.6\mathbf{i}] \text{ m/s}$

$\qquad \mathbf{v}_B = \mathbf{v}_A + \boldsymbol{\omega}_{AB} \times \mathbf{r}_{B/A}$

$\qquad v_B\mathbf{j} = (-3.6 \text{ m/s})\mathbf{i}$

$\qquad\qquad + (\omega_{AB}\mathbf{k}) \times (0.6 \cos 30° \mathbf{i} - 0.6 \sin 30° \mathbf{j}) \text{ m}$

$\qquad v_B\mathbf{j} = [\omega_{AB}(0.6 \sin 30°) - 3.6]\mathbf{i} + \omega_{AB}(0.6 \cos 30°)\mathbf{j}$

$\qquad 0 = \omega_{AB}(0.6 \sin 30°) - 3.6$ $\qquad\qquad$ (1)

$\qquad v_B = \omega_{AB}(0.6 \cos 30°)$ $\qquad\qquad$ (2)

$\qquad \omega_{AB} = 12 \text{ rad/s} \qquad v_B = 6.24 \text{ m/s} \uparrow$

F16–11. $\mathbf{v}_C = \mathbf{v}_B + \boldsymbol{\omega}_{BC} \times \mathbf{r}_{C/B}$

$\qquad v_C\mathbf{j} = (-60\mathbf{i}) \text{ ft/s}$

$\qquad\qquad + (-\omega_{BC}\mathbf{k}) \times (-2.5 \cos 30°\mathbf{i} + 2.5 \sin 30°\mathbf{j}) \text{ ft}$

$\qquad v_C\mathbf{j} = (-60)\mathbf{i} + 2.165\omega_{BC}\mathbf{j} + 1.25\omega_{BC}\mathbf{i}$

$\qquad 0 = -60 + 1.25\omega_{BC}$

$\qquad v_C = 2.165\,\omega_{BC}$

$\qquad \omega_{BC} = 48 \text{ rad/s}$ $\qquad\qquad$ *Ans.*

$\qquad v_C = 104 \text{ ft/s}$

F16–12. $\mathbf{v}_B = \mathbf{v}_A + \boldsymbol{\omega} \times \mathbf{r}_{B/A}$

$\qquad -v_B \cos 30° \mathbf{i} + v_B \sin 30° \mathbf{j} = (-3 \text{ m/s})\mathbf{j} +$

$\qquad\qquad (-\omega\mathbf{k}) \times (-2 \sin 45°\mathbf{i} - 2 \cos 45°\mathbf{j}) \text{ m}$

$\qquad -0.8660v_B\mathbf{i} + 0.5v_B\mathbf{j}$

$\qquad\qquad = -1.4142\omega\mathbf{i} + (1.4142\omega - 3)\mathbf{j}$

$\qquad -0.8660v_B = -1.4142\omega$

$\qquad 0.5v_B = 1.4142\omega - 3$

$\qquad \omega = 5.02 \text{ rad/s} \qquad v_B = 8.20 \text{ m/s}$ $\qquad$ *Ans.*

F16–13. $\omega_{AB} = \dfrac{v_A}{r_{A/IC}} = \dfrac{6}{3} = 2 \text{ rad/s}$ $\qquad$ *Ans.*

$\qquad r_{C/IC} = \sqrt{1.5^2 + 2^2} = 2.5 \text{ m}$

$\qquad \phi = \tan^{-1}\left(\tfrac{2}{1.5}\right) = 53.13°$

$\qquad v_C = \omega_{AB}r_{C/IC} = 2(2.5) = 5 \text{ m/s}$ $\qquad$ *Ans.*

$\qquad \theta = 90° - \phi = 90° - 53.13° = 36.9° \,\diagdown$ $\quad$ *Ans.*

F16–14. $v_B = \omega_{AB}\, r_{B/A} = 12(0.6) = 7.2 \text{ m/s} \downarrow$

$\qquad v_C = 0$ $\qquad\qquad\qquad\qquad\qquad\qquad$ *Ans.*

$\qquad \omega_{BC} = \dfrac{v_B}{r_{B/IC}} = \dfrac{7.2}{1.2} = 6 \text{ rad/s}$ $\qquad$ *Ans.*

F16–15. $\omega = \dfrac{v_O}{r_{O/IC}} = \dfrac{6}{0.3} = 20 \text{ rad/s}$ $\qquad$ *Ans.*

$\qquad r_{A/IC} = \sqrt{0.3^2 + 0.6^2} = 0.6708 \text{ m}$

$\qquad \phi = \tan^{-1}\left(\tfrac{0.3}{0.6}\right) = 26.57°$

$\qquad v_A = \omega r_{A/IC} = 20(0.6708) = 13.4 \text{ m/s}$ $\quad$ *Ans.*

$\qquad \theta = 90° - \phi = 90° - 26.57° = 63.4° \,\diagup$ $\quad$ *Ans.*

F16–16. The location of *IC* can be determined using similar triangles.

$\qquad \dfrac{0.5 - r_{C/IC}}{3} = \dfrac{r_{C/IC}}{1.5} \qquad r_{C/IC} = 0.1667 \text{ m}$

$\qquad \omega = \dfrac{v_C}{r_{C/IC}} = \dfrac{1.5}{0.1667} = 9 \text{ rad/s}$ $\qquad$ *Ans.*

$\qquad$ Also, $r_{O/IC} = 0.3 - r_{C/IC} = 0.3 - 0.1667$

$\qquad\qquad = 0.1333 \text{ m}.$

$\qquad v_O = \omega r_{O/IC} = 9(0.1333) = 1.20 \text{ m/s}$ $\qquad$ *Ans.*

F16–17. $v_B = \omega r_{B/A} = 6(0.2) = 1.2 \text{ m/s}$

$\qquad r_{B/IC} = 0.8 \tan 60° = 1.3856 \text{ m}$

$\qquad r_{C/IC} = \dfrac{0.8}{\cos 60°} = 1.6 \text{ m}$

$\qquad \omega_{BC} = \dfrac{v_B}{r_{B/IC}} = \dfrac{1.2}{1.3856} = 0.8660 \text{ rad/s}$

$\qquad\qquad = 0.866 \text{ rad/s}$ $\qquad\qquad\qquad\qquad$ *Ans.*

$\qquad$ Then,

$\qquad v_C = \omega_{BC}\, r_{C/IC} = 0.8660(1.6) = 1.39 \text{ m/s}$ $\quad$ *Ans.*

F16–18. $v_B = \omega_{AB}\, r_{B/A} = 10(0.2) = 2 \text{ m/s}$

$\qquad v_C = \omega_{CD}\, r_{C/D} = \omega_{CD}(0.2) \rightarrow$

$\qquad r_{B/IC} = \dfrac{0.4}{\cos 30°} = 0.4619 \text{ m}$

$\qquad r_{C/IC} = 0.4 \tan 30° = 0.2309 \text{ m}$

$\qquad \omega_{BC} = \dfrac{v_B}{r_{B/IC}} = \dfrac{2}{0.4619} = 4.330 \text{ rad/s}$

$\qquad\qquad = 4.33 \text{ rad/s}$ $\qquad\qquad\qquad\qquad$ *Ans.*

$\qquad v_C = \omega_{BC}\, r_{C/IC}$

$\qquad \omega_{CD}(0.2) = 4.330(0.2309)$

$\qquad \omega_{CD} = 5 \text{ rad/s}$ $\qquad\qquad\qquad\qquad$ *Ans.*

F16–19. $\omega = \dfrac{v_A}{r_{A/IC}} = \dfrac{6}{3} = 2 \text{ rad/s}$

$\qquad \mathbf{a}_B = \mathbf{a}_A + \boldsymbol{\alpha} \times \mathbf{r}_{B/A} - \omega^2\mathbf{r}_{B/A}$

$\qquad a_B\mathbf{i} = -5\mathbf{j} + (\alpha\mathbf{k}) \times (3\mathbf{i} - 4\mathbf{j}) - 2^2(3\mathbf{i} - 4\mathbf{j})$

$\qquad a_B\mathbf{i} = (4\alpha - 12)\mathbf{i} + (3\alpha + 11)\mathbf{j}$

$\qquad a_B = 4\alpha - 12$

$\qquad 0 = 3\alpha + 11$

$\qquad \alpha = -3.67 \text{ rad/s}^2$ $\qquad\qquad\qquad\qquad$ *Ans.*

$\qquad a_B = -26.7 \text{ m/s}^2$ $\qquad\qquad\qquad\qquad$ *Ans.*

F16–20. $\mathbf{a}_A = \mathbf{a}_O + \boldsymbol{\alpha} \times \mathbf{r}_{A/O} - \omega^2\mathbf{r}_{A/O}$ $\qquad$ *Ans.*

$\qquad = 1.8\mathbf{i} + (-6\mathbf{k}) \times (0.3\mathbf{i}) - 12^2(0.3\mathbf{j})$

$\qquad = \{3.6\mathbf{i} - 43.2\mathbf{j}\} \text{ m/s}^2$ $\qquad\qquad\qquad$ *Ans.*

F16–21. $\mathbf{a}_A = \mathbf{a}_B + \boldsymbol{\alpha} \times \mathbf{r}_{A/B} - \omega^2 \mathbf{r}_{A/B}$

$3\mathbf{i} = a_B\mathbf{j} + (-\alpha\mathbf{k}) \times 0.3\mathbf{j} - 20^2(0.3\mathbf{j})$

$3\mathbf{i} = 0.3\alpha\mathbf{i} + (a_B - 120)\mathbf{j}$

$3 = 0.3\alpha \qquad \alpha = 10 \text{ rad/s}^2$ *Ans.*

$\mathbf{a}_A = \mathbf{a}_O + \boldsymbol{\alpha} \times \mathbf{r}_{A/O} - \omega^2\mathbf{r}_{A/O}$

$= 3\mathbf{i} + (-10\mathbf{k}) \times (-0.6\mathbf{i}) - 20^2(-0.6\mathbf{j})$

$= \{243\mathbf{i} + 6\mathbf{j}\} \text{ m/s}^2$ *Ans.*

F16–22. $\dfrac{r_{A/IC}}{3} = \dfrac{0.5 - r_{A/IC}}{1.5}; \qquad r_{A/IC} = 0.3333 \text{ m}$

$\omega = \dfrac{v_A}{r_{A/IC}} = \dfrac{3}{0.3333} = 9 \text{ rad/s}$

$\mathbf{a}_A = \mathbf{a}_C + \boldsymbol{\alpha} \times \mathbf{r}_{A/C} - \omega^2\mathbf{r}_{A/C}$

$1.5\mathbf{i} - (a_A)_n\mathbf{j} = -0.75\mathbf{i} + (a_C)_n\mathbf{j}$
$\qquad\qquad + (-\alpha\mathbf{k}) \times 0.5\mathbf{j} - 9^2(0.5\mathbf{j})$

$1.5\mathbf{i} - (a_A)_n\mathbf{j} = (0.5\alpha - 0.75)\mathbf{i} + \left[(a_C)_n - 40.5\right]\mathbf{j}$

$1.5 = 0.5\alpha - 0.75$

$\alpha = 4.5 \text{ rad/s}^2$ *Ans.*

F16–23. $v_B = \omega \, r_{B/A} = 12(0.3) = 3.6 \text{ m/s}$

$\omega_{BC} = \dfrac{v_B}{r_{B/IC}} = \dfrac{3.6}{1.2} = 3 \text{ rad/s}$

$\mathbf{a}_B = \boldsymbol{\alpha} \times \mathbf{r}_{B/A} - \omega^2\mathbf{r}_{B/A}$

$= (-6\mathbf{k}) \times (0.3\mathbf{i}) - 12^2(0.3\mathbf{i})$

$= \{-43.2\mathbf{i} - 1.8\mathbf{j}\} \text{ m/s}$

$\mathbf{a}_C = \mathbf{a}_B + \boldsymbol{\alpha}_{BC} \times \mathbf{r}_{C/B} - \omega_{BC}^2\mathbf{r}_{C/B}$

$a_C\mathbf{i} = (-43.2\mathbf{i} - 1.8\mathbf{j})$
$\qquad\qquad + (\alpha_{BC}\,\mathbf{k}) \times (1.2\mathbf{i}) - 3^2(1.2\mathbf{i})$

$a_C\mathbf{i} = -54\mathbf{i} + (1.2\alpha_{BC} - 1.8)\mathbf{j}$

$a_C = -54 \text{ m/s}^2 = 54 \text{ m/s}^2 \leftarrow$ *Ans.*

$0 = 1.2\alpha_{BC} - 1.8 \qquad \alpha_{BC} = 1.5 \text{ rad/s}^2$ *Ans.*

F16–24. $v_B = \omega \, r_{B/A} = 6(0.2) = 1.2 \text{ m/s} \rightarrow$

$r_{B/IC} = 0.8 \tan 60° = 1.3856 \text{ m}$

$\omega_{BC} = \dfrac{v_B}{r_{B/IC}} = \dfrac{1.2}{1.3856} = 0.8660 \text{ rad/s}$

$\mathbf{a}_B = \boldsymbol{\alpha} \times \mathbf{r}_{B/A} - \omega^2\mathbf{r}_{B/A}$

$= (-3\mathbf{k}) \times (0.2\mathbf{j}) - 6^2(0.2\mathbf{j})$

$= [0.6\mathbf{i} - 7.2\mathbf{j}] \text{ m/s}$

$\mathbf{a}_C = \mathbf{a}_B + \boldsymbol{\alpha}_{BC} \times \mathbf{r}_{C/B} - \omega^2\mathbf{r}_{C/B}$

$a_C \cos 30°\mathbf{i} + a_C \sin 30°\mathbf{j}$

$= (0.6\mathbf{i} - 7.2\mathbf{j}) + (\alpha_{BC}\,\mathbf{k} \times 0.8\mathbf{i}) - 0.8660^2(0.8\mathbf{i})$

$0.8660a_C\mathbf{i} + 0.5a_C\mathbf{j} = (0.8\alpha_{BC} - 7.2)\mathbf{j}$

$0.8660a_C = 0$

$0.5a_C = 0.8\alpha_{BC} - 7.2$

$a_C = 0 \qquad\qquad \alpha_{BC} = 9 \text{ rad/s}^2$ *Ans.*

Chapter 17

F17–1. $\xrightarrow{+} \Sigma F_x = m(a_G)_x; \quad 100\left(\tfrac{4}{5}\right) = 100a$

$a = 0.8 \text{ m/s}^2 \rightarrow$ *Ans.*

$+\uparrow\Sigma F_y = m(a_G)_y;$

$N_A + N_B - 100\left(\tfrac{3}{5}\right) - 100(9.81) = 0$ (1)

$\zeta + \Sigma M_G = 0;$

$N_A(0.6) + 100\left(\tfrac{3}{5}\right)(0.7)$

$\qquad - N_B(0.4) - 100\left(\tfrac{4}{5}\right)(0.7) = 0$ (2)

$N_A = 430.4 \text{ N} = 430 \text{ N}$ *Ans.*

$N_B = 610.6 \text{ N} = 611 \text{ N}$ *Ans.*

F17–2. $\Sigma F_{x'} = m(a_G)_{x'}; \qquad 80(9.81) \sin 15° = 80a$

$a = 2.54 \text{ m/s}^2$ *Ans.*

$\Sigma F_{y'} = m(a_G)_{y'};$

$N_A + N_B - 80(9.81) \cos 15° = 0$ (1)

$\zeta + \Sigma M_G = 0;$

$N_A(0.5) - N_B(0.5) = 0$ (2)

$N_A = N_B = 379 \text{ N}$ *Ans.*

F17–3. $\zeta + \Sigma M_A = \Sigma(\mathcal{M}_k)_A; \quad 10\left(\tfrac{3}{5}\right)(7) = \tfrac{20}{32.2}a(3.5)$

$a = 19.3 \text{ ft/s}^2$ *Ans.*

$\xrightarrow{+}\Sigma F_x = m(a_G)_x; \, A_x + 10\left(\tfrac{3}{5}\right) = \tfrac{20}{32.2}(19.32)$

$A_x = 6 \text{ lb}$ *Ans.*

$+\uparrow\Sigma F_y = m(a_G)_y; \quad A_y - 20 + 10\left(\tfrac{4}{5}\right) = 0$

$A_y = 12 \text{ lb}$ *Ans.*

F17–4. $F_A = \mu_s N_A = 0.2N_A \qquad F_B = \mu_s N_B = 0.2N_B$

$\xrightarrow{+}\Sigma F_x = m(a_G)_x;$

$0.2N_A + 0.2N_B = 100a$ (1)

$+\uparrow\Sigma F_y = m(a_G)_y;$

$N_A + N_B - 100(9.81) = 0$ (2)

$\zeta + \Sigma M_G = 0;$

$0.2N_A(0.75) + N_A(0.9) + 0.2N_B(0.75)$

$\qquad\qquad - N_B(0.6) = 0$ (3)

Solving Eqs. (1), (2), and (3),

$N_A = 294.3 \text{ N} = 294 \text{ N}$

$N_B = 686.7 \text{ N} = 687 \text{ N}$

$a = 1.96 \text{ m/s}^2$ *Ans.*

Since N_A is positive, the table will indeed slide before it tips.

F17–5. $(a_G)_t = \alpha r = \alpha(1.5 \text{ m})$

$(a_G)_n = \omega^2 r = (5 \text{ rad/s})^2(1.5 \text{ m}) = 37.5 \text{ m/s}^2$

$\Sigma F_t = m(a_G)_t; \quad 100 \text{ N} = 50 \text{ kg}[\alpha(1.5 \text{ m})]$

$\alpha = 1.33 \text{ rad/s}^2$ *Ans.*

$\Sigma F_n = m(a_G)_n; \quad T_{AB} + T_{CD} - 50(9.81) \text{ N}$

$\qquad = 50 \text{ kg}(37.5 \text{ m/s}^2)$

$T_{AB} + T_{CD} = 2365.5$

$\zeta + \Sigma M_G = 0; \quad T_{CD}(1 \text{ m}) - T_{AB}(1 \text{ m}) = 0$

$T_{AB} = T_{CD} = 1182.75 \text{ N} = 1.18 \text{ kN}$ *Ans.*

F17–6. $\zeta + \Sigma M_C = 0;$

$D_y(0.6) - 450 = 0 \quad D_y = 750 \text{ N}$ *Ans.*

$(a_G)_n = \omega^2 r = 6^2(0.6) = 21.6 \text{ m/s}^2$

$(a_G)_t = \alpha r = \alpha(0.6)$

$+\uparrow \Sigma F_t = m(a_G)_t;$

$750 - 50(9.81) = 50[\alpha(0.6)]$

$\alpha = 8.65 \text{ rad/s}^2$ *Ans.*

$\overset{+}{\rightarrow} \Sigma F_n = m(a_G)_n;$

$F_{AB} + D_x = 50(21.6)$ (1)

$\zeta + \Sigma M_G = 0;$

$D_x(0.4) + 750(0.1) - F_{AB}(0.4) = 0$ (2)

$D_x = 446.25 \text{ N} = 446 \text{ N}$ *Ans.*

$F_{AB} = 633.75 \text{ N} = 634 \text{ N}$ *Ans.*

F17–7. $I_O = mk_O^2 = 100(0.5^2) = 25 \text{ kg} \cdot \text{m}^2$

$\zeta + \Sigma M_C = I_O \alpha; \quad -100(0.6) = -25\alpha$

$\alpha = 2.4 \text{ rad/s}^2$

$\omega = \omega_0 + \alpha_c t$

$\omega = 0 + 2.4(3) = 7.2 \text{ rad/s}$ *Ans.*

F17–8. $I_O = \frac{1}{2} mr^2 = \frac{1}{2}(50)(0.3^2) = 2.25 \text{ kg} \cdot \text{m}^2$

$\zeta + \Sigma M_C = I_O \alpha;$

$\qquad -9t = -2.25\alpha \quad \alpha = (4t) \text{ rad/s}^2$

$d\omega = \alpha \, dt$

$\int_0^\omega d\omega = \int_0^t 4t \, dt$

$\omega = (2t^2) \text{ rad/s}$

$\omega = 2(4^2) = 32 \text{ rad/s}$ *Ans.*

F17–9. $(a_G)_t = \alpha r_G = \alpha(0.15)$

$(a_G)_n = \omega^2 r_G = 6^2(0.15) = 5.4 \text{ m/s}^2$

$I_O = I_G + md^2 = \frac{1}{12}(30)(0.9^2) + 30(0.15^2)$

$\qquad = 2.7 \text{ kg} \cdot \text{m}^2$

$\zeta + \Sigma M_O = I_O \alpha; \quad 60 - 30(9.81)(0.15) = 2.7\alpha$

$\alpha = 5.872 \text{ rad/s}^2 = 5.87 \text{ rad/s}^2$ *Ans.*

$\overset{+}{\rightarrow} \Sigma F_n = m(a_G)_n; \quad O_n = 30(5.4) = 162 \text{ N}$ *Ans.*

$+\uparrow \Sigma F_t = m(a_G)_t;$

$O_t - 30(9.81) = 30[5.872(0.15)]$

$O_t = 320.725 \text{ N} = 321 \text{ N}$ *Ans.*

F17–10. $(a_G)_t = \alpha r_G = \alpha(0.3)$

$(a_G)_n = \omega^2 r_G = 10^2(0.3) = 30 \text{ m/s}^2$

$I_O = I_G + md^2 = \frac{1}{2}(30)(0.3^2) + 30(0.3^2)$

$\qquad = 4.05 \text{ kg} \cdot \text{m}^2$

$\zeta + \Sigma M_O = I_O \alpha;$

$50\left(\frac{3}{5}\right)(0.3) + 50\left(\frac{4}{5}\right)(0.3) = 4.05\alpha$

$\alpha = 5.185 \text{ rad/s}^2 = 5.19 \text{ rad/s}^2$ *Ans.*

$+\uparrow \Sigma F_n = m(a_G)_n;$

$O_n + 50\left(\frac{3}{5}\right) - 30(9.81) = 30(30)$

$O_n = 1164.3 \text{ N} = 1.16 \text{kN}$ *Ans.*

$\overset{+}{\rightarrow} \Sigma F_t = m(a_G)_t;$

$O_t + 50\left(\frac{4}{5}\right) = 30[5.185(0.3)]$

$O_t = 6.67 \text{ N}$ *Ans.*

F17–11. $I_G = \frac{1}{12}ml^2 = \frac{1}{12}(15 \text{ kg})(0.9 \text{ m})^2 = 1.0125 \text{ kg} \cdot \text{m}^2$

$(a_G)_n = \omega^2 r_G = 0$

$(a_G)_t = \alpha(0.15 \text{ m})$

$I_O = I_G + md_{OG}^2$

$\qquad = 1.0125 \text{ kg} \cdot \text{m}^2 + 15 \text{ kg}(0.15 \text{ m})^2$

$\qquad = 1.35 \text{ kg} \cdot \text{m}^2$

$\zeta + \Sigma M_O = I_O \alpha;$

$[15(9.81) \text{ N}](0.15 \text{ m}) = (1.35 \text{ kg} \cdot \text{m}^2)\alpha$

$\alpha = 16.35 \text{ rad/s}^2$ *Ans.*

$+\downarrow \Sigma F_t = m(a_G)_t; \quad -O_t + 15(9.81)\text{N}$

$\qquad = (15 \text{ kg})[16.35 \text{ rad/s}^2(0.15 \text{ m})]$

$O_t = 110.36 \text{ N} = 110 \text{ N}$ *Ans.*

$\overset{+}{\rightarrow} \Sigma F_n = m(a_G)_n; \quad O_n = 0$ *Ans.*

F17–12. $(a_G)_t = \alpha r_G = \alpha(0.45)$

$(a_G)_n = \omega^2 r_G = 6^2(0.45) = 16.2 \text{ m/s}^2$

$I_O = \frac{1}{3}ml^2 = \frac{1}{3}(30)(0.9^2) = 8.1 \text{ kg} \cdot \text{m}^2$

$\zeta + \Sigma M_O = I_O \alpha;$

$300\left(\frac{4}{5}\right)(0.6) - 30(9.81)(0.45) = 8.1\alpha$

$\alpha = 1.428 \text{ rad/s}^2 = 1.43 \text{ rad/s}^2$ *Ans.*

$\overset{+}{\leftarrow} \Sigma F_n = m(a_G)_n; \quad O_n + 300\left(\frac{3}{5}\right) - 30(16.2)$

$\qquad\qquad\qquad\qquad O_n = 306 \text{ N}$ *Ans.*

$+\uparrow \Sigma F_t = m(a_G)_t; \quad O_t + 300\left(\frac{4}{5}\right) - 30(9.81)$

$\qquad\qquad\qquad\qquad = 30[1.428(0.45)]$

$O_t = 73.58 \text{ N} = 73.6 \text{ N}$ *Ans.*

F17–13. $I_G = \frac{1}{12}ml^2 = \frac{1}{12}(60)(3^2) = 45 \text{ kg} \cdot \text{m}^2$

$+\uparrow \Sigma F_y = m(a_G)_y;$

$80 - 20 = 60a_G \quad a_G = 1 \text{ m/s}^2 \uparrow$

$\zeta + \Sigma M_G = I_G\alpha; \quad 80(1) + 20(0.75) = 45\alpha$

$\alpha = 2.11 \text{ rad/s}^2$ *Ans.*

F17–14. $\zeta + \Sigma M_A = (\mathcal{M}_k)_A;$

$-200(0.3) = -100a_G(0.3) - 4.5\alpha$

$30a_G + 4.5\alpha = 60 \quad (1)$

$a_G = \alpha r = \alpha(0.3) \quad (2)$

$\alpha = 4.44 \text{ rad/s}^2 \quad a_G = 1.33 \text{ m/s}^2 \rightarrow$ *Ans.*

F17–15. $+\uparrow \Sigma F_y = m(a_G)_y;$

$N - 20(9.81) = 0 \quad N = 196.2 \text{ N}$

$\xrightarrow{+} \Sigma F_x = m(a_G)_x; \quad 0.5(196.2) = 20a_O$

$a_O = 4.905 \text{ m/s}^2 \rightarrow$ *Ans.*

$\zeta + \Sigma M_O = I_O\alpha;$

$0.5(196.2)(0.4) - 100 = -1.8\alpha$

$\alpha = 33.8 \text{ rad/s}^2$ *Ans.*

F17–16. $\zeta + \Sigma M_A = (\mathcal{M}_k)_A;$

$20(9.81)\sin30^0 (0.15) = 0.18\alpha + (20a_G)(0.15)$

$0.18\alpha + 3a_G = 14.715$

$a_G = \alpha r = \alpha(0.15)$

$\alpha = 23.36 \text{ rad/s}^2 = 23.4 \text{ rad/s}^2$ *Ans.*

$a_G = 3.504 \text{ m/s}^2 = 3.50 \text{ m/s}^2$ *Ans.*

F17–17. $+\uparrow \Sigma F_y = m(a_G)_y;$

$N - 200(9.81) = 0 \quad N = 1962 \text{ N}$

$\xrightarrow{+} \Sigma F_x = m(a_G)_x;$

$T - 0.2(1962) = 200a_G \quad (1)$

$\zeta + \Sigma M_A = (\mathcal{M}_k)_A; \quad 450 - 0.2(1962)(1)$

$= 18\alpha + 200a_G(0.4) \quad (2)$

$(a_A)_t = 0 \quad a_A = (a_A)_n$

$\mathbf{a}_G = \mathbf{a}_A + \boldsymbol{\alpha} \times \mathbf{r}_{G/A} - \omega^2\mathbf{r}_{G/A}$

$a_G\mathbf{i} = -a_A\mathbf{j} + \alpha\mathbf{k} \times (0.4\mathbf{j}) - \omega^2(-0.4\mathbf{j})$

$a_G\mathbf{i} = 0.4\alpha\mathbf{i} + (0.4\omega^2 - a_A)\mathbf{j}$

$a_G = 0.4\alpha \quad (3)$

Solving Eqs. (1), (2), and (3),

$\alpha = 1.15 \text{ rad/s}^2 \quad a_G = 0.461 \text{ m/s}^2$

$T = 485 \text{ N}$ *Ans.*

F17–18. $\xrightarrow{+} \Sigma F_x = m(a_G)_x; \quad 0 = 12(a_G)_x \quad (a_G)_x = 0$

$\zeta + \Sigma M_A = (\mathcal{M}_k)_A$

$-12(9.81)(0.3) = 12(a_G)_y(0.3) - \frac{1}{12}(12)(0.6)^2\alpha$

$0.36\alpha - 3.6(a_G)_y = 35.316 \quad (1)$

$\omega = 0$

$\mathbf{a}_G = \mathbf{a}_A + \boldsymbol{\alpha} \times \mathbf{r}_{G/A} - \omega^2\mathbf{r}_{G/A}$

$(a_G)_y\mathbf{j} = a_A\mathbf{i} + (-\alpha\mathbf{k}) \times (0.3\mathbf{i}) - \mathbf{0}$

$(a_G)_y\mathbf{j} = (a_A)\mathbf{i} - 0.3\mathbf{j}$

$a_A = 0$ *Ans.*

$(a_G)_y = -0.3\alpha \quad (2)$

Solving Eqs. (1) and (2)

$\alpha = 24.5 \text{ rad/s}^2$

$(a_G)_y = -7.36 \text{ m/s}^2 = 7.36 \text{ m/s}^2 \downarrow$ *Ans.*

Chapter 18

F18–1. $I_O = mk_O^2 = 80(0.4^2) = 12.8 \text{ kg} \cdot \text{m}^2$

$T_1 = 0$

$T_2 = \frac{1}{2}I_O\omega^2 = \frac{1}{2}(12.8)\omega^2 = 6.4\omega^2$

$s = \theta r = 20(2\pi)(0.6) = 24\pi \text{ m}$

$T_1 + \Sigma U_{1-2} = T_2$

$0 + 50(24\pi) = 6.4\omega^2$

$\omega = 24.3 \text{ rad/s}$ *Ans.*

F18–2. $T_1 = 0$

$T_2 = \frac{1}{2}m(v_G)_2^2 + \frac{1}{2}I_G\omega_2^2$

$= \frac{1}{2}\left(\frac{50}{32.2} \text{ slug}\right)(2.5\omega_2)^2$

$+ \frac{1}{2}\left[\frac{1}{12}\left(\frac{50}{32.2} \text{ slug}\right)(5 \text{ ft})^2\right]\omega_2^2$

$T_2 = 6.4700\omega_2^2$

Or,

$I_O = \frac{1}{3}ml^2 = \frac{1}{3}\left(\frac{50}{32.2} \text{ slug}\right)(5 \text{ ft})^2$

$= 12.9400 \text{ slug} \cdot \text{ft}^2$

So that

$T_2 = \frac{1}{2}I_O\omega_2^2 = \frac{1}{2}(12.9400 \text{ slug} \cdot \text{ft}^2)\omega_2^2$

$= 6.4700\omega_2^2$

$T_1 + \Sigma U_{1-2} = T_2$

$T_1 + [-Wy_G + M\theta] = T_2$

$0 + [-(50 \text{ lb})(2.5 \text{ ft}) + (100 \text{ lb}\cdot\text{ft})(\frac{\pi}{2})]$

$= 6.4700\omega_2^2$

$\omega_2 = 2.23 \text{ rad/s}$ *Ans.*

F18–3. $(v_G)_2 = \omega_2 r_{G/IC} = \omega_2(2.5)$

$I_G = \frac{1}{12} ml^2 = \frac{1}{12}(50)(5^2) = 104.17 \text{ kg} \cdot \text{m}^2$

$T_1 = 0$

$T_2 = \frac{1}{2} m(v_G)_2^2 + \frac{1}{2} I_G \omega_2^2$

$\quad = \frac{1}{2}(50)[\omega_2(2.5)]^2 + \frac{1}{2}(104.17)\omega_2^2 = 208.33\omega_2^2$

$U_P = Ps_P = 600(3) = 1800 \text{ J}$

$U_W = -Wh = -50(9.81)(2.5 - 2) = -245.25 \text{ J}$

$T_1 + \Sigma U_{1-2} = T_2$

$0 + 1800 + (-245.25) = 208.33\omega_2^2$

$\omega_2 = 2.732 \text{ rad/s} = 2.73 \text{ rad/s}$ *Ans.*

F18–4. $T = \frac{1}{2} mv_O^2 + \frac{1}{2} I_O \omega^2$

$\quad = \frac{1}{2}(50 \text{ kg})(0.4\omega)^2 + \frac{1}{2}[50 \text{ kg}(0.3 \text{ m})^2]\omega^2$

$\quad = 6.25\omega^2 \text{ J}$

Or,

$T = \frac{1}{2} I_{IC} \omega^2$

$\quad = \frac{1}{2}[50 \text{ kg}(0.3 \text{ m})^2 + 50 \text{ kg}(0.4 \text{ m})^2]\omega^2$

$\quad = 6.25\omega^2 \text{ J}$

$s_O = \theta r = 10(2\pi \text{ rad})(0.4 \text{ m}) = 8\pi \text{ m}$

$T_1 + \Sigma U_{1-2} = T_2$

$T_1 + P \cos 30° \, s_O = T_2$

$0 + (50 \text{ N})\cos 30°(8\pi \text{ m}) = 6.25\omega^2 \text{ J}$

$\omega = 13.2 \text{ rad/s}$ *Ans.*

F18–5. $I_G = \frac{1}{12} ml^2 = \frac{1}{12}(30)(3^2) = 22.5 \text{ kg} \cdot \text{m}^2$

$T_1 = 0$

$T_2 = \frac{1}{2} mv_G^2 + \frac{1}{2} I_G \omega^2$

$\quad = \frac{1}{2}(30)[\omega(0.5)]^2 + \frac{1}{2}(22.5)\omega^2 = 15\omega^2$

$I_O = I_G + md^2 = \frac{1}{12}(30)(3^2) + 30(0.5^2)$

$\quad = 30 \text{ kg} \cdot \text{m}^2$

Or,

$T_2 = \frac{1}{2} I_O \omega^2 = \frac{1}{2}(30)\omega^2 = 15\omega^2$

$s_1 = \theta r_1 = 8\pi(0.5) = 4\pi \text{ m}$

$s_2 = \theta r_2 = 8\pi(1.5) = 12\pi \text{m}$

$U_{P_2} = P_2 s_2 = 20(12\pi) = 240\pi \text{ J}$

$U_M = M\theta = 20[4(2\pi)] = 160\pi \text{ J}$

$T_1 + \Sigma U_{1-2} = T_2$

$0 + 120\pi + 240\pi + 160\pi = 15\omega^2$

$\omega = 10.44 \text{ rad/s} = 10.4 \text{ rad/s}$ *Ans.*

F18–6. $v_O = \omega r = \omega(0.4)$

$I_O = mk_O{}^2 = 20(0.3^2) = 1.8 \text{ kg} \cdot \text{m}^2$

$T_1 = 0$

$T_2 = \frac{1}{2} mv_G{}^2 + \frac{1}{2} I_G \omega^2$

$\quad = \frac{1}{2}(20)[\omega(0.4)]^2 + \frac{1}{2}(1.8)\omega^2$

$\quad = 2.5\omega^2$

$U_M = M\theta = M\left(\dfrac{s_O}{r}\right) = 50\left(\dfrac{20}{0.4}\right) = 2500 \text{ J}$

$T_1 + \Sigma U_{1-2} = T_2$

$0 + 2500 = 2.5\omega^2$

$\omega = 31.62 \text{ rad/s} = 31.6 \text{ rad/s}$ *Ans.*

F18–7. $v_G = \omega r = \omega(0.3)$

$I_G = \frac{1}{2} mr^2 = \frac{1}{2}(30)(0.3^2) = 1.35 \text{ kg} \cdot \text{m}^2$

$T_1 = 0$

$T_2 = \frac{1}{2} m(v_G)_2^2 + \frac{1}{2} I_G \omega_2^2$

$\quad = \frac{1}{2}(30)[\omega_2(0.3)]^2 + \frac{1}{2}(1.35)\omega_2^2 = 2.025\omega_2^2$

$(V_g)_1 = W y_1 = 0$

$(V_g)_2 = -W y_2 = -30(9.81)(0.3) = -88.92 \text{ J}$

$T_1 + V_1 = T_2 + V_2$

$0 + 0 = 2.025\omega_2^2 + (-88.29)$

$\omega_2 = 6.603 \text{ rad/s} = 6.60 \text{ rad/s}$ *Ans.*

F18–8. $v_O = \omega r_{O/IC} = \omega(0.2)$

$I_O = mk_O{}^2 = 50(0.3^2) = 4.5 \text{ kg} \cdot \text{m}^2$

$T_1 = 0$

$T_2 = \frac{1}{2} m(v_O)_2^2 + \frac{1}{2} I_O \omega_2^2$

$\quad = \frac{1}{2}(50)[\omega_2(0.2)]^2 + \frac{1}{2}(4.5)\omega_2^2$

$\quad = 3.25\omega_2^2$

$(V_g)_1 = W y_1 = 0$

$(V_g)_2 = -W y_2 = -50(9.81)(6 \sin 30°)$

$\quad = -1471.5 \text{J}$

$T_1 + V_1 = T_2 + V_2$

$0 + 0 = 3.25\omega_2^2 + (-1471.5)$

$\omega_2 = 21.28 \text{ rad/s} = 21.3 \text{ rad/s}$ *Ans.*

F18–9. $v_G = \omega r_G = \omega(1.5)$

$I_G = \frac{1}{12}(60)(3^2) = 45 \text{ kg} \cdot \text{m}^2$

$T_1 = 0$

$T_2 = \frac{1}{2}m(v_G)_2^2 + \frac{1}{2}I_G\omega_2^2$

$\quad = \frac{1}{2}(60)[\omega_2(1.5)]^2 + \frac{1}{2}(45)\omega_2^2$

$\quad = 90\omega_2^2$

Or,

$T_2 = \frac{1}{2}I_O\omega_2^2 = \frac{1}{2}\left[45 + 60(1.5^2)\right]\omega_2^2 = 90\omega_2^2$

$(V_g)_1 = Wy_1 = 0$

$(V_g)_2 = -Wy_2 = -60(9.81)(1.5 \sin 45°)$

$\quad = -624.30 \text{ J}$

$(V_e)_1 = \frac{1}{2}ks_1^2 = 0$

$(V_e)_2 = \frac{1}{2}ks_2^2 = \frac{1}{2}(150)(3 \sin 45°)^2 = 337.5 \text{ J}$

$T_1 + V_1 = T_2 + V_2$

$0 + 0 = 90\omega_2^2 + [-624.39 + 337.5]$

$\omega_2 = 1.785 \text{ rad/s} = 1.79 \text{ rad/s}$ *Ans.*

F18–10. $v_G = \omega r_G = \omega(0.75)$

$I_G = \frac{1}{12}(30)(1.5^2) = 5.625 \text{ kg} \cdot \text{m}^2$

$T_1 = 0$

$T_2 = \frac{1}{2}m(v_G)_2^2 + \frac{1}{2}I_G\omega_2^2$

$\quad = \frac{1}{2}(30)[\omega(0.75)]^2 + \frac{1}{2}(5.625)\omega_2^2 = 11.25\omega_2^2$

Or,

$T_2 = \frac{1}{2}I_O\omega_2^2 = \frac{1}{2}\left[5.625 + 30(0.75^2)\right]\omega_2^2$

$\quad = 11.25\omega_2^2$

$(V_g)_1 = Wy_1 = 0$

$(V_g)_2 = -Wy_2 = -30(9.81)(0.75)$

$\quad = -220.725 \text{ J}$

$(V_e)_1 = \frac{1}{2}ks_1^2 = 0$

$(V_e)_1 = \frac{1}{2}ks_2^2 = \frac{1}{2}(80)\left(\sqrt{2^2 + 1.5^2} - 0.5\right)^2 = 160 \text{ J}$

$T_1 + V_1 = T_2 + V_2$

$0 + 0 = 11.25\omega_2^2 + (-220.725 + 160)$

$\omega_2 = 2.323 \text{ rad/s} = 2.32 \text{ rad/s}$ *Ans.*

F18–11. $(v_G)_2 = \omega_2 r_{G/IC} = \omega_2(0.75)$

$I_G = \frac{1}{12}(30)(1.5^2) = 5.625 \text{ kg} \cdot \text{m}^2$

$T_1 = 0$

$T_2 = \frac{1}{2}m(v_G)_2^2 + \frac{1}{2}I_G\omega_2^2$

$\quad = \frac{1}{2}(30)[\omega_2(0.75)]^2 + \frac{1}{2}(5.625)\omega_2^2 = 11.25\omega_2^2$

$(V_g)_1 = Wy_1 = 30(9.81)(0.75 \sin 45°) = 156.08 \text{ J}$

$(V_g)_1 = -Wy_2 = 0$

$(V_e)_1 = \frac{1}{2}ks_1^2 = 0$

$(V_e)_1 = \frac{1}{2}ks_2^2 = \frac{1}{2}(300)(1.5 - 1.5 \cos 45°)^2$

$\quad = 28.95 \text{ J}$

$T_1 + V_1 = T_2 + V_2$

$0 + (156.08 + 0) = 11.25\omega_2^2 + (0 + 28.95)$

$\omega_2 = 3.362 \text{ rad/s} = 3.36 \text{ rad/s}$ *Ans.*

F18–12. $(V_g)_1 = -Wy_1 = -[20(9.81) \text{ N}](1 \text{ m}) = -196.2 \text{ J}$

$(V_g)_2 = 0$

$(V_e)_1 = \frac{1}{2}ks_1^2$

$\quad = \frac{1}{2}(100 \text{ N/m})\left(\sqrt{(3 \text{ m})^2 + (2 \text{ m})^2} - 0.5 \text{ m}\right)^2$

$\quad = 482.22 \text{ J}$

$(V_e)_2 = \frac{1}{2}ks_2^2 = \frac{1}{2}(100 \text{ N/m})(1 \text{ m} - 0.5 \text{ m})^2$

$\quad = 12.5 \text{ J}$

$T_1 = 0$

$T_2 = \frac{1}{2}I_A\omega^2 = \frac{1}{2}\left[\frac{1}{3}(20 \text{ kg})(2 \text{ m})^2\right]\omega^2$

$\quad = 13.3333\omega^2$

$T_1 + V_1 = T_2 + V_2$

$0 + [-196.2 \text{ J} + 482.22 \text{ J}]$

$\quad = 13.3333\omega_2^2 + [0 + 12.5 \text{ J}]$

$\omega_2 = 4.53 \text{ rad/s}$ *Ans.*

Chapter 19

F19–1. $\zeta + I_O\omega_1 + \Sigma \int_{t_1}^{t_2} M_O\, dt = I_O\omega_2$

$0 + \int_0^{4 \text{ s}} 3t^2\, dt = \left[60(0.3)^2\right]\omega_2$

$\omega_2 = 11.85 \text{ rad/s} = 11.9 \text{ rad/s}$ *Ans.*

F19–2. $\zeta + (H_A)_1 + \Sigma \int_{t_1}^{t_2} M_A\, dt = (H_A)_2$

$0 + 300(6) = 300(0.4^2)\omega_2 + 300[\omega(0.6)](0.6)$

$\omega_2 = 11.54 \text{ rad/s} = 11.5 \text{ rad/s}$ *Ans.*

$\overset{+}{\rightarrow} \quad m(v_1)_x + \Sigma \int_{t_1}^{t_2} F_x\, dt = m(v_2)_x$

$0 + F_f(6) = 300[11.54(0.6)]$

$F_f = 346 \text{ N}$ *Ans.*

F19–3. $v_A = \omega_A r_{A/IC} = \omega_A(0.15)$

$\zeta + \Sigma M_O = 0; \quad 9 - A_t(0.45) = 0 \quad A_t = 20 \text{ N}$

$\zeta + (H_C)_1 + \Sigma \int_{t_1}^{t_2} M_C dt = (H_C)_2$

$0 + [20(5)](0.15)$

$\quad = 10[\omega_A(0.15)](0.15)$

$\quad + \left[10(0.1^2)\right]\omega_A$

$\omega_A = 46.2 \text{ rad/s} \hspace{3cm} Ans.$

F19–4. $I_A = mk_A^2 = 10(0.08^2) = 0.064 \text{ kg} \cdot \text{m}^2$

$I_B = mk_B^2 = 50(0.15^2) = 1.125 \text{ kg} \cdot \text{m}^2$

$\omega_A = \left(\dfrac{r_B}{r_A}\right)\omega_B = \left(\dfrac{0.2}{0.1}\right)\omega_B = 2\omega_B$

$\zeta + I_A(\omega_A)_1 + \Sigma \int_{t_1}^{t_2} M_A \, dt = I_A(\omega_A)_2$

$0 + 10(5) - \int_0^{5s} F(0.1)dt = 0.064[2(\omega_B)_2]$

$\int_0^{5\,s} F dt = 500 - 1.28(\omega_B)_2 \hspace{2cm} (1)$

$\zeta + I_B(\omega_B)_1 + \Sigma \int_{t_1}^{t_2} M_B \, dt = I_B(\omega_B)_2$

$0 + \int_0^{5s} F(0.2)dt = 1.125(\omega_B)_2$

$\int_0^{5\,s} F dt = 5.625(\omega_B)_2 \hspace{2.3cm} (2)$

Equating Eqs. (1) and (2),

$500 - 1.28(\omega_B)_2 = 5.625(\omega_B)_2$

$(\omega_B)_2 = 72.41 \text{ rad/s} = 72.4 \text{ rad/s} \hspace{1cm} Ans.$

F19–5. $(\overset{+}{\rightarrow}) \quad m[(v_O)_x]_1 + \Sigma \int F_x \, dt = m[(v_O)_x]_2$

$0 + (150 \text{ N})(3 \text{ s}) + F_A(3 \text{ s})$

$\quad = (50 \text{ kg})(0.3\omega_2)$

$\zeta + I_G\omega_1 + \Sigma \int M_G \, dt = I_G\omega_2$

$0 + (150 \text{ N})(0.2 \text{ m})(3 \text{ s}) - F_A(0.3 \text{ m})(3 \text{ s})$

$\quad = [(50 \text{ kg})(0.175 \text{ m})^2]\omega_2$

$\omega_2 = 37.3 \text{ rad/s} \hspace{2.5cm} Ans.$

$F_A = 36.53 \text{ N}$

Also,

$I_{IC}\omega_1 + \Sigma \int M_{IC} \, dt = I_{IC}\omega_2$

$0 + [(150 \text{ N})(0.2 + 0.3) \text{ m}](3 \text{ s})$

$\quad = [(50 \text{ kg})(0.175 \text{ m})^2 + (50 \text{ kg})(0.3 \text{ m})^2]\omega_2$

$\omega_2 = 37.3 \text{ rad/s} \hspace{2.5cm} Ans.$

F19–6. $(+\uparrow) \, m[(v_G)_1]_y + \Sigma \int F_y \, dt = m[(v_G)_2]_y$

$0 + N_A(3 \text{ s}) - (150 \text{ lb})(3 \text{ s}) = 0$

$N_A = 150 \text{ lb}$

$\zeta + (H_{IC})_1 + \Sigma \int M_{IC} \, dt = (H_{IC})_2$

$0 + (25 \text{ lb} \cdot \text{ft})(3 \text{ s}) - [0.15(150 \text{ lb})(3 \text{ s})](0.5 \text{ ft})$

$= \left[\tfrac{150}{32.2} \text{ slug}(1.25 \text{ ft})^2\right]\omega_2 + \left(\tfrac{150}{32.2} \text{ slug}\right)\left[\omega_2(1 \text{ ft})\right](1 \text{ ft})$

$\omega_2 = 3.46 \text{ rad/s} \hspace{2.5cm} Ans.$

Answers to Selected Problems

Chapter 12

12–1. $v^2 = v_0^2 + 2a_c(s - s_0)$
$a_c = 0.5625 \text{ m/s}^2$
$v = v_0 + a_c t$
$t = 26.7 \text{ s}$

12–2. $v = 0 + 1(30) = 30 \text{ m/s}$
$s = 450 \text{ m}$

12–3. $t = 3 \text{ s}$
$s = 22.5 \text{ ft}$

12–5. $dv = a \, dt$
$v = \left(6t^2 - 2t^{3/2}\right) \text{ ft/s}$
$ds = v \, dt$
$s = \left(2t^3 - \frac{4}{5}t^{5/2} + 15\right) \text{ ft}$

12–6. $h = 127 \text{ ft}$
$v = -90.6 \text{ ft/s} = 90.6 \text{ ft/s} \downarrow$

12–7. $v = 13 \text{ m/s}$
$\Delta s = 76 \text{ m}$
$t = 8.33 \text{ s}$

12–9. $dt = \dfrac{dv}{a}$
$v = \sqrt{2kt + v_0^2}$

12–10. $s_A = 3200 \text{ ft}$

12–11. $a = -24 \text{ m/s}^2$
$\Delta s = -880 \text{ m}$
$s_T = 912 \text{ m}$

12–13. $\Delta s = 2 \text{ m}$
$s_T = 6 \text{ m}$
$v_{\text{avg}} = 0.333 \text{ m/s}$
$(v_{sp})_{\text{avg}} = 1 \text{ m/s}$

12–14. $v_{\text{avg}} = 0.222 \text{ m/s}$
$(v_{sp})_{\text{avg}} = 2.22 \text{ m/s}$

12–15. $d = 517 \text{ ft}$
$d = 616 \text{ ft}$

12–17. $h = 5t' - 4.905(t')^2 + 10$
$h = 19.81t' - 4.905(t')^2 - 14.905$
$t' = 1.682 \text{ m}$
$h = 4.54 \text{ m}$

12–18. $s = 1708 \text{ m}$
$v_{\text{avg}} = 22.3 \text{ m/s}$

12–19. $a|_{t=4} = 1.06 \text{ m/s}^2$

12–21. $v_A = (3t^2 - 3t) \text{ ft/s}$
$v_B = (4t^3 - 8t) \text{ ft/s}$
$t = 0 \text{ s and} = 1 \text{ s}$
B stops

$t = 0 \text{ s}$
$t = \sqrt{2} \text{ s}$
$s_{AB}|_{t=4\,\text{s}} = 152 \text{ ft}$
$(s_T)_A = 41 \text{ ft}$
$(s_T)_B = 200 \text{ ft}$

12–22. Choose the root greater than 10 m $s = 11.9 \text{ m}$
$v = 0.250 \text{ m/s}$

12–23. $v = \left(20e^{-2t}\right) \text{ m/s}$
$a = \left(-40e^{-2t}\right) \text{ m/s}^2$
$s = 10\left(1 - e^{-2t}\right) \text{ m}$

12–25. $s = \dfrac{1}{2k} \ln\left(\dfrac{g + kv_0^2}{g + kv^2}\right)$

$h_{\max} = \dfrac{1}{2k} \ln\left(1 + \frac{k}{g}v_0^2\right)$

12–26. $v = 4.11 \text{ m/s}$
$a = 4.13 \text{ m/s}^2$

12–27. $v = 1.29 \text{ m/s}$

12–29. $s|_{t=6\,\text{s}} = -27.0 \text{ ft}$
$v = 4.50t^2 - 27.0t + 22.5$
The times when the particle stops are
$t = 1 \text{ s and } t = 5 \text{ s.}$
$s_{\text{tot}} = 69.0 \text{ ft}$

12–30. $s = \dfrac{v_0}{k}\left(1 - e^{-kt}\right)$
$a = -kv_0 e^{-kt}$

12–31. $t = \dfrac{v_f}{2g} \ln\left(\dfrac{v_f + v}{v_f - v}\right)$

12–33. Distance between motorcycle and car 5541.67 ft
$t = 77.6 \text{ s}$
$s_m = 3.67(10)^3 \text{ ft}$

12–34. $a = 80 \text{ km/s}^2$
$t = 6.93 \text{ ms}$

12–35. $v_{\text{avg}} = 10 \text{ m/s} \leftarrow$
$a_{\text{avg}} = 6 \text{ m/s}^2 \leftarrow$

12–37. ball A
$h = v_0 t' - \dfrac{g}{2}t'^2$
$v_A = v_0 - gt'$
$h = v_0(t' - t) - \dfrac{g}{2}(t' - t)^2$
$v_B = v_0 - g(t' - t)$
$t' = \dfrac{2v_0 + gt}{2g}$

699

$v_A = \frac{1}{2}gt \downarrow$

$v_B = \frac{1}{2}gt \uparrow$

12–38. $v = 11.2$ km/s

12–39. $v = 3.02$ km/s $\downarrow$

12–41. $v = -30t + 15t^2$ m/s

At rest at $t = 0$ and $t = 2$ s

$s_{tot} = 30$ m

$v_{avg} = 15$ m/s

12–42. $s_T = 980$ m

12–45. $v = \frac{2\pi}{5}\cos\frac{\pi}{5}t$

$a = -\frac{2\pi^2}{25}\sin\frac{\pi}{5}t$

12–46. $v_{max} = 16.7$ m/s

12–49. $v = 3t^2 - 6t + 2$

$a = 6t - 6$

12–51. $s|_{t=90\,s} = 1350$ m

12–53. $s = \left(\frac{1}{5}t^2\right)$ m and $s = (12t - 180)$ m

$a = 0.4$ m/s^2 and $a = 0$

12–54. $t = 9.88$ s

12–55. $t' = 8.75$ s

$s|_{t=8.75\,s} = 272$ m

12–57. $v = \left(\sqrt{0.1s^2 + 10s}\right)$ m/s and

$v = \left(\sqrt{-30s + 12\,000}\right)$ m/s

$s' = 400$ m

12–58. $s' = 2500$ ft

12–59. $s = 917$ m

12–61. $s = 2t^2$

$s = 20t - 50$

$s = -t^2 + 60t - 450$

12–62. $v_{max} = 36.7$ m/s

$s' = 319$ m

12–63. $v = 4t^{3/2}$ and $v = 2t^2 - 18t + 108$

$s = \frac{8}{5}t^{5/2}$ and $s = \frac{2}{3}t^3 - 9t^2 + 108t - 340$

12–65. $v = \sqrt{0.04s^2 + 4s}$ ft/s

$v = \sqrt{20s - 1600}$ ft/s

12–66. $t = 16.9$ s

$v = 0.8t, v = 24.0$

$a = 0.8, a = 0$

12–69. $v = \left(0.4t^2\right)$ m/s

$v = (8t - 40)$ m/s

$t' = 16.25$ s

$s|_{t=16.25\,s} = 540$ m

12–70. $t' = 133$ s, $s = 8857$ m

12–71. $v = 36.1$ m/s

$a = 36.5$ m/s^2

12–73. $x = \sqrt{\dfrac{c}{3b}}t^{3/2}$

$a_x = \dfrac{3}{4}\sqrt{\dfrac{c}{3b}}\dfrac{1}{\sqrt{t}}$

$a_y = 2ct$

12–74. $a = 80.2$ m/s^2

$(42.7, 16.0, 14.0)$ m

12–75. $a_x = \pm 4r\cos 2t$

$a_y = -4r\sin 2t$

12–77. $\mathbf{v} = \{-10\sin 2t\mathbf{i} + 8\cos 2t\mathbf{j}\}$ m/s

$\mathbf{a} = \{-20\cos 2t\mathbf{i} - 16\sin 2t\mathbf{j}\}$ m/s^2

$v = 9.68$ m/s

$a = 16.8$ m/s^2

12–78. $v = 10.4$ m/s

$a = 38.5$ m/s^2

12–79. $v_x = 3.58$ m/s, $v_y = 1.79$ m/s

$a_x = 0.32$ m/s^2 $a_y = 0.64$ m/s^2 $\downarrow$

12–81. $\mathbf{r}_B = \{21.21\mathbf{i} - 21.21\mathbf{j}\}$ m

$\mathbf{r}_C = \{28.98\mathbf{i} - 7.765\mathbf{j}\}$ m

$(\mathbf{v}_{BC})_{avg} = \{3.88\mathbf{i} + 6.72\mathbf{j}\}$ m/s

12–82. $s = 9$ km

$\Delta r = 3.61$ km

$v_{avg} = 2.61$ m/s

$(v_{sp})_{avg} = 6.52$ m/s

12–83. $v = \sqrt{c^2 k^2 + b^2}$

$a = ck^2$

12–85. $v_y = v_x - \dfrac{x}{200}v_x$

$v = 2.69$ ft/s

$a_y = a_x - \dfrac{1}{200}\left(v_x^2 + xa_x\right)$

$a = 0.0200$ ft/s^2

12–86. $v_x = v_0\left[1 + \left(\frac{\pi}{L}c\right)^2\cos^2\left(\frac{\pi}{L}x\right)\right]^{-\frac{1}{2}}$

$v_y = \dfrac{v_0\,\pi c}{L}\left(\cos\frac{\pi}{L}x\right)\left[1 + \left(\frac{\pi}{L}c\right)^2\cos^2\left(\frac{\pi}{L}x\right)\right]^{-\frac{1}{2}}$

12–87. $v_A = 6.49$ m/s

$t = 0.890$ s

12–89. $v_A\cos\theta = 20$

$v_A\sin\theta = 23.3$

$\theta = 49.4°$

$v_A = 30.7$ ft/s

$v_B = 76.0$ ft/s

$\theta = 57.6°$

$x = 222$ m

$y = 116$ m

12–91. $s = 8.68$ ft

$s = 34.4$ ft

12–93. $t = \dfrac{1}{\cos \theta_A}$

$4.905t^2 - 30 \sin \theta_A t - 1.2 = 0$

Solve by trial and error.

$\theta_A = 7.19°$ and $80.5°$

12–94. $\theta_A = 30.5°$ $v_A = 23.2$ m/s

12–95. $h = 14.7$ ft

12–97. $y = 0 + v_0 \sin \theta_1 t_1 + \frac{1}{2}(-g)t_1^2$

$y = 0 + v_0 \sin \theta_2 t_2 + \frac{1}{2}(-g)t_2^2$

$x = 0 + v_0 \cos \theta_1 t_1$

$x = 0 + v_0 \cos \theta_2 t_2$

$\Delta t = \dfrac{2v_0 \sin(\theta_1 - \theta_2)}{g(\cos \theta_2 + \cos \theta_1)}$

12–98. $d = 94.1$ m

12–99. $v_A = 76.7$ ft/s

$s = 22.9$ ft

12–101. $20 = 0 + v_A \cos 30° \, t$

$10 = 1.8 + v_A \sin 30°(t) + \frac{1}{2}(-9.81)(t)^2$

$v_A = 28.0$ m/s

12–102. $d = 166$ ft

12–103. Since $H > 15$ ft, the football is kicked **over**

the goalpost.

$h = 22.0$ ft

12–105. $0 = 15 \sin \theta_A + (-9.81)t$

$8 = 1 + 15 \sin \theta_A t + \frac{1}{2}(-9.81)t^2$

$\theta_A = 51.4°$

$d = 7.18$ m

12–106. $v_A = 18.2$ m/s

$t = 1.195$ s

$d = 12.7$ m

12–107. $\theta_1 = 25.0°$ ↰

$\theta_2 = 85.2°$ ↲

12–109. $3 = 7.5 + 0 + \frac{1}{2}(-32.2)t_1^2$

$0 = 7.5 + 0 + \frac{1}{2}(-32.2)t_2^2$

$21 = 0 + v_A (0.5287)$

$v_A = 39.7$ ft/s

$s_x = (s_0)_x + (v_0)_x \, t$

$s = 6.11$ ft

12–110. $v_A = 19.4$ m/s

$t_{AB} = 4.54$ s

12–111. $\rho = 208$ m

12–113. $7.5 = \dfrac{v^2}{200}$

$v = 38.7$ m/s

12–114. $v = 63.2$ ft/s

12–115. $a = 0.488$ m/s²

12–117. $t = 7.071$ s

$v = 5.66$ m/s

$a_t = \dot{v} = 0.8$ m/s²

$a_n = 0.640$ m/s²

$a = 1.02$ m/s²

12–118. $v = 1.80$ m/s

$a = 1.20$ m/s²

12–119. $a = 15.1$ ft/s²

$\Delta s = 14$ ft

12–121. $\rho = 3808.96$ m

$a = 0.511$ m/s²

12–122. $a = 0.309$ m/s²

12–123. $a = 2.75$ m/s²

12–125. $v = \left(25 - \frac{1}{6}t^{3/2}\right)$ m/s

When the car reaches C

$t = 15.942$ s

$a = 1.30$ m/s²

12–126. $a = 0.730$ m/s²

12–127. $a = 7.85$ ft/s²

12–129. $\rho = 79.30$ m

$a = 8.43$ m/s²

$\theta = 38.2°$

12–130. $a = 6.03$ m/s²

12–131. $a = 0.824$ m/s²

12–133. $v = \left(\sqrt{400 - 0.25s^2}\right)$ m/s

$t = 2 \sin^{-1}\left(\frac{s}{40}\right)$

When $t = 2$ s,

$s = 33.7$ m

$a_t = -8.42$ m/s²

$a_n = 5.84$ m/s²

$a = 10.2$ m/s²

12–134. $a_A = 4.44$ m/s²

12–135. $a_B = 0.556$ m/s²

12–137. $\mathbf{v} = \{3t^2\mathbf{i} + 6t\mathbf{j} + 8\mathbf{k}\}$ m/s

$v = 18.8$ m/s

$\mathbf{a} = \{6t\mathbf{i} + 6\mathbf{j}\}$ m/s²

$a = 13.4$ m/s²

$\rho = 51.1$ m

12–138. $v = 3.68$ m/s

$a = 4.98$ m/s²

12–139. $v = 3.19$ m/s

$a = 4.22$ m/s²

12–141. $dv = a \, dt, v = 7.20$ m/s

$a_n = 1.037$ m/s², $a = 1.91$ m/s²

12–142. $d = 106$ ft
$\quad a_A = 9.88$ ft/s^2
$\quad a_B = 1.28$ ft/s^2

12–143. $a = 3.05$ m/s^2

12–145. $\rho = 449.4$ m, $a_n = a = 26.9$ m/s^2

12–146. $a = 0.897$ ft/s^2

12–147. $a = 8.61$ m/s^2

12–149. $v_A = 0.6325\sqrt{s_A^2 + 160}$
$\quad s_A = 8.544$ m
$\quad d = 17.0$ m
$\quad (a_n)_A = 18.64$ m/s^2
$\quad (a_n)_B = 12.80$ m/s^2
$\quad a_A = 19.0$ m/s^2
$\quad a_B = 12.8$ m/s^2

12–150. $t = 2.51$ s
$\quad a_A = 22.2$ m/s^2
$\quad a_B = 65.1$ m/s^2

12–151. $t = 10.1$ s
$\quad v = 47.6$ m/s
$\quad a = 11.8$ m/s^2

12–153. $x = 0 + 6.128t$
$\quad y = 0 + 5.143t + \frac{1}{2}(-9.81)(t^2)$
$\quad y = \{0.839x - 0.131x^2\}$ m
$\quad a_t = 3.94$ m/s^2
$\quad a_n = 8.98$ m/s^2

12–154. $v_n = 0$
$\quad v_t = 7.21$ m/s
$\quad a_n = 0.555$ m/s^2
$\quad a_t = 2.77$ m/s^2

12–155. $a_{max} = \dfrac{a}{b^2}v^2$

12–157. $\theta = \left(t^3\right)$ rad
$\quad \dot{r} = \ddot{r} = 0$
$\quad \dot{\theta} = 2.554$ rad/s $\quad \ddot{\theta} = 5.536$ rad/s^2
$\quad v = 0.766$ m/s
$\quad a = 2.57$ m/s^2

12–158. $a = 3.66$ ft/s^2

12–159. $v = 30.1$ m/s
$\quad a = 85.3$ m/s^2

12–161. $v_{Pl} = 293.3$ ft/s
$\quad a_{Pl} = 0.001\,22$ ft/s^2
$\quad v = 464$ ft/s
$\quad a_{pr} = 43\,200$ ft/s^2
$\quad a = 43.2(10^3)$ ft/s^2

12–162. $a = 14.3$ in./s^2

12–163. $v_r = a\sin\theta\,\dot{\theta}$
$\quad v_\theta = (b - a\cos\theta)\dot{\theta}$

$\quad a_r = (2a\cos\theta - b)\,\dot{\theta}^2 + a\sin\theta\,\ddot{\theta}$
$\quad a_\theta = (b - a\cos\theta)\ddot{\theta} + 2a\dot{\theta}^2\sin\theta$

12–165. $v_r = 0 \quad v_\theta = 120$ ft/s
$\quad v = 120$ ft/s
$\quad a_r = -48.0$ ft/s^2
$\quad a_\theta = 60.0$ ft/s^2
$\quad a = 76.8$ ft/s^2

12–166. $v = 2a\dot{\theta}$
$\quad a = 4a\dot{\theta}^2$

12–167. $v = 2a\dot{\theta}$
$\quad a = 2a\sqrt{4\dot{\theta}^4 + \ddot{\theta}^2}$

12–169. $v_r = 0 \quad v_\theta = 400\left(\dot{\theta}\right)$
$\quad \theta = 0.075$ rad/s
$\quad a_r = -2.25$ ft/s^2
$\quad a_\theta = 0$
$\quad a = 2.25$ ft/s^2

12–170. $v_r = 1.50$ m/s
$\quad v_\theta = 0.450$ m/s
$\quad a_r = 0.410$ m/s^2
$\quad a_\theta = 0.600$ m/s^2

12–171. $\mathbf{v} = \{-116\mathbf{u}_r - 163\mathbf{u}_z\}$ mm/s
$\quad \mathbf{a} = \{-5.81\mathbf{u}_r - 8.14\mathbf{u}_z\}$ mm/s^2

12–173. $v_r = 2.149$ m/s $\quad v_\theta = 3.722$ m/s
$\quad v = 4.30$ m/s
$\quad a_r = -23.20$ m/s^2
$\quad a_\theta = 11.39$ m/s^2
$\quad a = 25.8$ m/s^2

12–174. $v_r = 0$
$\quad v_\theta = 0.8$ m/s
$\quad v_z = -0.0932$ m/s
$\quad a_r = -0.16$ m/s^2
$\quad a_\theta = 0$
$\quad a_z = -0.00725$ m/s^2

12–175. $v = 8.49$ m/s
$\quad a = 88.2$ m/s^2

12–177. $\dot{r} = (-200\sin 2\theta\,\dot{\theta})$ m/s
$\quad \dot{\theta} = 0.302$ rad/s

12–178. $\dot{\theta} = 0.378$ rad/s

12–179. $\dot{r} = -250$ mm/s
$\quad \ddot{r} = -2165$ mm/s^2

12–181. $v_r = 0$
$\quad v_\theta = 1.473$ m/s
$\quad v_z = -0.2814$ m/s
$\quad a_r = -0.217$
$\quad a_\theta = 0$
$\quad a_z = 0$
$\quad a = 0.217$ m/s^2

12–182. $a = 7.26 \text{ m/s}^2$

12–183. $v = 4.16 \text{ m/s}$
$a = 33.1 \text{ m/s}^2$

12–185. $v_r = 5.405 \text{ m/s}$
$v_\theta = 5.660 \text{ m/s}$
$v = 7.83 \text{ m/s}$
$a_r = -5.998 \text{ m/s}^2$
$a_\theta = 38.95 \text{ m/s}^2$
$a = 39.4 \text{ m/s}^2$

12–186. $v_r = a\dot{\theta}$
$v_\theta = a\theta\dot{\theta}$
$a_r = -a\theta\dot{\theta}^2$
$a_\theta = 2a\dot{\theta}^2$

12–187. $v_r = 6.00 \text{ ft/s}$
$v_\theta = 18.3 \text{ ft/s}$
$a_r = -67.1 \text{ ft/s}^2$
$a_\theta = 66.3 \text{ ft/s}^2$

12–189. $v_r = 8.2122 \text{ mm/s}$
$v_\theta = 164.24 \text{ mm/s}$
$v = 164 \text{ mm/s}$
$a_r = -651.2 \text{ mm/s}^2$
$a_\theta = 147.82 \text{ mm/s}^2$
$a = 668 \text{ mm/s}^2$

12–190. $v_r = 32.0 \text{ ft/s}$
$v_\theta = 50.3 \text{ ft/s}$
$a_r = -201 \text{ ft/s}^2$
$a_\theta = 256 \text{ ft/s}^2$

12–191. $v_r = 32.0 \text{ ft/s}$
$v_\theta = 50.3 \text{ ft/s}$
$a_r = -161 \text{ ft/s}^2$
$a_\theta = 319 \text{ ft/s}^2$

12–193. $v_r = -\dfrac{1800}{\pi^2}\dot{\theta}$ $v_\theta = \dfrac{600}{\pi}\dot{\theta}$
$v_r = -24.2 \text{ ft/s}$
$v_\theta = 25.3 \text{ ft/s}$

12–194. $v_r = -306 \text{ m/s}$
$v_\theta = 177 \text{ m/s}$
$a_r = -128 \text{ m/s}^2$
$a_\theta = 67.7 \text{ m/s}^2$

12–195. $v_P = 6 \text{ m/s} \nearrow$

12–197. $2v_H = -v_A$
$v_A = 4 \text{ ft/s} \leftarrow$

12–198. $v_B = 20 \text{ m/s} \uparrow$

12–199. $v_E = 2.14 \text{ m/s} \uparrow$

12–201. $3v_A + v_M = 0$
$v_A = 1.67 \text{ m/s} \uparrow$

12–202. $v_B = 0.5 \text{ m/s} \uparrow$

12–203. $v_B = 1 \text{ m/s} \uparrow$

12–205. $2v_A = v_C$
$v_B = 4v_A$
$v_A = 1 \text{ ft/s} \uparrow$
$a_A = 0.5 \text{ ft/s}^2 \downarrow$

12–206. $v_B = 1 \text{ ft/s} \uparrow$

12–207. $v_B = 12 \text{ ft/s} \uparrow$

12–209. $v_A = -2v_D$
$2v_C - v_D + v_B = 0$
$t = 5.43 \text{ s}$
$v_C = 2.21 \text{ m/s} \uparrow$

12–210. $t = 1.07 \text{ s}$
$v_A = 0.605 \text{ m/s}$
$v_B = 5.33 \text{ m/s}$

12–211. $\dot{s}_B = 1.20 \text{ ft/s} \downarrow$
$\ddot{s}_B = 1.11 \text{ ft/s}^2 \uparrow$

12–213. $y_B = 16 - \sqrt{x_A^2 + 64}$
$v_B = -\dfrac{x_A}{\sqrt{x_A^2 + 64}}\, v_A$
$v_B = 1.41 \text{ m/s} \uparrow$

12–214. $v_C = (6 \sec \theta) \text{ ft/s} \rightarrow$

12–215. $v_{B/A} = 11.2 \text{ m/s}$
$\theta_v = 50.3°$

12–217. $v_{BC} = 18.6 \text{ m/s}$
$\theta_v = 66.2°$
$\mathbf{v}_{B/C} = \{7.5\mathbf{i} + 17.01\mathbf{j}\} \text{ m/s}$
$(\mathbf{a}_B)_t\{-2\cos 60°\mathbf{i} + 2\sin 60°\mathbf{j}\} \text{ m/s}^2$
$\mathbf{a}_{B/C} = [0.9486\mathbf{i} - 0.1429\mathbf{j}]$
$a_{B/C} = 0.959 \text{ m/s}^2$
$\theta_a = 8.57°$

12–218. $v_{w/s} = 19.9 \text{ m/s}$
$\theta = 74.0°$

12–219. $\theta = 9.58°$
$v_{r/c} = 19.9 \text{ m/s}$
$\theta = 9.58°$

12–221. $-20\sin 30° = -30 + (v_{B/A})_x$
$20\cos 30° = (v_{B/A})_y$
$v_{B/A} = 26.5 \text{ mi/h}$
$\theta = 40.9° \measuredangle$
$-1200\sin 30° + 1333.3\cos 30° = (a_{B/A})_x$
$1200\cos 30° + 1333.3\sin 30° = (a_{B/A})_y$
$a_{B/A} = 1.79(10^3) \text{ mi/h}^2$
$\theta = 72.0° \measuredangle$

12–222. $v_{B/A} = 26.5 \text{ mi/h}$
$\theta_v = 40.90° \measuredangle$
$a_{B/A} = 1955 \text{ mi/h}^2$
$\theta_a = 0.767° \searrow$

12–223. $v_{A/B} = 21.7$ ft/s
$\theta = 18.0°$ ⤷
$t = 36.9$ s

12–225. $\mathbf{a}_{B/A} = \{2392.95\mathbf{i} - 3798.15\mathbf{j}\}$ mi/h^2
$a_{B/A} = 4489$ mi/h^2
$\phi = 57.8°$ ⤵

12–226. $v_{A/B} = 49.1$ km/h
$\theta = 67.2°$ ⤷

12–227. $v_W = 58.3$ km/h
$\theta = 59.0°$ ⤶

12–229. $\mathbf{v}_{A/B} = (v\sin\theta - v)\mathbf{i} + v\cos\theta\mathbf{j}$
$v_{A/B} = v\sqrt{2(1 - \sin\theta)}$

12–230. $v_{r/m} = 16.6$ km/h,
$\theta = 25.0°$ ⤵

12–231. $v_b = 6.21$ m/s
$t = 11.4$ s

Chapter 13

13–1. $a = 0.6667$ m/s^2
$F_{AB} = F_{AC} = 18.1$ kN

13–2. $a = -0.505$ m/s^2

13–3. $v = 22.4$ m/s

13–5. $40.55 - F = 10a$
$F + 14.14 = 6a$
$a = 3.42$ m/s^2
$F = 6.37$ N

13–6. $a_C = 2.5$ ft/s^2 ↑
$T = 162$ lb

13–7. $F = 85.7$ N

13–9. $F = 7.50$ kN
$a = 0.0278$ m/s^2

13–10. $a = 1.66$ m/s^2

13–11. $a = 1.75$ m/s^2

13–13. $a = 3.61$ ft/s^2
$T = 5.98$ kip

13–14. $T_{CA} = T_{CB} = 27.9$ kN

13–15. $s = 12.9$ m

13–17. $a_A = 32.2$ ft/s^2
$s = 64.4$ ft

13–18. $F = 13.1$ lb

13–19. (a) $x = 0$
(b) $x = 0.955$ m

13–21. $N_B = mg\cos\theta$
$-T + N_B\sin\theta = 0$
$T = \left(\dfrac{mg}{2}\right)\sin 2\theta$

13–22. $T = mg\cos\theta(\sin\theta - \mu_k\cos\theta)$

13–23. $v = 30$ m/s

13–25. $2a_C - a_P = 0$
$T = 1131$ N
$B_y = 1.92$ kN
$A_x = 0$
$A_y = 2.11$ kN

13–26. $a_E = 0.75$ m/s^2 ↑
$T = 1.32$ kN

13–27. $m_A = 13.7$ kg

13–29. $12 - s_B + \sqrt{s_A^2 + (12)^2} = 24$
$T = 1.63$ kN

13–30. $T = 1.80$ kN

13–31. $a_A = 0.195$ m/s^2 ↓
$T = 769$ N

13–33. $F_s = 4\left(\sqrt{1 + s^2} - 1\right)$
$v = 14.6$ ft/s

13–34. $d = \dfrac{eVLl}{v_0^2 wm}$

13–35. (a), (b) $a_C = 6.94$ m/s^2
(c) $a_C = 7.08$ m/s^2
$\theta = 56.5°$ ⤵

13–37. $N\cos\theta - \mu_s N\sin\theta - mg = 0$
$N\sin\theta + \mu_s N\cos\theta = ma$
$P = 2mg\left(\dfrac{\sin\theta + \mu_s\cos\theta}{\cos\theta - \mu_s\sin\theta}\right)$

13–38. $t = 1.08$ s

13–39. $t = 42.1$ min
$v_{max} = 2.49$ km/s

13–41. $a_B = 5.68$ ft/s^2
$a_A = 21.22$ ft/s^2 $N_B = 18.27$ lb

13–42. $x = d$
$v = \sqrt{\dfrac{kd^2}{(m_A + m_B)}}$

13–43. $N = 0$, then $x = d$ for separation.

13–45. $a = (2.19 - 0.2v)$ m/s^2
$v = 10.95(1 - e^{-t/5})$
$v_{max} = 10.95$ m/s

13–46. $v_{max} = \sqrt{\dfrac{mg}{C}}$

13–47. $v = \sqrt{v_0^2 - 2gr_0\left(1 - \dfrac{r_0}{r}\right)}$
$r_{max} = \dfrac{2gr_0^2}{2gr_0 - v_0^2}$
$v_{esc} = \sqrt{2gr_0}$
$t = \dfrac{2}{3r_0\sqrt{2g}}\left(r_{max}^{\frac{3}{2}} - r_0^{\frac{3}{2}}\right)$

13–49. $a_n = \dfrac{v^2}{1.5}$

$v = 10.5$ m/s

13–50. $a_t = -4.905$ m/s^2

$\rho = 188$ m

13–51. $v = 41.2$ m/s

13–53. $N = 19\,140.6$ N

$v_{max} = 24.4$ m/s

13–54. $v_{min} = 12.2$ m/s

13–55. $v = 9.90$ m/s

13–57. $a_n = 0$

$T_{CD} = mg \sin \theta$

13–58. $T = 2\pi \sqrt{\dfrac{r^3}{GM_e}}$

13–59. $\theta = 78.1°$

13–61. $T = 414$ N

$a_t = -9.81 \sin \theta$

$a_t \, ds = v \, dv$

$\theta = 37.2°$

13–62. $a_t = 3.36$ m/s^2 $\swarrow$

$T = 361$ N

13–63. $\theta = 26.7°$

13–65. $\rho = 0.120$ m

$T = 1.82$ N

$N_B = 0.844$ N

13–66. $\mu_s = 0.252$

13–67. $v = 22.1$ m/s

13–69. $v = \sqrt{gr}$

$a_n = g$

$N = 2mg$

13–70. $\theta = 17.8°$ $T = 51.5$ kN

13–71. $L = 50.8$ kN

$r = 3.60$ km

13–73. $\theta = -26.57°$

$\rho = 223.61$ m

$F_f = 1.11$ kN

$N = 6.73$ kN

13–74. $N = 11.2$ N

$a_t = 6.35$ m/s^2

13–77. $\theta = 0°$

$\rho = 10.0$ m

$a_t = -9.81 \sin \theta$

$v^2 = 98.1$ m^2/s^2

$N = 1.02$ kN

13–78. $\theta = 112°$

13–79. $v_{min} = 25.4$ ft/s

$v_B = 12.8$ ft/s

13–81. $\rho_A = 354.05$ m

$v = 22.22$ m/s, $a_n = 1.395$ m/s^2

$N = 19.3$ kN

13–82. $v = 31.3$ m/s

$N_B = 840$ N

13–83. $N = 33.8$ lb, $a = 59.8$ ft/s^2

13–85. $a_r = 0$

$a_\theta = 42$ m/s^2

$F = 210$ N

13–86. $F_r = -2$ N

$F_\theta = 16$ N

13–87. $F_r = -2$ N

$F_\theta = 36$ N

$F_z = 11.6$ N

13–89. $a_r = -2.4$ m/s^2

$a_\theta = 1.2$ m/s^2

$N_B = 1.20$ N

$F_{AB} = 0.6$ N

13–90. $F_z = 18.6$ N

13–91. $(F_z)_{min} = 18.6$ N

$(F_z)_{max} = 20.6$ N

13–93. $N = 17.34m$

$a_r = -14.715$ m/s^2

$\dot{\theta} = 7.00$ rad/s

13–94. $\dot{\theta} = 5.70$ rad/s

13–95. $r = 816$ mm

13–97. $a_r = -7.128$ ft/s^2

$a_\theta = -3.6536$ ft/s^2

$F = 0.0108$ lb

$N = 0.1917$ N

13–98. $F = 0.444$ lb

13–99. $F = -0.0155$ lb

13–101. $a_r = -4.235$ ft/s^2

$a_\theta = -1.919$ ft/s^2

$N = 0.267$ lb

$F = 0.163$ lb

13–102. $F_r = -131$ N

$F_\theta = -38.4$ N

$F_z = 215$ N

13–103. $N = 2.86$ kN

13–105. $a_r = 34.641$ m/s^2

$a_\theta = 20$ m/s^2

$F = 7.67$ N

$N = 12.1$ N

13–106. $F = 7.82$ N

13–107. $F_{OA} = 12.7$ N

13–109. $\psi = 84.3°$

$a_t = 12$ m/s^2

13–110. $v_r = 2.50$ m/s
$v_\theta = 2$ m/s
13–111. $N = 113$ lb
13–113. $a_r = -4r_c \cos\theta\, \dot\theta_0^2$

$$\theta = \tan^{-1}\left(\frac{4r_c\dot\theta_0^2}{g}\right)$$

13–114. $N = 9.66$ N
$F = 19.3$ N
13–115. $N = 10.4$ N
$F_{OA} = 20.9$ N
13–117. $r_0 = 11.1$ Mm
$v_{A'} = 1964.19$ m/s
$\Delta v_A = 814$ m/s
13–118. $v_p = 7.76$ km/s
$v_A = 4.52$ km/s
$T = 3.35$ hr
13–119. $v_B = 7.71$ km/s
$v_A = 4.63$ km/s
13–121. $v_O = 6899.15$ m/s
$v_p = 7755.54$ m/s
$\Delta v_p = 856$ m/s
$v_A = 4.52$ km/s
13–122. $v_0 = 23.9(10^3)$ ft/s
13–123. $v_{A'} = 7.30(10^3)$ ft/s
$t = 1.69$ h
13–125. $v_0 = \sqrt{\dfrac{66.73(10^{-12})(5.976)(10^{24})}{(800 + 6378)(10^3)}}$
7.45 km/s
13–126. $v_0 = 30.8$ km/s
$\frac{1}{r} = 0.502(10^{-12})\cos\theta + 6.11(10^{-12})$
13–127. $\Delta v = \sqrt{\dfrac{GM_e}{r_0}}\left(\sqrt{2} - \sqrt{1+e}\right)$
The change in speed should occur at perigee.
13–129.

$$9(10^6) = \frac{6(10^6)}{\left(\dfrac{2(66.73)(10^{-12})(0.7)[5.976(10^{24})]}{6(10^6)\,v_p^2}\right) - 1}$$

$v_P = 7.47$ km/s
13–130. $v_a = 3.94$ km/s
$t = 46.1$ min
13–131. $v_A = 3.44$ km/s
13–133. $h = 101.575(10^9)$ m²/s
$r_P = 14.6268(10^6)$ m
$T = 119$ h
13–134. $r = 317$ Mm
$r = 640$ Mm

317 Mm $< r < 640$ Mm
$r > 640$ Mm
13–135. $v_A = 6.11$ km/s
$\Delta v_B = -2.37$ km/s
13–137. $v = \sqrt{G\dfrac{m_e}{r}} = 5.16$ km/s

Chapter 14

14–1. $N = 1307$ lb
$T = 744$ lb
$U_T = 18.0(10^3)$ ft · lb
14–2. $s = 1.05$ ft
14–3. $v = 0.365$ ft/s
14–5. $\frac{1}{2}(1.5)(4^2) + \left[-\displaystyle\int_0^{0.2\,\text{m}} 900s^2\, ds\right] = \frac{1}{2}(1.5)v^2$
$v = 3.58$ m/s
14–6. $d = 192$ m
14–7. $s = 7.59$ in
14–9. $0 + 150\cos 30°(0.2) + \left[-\frac{1}{2}(300)(0.2^2)\right]$
$+ \left[-\frac{1}{2}(200)(0.2^2)\right] = \frac{1}{2}(2)v^2$
$v = 4.00$ m/s
14–10. $s = 178$ m
14–11. $\mu_k = 0.255$
14–13. $F_A = 3$ lb
$F_B = 3.464$ lb
$N_B = 34.64$ lb
$v_A = 0.771$ ft/s
14–14. $s = 3.41$ m
14–15. $v = 3.77$ m/s
14–17. $0 + \displaystyle\int_0^{0.05\,\text{ft}} 100s^{1/3}\, ds - 20(0.05) = \frac{1}{2}\left(\frac{20}{32.2}\right)v^2$
$v = 1.11$ ft/s
14–18. $v_C = 1.37$ m/s
14–19. $h = 47.5$ m
14–21. $s = 179$ mm
14–22. $v_B = 24.0$ ft/s
$N_B = 7.18$ lb
$v_C = 16.0$ ft/s
$N_C = 1.18$ lb
14–23. $v_B = 7.22$ ft/s
$N_B = 27.1$ lb
$v_C = 17.0$ ft/s
$N_C = 133$ lb
$v_D = 18.2$ ft/s
14–25. $v_B = 30.0$ m/s
$s\cos 30° = 0 + 30.04t$

$s \sin 30° + 4 = 0 + 0 + \frac{1}{2}(9.81)t^2$

$s = 130$ m

14–26. $s = 1.35$ m

14–27. $v_B = 31.5$ ft/s

$d = 22.6$ ft

$v_C = 54.1$ ft/s

14–29. $F_s = 1284.85$ lb

$k = 642$ lb/ft

$v_2 = 18.0$ ft/s

14–30. $h_A = 22.5$ m

$h_C = 12.5$ m

14–31. $R = 2.83$ m

$v_C = 7.67$ m/s

14–33. $N = 693.67$ N

$F_f = 173.42$ N

$x = 2.57$ m

14–34. $v = 8.64$ m/s

14–35. $l_0 = 2.77$ ft

14–37. $s = 3.675$ m $N = 1.25$ kN

$v_B = 5.42$ m/s

14–38. $F = 367$ N

14–39. $v_B = 14.9$ m/s, $N = 1.25$ kN

14–41. $v^2 = gr\left(\frac{9}{4} - 2 \cos \theta\right)$

$N = mg\left(3 \cos \theta - \frac{9}{4}\right)$

$\theta = 41.4°$

14–42. $P_{\text{avg}} = 200$ kW

14–43. power input $= 4.20$ hp

14–45. $P = 5200(600)\left(\dfrac{88 \text{ ft/s}}{60 \text{ mi/h}}\right)\dfrac{1}{550} = 8.32\left(10^3\right)$ hp

14–46. $v = 63.2$ ft/s

14–47. $P_{max} = 119$ hp

14–49. $v_y = 0.2683$ m/s

$t = 7.454$ s

$P = 12.6$ kW

14–50. $P_{max} = 1.02$ hp

$t = 30.5$ s

14–51. $P_{\text{in}} = 19.5$ kW

14–53. $a_c = 0.8333$ m/s^2

$F = 3618.93$ N

$P_{\text{in}} = 113$ kW

$(P_{\text{in}})_{\text{avg}} = 56.5$ kW

14–54. $v = 22.3$ ft/s

14–55. $v = 56.5$ ft/s

14–57. $F = 1500\left(v\frac{dv}{ds}\right)$

$v = 18.7$ m/s

14–58. $P_{\text{out}} = 42.2$ kW

14–59. $v = 13.1$ m/s

14–61. $a = 7.20$ ft/s^2

$2\,v_C = v_P$

$P_{\text{in}} = 2.05$ hp

14–62. $\epsilon = 0.460$

14–63. $P = [400(10^3)t]$ W

14–65. $T = 1968.33$ N

$v_P = 18$ m/s

$P_o = 35.4$ kW

14–66. $P = 8.31t$ MW

14–67. $P = 1.12$ kW

14–69. $F = 308.68$ N

$v = 4.86$ m/s

14–70. power input $= 1.60$ kW

14–71. power input $= 2.28$ kW

14–73. $0 + 6(2) = 0 + \frac{1}{2}(5)(12)(x)^2$

$x = 7.59$ in.

14–74. $v = 1.37$ m/s

14–75. $v = 1.37$ m/s

14–77. $0 + (2)\left(\frac{1}{2}\right)(50)\left[\sqrt{(0.05)^2 + (0.240)^2} - 0.2\right]^2$

$= \frac{1}{2}(0.025)v^2$

$v = 2.86$ m/s

14–78. $h = 416$ mm

14–79. $v_2 = 106$ ft/s

14–81. $0 + \frac{1}{2}(200)(4)^2 + \frac{1}{2}(100)(6)^2 = h(3)$

$h = 133$ in.

14–82. $v_2 = 2.15$ m/s

14–83. $v_C = 2.09$ m/s

14–85. final elastic potential energy $= 103.11$ J

$v = 6.97$ m/s

14–86. $v_C = 7.58$ m/s

$T = 1.56$ kN

$T = 2.90$ kN

14–87. $h = 24.5$ m

$N_B = 0$

$N_C = 16.8$ kN

14–89. $v_B^2 = \rho_B\, g$

$v_A = \sqrt{\rho_B\, g + 2gh}$

$N_C = \dfrac{mg}{\rho_C}(\rho_B + \rho_C + 2h)$

14–90. $v = 32.3$ ft/s

14–91. $k = 8.57$ lb/ft

14–93. $\Delta s_P + 2\Delta s_A = 0$

$(v_A)_2 = 1.42$ m/s

$\Delta s_A = 617.5$ mm

14–94. $d = 1.34$ m

14–95. $v_A = 11.0$ m/s

14–97. $(V_g)_A = 110\,362.5\,\text{J}$ $(V_g)_B = 0$
$(V_e)_A = 0$ $(V_e)_B = 1500(150 - l_0)^2$
$l_0 = 141\,\text{m}$

14–98. $x = 453\,\text{mm}$

14–99. $v_B = 32.1\,\text{ft/s}$

14–101. $(V_g)_1 = \left(\frac{\pi-2}{2}\right)m_0\,r^2g$ $(V_g)_2 = 0$
$v_2 = \sqrt{\frac{2}{\pi}(\pi - 2)gr}$

14–102. $x = \frac{2}{3}r$

14–103. $v_C = \sqrt{\frac{7}{3}gr}$
$T = 6mg$

14–105. $v_A = 11\,111.1\,\text{m/s}$
$v_B = 34.8\,\text{Mm/h}$

14–106. $s_A = 1.29\,\text{ft}$

Chapter 15

15–1. $\frac{5}{32.2}(10) + (-5\sin 45°)t = 0$
$t = 0.439\,\text{s}$

15–2. $v = 16.1\,\text{m/s}$
$s = 48.4\,\text{m}$

15–3. $I = 90.0\,\text{lb}\cdot\text{s}$

15–5. $(v_B)_1 = 2\,\text{m/s}\uparrow$
$(v_A)_2 = 1.27\,\text{m/s}\uparrow$
$(v_B)_2 = 1.27\,\text{m/s}\downarrow$
$T = 43.6\,\text{N}$

15–6. $F = 19.44\,\text{kN}$
$T = 12.5\,\text{kN}$

15–7. $v_{max} = 90\,\text{m/s}$

15–9. $0 + \int_0^{10\,\text{s}} 30\left(10^6\right)\left(1 - e^{-0.1t}\right)\,dt = 0.130\left(10^9\right)v$
$v = 0.849\,\text{m/s}$

15–10. $t = 4.64\,\text{s}$

15–11. $v = 21.0\,\text{ft/s}$

15–13. $0 + 2(T\cos 30°)(0.3) - 600(0.3) = \left(\frac{600}{32.2}\right)(5)$
$T = 526\,\text{lb}$

15–14. $v = 4.50\,\text{m/s}$

15–15. $T = 520.5\,\text{N}$

15–17. $0 + 12\left(10^3\right)(3) - F(1.5) = 0 + 0$
$F = 24\,\text{kN}$
$12\left(10^3\right)(3) - T(1.5) = 0$
$T = 24\,\text{kN}$

15–18. $v_2 = \frac{2C\,t'}{\pi m}, s = \frac{C\,t'^2}{\pi m}$

15–19. $(v_x)_2 = 91.4\,\text{ft/s}\leftarrow$

15–21. $40(1.5) + 4[(30)4 + 10(6 - 4)]$
$- [10(2) + 20(4 - 2) + 40(6 - 4)] = 40v_2$
$v_2 = 12.0\,\text{m/s}(\rightarrow)$

15–22. $v = 26.4\,\text{ft/s}$

15–23. $v = 8.07\,\text{m/s}$
$\theta = 48.1°\,\searrow$

15–25. $63\,000(0) + 30(10^3)(30) = 63\,000v$
$v = 14.3\,\text{m/s}$
$33\,000(0) + F_D(30) = 33\,000(14.29)$
$F_D = 15.7\,\text{kN}$

15–26. $v = 4.14\,\text{m/s}$

15–27. $v_2 = 21.8\,\text{m/s}$

15–29. $v = 136.35\,\text{ft/s}$
$F_{avg} = 847\,\text{lb}$

15–30. $F_{avg} = 12.7\,\text{kN}$

15–31. $v_2 = 1.92\,\text{m/s}$

15–33. $v_2 = 0.5\,\text{m/s}$
$T_1 = 20.25\,\text{kJ}$
$T_2 = 3.375\,\text{kJ}$
$\Delta T = 16.9\,\text{kJ}$

15–34. $v = 0.6\,\text{ft/s}$

15–35. $\theta = \phi = 9.52°$

15–37. $v_A = -v_B + 2$
$v_B = 1.33\,\text{m/s}\leftarrow v_A = 0.667\,\text{m/s}\rightarrow$
$t = 2.5\,\text{s}$

15–38. $v_3 = 2.31\,\text{m/s}$
$s_{max} = 163\,\text{mm}$

15–39. $v_A = 29.8\,\text{m/s}$
$v_B = 11.9\,\text{m/s}$

15–41. $v_A = 3.09(10^3)\,\text{m/s}$
$v_B = 2.62(10^3)\,\text{m/s}$
$t_{BD} = 0.04574\,\text{s}$
$d_B = 104\,\text{m}$

15–42. $v = 0.720\,\text{m/s}\leftarrow$

15–43. $s_P = 0$
$t = 0.408\,\text{s}$

15–45. $v_C = 1.443\,\text{m/s}\leftarrow$ $(v_B)_x = 2.887\,\text{m/s}$
$s_C = 0.577\,\text{m}\leftarrow$

15–46. $(v_c)_2 = 0.800\,\text{ft/s}\leftarrow$
$v_3 = 0$

15–47. $v_r = 8.93\,\text{ft/s}$

15–49. $v_G = 0.8660\,\text{m/s}\leftarrow$
$(v_B)_y = 2.5\,\text{m/s}$
$t = 0.5097\,\text{s}$
$s = 2.207\,\text{m}\rightarrow$
$d = 2.65\,\text{m}$

15–50. $1.36\,\text{m}$

15–51. $v_M = 0.178\,\text{m/s}$
$N = 771\,\text{N}$

15–53. $\left(\frac{10}{32.2}\right)(10) + 0 = \left(\frac{10 + 20}{32.2}\right)v$

$v = 3.33$ ft/s

$t = 0.518$ s

15–54. $t = 0.518$ s

$s = 0.863$ ft

15–55. $t = 0.226$ s

15–57. $(v_A)_2 = \dfrac{v(1 - e)}{2}$ $(v_B)_2 = \dfrac{v(1 + e)}{2}$

$(v_C)_2 = \dfrac{v(1 + e)^2}{1}$

$(v_C)_2 = \dfrac{v(1 + e)^2}{4}$

15–58. $(v_A)_1 = 19.7$ ft/s

$(v_A)_2 = 9.44$ ft/s $\leftarrow$

$(v_B)_2 = 15.3$ ft/s $\leftarrow$

$s_{B_*} = 9.13$ ft

15–59. $h = 21.8$ mm

15–61. $(v_A)_2 = 2.40$ m/s

$(v_B)_2 = 5.60$ m/s

$s_{max} = 1.53$ m

15–62. $F_{avg} = 1.68$ kN

15–63. $(v_P)_2 = 0.940$ m/s

15–65. $(v_2)_y = 11.12$ ft/s

$(v_x)_2 = 8$ ft/s

$v_2 = 13.7$ ft/s

$\theta = 54.3° \,\measuredangle$

$h = 1.92$ ft

15–66. $e = 0.261$

$\int F\,dt = 1.99$ kip $\cdot$ s

15–67. $(v_A)_3 = 0$

$(v_B)_3 = 13.9$ ft/s

$s_{max} = 1.41$ ft

15–69. $(v_B)_1 = 13.900$ ft/s

$\phi = 6.34°$

$v_{Ay} = 11.434$ ft/s

$t = 0.3119$ s

$v_{Ax} = 12.510$ ft/s

$s = 1.90$ ft

$v_A = 16.9$ ft/s

15–70. $(v_A)_3 = \left(\frac{1 - e}{2}\right)\sqrt{gL}$

$(v_B)_3 = \left(\frac{1 + e}{2}\right)\sqrt{gL}$

$\theta = \cos^{-1}\left[1 - \dfrac{(1 + e)^2}{8}\right]$

15–71. $e = 0.75$

$\Delta E = 9.65$ kJ

15–73. $v_2' = \left(\frac{1 + e}{2}\right)v_1$

$v_3' = \left(\frac{1 + e}{2}\right)^2 v_1$

$v_n' = \left(\frac{1 + e}{2}\right)^{n-1}v_1$

15–74. $\phi = \cos^{-1}\left[1 - \dfrac{(1 + e)^4}{16}(1 - \cos\theta)\right]$

15–75. $(v_B)_3 = 3.24$ m/s $\theta = 43.9°$

15–77. $(v_B')_x = 21.65$ m/s $\leftarrow$

$(v_B')_y = 5$ m/s $\uparrow$

$v_B' = 22.2$ m/s

$\theta = 13.0°$

15–78. $v_B' = 31.8$ ft/s

15–79. $(v_A)_2 = 4.60$ m/s

$(v_B)_2 = 3.16$ m/s

$d = 0.708$ m

15–81. $(v_A)_2 = 9.829$ ft/s

$(v_A)_1 = 44$ ft/s

$(v_B)_2 = 43.51$ ft/s

$(v_B)_1 = 29.3$ ft/s

15–82. $v_A' = 5.07$ m/s $\theta_A = 80.2° \,\measuredangle$

$v_B' = 7.79$ m/s $\leftarrow$

15–83. $(v_B)_2 = 2.88$ ft/s

$(v_A)_2 = 1.77$ ft/s

15–85. $15v_A'\cos\phi_A + 10v_B'\cos\phi_B = 42$

$v_A'\sin\phi_A = 8$

$v_B'\sin\phi_B = 6.4$

$v_A' = 8.19$ m/s

$\phi_A = 102.52°$

$v_B' = 9.38$ m/s

$\phi_B = 42.99°$

15–86. $v_A' = 9.68$ m/s

$\phi_A = 86.04°$

$v_B' = 4.94$ m/s

$\phi_B = 61.16°$

15–87. $v_A' = 12.6$ ft/s

$\phi_A = 72.86°$

$v_B' = 14.7$ ft/s

$\phi_B = 42.80°$

15–89. $(v_{Ax})_2 = 0.550$ ft/s

$(v_{Bx})_2 = 1.95$ ft/s

$(v_{A_y})_2 = -2.40$ ft/s

$(v_{B_y})_2 = -2.40$ ft/s

$(v_A)_2 = 2.46$ ft/s

$(v_B)_2 = 3.09$ ft/s

15–90. $v = 95.6$ ft/s

15–91. $v = 3.33$ m/s

Geometric Properties of Line and Area Elements

Centroid Location	Centroid Location	Area Moment of Inertia

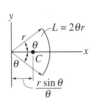

$L = 2\theta r$

$r \sin \theta / \theta$

Circular arc segment

$A = \theta r^2$

$\frac{2}{3} \frac{r \sin \theta}{\theta}$

Circular sector area

$I_x = \frac{1}{4} r^4 (\theta - \frac{1}{2} \sin 2\theta)$

$I_x = \frac{1}{4} r^4 (\theta + \frac{1}{2} \sin 2\theta)$

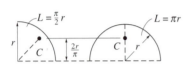

$L = \frac{\pi}{2} r$ $L = \pi r$

$\frac{2r}{\pi}$

Quarter and semicircle arcs

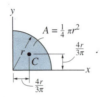

$A = \frac{1}{4} \pi r^2$

$\frac{4r}{3\pi}$

$\frac{4r}{3\pi}$

Quarter circle area

$I_x = \frac{1}{16} \pi r^4$

$I_y = \frac{1}{16} \pi r^4$

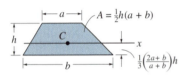

$A = \frac{1}{2} h(a + b)$

$\frac{1}{3}\left(\frac{2a+b}{a+b}\right)h$

Trapezoidal area

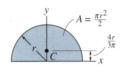

$A = \frac{\pi r^2}{2}$

$\frac{4r}{3\pi}$

Semicircular area

$I_x = \frac{1}{8} \pi r^4$

$I_y = \frac{1}{8} \pi r^4$

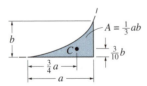

$A = \frac{2}{3} ab$

$\frac{3}{5} a$

$\frac{3}{8} b$

Semiparabolic area

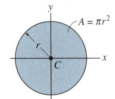

$A = \pi r^2$

Circular area

$I_x = \frac{1}{4} \pi r^4$

$I_y = \frac{1}{4} \pi r^4$

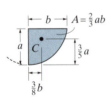

$A = \frac{1}{3} ab$

$\frac{3}{4} a$

$\frac{3}{10} b$

Exparabolic area

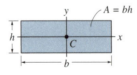

$A = bh$

Rectangular area

$I_x = \frac{1}{12} bh^3$

$I_y = \frac{1}{12} hb^3$

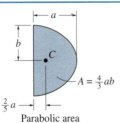

$A = \frac{4}{3} ab$

$\frac{2}{5} a$

Parabolic area

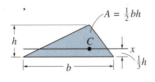

$A = \frac{1}{2} bh$

$\frac{1}{3} h$

Triangular area

$I_x = \frac{1}{36} bh^3$